Table of Normal Curve Areas	z	.00	.01	.02	.03	.04	.05	.06	.07	.08	.09
	.0	.0000	.0040	.0080	.0120	.0160	.0199	.0239	.0279	.0319	.0359
	.1	.0398	.0438	.0478	.0517	.0557	.0596	.0636	.0675	.0714	.0753
	.2	.0793	.0832	.0871	.0910	.0948	.0987	.1026	.1064	.1103	.1141
	.3	.1179	.1217	.1255	.1293	.1331	.1368	.1406	.1443	.1480	.1517
	.4	.1554	.1591	.1628	.1664	.1700	.1736	.1772	.1808	.1844	.1879
	.5	.1915	.1950	.1985	.2019	.2054	.2088	.2123	.2157	.2190	.2224
	.6	.2257	.2291	.2324	.2357	.2389	.2422	.2454	.2486	.2517	.2549
	.7	.2580	.2611	.2642	.2673	.2704	.2734	.2764	.2794	.2823	.2852
	.8	.2881	.2910	.2939	.2967	.2995	.3023	.3051	.3078	.3106	.3133
	.9	.3159	.3186	.3212	.3238	.3264	.3289	.3315	.3340	.3365	.3389
	1.0	.3413	.3438	.3461	.3485	.3508	.3531	.3554	.3577	.3599	.3621
	1.1	.3643	.3665	.3686	.3708	.3729	.3749	.3770	.3790	.3810	.3830
	1.2	.3849	.3869	.3888	.3907	.3925	.3944	.3962	.3980	.3997	.4015
	1.3	.4032	.4049	.4066	.4082	.4099	.4115	.4131	.4147	.4162	.4177
	1.4	.4192	.4207	.4222	.4236	.4251	.4265	.4279	.4292	.4306	.4319
	1.5	.4332	.4345	.4357	.4370	.4382	.4394	.4406	.4418	.4429	.4441
	1.6	.4452	.4463	.4474	.4484	.4495	.4505	.4515	.4525	.4535	.4545
	1.7	.4554	.4564	.4573	.4582	.4591	.4599	.4608	.4616	.4625	.4633
	1.8	.4641	.4649	.4656	.4664	.4671	.4678	.4686	.4693	.4699	.4706
	1.9	.4713	.4719	.4726	.4732	.4738	.4744	.4750	.4756	.4761	.4767
	2.0	.4772	.4778	.4783	.4788	.4793	.4798	.4803	.4808	.4812	.4817
	2.1	.4821	.4826	.4830	.4834	.4838	.4842	.4846	.4850	.4854	.4857
	2.2	.4861	.4864	.4868	.4871	.4875	.4878	.4881	.4884	.4887	.4890
	2.3	.4893	.4896	.4898	.4901	.4904	.4906	.4909	.4911	.4913	.4916
	2.4	.4918	.4920	.4922	.4925	.4927	.4929	.4931	.4932	.4934	.4936
	2.5	.4938	.4940	.4941	.4943	.4945	.4946	.4948	.4949	.4951	.4952
	2.6	.4953	.4955	.4956	.4957	.4959	.4960	.4961	.4962	.4963	.4964
	2.7	.4965	.4966	.4967	.4968	.4969	.4970	.4971	.4972	.4973	.4974
	2.8	.4974	.4975	.4976	.4977	.4977	.4978	.4979	.4979	.4980	.4981
	2.9	.4981	.4982	.4982	.4983	.4984	.4984	.4985	.4985	.4986	.4986
	3.0	.4987	.4987	.4987	.4988	.4988	.4989	.4989	.4989	.4990	.4990

SOURCE: This table is abridged from Table 1 of *Statistical Tables and Formulas* by A. Hald (New York: Wiley, 1952). Reproduced by permission of A. Hald and the publisher, John Wiley & Sons, Inc.

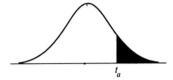

$$t_a$$

Table of Critical Values of t	*df*	$t_{.100}$	$t_{.050}$	$t_{.025}$	$t_{.010}$	$t_{.005}$	*df*
	1	3.078	6.314	12.706	31.821	63.657	1
	2	1.886	2.920	4.303	6.965	9.925	2
	3	1.638	2.353	3.182	4.541	5.841	3
	4	1.533	2.132	2.776	3.747	4.604	4
	5	1.476	2.015	2.571	3.365	4.032	5
	6	1.440	1.943	2.447	3.143	3.707	6
	7	1.415	1.895	2.365	2.998	3.499	7
	8	1.397	1.860	2.306	2.896	3.355	8
	9	1.383	1.833	2.262	2.821	3.250	9
	10	1.372	1.812	2.228	2.764	3.169	10
	11	1.363	1.796	2.201	2.718	3.106	11
	12	1.356	1.782	2.179	2.681	3.055	12
	13	1.350	1.771	2.160	2.650	3.012	13
	14	1.345	1.761	2.145	2.624	2.977	14
	15	1.341	1.753	2.131	2.602	2.947	15
	16	1.337	1.746	2.120	2.583	2.921	16
	17	1.333	1.740	2.110	2.567	2.898	17
	18	1.330	1.734	2.101	2.552	2.878	18
	19	1.328	1.729	2.093	2.539	2.861	19
	20	1.325	1.725	2.086	2.528	2.845	20
	21	1.323	1.721	2.080	2.518	2.831	21
	22	1.321	1.717	2.074	2.508	2.819	22
	23	1.319	1.714	2.069	2.500	2.807	23
	24	1.318	1.711	2.064	2.492	2.797	24
	25	1.316	1.708	2.060	2.485	2.787	25
	26	1.315	1.706	2.056	2.479	2.779	26
	27	1.314	1.703	2.052	2.473	2.771	27
	28	1.313	1.701	2.048	2.467	2.763	28
	29	1.311	1.699	2.045	2.462	2.756	29
	∞	1.282	1.645	1.960	2.326	2.576	∞

SOURCE: From "Table of Percentage Points of the *t*-Distribution," *Biometrika* 32 (1941):300. Reproduced by permission of the *Biometrika* Trustees.

Introduction to Probability & Statistics

William Mendenhall

Robert J. Beaver

Barbara M. Beaver

THOMSON

™

BROOKS/COLE

Australia · Canada · Mexico · Singapore · Spain · United Kingdom · United States

Introduction to Probability & Statistics
Mendenhall, Beaver, & Beaver

Custom Editor:
Marc Bove

Project Development Editor:
Lea Riddle

Marketing Coordinator:
Sara Mercurio

Production/Manufacturing Supervisor:
Donna M. Brown

Project Coordinator:
Rebecca A. Walker

Pre-Media Services Supervisor:
Dan Plofchan

Rights and Permissions Specialist:
Kalina Hintz

Senior Prepress Specialist:
Kim Fry

Cover Design:
Krista Pierson

Printer:
Globus

Printed in the United States of America
1 2 3 4 5 6 7 8 9 10 11 12 13 14 09 08 07 06

For information about our products, contact us at:
Thomson Learning Academic Resource Center
(800) 423-0563

For permission to use material from this text or product, submit a request online at **http://www.thomsonrights.com**. Any additional questions about permissions can be submitted by email to **thomsonrights@thomson.com**.

The Adaptable Courseware Program consists of products and additions to existing Wadsworth products that are produced from camera-ready copy. Peer review, class testing, and accuracy are primarily the responsibility of the author(s).

Student Edition: ISBN 0-534-61594-5

Thomson Custom Solutions
5191 Natorp Boulevard
Mason, OH 45040
www.thomsoncustom.com

Thomson Higher Education
10 Davis Drive
Belmont, CA 94002-3098
USA

Asia (Including India):
Thomson Learning
60 Albert Street, #15-01
Albert Complex
Singapore 189969
Tel 65 336-6411
Fax 65 336-7411

Australia/New Zealand:
Thomson Learning Australia
102 Dodds Street
Southbank, Victoria 3006
Australia

Latin America:
Thomson Learning
Seneca 53
Colonia Polano
11560 Mexico, D.F., Mexico
Tel (525) 281-2906
Fax (525) 281-2656

Canada:
Thomson Nelson
1120 Birchmount Road
Toronto, Ontario
Canada M1K 5G4
Tel (416) 752-9100
Fax (416) 752-8102

UK/Europe/Middle East/Africa:
Thomson Learning
High Holborn House
50-51 Bedford Row
London, WC1R 4L$
United Kingdom
Tel 44 (020) 7067-2500
Fax 44 (020) 7067-2600

Spain (Includes Portugal):
Thomson Paraninfo
Calle Magallanes 25
28015 Madrid
España
Tel 34 (0)91 446-3350
Fax 34 (0)91 445-6218

Custom Table of Contents

Describing Data with Graphs

© Picture Arts/Royalty-Free/CORBIS

GENERAL OBJECTIVES

Many sets of measurements are samples selected from larger populations. Other sets constitute the entire population, as in a national census. In this chapter, you will learn what a *variable* is, how to classify variables into several types, and how measurements or data are generated. You will then learn how to use graphs to describe data sets.

CHAPTER INDEX

MY | PERSONAL TRAINER

How Do I Construct a Stem and Leaf Plot?
How Do I Construct a Relative Frequency Histogram?

How Is Your Blood Pressure?

Is your blood pressure normal, or is it too high or too low? The case study at the end of this chapter examines a large set of blood pressure data. You will use graphs to describe these data and compare your blood pressure with that of others of your same age and gender.

VARIABLES AND DATA

1.1

In Chapters 1 and 2, we will present some basic techniques in *descriptive statistics—* the branch of statistics concerned with describing sets of measurements, both *samples* and *populations*. Once you have collected a set of measurements, how can you display this set in a clear, understandable, and readable form? First, you must be able to define what is meant by measurements or "data" and to categorize the types of data that you are likely to encounter in real life. We begin by introducing some definitions—new terms in the statistical language that you need to know.

Definition   A **variable** is a characteristic that changes or varies over time and/or for different individuals or objects under consideration.

For example, body temperature is a variable that changes over time within a single individual; it also varies from person to person. Religious affiliation, ethnic origin, income, height, age, and number of offspring are all variables—characteristics that vary depending on the individual chosen.

In the Introduction, we defined an *experimental unit* or an *element of the sample* as the object on which a measurement is taken. Equivalently, we could define an experimental unit as the object on which a variable is measured. When a variable is actually measured on a set of experimental units, a set of measurements or **data** result.

Definition   An **experimental unit** is the individual or object on which a variable is measured. A single **measurement** or data value results when a variable is actually measured on an experimental unit.

If a measurement is generated for every experimental unit in the entire collection, the resulting data set constitutes the *population* of interest. Any smaller subset of measurements is a *sample*.

Definition   A **population** is the set of all measurements of interest to the investigator.

Definition   A **sample** is a subset of measurements selected from the population of interest.

EXAMPLE   **1.1**

A set of five students is selected from all undergraduates at a large university, and measurements are entered into a spreadsheet as shown in Figure 1.1. Identify the various elements involved in generating this set of measurements.

Solution   There are several *variables* in this example. The *experimental unit* on which the variables are measured is a particular undergraduate student on the campus, identified in column C1. Five variables are measured for each student: grade point average (GPA), gender, year in college, major, and current number of units enrolled. Each of these characteristics varies from student to student. If we consider the GPAs of all students at this university to be the population of interest, the five GPAs in column C2 represent a *sample* from this population. If the GPA of each undergraduate student at the university had been measured, we would have generated the entire *population* of measurements for this variable.

FIGURE 1.1

Measurements on five
undergraduate students

Worksheet 1 ***

	C1	C2	C3-T	C4-T	C5-T	C6
	Student	GPA	Gender	Year	Major	Number of Units
1	1	2.0	F	Fr	Psychology	16
2	2	2.3	F	So	Mathematics	15
3	3	2.9	M	So	English	17
4	4	2.7	M	Fr	English	15
5	5	2.6	F	Jr	Business	14

The second variable measured on the students is gender, in column C3-T. This variable can take only one of two values—male (M) or female (F). It is not a numerically valued variable and hence is somewhat different from GPA. The population, if it could be enumerated, would consist of a set of Ms and Fs, one for each student at the university. Similarly, the third and fourth variables, year and major, generate nonnumerical data. Year has four categories (Fr, So, Jr, Sr), and major has one category for each undergraduate major on campus. The last variable, current number of units enrolled, is numerically valued, generating a set of numbers rather than a set of qualities or characteristics.

Although we have discussed each variable individually, remember that we have measured each of these five variables on a single experimental unit: the student. Therefore, in this example, a "measurement" really consists of five observations, one for each of the five measured variables. For example, the measurement taken on student 2 produces this observation:

(2.3, F, So, Mathematics, 15)

You can see that there is a difference between a *single* variable measured on a single experimental unit and *multiple* variables measured on a single experimental unit as in Example 1.1.

Definition   **Univariate data** result when a single variable is measured on a single experimental unit.

Definition   **Bivariate data** result when two variables are measured on a single experimental unit. **Multivariate data** result when more than two variables are measured.

If you measure the body temperatures of 148 people, the resulting data are *univariate*. In Example 1.1, five variables were measured on each student, resulting in *multivariate* data.

1.2 TYPES OF VARIABLES

Variables can be classified into one of two categories: **qualitative** or **quantitative.**

Definition **Qualitative variables** measure a quality or characteristic on each experimental unit. **Quantitative variables** measure a numerical quantity or amount on each experimental unit.

Qualitative variables produce data that can be categorized according to similarities or differences in kind; hence, they are often called **categorical data.** The variables gender, year, and major in Example 1.1 are qualitative variables that produce categorical data. Here are some other examples:

 MY TIP

Qualitative ⇔ "quality"
or characteristic

Quantitative ⇔
"quantity" or number

- Political affiliation: Republican, Democrat, Independent
- Taste ranking: excellent, good, fair, poor
- Color of an M&M® candy: brown, yellow, red, orange, green, blue

Quantitative variables, often represented by the letter x, produce numerical data, such as those listed here:

- x = Prime interest rate
- x = Number of unregistered taxicabs in a city
- x = Weight of a package ready to be shipped
- x = Volume of orange juice in a glass

Notice that there is a difference in the types of numerical values that these quantitative variables can assume. The number of unregistered taxicabs, for example, can take on only the values $x = 0, 1, 2, \ldots$, whereas the weight of a package can take on any value greater than zero, or $0 < x < \infty$. To describe this difference, we define two types of quantitative variables: **discrete** and **continuous.**

Definition A **discrete variable** can assume only a finite or countable number of values. A **continuous variable** can assume the infinitely many values corresponding to the points on a line interval.

MY TIP

Discrete ⇔ "listable"

Continuous ⇔
"unlistable"

The name *discrete* relates to the discrete gaps between the possible values that the variable can assume. Variables such as number of family members, number of new car sales, and number of defective tires returned for replacement are all examples of discrete variables. On the other hand, variables such as height, weight, time, distance, and volume are *continuous* because they can assume values at any point along a line interval. For any two values you pick, a third value can always be found between them!

EXAMPLE 1.2 Identify each of the following variables as qualitative or quantitative:

1. The most frequent use of your microwave oven (reheating, defrosting, warming, other)
2. The number of consumers who refuse to answer a telephone survey
3. The door chosen by a mouse in a maze experiment (A, B, or C)
4. The winning time for a horse running in the Kentucky Derby
5. The number of children in a fifth-grade class who are reading at or above grade level

Discrete variables often involve the "number of" items in a set.

Solution Variables 1 and 3 are both *qualitative* because only a quality or characteristic is measured for each individual. The categories for these two variables are shown in parentheses. The other three variables are *quantitative*. Variable 2, the number of consumers, is a *discrete* variable that can take on any of the values $x = 0, 1, 2, \ldots$, with a maximum value depending on the number of consumers called. Similarly, variable 5, the number of children reading at or above grade level, can take on any of the values $x = 0, 1, 2, \ldots$, with a maximum value depending on the number of children in the class. Variable 4, the winning time for a Kentucky Derby horse, is the only *continuous* variable in the list. The winning time, if it could be measured with sufficient accuracy, could be 121 seconds, 121.5 seconds, 121.25 seconds, or any values between any two times we have listed.

Figure 1.2 depicts the types of data we have defined. Why should you be concerned about different kinds of variables and the data that they generate? The reason is that the methods used to describe data sets depend on the type of data you have collected. For each set of data that you collect, the key will be to determine what type of data you have and how you can present them most clearly and understandably to your audience!

FIGURE 1.2

Types of data

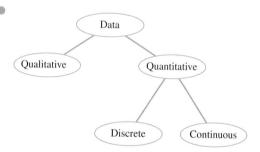

GRAPHS FOR CATEGORICAL DATA

1.3

After the data have been collected, they can be consolidated and summarized to show the following information:

- What values of the variable have been measured
- How often each value has occurred

For this purpose, you can construct a *statistical table* that can be used to display the data graphically as a data distribution. The type of graph you choose depends on the type of variable you have measured.

When the variable of interest is *qualitative,* the statistical table is a list of the categories being considered along with a measure of how often each value occurred. You can measure "how often" in three different ways:

- The **frequency,** or number of measurements in each category
- The **relative frequency,** or proportion of measurements in each category
- The **percentage** of measurements in each category

For example, if you let n be the total number of measurements in the set, you can find the relative frequency and percentage using these relationships:

$$\text{Relative frequency} = \frac{\text{Frequency}}{n}$$

$$\text{Percent} = 100 \times \text{Relative frequency}$$

You will find that the sum of the frequencies is always n, the sum of the relative frequencies is 1, and the sum of the percentages is 100%.

The categories for a qualitative variable should be chosen so that

- a measurement will belong to one and only one category
- each measurement has a category to which it can be assigned

MY TIP

Three steps to a data distribution:
(1) raw data ⇒
(2) statistical table ⇒
(3) graph

For example, if you categorize meat products according to the type of meat used, you might use these categories: beef, chicken, seafood, pork, turkey, other. To categorize ranks of college faculty, you might use these categories: professor, associate professor, assistant professor, instructor, lecturer, other. The "other" category is included in both cases to allow for the possibility that a measurement cannot be assigned to one of the earlier categories.

Once the measurements have been categorized and summarized in a *statistical table,* you can use either a pie chart or a bar chart to display the distribution of the data. A **pie chart** is the familiar circular graph that shows how the measurements are distributed among the categories. A **bar chart** shows the same distribution of measurements in categories, with the height of the bar measuring how often a particular category was observed.

EXAMPLE 1.3

In a survey concerning public education, 400 school administrators were asked to rate the quality of education in the United States. Their responses are summarized in Table 1.1. Construct a pie chart and a bar chart for this set of data.

Solution To construct a pie chart, assign one sector of a circle to each category. The angle of each sector should be proportional to the proportion of measurements (or *relative frequency*) in that category. Since a circle contains 360°, you can use this equation to find the angle:

$$\text{Angle} = \text{Relative frequency} \times 360°$$

TABLE 1.1

U.S. Education Rating by 400 Educators

Rating	Frequency
A	35
B	260
C	93
D	12
Total	400

MY TIP

Proportions add to 1.
Percents add to 100.
Sector angles add to 360°.

Table 1.2 shows the ratings along with the frequencies, relative frequencies, percentages, and sector angles necessary to construct the pie chart. Figure 1.3 shows the pie chart constructed from the values in the table. While pie charts use percentages to determine the relative sizes of the "pie slices," bar charts usually plot frequency against the categories. A bar chart for these data is shown in Figure 1.4.

TABLE 1.2   ● **Calculations for the Pie Chart in Example 1.3**

Rating	Frequency	Relative Frequency	Percent	Angle
A	35	35/400 = .09	9%	$.09 \times 360 = 32.4°$
B	260	260/400 = .65	65%	234.0°
C	93	93/400 = .23	23%	82.8°
D	12	12/400 = .03	3%	10.8°
Total	400	1.00	100%	360°

The visual impact of these two graphs is somewhat different. The pie chart is used to display the relationship of the parts to the whole; the bar chart is used to emphasize the actual quantity or frequency for each category. Since the categories in this example are ordered "grades" (A, B, C, D), we would not want to rearrange the bars in the chart to change its *shape*. In a pie chart, the order of presentation is irrelevant.

FIGURE 1.3

Pie chart for Example 1.3

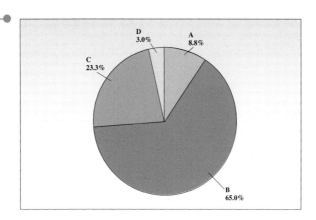

FIGURE 1.4

Bar chart for Example 1.3

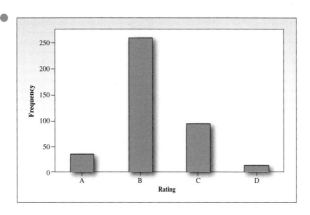

EXAMPLE 1.4   A snack size bag of peanut M&M® candies contains 21 candies with the colors listed in Table 1.3. The variable "color" is *qualitative,* so Table 1.4 lists the six categories along with a tally of the number of candies of each color. The last three columns of Table 1.4 give the three different measures of how often each category occurred. Since the categories are colors and have no particular order, you could construct bar charts with many different *shapes* just by reordering the bars. To emphasize that brown is the most frequent color, followed by blue, green, and orange, we order the bars from largest to smallest and generate the bar chart using *MINITAB* in Figure 1.5. A bar chart in which the bars are ordered from largest to smallest is called a **Pareto chart.**

TABLE 1.3 ● **Raw Data: Colors of 21 Candies**

Brown	Green	Brown	Blue
Red	Red	Green	Brown
Yellow	Orange	Green	Blue
Brown	Blue	Blue	Brown
Orange	Blue	Brown	Orange
Yellow			

TABLE 1.4 ● **Statistical Table: M&M Data for Example 1.4**

Category	Tally	Frequency	Relative Frequency	Percent
Brown	ⅢⅠ	6	6/21	28%
Green	Ⅲ	3	3/21	14
Orange	Ⅲ	3	3/21	14
Yellow	Ⅱ	2	2/21	10
Red	Ⅱ	2	2/21	10
Blue	Ⅲ	5	5/21	24
Total		21	1	100%

FIGURE 1.5

MINITAB bar chart for Example 1.4

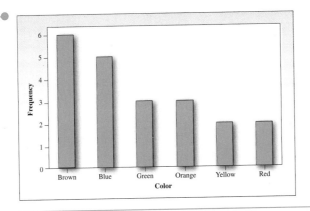

EXERCISES

UNDERSTANDING THE CONCEPTS

1.1 Experimental Units Identify the experimental units on which the following variables are measured:

a. Gender of a student

b. Number of errors on a midterm exam

c. Age of a cancer patient

d. Number of flowers on an azalea plant

e. Color of a car entering the parking lot

1.2 Qualitative or Quantitative? Identify each variable as quantitative or qualitative:

a. Amount of time it takes to assemble a simple puzzle

b. Number of students in a first-grade classroom

c. Rating of a newly elected politician (excellent, good, fair, poor)

d. State in which a person lives

1.3 Discrete or Continuous? Identify the following quantitative variables as discrete or continuous:

a. Population in a particular area of the United States

b. Weight of newspapers recovered for recycling on a single day

c. Time to complete a sociology exam

d. Number of consumers in a poll of 1000 who consider nutritional labeling on food products to be important

1.4 Discrete or Continuous? Identify each quantitative variable as discrete or continuous.

a. Number of boating accidents along a 50-mile stretch of the Colorado River

b. Time required to complete a questionnaire

c. Choice of color for a new refrigerator

d. Number of brothers and sisters you have

e. Yield in kilograms of wheat from a 1-hectare plot in a wheat field

1.5 Parking on Campus Six vehicles are selected from the vehicles that are issued campus parking permits, and the following data are recorded:

Vehicle	Type	Make	Carpool?	One-way commute distance (miles)	Age of vehicle (years)
1	Car	Honda	No	23.6	6
2	Car	Toyota	No	17.2	3
3	Truck	Toyota	No	10.1	4
4	Van	Dodge	Yes	31.7	2
5	Motor-cycle	Harley-Davidson	No	25.5	1
6	Car	Chevrolet	No	5.4	9

a. What are the experimental units?

b. What are the variables being measured? What types of variables are they?

c. Is this univariate, bivariate, or multivariate data?

1.6 Past U.S. Presidents A data set consists of the ages at death for each of the 38 past presidents of the United States.

a. Is this set of measurements a population or a sample?

b. What is the variable being measured?

c. Is the variable in part b quantitative or qualitative?

1.7 Voter Attitudes You are a candidate for your state legislature, and you want to survey voter attitudes regarding your chances of winning. Identify the population that is of interest to you and from which you would like to select your sample. How is this population dependent on time?

1.8 Cancer Survival Times A medical researcher wants to estimate the survival time of a patient after the onset of a particular type of cancer and after a particular regimen of radiotherapy.

a. What is the variable of interest to the medical researcher?

b. Is the variable in part a qualitative, quantitative discrete, or quantitative continuous?

c. Identify the population of interest to the medical researcher.

d. Describe how the researcher could select a sample from the population.

e. What problems might arise in sampling from this population?

1.9 New Teaching Methods An educational researcher wants to evaluate the effectiveness of a new method for teaching reading to deaf students. Achievement at the end of a period of teaching is measured by a student's score on a reading test.

a. What is the variable to be measured? What type of variable is it?

b. What is the experimental unit?

c. Identify the population of interest to the experimenter.

BASIC TECHNIQUES

1.10 Fifty people are grouped into four categories—A, B, C, and D—and the number of people who fall into each category is shown in the table:

Category	Frequency
A	11
B	14
C	20
D	5

a. What is the experimental unit?

b. What is the variable being measured? Is it qualitative or quantitative?

c. Construct a pie chart to describe the data.

d. Construct a bar chart to describe the data.

e. Does the shape of the bar chart in part d change depending on the order of presentation of the four categories? Is the order of presentation important?

f. What *proportion* of the people are in category B, C, or D?

g. What *percentage* of the people are *not* in category B?

1.11 Jeans A manufacturer of jeans has plants in California, Arizona, and Texas. A group of 25 pairs of jeans is randomly selected from the computerized database, and the state in which each is produced is recorded:

CA	AZ	AZ	TX	CA
CA	CA	TX	TX	TX
AZ	AZ	CA	AZ	TX
CA	AZ	TX	TX	TX
CA	AZ	AZ	CA	CA

a. What is the experimental unit?

b. What is the variable being measured? Is it qualitative or quantitative?

c. Construct a pie chart to describe the data.

d. Construct a bar chart to describe the data.

e. What *proportion* of the jeans are made in Texas?

f. What state produced the most jeans in the group?

g. If you want to find out whether the three plants produced equal numbers of jeans, or whether one produced more jeans than the others, how can you use the charts from parts c and d to help you? What conclusions can *you* draw from these data?

APPLICATIONS

1.12 Election 2004 During the months leading up to the Democratic convention in the summer of 2004, the news media regularly presented opinion polls that tracked the fortunes of the major candidates hoping to become the Democratic Party's nominee for president of the United States. One such poll, taken for *Time*/CNN, showed the following results:[1]

> "Suppose the 2004 election for president were being held today and you had to choose between [see below], the Democrat, and George W. Bush, the Republican. For whom would you vote: [see below]?"

George W. Bush %	John Kerry %	Not Sure %
50	48	2

George W. Bush %	John Edwards %	Not Sure %
52	46	2

George W. Bush %	Wesley Clark %	Not Sure %
55	41	4

Source: www.pollingreport.com

The results were based on a sample taken February 5 and 6, 2004, of 580 likely voters nationwide.

a. If the pollsters were planning to use these results to predict the outcome of the 2004 presidential election, describe the population of interest to them.

b. Describe the actual population from which the sample was drawn.

c. What is the difference between "likely voters" and "registered voters"? Why is this important?

d. Is the sample selected by the pollsters representative of the population described in part a? Explain.

1.13 Want to Be President? Would you want to be the president of the United States? Although many teenagers think that they could grow up to be the president, most don't want the job. In an opinion poll conducted by *ABC News,* nearly 80% of the teens were not interested in the job.[2] When asked "What's the main reason you would not want to be president?" they gave these responses:

Other career plans/no interest	40%
Too much pressure	20%
Too much work	15%
Wouldn't be good at it	14%
Too much arguing	5%

a. Are all of the reasons accounted for in this table? Add another category if necessary.

b. Would you use a pie chart or a bar chart to graphically describe the data? Why?

c. Draw the chart you chose in part b.

d. If you were the person conducting the opinion poll, what other types of questions might you want to investigate?

1.14 Race Distributions in the Armed Forces The four branches of the armed forces in the United States are quite different in their makeup with regard to gender, race, and age distributions. The table below shows the racial breakdown of the members of the United States Army and the United States Air Force.[3]

	Army	Air Force
White	58.4%	75.5%
Black	26.3%	16.2%
Hispanic	8.9%	5.0%
Other	6.4%	3.3%

Source: Time Magazine

a. Define the variable that has been measured in this table.

b. Is the variable quantitative or qualitative?

c. What do the numbers represent?

d. Construct a pie chart to describe the racial breakdown in the U.S. Army.

e. Construct a bar chart to describe the racial breakdown in the U.S. Air Force.

f. What percentage of the members of the U.S. Army are minorities—that is, not white? What is this percentage in the U.S. Air Force?

1.15 Back to Work How long does it take you to adjust to your normal work routine after coming back from vacation? A bar graph with data from the Snapshots section of *USA Today* is shown below:[4]

a. Are all of the opinions accounted for in the table? Add another category if necessary.

b. Is the bar chart drawn accurately? That is, are the three bars in the correct proportion to each other?

c. Use a pie chart to describe the opinions. Which graph is more interesting to look at?

GRAPHS FOR QUANTITATIVE DATA

(1.4)

Quantitative variables measure an amount or quantity on each experimental unit. If the variable can take only a finite or countable number of values, it is a *discrete* variable. A variable that can assume an infinite number of values corresponding to points on a line interval is called *continuous*.

Pie Charts and Bar Charts

Sometimes information is collected for a quantitative variable measured on different segments of the population, or for different categories of classification. For example, you might measure the average incomes for people of different age groups, different genders, or living in different geographic areas of the country. In such cases, you can use pie charts or bar charts to describe the data, using the amount measured in each category rather than the frequency of occurrence of each category. The *pie chart* displays how the total quantity is distributed among the categories, and the *bar chart* uses the height of the bar to display the amount in a particular category.

EXAMPLE   1.5   The amount of money expended in fiscal year 2000 by the U.S. Department of Defense in various categories is shown in Table 1.5.[5] Construct both a pie chart and a bar chart to describe the data. Compare the two forms of presentation.

TABLE 1.5

Expenses by Category

Category	Amount (in billions)
Military personnel	$86.8
Operation and maintenance	130.2
Procurement	62.5
Research and development	44.4
Military construction	5.1
Other	3.1
Total	$332.1

Source: The World Almanac and Book of Facts, 2004

Solution   Two variables are being measured: the category of expenditure (*qualitative*) and the amount of the expenditure (*quantitative*). The bar chart in Figure 1.6 displays the categories on the horizontal axis and the amounts on the vertical axis. For

FIGURE 1.6

Bar chart for Example 1.5

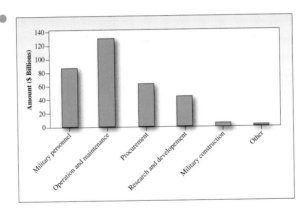

the pie chart in Figure 1.7, each "pie slice" represents the proportion of the total expenditures ($332.1 billion) corresponding to its particular category. For example, for the research and development category, the angle of the sector is

$$\frac{44.4}{332.1} \times 360° = 48.1°$$

FIGURE 1.7

Pie chart for Example 1.5

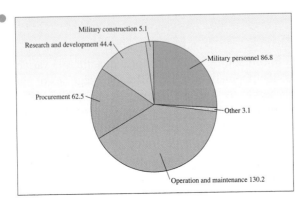

Both graphs show that the largest amounts of money were spent on personnel and operations. Since there is no inherent order to the categories, you are free to rearrange the bars or sectors of the graphs in any way you like. The *shape* of the bar chart has no bearing on its interpretation.

Line Charts

When a quantitative variable is recorded over time at equally spaced intervals (such as daily, weekly, monthly, quarterly, or yearly), the data set forms a **time series.** Time series data are most effectively presented on a **line chart** with time as the horizontal axis. The idea is to try to discern a pattern or **trend** that will likely continue into the future, and then to use that pattern to make accurate predictions for the immediate future.

EXAMPLE 1.6

In the year 2025, the oldest "baby boomers" (born in 1946) will be 79 years old, and the oldest "Gen-Xers" (born in 1965) will be two years from Social Security eligibility. How will this affect the consumer trends in the next 25 years? As the population becomes older and more culturally diverse, the largest population growth will occur in areas with a high "diversity quotient" or with a large senior population. *American Demographics* gives projections for one such area, Naples, Florida, as shown in Table 1.6 below.[6] Construct a line chart to illustrate the data. What is the effect of stretching and shrinking the vertical axis on the line chart?

TABLE 1.6 ● **Population Growth Projections**

Year	2000	2005	2010	2015	2020	2025
Population (thousands)	265.4	287.6	333.3	348.7	382.4	412.4

MY TIP

Beware of stretching or shrinking axes when you look at a graph!

Solution The quantitative variable "percent" is measured over five time intervals, creating a *time series* that you can graph with a line chart. The time intervals are marked on the horizontal axis and the percentages on the vertical axis. The data points are then connected by line segments to form the line charts in Figure 1.8. Notice the marked difference in the vertical scales of the two graphs. *Shrinking* the scale on the vertical axis causes large changes to appear small, and vice versa. To avoid misleading conclusions, you must look carefully at the scales of the vertical and horizontal axes. However, from both graphs you get a clear picture of the steadily increasing percentages in the early years of the new millennium.

FIGURE 1.8

Line charts for Example 1.6

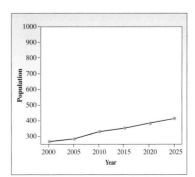

Dotplots

Many sets of quantitative data consist of numbers that cannot easily be separated into categories or intervals of time. You need a different way to graph this type of data!

The simplest graph for quantitative data is the **dotplot.** For a small set of measurements—for example, the set 2, 6, 9, 3, 7, 6—you can simply plot the measurements as points on a horizontal axis. This dotplot, generated by *MINITAB,* is shown in Figure 1.9(a). For a large data set, however, such as the one in Figure 1.9(b), the dotplot can be uninformative and tedious to interpret.

FIGURE 1.9

Dotplots for small and large data sets

Stem and Leaf Plots

Another simple way to display the distribution of a quantitative data set is the **stem and leaf plot.** This plot presents a graphical display of the data using the actual numerical values of each data point.

MY PERSONAL TRAINER

How Do I Construct a Stem and Leaf Plot?

1. Divide each measurement into two parts: the **stem** and the **leaf.**
2. List the stems in a column, with a vertical line to their right.
3. For each measurement, record the leaf portion in the same row as its corresponding stem.
4. Order the leaves from lowest to highest in each stem.
5. Provide a key to your stem and leaf coding so that the reader can re-create the actual measurements if necessary.

EXAMPLE 1.7

Table 1.7 lists the prices (in dollars) of 19 different brands of walking shoes. Construct a stem and leaf plot to display the distribution of the data.

TABLE 1.7

Prices of Walking Shoes

90	70	70	70	75	70
65	68	60	74	70	95
75	70	68	65	40	65
70					

Solution To create the stem and leaf, you could divide each observation between the ones and the tens place. The number to the left is the stem; the number to the right is the leaf. Thus, for the shoes that cost $65, the stem is 6 and the leaf is 5. The stems, ranging from 4 to 9, are listed in Figure 1.10, along with the leaves for each of the 19 measurements. If you indicate that the leaf unit is 1, the reader will realize that the stem and leaf 6 and 8, for example, represent the number 68, recorded to the nearest dollar.

FIGURE 1.10

Stem and leaf plot for the data in Table 1.7

```
4 | 0                  Leaf unit = 1    4 | 0
5 |                                     5 |
6 | 5 8 0 8 5 5        Reordering ──→   6 | 0 5 5 5 8 8
7 | 0 0 0 5 0 4 0 5 0 0                 7 | 0 0 0 0 0 0 0 4 5 5
8 |                                     8 |
9 | 0 5                                 9 | 0 5
```

MY TIP

stem | leaf

Sometimes the available stem choices result in a plot that contains too few stems and a large number of leaves within each stem. In this situation, you can stretch the stems by dividing each one into several lines, depending on the leaf values assigned to them. Stems are usually divided in one of two ways:

- Into two lines, with leaves 0–4 in the first line and leaves 5–9 in the second line

- Into five lines, with leaves 0–1, 2–3, 4–5, 6–7, and 8–9 in the five lines, respectively

EXAMPLE 1.8

The data in Table 1.8 are the GPAs of 30 Bucknell University freshmen, recorded at the end of the freshman year. Construct a stem and leaf plot to display the distribution of the data.

TABLE 1.8

Grade Point Averages of 30 Bucknell University Freshmen

2.0	3.1	1.9	2.5	1.9
2.3	2.6	3.1	2.5	2.1
2.9	3.0	2.7	2.5	2.4
2.7	2.5	2.4	3.0	3.4
2.6	2.8	2.5	2.7	2.9
2.7	2.8	2.2	2.7	2.1

Solution The data, though recorded to an accuracy of only one decimal place, are measurements of the continuous variable x = GPA, which can take on values in the interval 0–4.0. By examining Table 1.8, you can quickly see that the highest and lowest GPAs are 3.4 and 1.9, respectively. But how are the remaining GPAs distributed?

If you use the decimal point as the dividing line between the stem and the leaf, you have only three stems, which does not produce a very good picture. Even if you divide each stem into two lines, there are only four stems, since the first line of stem 1 and the second line of stem 4 are empty! Dividing each stem into five lines produces the most descriptive plot, as shown in Figure 1.11. For these data, the leaf unit is .1, and the reader can infer that the stem and leaf 2 and 6, for example, represent the measurement $x = 2.6$.

FIGURE 1.11

Stem and leaf plot for the data in Table 1.8

1	9 9			1	9 9
2	0 1 1			2	0 1 1
2	3 2			2	2 3
2	5 4 5 5 5 5 4	Reordering →		2	4 4 5 5 5 5 5
2	7 6 7 6 7 7 7			2	6 6 7 7 7 7 7
2	9 8 8 9			2	8 8 9 9
3	1 0 1 0			3	0 0 1 1
3				3	
3	4	Leaf unit = .1		3	4

If you turn the stem and leaf plot sideways, so that the vertical line is now a horizontal axis, you can see that the data have "piled up" or been "distributed" along the axis in a pattern that can be described as "mound-shaped"—much like a pile of sand on the beach. One GPA was somewhat higher than the rest ($x = 3.4$), and the gap in the distribution shows that no GPAs were between 3.1 and 3.4.

Interpreting Graphs with a Critical Eye

Once you have created a graph or graphs for a set of data, what should you look for as you attempt to describe the data?

- First, check the horizontal and vertical **scales,** so that you are clear about what is being measured.
- Examine the **location** of the data distribution. Where on the horizontal axis is the center of the distribution? If you are comparing two distributions, are they both centered in the same place?
- Examine the **shape** of the distribution. Does the distribution have one "peak," a point that is higher than any other? If so, this is the most frequently occurring measurement or category. Is there more than one peak? Are there an approximately equal number of measurements to the left and right of the peak?
- Look for any unusual measurements or **outliers.** That is, are any measurements much bigger or smaller than all of the others? These outliers may not be representative of the other values in the set.

Distributions are often described according to their shapes.

Definition A distribution is **symmetric** if the left and right sides of the distribution, when divided at the middle value, form mirror images.

A distribution is **skewed to the right** if a greater proportion of the measurements lie to the right of the peak value. Distributions that are **skewed right** contain a few unusually large measurements.

A distribution is **skewed to the left** if a greater proportion of the measurements lie to the left of the peak value. Distributions that are **skewed left** contain a few unusually small measurements.

A distribution is **unimodal** if it has one peak; a **bimodal** distribution has two peaks. Bimodal distributions often represent a mixture of two different populations in the data set.

EXAMPLE 1.9

Examine the three dotplots generated by *MINITAB* and shown in Figure 1.12. Describe these distributions in terms of their locations and shapes.

FIGURE 1.12

Shapes of data distributions for Example 1.9

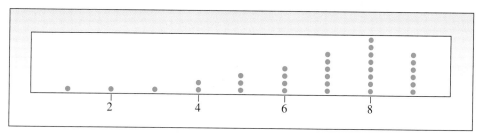

MY **TIP**

Symmetric ⇔ mirror images
Skewed right ⇔ long right tail
Skewed left ⇔ long left tail

Solution   The first dotplot shows a *relatively symmetric* distribution with a single peak located at $x = 4$. If you were to fold the page at this peak, the left and right halves would *almost* be mirror images. The second dotplot, however, is far from symmetric. It has a long "right tail," meaning that there are a few unusually large observations. If you were to fold the page at the peak, a larger proportion of measurements would be on the right side than on the left. This distribution is *skewed to the right*. Similarly, the third dotplot with the long "left tail" is *skewed to the left*.

EXAMPLE 1.10

A quality control analyst is interested in monitoring the weights of a particular style of walking sneaker. She enters the weights (in ounces) of eight randomly selected shoes into the database but accidentally misplaces the decimal point in the last entry:

9.72   9.74   9.70   9.71   9.71   9.73   9.72   .972

Use a dotplot to describe the data and uncover the analyst's mistake.

Solution   The dotplot of this small data set is shown in Figure 1.13(a). You can clearly see the *outlier* or unusual observation caused by the analyst's data entry error. Once the error has been corrected, as in Figure 1.13(b), you can see the correct distribution of the data set. Since this is a very small set, it is difficult to describe the shape of the distribution, although it seems to have a peak value around 9.72 and it appears to be relatively symmetric.

FIGURE 1.13

Distributions of weights for Example 1.10

 TIP

Outliers lie out, away from the main body of data.

When comparing graphs created for two data sets, you should compare their *scales of measurement, locations,* and *shapes,* and look for unusual measurements or outliers. Remember that outliers are not always caused by errors or incorrect data entry. Sometimes they provide very valuable information that should not be ignored. You may need additional information to decide whether an outlier is a valid measurement that is simply unusually large or small, or whether there has been some sort of mistake in the data collection. If the scales differ widely, be careful about making comparisons or drawing conclusions that might be inaccurate!

1.5 RELATIVE FREQUENCY HISTOGRAMS

A relative frequency histogram resembles a bar chart, but it is used to graph quantitative rather than qualitative data. The data in Table 1.9 are the GPAs of 30 Bucknell University freshmen, reproduced from Example 1.8 and shown as a dotplot in Figure 1.14(a). First, divide the interval from the smallest to the largest measurements into subintervals or *classes of equal length.* If you stack up the dots in each subinterval (Figure 1.14(b)), and draw a bar over each stack, you will have created a **frequency histogram** or a **relative frequency histogram,** depending on the scale of the vertical axis.

TABLE 1.9 **Grade Point Averages of 30 Bucknell University Freshmen**

2.0	3.1	1.9	2.5	1.9
2.3	2.6	3.1	2.5	2.1
2.9	3.0	2.7	2.5	2.4
2.7	2.5	2.4	3.0	3.4
2.6	2.8	2.5	2.7	2.9
2.7	2.8	2.2	2.7	2.1

FIGURE 1.14

How to construct a histogram

Definition

A **relative frequency histogram** for a quantitative data set is a bar graph in which the height of the bar shows "how often" (measured as a proportion or relative frequency) measurements fall in a particular class or subinterval. The classes or subintervals are plotted along the horizontal axis.

As a rule of thumb, the number of classes should range from 5 to 12; the more data available, the more classes you need.[†] The classes must be chosen so that each measurement falls into one and only one class. For the GPAs in Table 1.9, we decided to use eight intervals of equal length. Since the total span of the GPAs is

$$3.4 - 1.9 = 1.5$$

the minimum class width necessary to cover the range of the data is $(1.5 \div 8) = .1875$. For convenience, we round this approximate width up to .2. Beginning the first interval at the lowest value, 1.9, we form subintervals from 1.9 up to *but not including* 2.1, 2.1 up to *but not including* 2.3, and so on. By using the **method of left inclusion,** and including the left class boundary point but not the right boundary point in the class, we eliminate any confusion about where to place a measurement that happens to fall on a class boundary point.

Table 1.10 shows the eight classes, labeled from 1 to 8 for identification. The boundaries for the eight classes, along with a tally of the number of measurements that fall in each class, are also listed in the table. As with the charts in Section 1.3, you can now measure *how often* each class occurs using *frequency* or *relative frequency*.

[†]You can use this table as a guide for selecting an appropriate number of classes. Remember that this is only a guide; you may use more or fewer classes than the table recommends if it makes the graph more descriptive.

Sample Size	25	50	100	200	500
Number of Classes	6	7	8	9	10

To construct the relative frequency histogram, plot the class boundaries along the horizontal axis. Draw a bar over each class interval, with height equal to the relative frequency for that class. The relative frequency histogram for the GPA data, Figure 1.15, shows at a glance how GPAs are distributed over the interval 1.9 to 3.5.

TABLE 1.10 ● **Relative Frequencies for Data of Table 1.9**

Class	Class Boundaries	Tally	Class Frequency	Class Relative Frequency
1	1.9 to <2.1	III	3	3/30
2	2.1 to <2.3	III	3	3/30
3	2.3 to <2.5	III	3	3/30
4	2.5 to <2.7	ЖII	7	7/30
5	2.7 to <2.9	ЖII	7	7/30
6	2.9 to <3.1	IIII	4	4/30
7	3.1 to <3.3	II	2	2/30
8	3.3 to <3.5	I	1	1/30

MY TIP

Relative frequencies add to 1—
Frequencies add to *n*.

FIGURE 1.15
Relative frequency histogram

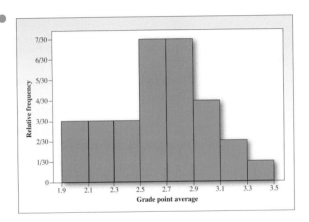

EXAMPLE 1.11 — Twenty-five households are polled in a marketing survey, and Table 1.11 lists the numbers of quarts of milk purchased during a particular week. Construct a relative frequency histogram to describe the data.

TABLE 1.11 ● **Quarts of Milk Purchased by 25 Households**

0	3	5	4	3
2	1	3	1	2
1	1	2	0	1
4	3	2	2	2
2	2	2	3	4

Solution The variable being measured is "number of quarts of milk," which is a discrete variable that takes on only integer values. In this case, it is simplest to choose the classes or subintervals as the integer values over the range of observed values: 0, 1, 2, 3, 4, and 5. Table 1.12 shows the classes and their corresponding frequencies and relative frequencies. The relative frequency histogram, generated using *MINITAB*, is shown in Figure 1.16.

TABLE 1.12 ● **Frequency Table for Example 1.11**

Number of Quarts	Frequency	Relative Frequency
0	2	.08
1	5	.20
2	9	.36
3	5	.20
4	3	.12
5	1	.04

FIGURE 1.16 ●

MINITAB histogram for Example 1.11

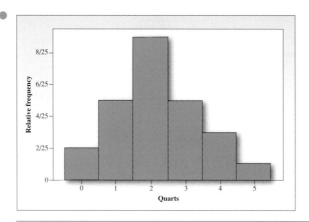

MY PERSONAL TRAINER

How Do I Construct a Relative Frequency Histogram?

1. Choose the number of classes, usually between 5 and 12. The more data you have, the more classes you should use.

2. Calculate the approximate class width by dividing the difference between the largest and smallest values by the number of classes.

3. Round the approximate class width up to a convenient number.

4. If the data are discrete, you might assign one class for each integer value taken on by the data. For a large number of integer values, you may need to group them into classes.

5. Locate the class boundaries. The lowest class must include the smallest measurement. Then add the remaining classes using the left inclusion method.

6. Construct a statistical table containing the classes, their frequencies, and their relative frequencies.

7. Construct the histogram like a bar graph, plotting class intervals on the horizontal axis and relative frequencies as the heights of the bars.

(continued)

Exercise Reps

A. For the following data sets, find the range, the minimum class width, and a convenient class width.

Number of Measurements	Smallest and Largest Values	Number of Classes	Range	Minimum Class Width	Convenient Class Width
50	10 to 100	7			
25	0.1 to 6.0	6			
100	500 to 700	8			

B. For the same data sets, select a convenient starting point, and list the class boundaries for the first two classes.

Number of Measurements	Smallest and Largest Values	Convenient Starting Point	First Two Classes
50	10 to 100		
25	0.1 to 6.0		
100	500 to 700		

Progress Report

- Still having trouble? Try again using the Exercise Reps at the end of this section.
- Mastered relative frequency histograms? You can skip the Exercise Reps and go straight to the Basic Techniques Exercises at the end of this section.

Answers are located on the perforated card at the back of this book.

A relative frequency histogram can be used to describe the distribution of a set of data in terms of its *location* and *shape,* and to check for *outliers* as you did with other graphs. For example, both the GPA and the "milk" data were relatively symmetric, with no unusual measurements. Since the bar constructed above each class represents the *relative frequency* or proportion of the measurements in that class, these heights can be used to give us further information:

- The proportion of the measurements that fall in a particular class or group of classes
- The probability that a measurement drawn at random from the set will fall in a particular class or group of classes

Consider the relative frequency histogram for the GPA data in Figure 1.15. What proportion of the students had GPAs of 2.7 or higher? This involves all classes beyond 2.7 in Table 1.10. Because there are 14 students in those classes, the proportion who have GPAs of 2.7 or higher is 14/30, or approximately 47%. This is also the percentage of the total area under the histogram in Figure 1.15 that lies to the right of 2.7.

Suppose you wrote each of the 30 GPAs on a piece of paper, put them in a hat, and drew one at random. What is the chance that this piece of paper contains a GPA of 2.7 or higher? Since 14 of the 30 pieces of paper fall in this category, you have 14 chances out of 30; that is, the probability is 14/30. The word *probability* is not unfamiliar to you; we will discuss it in more detail in Chapter 4.

Although we are interested in describing the set of $n = 30$ measurements, we might also be interested in the population from which the sample was drawn, which is the set of GPAs of all freshmen currently in attendance at Bucknell University. Or, if we are interested in the academic achievement of college freshmen in general, we might consider our sample as representative of the population of GPAs for freshmen attending Bucknell or colleges *similar* to Bucknell. A sample histogram provides valuable information about the population histogram—the graph that describes the distribution of the entire population. Remember, though, that different samples from the same population will produce *different* histograms, even if you use the same class boundaries. However, you can expect that the sample and population histograms will be similar. As you add more and more data to the sample, the two histograms become more and more alike. If you enlarge the sample to include the entire population, the two histograms are identical!

1.5 EXERCISES

EXERCISE REPS

1.16 For the following data sets, find the range, the minimum class width, and a convenient class width.

Number of Measurements	Smallest and Largest Values	Number of Classes	Range	Minimum Class Width	Convenient Class Width
75	0.5 to 1.0	8			
25	0 to 100	6			
200	1,200 to 1,500	9			

1.17 Refer to Exercise 1.16. For the same data sets, select a convenient starting point, and list the class boundaries for the first two classes.

Number of Measurements	Smallest and Largest Values	Convenient Starting Point	First Two Classes
75	0.5 to 1.0		
25	0 to 100		
200	1,200 to 1,500		

BASIC TECHNIQUES

MY CD **1.18** Construct a stem and leaf plot for these 50

EX0118 measurements:

3.1	4.9	2.8	3.6	2.5	4.5	3.5	3.7	4.1	4.9
2.9	2.1	3.5	4.0	3.7	2.7	4.0	4.4	3.7	4.2
3.8	6.2	2.5	2.9	2.8	5.1	1.8	5.6	2.2	3.4
2.5	3.6	5.1	4.8	1.6	3.6	6.1	4.7	3.9	3.9
4.3	5.7	3.7	4.6	4.0	5.6	4.9	4.2	3.1	3.9

a. Describe the shape of the data distribution. Do you see any outliers?

b. Use the stem and leaf plot to find the smallest observation.

c. Find the eighth and ninth largest observations.

1.19 Refer to Exercise 1.18. Construct a relative frequency histogram for the data.

a. Approximately how many class intervals should you use?

b. Suppose you decide to use classes starting at 1.6 with a class width of .5 (i.e., 1.6 to <2.1, 2.1 to <2.6). Construct the relative frequency histogram for the data.

c. What fraction of the measurements are less than 5.1?

d. What fraction of the measurements are larger than 3.6?

e. Compare the relative frequency histogram with the stem and leaf plot in Exercise 1.18. Are the shapes similar?

MY CD **1.20** Consider this set of data:

EX0120

4.5	3.2	3.5	3.9	3.5	3.9
4.3	4.8	3.6	3.3	4.3	4.2
3.9	3.7	4.3	4.4	3.4	4.2
4.4	4.0	3.6	3.5	3.9	4.0

a. Construct a stem and leaf plot by using the leading digit as the stem.

b. Construct a stem and leaf plot by using each leading digit twice. Does this technique improve the presentation of the data? Explain.

1.21 A discrete variable can take on only the values 0, 1, or 2. A set of 20 measurements on this variable is shown here:

1	2	1	0	2
2	1	1	0	0
2	2	1	1	0
0	1	2	1	1

a. Construct a relative frequency histogram for the data.

b. What proportion of the measurements are greater than 1?

c. What proportion of the measurements are less than 2?

d. If a measurement is selected at random from the 20 measurements shown, what is the probability that it is a 2?

e. Describe the shape of the distribution. Do you see any outliers?

1.22 Refer to Exercise 1.21.

a. Draw a dotplot to describe the data.

b. How could you define the stem and the leaf for this data set?

c. Draw the stem and leaf plot using your decision from part b.

d. Compare the dotplot, the stem and leaf plot, and the relative frequency histogram (Exercise 1.21). Do they all convey roughly the same information?

1.23 Navigating a Maze An experimental psychologist measured the length of time it took for a rat to successfully navigate a maze on each of five days. The results are shown in the table. Create a line chart to describe the data. Do you think that any learning is taking place?

Day	1	2	3	4	5
Time (sec.)	45	43	46	32	25

MY CD **1.24** Measuring over Time The value of a

EX0124 quantitative variable is measured once a year for a 10-year period. Here are the data:

Year	Measurement	Year	Measurement
1	61.5	6	58.2
2	62.3	7	57.5
3	60.7	8	57.5
4	59.8	9	56.1
5	58.0	10	56.0

a. Create a line chart to describe the variable as it changes over time.

b. Describe the measurements using the chart constructed in part a.

MY CD **1.25** Test Scores The test scores on a

EX0125 100-point test were recorded for 20 students:

61	93	91	86	55	63	86	82	76	57
94	89	67	62	72	87	68	65	75	84

a. Use an appropriate graph to describe the data.

b. Describe the shape and location of the scores.

c. Is the shape of the distribution unusual? Can you think of any reason the distribution of the scores would have such a shape?

APPLICATIONS

(MY) CD **1.26** A Recurring Illness The length of time
EX0126 (in months) between the onset of a particular illness and its recurrence was recorded for $n = 50$ patients:

2.1	4.4	2.7	32.3	9.9	9.0	2.0	6.6	3.9	1.6
14.7	9.6	16.7	7.4	8.2	19.2	6.9	4.3	3.3	1.2
4.1	18.4	.2	6.1	13.5	7.4	.2	8.3	.3	1.3
14.1	1.0	2.4	2.4	18.0	8.7	24.0	1.4	8.2	5.8
1.6	3.5	11.4	18.0	26.7	3.7	12.6	23.1	5.6	.4

a. Construct a relative frequency histogram for the data.

b. Would you describe the shape as roughly symmetric, skewed right, or skewed left?

c. Give the fraction of recurrence times less than or equal to 10 months.

1.27 Education Pays Off! Education pays off, according to a snapshot provided in a report to the city of Riverside by the Riverside County Office of Education.[7] The average annual incomes for six different levels of education are shown in the table:

Educational Level	Average Annual Income
High school graduate	$26,795
Some college, no degree	29,095
Bachelor's degree	50,623
Master's degree	63,592
Doctorate	85,675
Professional (Doctor, Lawyer)	101,375

Source: U.S. Census Bureau

a. What graphical methods could you use to describe the data?

b. Select the method from part a that you think best describes the data.

c. How would you summarize the information that you see in the graph regarding educational levels and salary?

(MY) CD **1.28** Preschool The ages (in months) at
EX0128 which 50 children were first enrolled in a preschool are listed below.

38	40	30	35	39	40	48	36	31	36
47	35	34	43	41	36	41	43	48	40
32	34	41	30	46	35	40	30	46	37
55	39	33	32	32	45	42	41	36	50
42	50	37	39	33	45	38	46	36	31

a. Construct a stem and leaf display for the data.

b. Construct a relative frequency histogram for these data. Start the lower boundary of the first class at 30 and use a class width of 5 months.

c. Compare the graphs in parts a and b. Are there any significant differences that would cause you to choose one as the better method for displaying the data?

d. What proportion of the children were 35 months (2 years, 11 months) or older, but less than 45 months (3 years, 9 months) of age when first enrolled in preschool?

e. If one child were selected at random from this group of children, what is the probability that the child was less than 50 months old (4 years, 2 months) when first enrolled in preschool?

(MY) CD **1.29** Happy in the Air? Of the 6,229 con-
EX0129 sumer complaints against major U.S. airlines in a recent year, the distribution by airline is shown in the table.[8]

Airline	Number of Complaints	Airline	Number of Complaints
United	1172	American	1212
America West	318	U.S. Airways	532
Northwest	765	Alaska	129
Continental	563	American Eagle	71
Delta	1231	Southwest	236

Source: Time Almanac 2004

a. Construct a pie chart to describe the total number of complaints by airline.

b. Order the airlines from the smallest to the largest number of complaints. Construct a Pareto chart to describe the data. Which display is more effective?

c. Is there another variable that you could measure that might help to explain why some airlines have many more complaints than others? Explain.

(MY) CD **1.30** How Long Is the Line? To decide on
EX0130 the number of service counters needed for stores to be built in the future, a supermarket chain wanted to obtain information on the length of time (in minutes) required to service customers. To find the distribution of customer service times, a sample of 1,000 customers' service times was recorded. Sixty of these are shown here:

3.6	1.9	2.1	.3	.8	.2	1.0	1.4	1.8	1.6
1.1	1.8	.3	1.1	.5	1.2	.6	1.1	.8	1.7
1.4	.2	1.3	3.1	.4	2.3	1.8	4.5	.9	.7
.6	2.8	2.5	1.1	.4	1.2	.4	1.3	.8	1.3
1.1	1.2	.8	1.0	.9	.7	3.1	1.7	1.1	2.2
1.6	1.9	5.2	.5	1.8	.3	1.1	.6	.7	.6

a. Construct a stem and leaf plot for the data.

b. What fraction of the service times are less than or equal to 1 minute?

c. What is the smallest of the 60 measurements?

1.31 Service Times, continued Refer to Exercise 1.30. Construct a relative frequency histogram for the supermarket service times.

a. Describe the shape of the distribution. Do you see any outliers?

b. Assuming that the outliers in this data set are valid observations, how would you explain them to the management of the supermarket chain?

c. Compare the relative frequency histogram with the stem and leaf plot in Exercise 1.30. Do the two graphs convey the same information?

1.32 Calcium Content The calcium (Ca) content of a powdered mineral substance was analyzed ten times with the following percent compositions recorded:

.0271	.0282	.0279	.0281	.0268
.0271	.0281	.0269	.0275	.0276

a. Draw a dotplot to describe the data. (HINT: The scale of the horizontal axis should range from .0260 to .0290.)

b. Draw a stem and leaf plot for the data. Use the numbers in the hundredths and thousandths places as the stem.

c. Are any of the measurements inconsistent with the other measurements, indicating that the technician may have made an error in the analysis?

1.33 American Presidents Listed below are the ages at the time of death for the 38 American presidents from George Washington to Ronald Reagan:[5]

Washington	67	Garfield	49
J. Adams	90	Arthur	56
Jefferson	83	Cleveland	71
Madison	85	B. Harrison	67
Monroe	73	Cleveland	71
J. Q. Adams	80	McKinley	58
Jackson	78	T. Roosevelt	60
Van Buren	79	Taft	72
W. H. Harrison	68	Wilson	67
Tyler	71	Harding	57
Polk	53	Coolidge	60
Taylor	65	Hoover	90
Fillmore	74	F. D. Roosevelt	63
Pierce	64	Truman	88
Buchanan	77	Eisenhower	78
Lincoln	56	Kennedy	46
A. Johnson	66	L. Johnson	64
Grant	63	Nixon	81
Hayes	70	Reagan	93

a. Before you graph the data, try to visualize the distribution of the ages at death for the presidents. What shape do you think it will have?

b. Construct a stem and leaf plot for the data. Describe the shape. Does it surprise you?

c. The five youngest presidents at the time of death appear in the lower "tail" of the distribution. Three of the five youngest have one common trait. Identify the five youngest presidents at death. What common trait explains these measurements?

1.34 RBC Counts The red blood cell count of a healthy person was measured on each of 15 days. The number recorded is measured in 10^6 cells per microliter (μL).

5.4	5.2	5.0	5.2	5.5
5.3	5.4	5.2	5.1	5.3
5.3	4.9	5.4	5.2	5.2

a. Use an appropriate graph to describe the data.

b. Describe the shape and location of the red blood cell counts.

c. If the person's red blood cell count is measured today as $5.7 \times 10^6/\mu L$, would you consider this unusual? What conclusions might you draw?

1.35 Batting Champions The officials of major league baseball have crowned a batting champion in the National League each year since 1876. A sample of winning batting averages is listed in the table:[5]

Year	Name	Average
1876	Roscoe Barnes	.403
1893	Hugh Duffy	.378
1915	Larry Doyle	.320
1917	Edd Roush	.341
1934	Paul Waner	.362
1911	Honus Wagner	.334
1898	Willie Keeler	.379
1924	Roger Hornsby	.424
1963	Tommy Davis	.326
1992	Gary Sheffield	.330
1954	Willie Mays	.345
1975	Bill Madlock	.354
1958	Richie Ashburn	.350
1942	Ernie Lombardi	.330
1948	Stan Musial	.376
1971	Joe Torre	.363
1996	Tony Gwynn	.353
1961	Roberto Clemente	.351
1968	Pete Rose	.335
1885	Roger Connor	.371

a. Construct a relative frequency histogram to describe the batting averages for these 20 champions.

b. If you were to randomly choose one of the 20 names, what is the chance that you would choose a player whose average was above .400 for his championship year?

 1.36 Top 20 Movies The table below shows the weekend gross ticket sales for the top 20 movies during the week of January 23–25, 2004:[9]
EX0136

Movie	Weekend Gross ($ millions)
1. *The Butterfly Effect*	$17.1
2. *Along Came Polly*	16.4
3. *Win a Date with Tad Hamilton*	7.3
4. *Big Fish*	7.1
5. *Lord of the Rings: The Return of the King*	6.8
6. *Cheaper by the Dozen*	6.4
7. *Cold Mountain*	5.0
8. *Torque*	4.5
9. *Something's Gotta Give*	4.0
10. *Mystic River*	3.4
11. *Calendar Girls*	2.4
12. *The Last Samurai*	2.2
13. *Monster*	2.1
14. *My Baby's Daddy*	2.0
15. *Disney's Teacher's Pet*	1.4
16. *Peter Pan*	1.2
17. *Paycheck*	1.2
18. *The Cooler*	0.9
19. *Mona Lisa Smile*	0.9
20. *21 Grams*	0.9

Source: *Entertainment Weekly*

a. Draw a stem and leaf plot for the data. Describe the shape of the distribution. Are there any outliers?

b. Construct a dotplot for the data. Which of the two graphs is more informative? Explain.

 1.37 Hazardous Waste How safe is your
EX0137 neighborhood? Are there any hazardous waste

sites nearby? The table shows the number of hazardous waste sites in each of the 50 states and the District of Columbia in the year 2003:[5]

AL	15	HI	3	MA	32	NM	12	SD	2
AK	6	ID	9	MI	69	NY	91	TN	13
AZ	9	IL	45	MN	24	NC	29	TX	45
AR	12	IN	29	MS	4	ND	0	UT	19
CA	99	IA	14	MO	25	OH	35	VT	9
CO	18	KS	12	MT	15	OK	11	VA	30
CT	16	KY	14	NE	11	OR	12	WA	47
DE	15	LA	15	NV	1	PA	96	WV	9
DC	1	ME	12	NH	20	RI	12	WI	40
FL	52	MD	19	NJ	116	SC	25	WY	2
GA	15								

a. What variable is being measured? Is the variable discrete or continuous?

b. A stem and leaf plot generated by *MINITAB* is shown here. Describe the shape of the data distribution. Identify the unusually large measurements marked "HI" by state.

Stem-and-Leaf Display: Hazardous Waste

```
Stem-and-leaf of Hazardous Waste
N = 51   Leaf Unit = 1.0
    7   0  0112234
   12   0  69999
   23   1  11222222344
   (9)  1  555556899
   19   2  04
   17   2  5599
   13   3  02
   11   3  5
   10   4  0
    9   4  557
    6   5  2
        HI   69,   91,   96,   99,  116
```

c. Can you think of any reason these five states would have a large number of hazardous waste sites? What other variable might you measure to help explain why the data behave as they do?

As you continue to work through the exercises in this chapter, you will become more experienced in recognizing different types of data and in determining the most appropriate graphical method to use. Remember that the type of graphic you use is not as important as the interpretation that accompanies the picture. Look for these important characteristics:

- Location of the center of the data
- Shape of the distribution of data
- Unusual observations in the data set

Using these characteristics as a guide, you can interpret and compare sets of data using graphical methods, which are only the first of many statistical tools that you will soon have at your disposal.

CHAPTER REVIEW

Key Concepts

I. How Data Are Generated

1. Experimental units, variables, measurements
2. Samples and populations
3. Univariate, bivariate, and multivariate data

II. Types of Variables

1. Qualitative or categorical
2. Quantitative
   a. Discrete
   b. Continuous

III. Graphs for Univariate Data Distributions

1. Qualitative or categorical data
   a. Pie charts
   b. Bar charts

2. Quantitative data
   a. Pie and bar charts
   b. Line charts
   c. Dotplots
   d. Stem and leaf plots
   e. Relative frequency histograms

3. Describing data distributions
   a. Shapes—symmetric, skewed left, skewed right, unimodal, bimodal
   b. Proportion of measurements in certain intervals
   c. Outliers

 APPLET

Easy access to the Web has made it possible for you to understand statistical concepts using an interactive Web tool called an **applet.** These applets provide visual reinforcement for the concepts that have been presented in the chapter. Sometimes you will be able to perform statistical experiments, sometimes you will be able to interact with a statistical graph to change its form, and sometimes you will be able to use the applet as an interactive "statistical table." At the end of each chapter, you will find exercises designed specifically for use with a particular applet.

Rather than having to access these applets online, they have been included on the CD-ROM that accompanies each new copy of the text. When you insert the CD-ROM into the appropriate drive of your computer, the main menu will appear. Click on "MyApplet." If necessary, follow the instructions to download the latest web browser and/or Java plug-in, or just click the button to load the applets. Your web browser will open the index of applets, organized by chapter and name. When you click a particular applet title, the applet will appear in your browser. To return to the index of applets, simply click the link at the bottom of the page.

Dotplots

Click the Chapter 1 applet called **Building a Dotplot.** If you move your cursor over the applet marked **Dotplot Demo** you will see a green line with a value that changes as you move along the horizontal axis. When you left-click your mouse, a dot will appear at that point on the dotplot. If two measurements are identical, the dots will pile up on top of each other (Figure 1.17). Follow the directions in the **Dotplot Demo,** using the sample data given there. If you make a mistake, the applet will tell you. The second applet will not correct your mistakes; you can add as many dots as you want!

FIGURE 1.17

Building a Dotplot applet

Histograms

Click the Chapter 1 applet called **Building a Histogram.** If you scroll down to the applet marked **Histogram Demo,** you will see the interval boundaries (or interval midpoints) for the histogram along the horizontal axis. As you move the mouse across the graph, a light gray box will show you where the measurement will be added at your next mouse click. When you release the mouse, the box turns dark teal. The distribution in Figure 1.18 contains two 0s, one 1, three 2s, and one 3. Follow the directions in the **Histogram Demo** using the sample data given there. Click the link to compare your results to the correct histogram. The second applet will be used for some of the MyApplet Exercises.

Click the applet called **Flipping Fair Coins,** and scroll down to the applet marked **sample size = 3.** The computer will collect some data by "virtually" tossing 3 coins and recording the quantitative discrete variable

$$x = \text{number of heads observed}$$

Click on "New Coin Flip." You will see the result of your three tosses in the upper-left-hand corner, along with the value of x. For the experiment in Figure 1.19 we observed $x = 2$.

The applet begins to build a relative frequency histogram to describe the data set, which at this point contains only one observation. Click "New Coin Flip" a few more times. Watch the coins appear, along with the value of x, and watch the

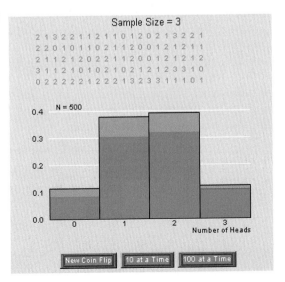

relative frequency histogram grow. The red area (light teal in Figures 1.19 and 1.20) represents the current data added to the histogram, and the dark teal area in Figure 1.20 is contributed from the previous coin flips. You can flip the three coins 10 at a time or 100 at a time to generate data more quickly.

Figure 1.20 shows the relative frequency histogram for 500 observations in our data set. Your data set will look a little different. However, it should have the same approximate shape—it should be relatively symmetric. For our histogram, we can say that the values $x = 0$ and $x = 3$ occurred about 12–13% of the time, while the values $x = 1$ and $x = 2$ occurred between 38% and 40% of the time. Does your histogram produce similar results?

 MINITAB

Introduction to *MINITAB*™

MINITAB is a computer software package that is available in many forms for different computer environments. The current version of *MINITAB* at the time of this printing is *MINITAB 14*, which is used in the Windows environment. We will assume that you are familiar with Windows. If not, perhaps a lab or teaching assistant can help you to master the basics.

Once you have started Windows, there are two ways to start *MINITAB*:

- If there is a *MINITAB* shortcut icon on the desktop double-click on the icon.
- Click the Start button on the taskbar. Follow the menus, highlighting **All Programs** → *MINITAB* **14 for Windows** → *MINITAB* **14.** Click on *MINITAB 14* to start the program.

When *MINITAB* is opened, the main *MINITAB* screen will be displayed (see Figure 1.21). It contains two windows: the Data window and the Session window. Clicking

FIGURE 1.21

![MINITAB screen showing the menu bar with File, Edit, Data, Calc, Stat, Graph, Editor, Tools, Window, Help; a Session window dated 3/1/2004 3:43:20 PM with "Welcome to Minitab, press F1 for help."; and Worksheet 1 with columns C1 through C8.]

anywhere on the window will make that window active so that you can either enter data or type commands. Although it is possible to manually type *MINITAB* commands in the Session window, we choose to use the Windows approach, which will be familiar to most of you. If you prefer to use the typed commands, consult the *MINITAB* manual for detailed instructions.

At the top of the Session window, you will see a Menu bar. Highlighting and clicking on any command on the Menu bar will cause a menu to drop down, from which you may then select the necessary command. We will use the standard notation to indicate a sequence of commands from the drop-down menus. For example, **File →
Open Worksheet** will allow you to retrieve a "worksheet"—a set of data from the Data window—which you have previously saved. To close the program, the command sequence is **File → Exit.**

MINITAB 14 allows multiple worksheets to be saved as "projects." When you are working on a project, you can add new worksheets or open worksheets from other projects to add to your current project. As you become more familiar with *MINITAB*, you will be able to organize your information into either "worksheets" or "projects," depending on the complexity of your task. If you are using an earlier version of *MINITAB*, you only need to open, use, and save stand-alone worksheets.

Graphing with *MINITAB*

The first data set to be graphed consists of qualitative data whose frequencies have already been recorded. The class status of 105 students in an introductory statistics class are listed in Table 1.13. Before you enter the data into the Minitab Data window, start a project called "Chapter 1" by highlighting **File → New.** A Dialog box called "New" will appear. Highlight **Minitab project** and click **OK.** Before you continue, let's save this project as "Chapter 1" using the series of commands **File → Save Project.** Type **Chapter 1** in the File Name box, and select a location using the white box marked "Save in:" at the top of the Dialog box. Click **Save.** In the Data window at the top of the screen, you will see your new project name, "Chapter 1.MPJ."

TABLE 1.13 ● **Status of Students in Statistics Class**

Status	Freshman	Sophomore	Junior	Senior	Grad Student
Frequency	5	23	32	35	10

To enter the data into the worksheet, click on the gray cell just below the name C1 in the Data window. You can enter your own descriptive name for the categories—possibly "Status." Now use the down arrow ↓ or your mouse to continue down column C1, entering the five status descriptions. Notice that the name **C1** has changed to **C1-T** because you are entering text rather than numbers. Continue by naming column 2 (C2) "Frequency," and enter the five numerical frequencies into C2. The Data window will appear as in Figure 1.22.

To construct a pie chart for these data, click on **Graph → Pie Chart,** and a Dialog box will appear (see Figure 1.23). In this box, you must specify how you want to create the chart. Click the radio button marked **Chart values from a table.** Then place your cursor in the box marked "Categorical variable." Either (1) highlight C1 in the list at the left and choose **Select,** (2) double-click on C1 in the list at the left, or (3) type C1 in the "Categorical variable" box. Similarly, place the cursor in the box marked

FIGURE 1.22

FIGURE 1.23

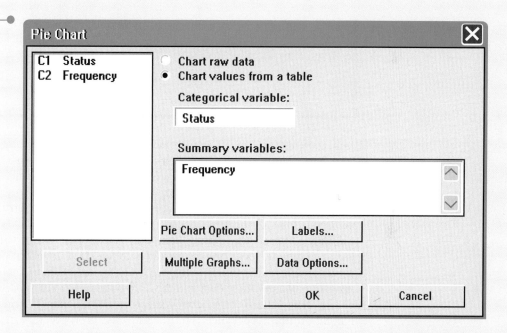

"Summary variables" and select C2. Click **Labels** and select the tab marked **Slice Labels.** Check the boxes marked "Category names" and "Percent." When you click **OK,** *MINITAB* will create the pie chart in Figure 1.24. We have removed the legend, by selecting and deleting it.

As you become more proficient at using the pie chart command, you may want to take advantage of some of the options available. Once the chart is created, *right-click* on the pie chart and select **Edit Pie.** You can change the colors and format of the chart, "explode" important sectors of the pie, and change the order of the categories. If you *right-click* on the pie chart and select **Update Graph Automatically,** the pie chart will automatically update when you change the data in columns C1 and C2 of the Minitab worksheet.

If you would rather construct a bar chart, use the command **Graph → Bar Chart.** In the Dialog box that appears, choose **Simple.** Choose an option in the "Bars represent" drop-down list, depending on the way that the data has been entered into the

FIGURE 1.24

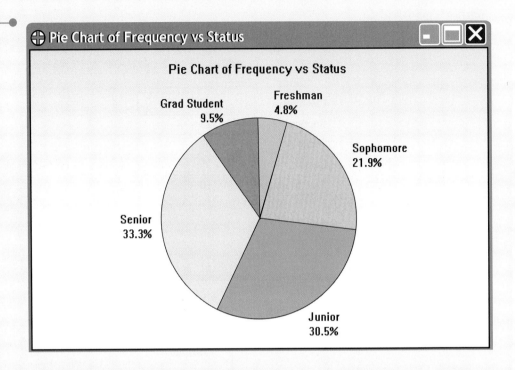

worksheet. For the data in Table 1.13, we choose "Values from a table" and click **OK.** When the Dialog box appears, place your cursor in the "Graph variables" box and **select** C2. Place your cursor in the "Categorical variable" box, and **select** C1. Click **OK** to finish the bar chart, shown in Figure 1.25. Once the chart is created, *right-click* on various parts of the bar chart and choose **Edit** to change the look of the chart.

MINITAB can create dotplots, stem and leaf plots, and histograms for quantitative data. The top 40 stocks on the over-the-counter (OTC) market, ranked by percentage of outstanding shares traded on a particular day, are listed in Table 1.14. Although we could simply enter these data into the third column (C3) of Worksheet 1 in the "Chapter 1" project, let's start a new worksheet within "Chapter 1" using **File → New,** highlighting **Minitab Worksheet,** and clicking **OK.** Worksheet 2 will appear on the screen. Enter the data into column C1 and name them "Stocks" in the gray cell just below the C1.

TABLE 1.14

Percentage of OTC Stocks Traded

11.88	6.27	5.49	4.81	4.40	3.78	3.44	3.11	2.88	2.68
7.99	6.07	5.26	4.79	4.05	3.69	3.36	3.03	2.74	2.63
7.15	5.98	5.07	4.55	3.94	3.62	3.26	2.99	2.74	2.62
7.13	5.91	4.94	4.43	3.93	3.48	3.20	2.89	2.69	2.61

To create a dotplot, use **Graph → Dotplot.** In the Dialog box that appears, choose **One Y → Simple** and click **OK.** To create a stem and leaf plot, use **Graph → Stem-and-Leaf.** For either graph, place your cursor in the "Graph variables" box, and select "Stocks" from the list to the left (see Figure 1.26).

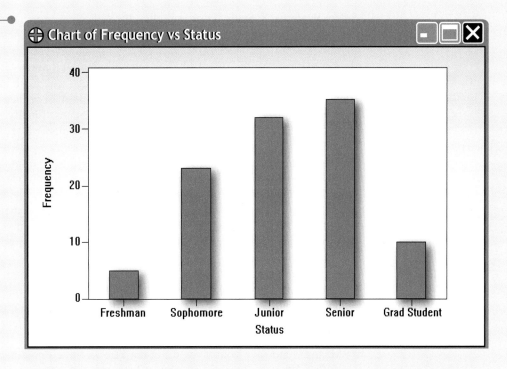

You can choose from a variety of formatting options before clicking **OK.** The dot-plot appears as a graph, while the stem and leaf plot appears in the Session window. To print either a Graph window or the Session window, click on the window to make it active and use **File** → **Print Graph** (or **Print Session Window**).

To create a histogram, use **Graph** → **Histogram.** In the Dialog box that appears, choose **Simple** and click **OK,** selecting "Stocks" for the "Graph variables" box.

Select **Scale** → **Y-Scale Type** and click the radio button marked "Frequency." (You can edit the histogram later to show relative frequencies.) Click **OK.** Once the histogram has been created, *right-click* on the *Y*-axis and choose **Edit Y Scale.** Under the tab marked "Scale," you can click the radio button marked "Position of ticks" and type in **0 5 10 15.** Then click the tab marked "Labels," the radio button marked "Specified" and type **0 5/40 10/40 15/40.** Click **OK.** This will reduce the number of ticks on the *y*-axis and change them to relative frequencies. Finally, double-click on the word "Frequency" along the *y*-axis. Change the box marked "Text" to read "Relative frequency" and click **OK.**

To adjust the type of boundaries for the histogram, *right-click* on the bars of the histogram and choose **Edit Bars.** Use the tab marked "Binning" to choose either "Cutpoints" or "Midpoints" for the histogram; you can specify the cutpoint or midpoint positions if you want. In this same **Edit** box, you can change the colors, fill type, and font style of the histogram. If you *right-click* on the bars and select **Update Graph Automatically,** the histogram will automatically update when you change the data in the "Stocks" column.

As you become more familiar with *MINITAB* for Windows, you can explore the various options available for each type of graph. It is possible to plot more than one variable at a time, to change the axes, to choose the colors, and to modify graphs in many ways. However, even with the basic default commands, it is clear that the distribution of OTC stocks is highly skewed to the right.

Make sure to save your work using the **File** → **Save Project** command before you exit *MINITAB*!

FIGURE 1.27

Supplementary Exercises

1.38 Quantitative or Qualitative? Identify each variable as quantitative or qualitative:

a. Ethnic origin of a candidate for public office

b. Score (0–100) on a placement examination

c. Fast-food establishment preferred by a student (McDonald's, Burger King, or Carl's Jr.)

d. Mercury concentration in a sample of tuna

1.39 Symmetric or Skewed? Do you expect the distributions of the following variables to be symmetric or skewed? Explain.

a. Size in dollars of nonsecured loans

b. Size in dollars of secured loans

c. Price of an 8-ounce can of peas

d. Height in inches of freshman women at your university

e. Number of broken taco shells in a package of 100 shells

f. Number of ticks found on each of 50 trapped cottontail rabbits

1.40 Continuous or Discrete? Identify each variable as continuous or discrete:

a. Number of homicides in Detroit during a one-month period

b. Length of time between arrivals at an outpatient clinic

c. Number of typing errors on a page of manuscript

d. Number of defective lightbulbs in a package containing four bulbs

e. Time required to finish an examination

1.41 Continuous or Discrete, again Identify each variable as continuous or discrete:

a. Weight of two dozen shrimp

b. A person's body temperature

c. Number of people waiting for treatment at a hospital emergency room

d. Number of properties for sale by a real estate agency

e. Number of claims received by an insurance company during one day

1.42 Continuous or Discrete, again Identify each variable as continuous or discrete:

a. Number of people in line at a supermarket checkout counter

b. Depth of a snowfall

c. Length of time for a driver to respond when faced with an impending collision

d. Number of aircraft arriving at the Atlanta airport in a given hour

(MY) CD **1.43 Aqua Running** Aqua running has been
EX0143 suggested as a method of cardiovascular conditioning for injured athletes and others who want a low-impact aerobics program. A study reported in the *Journal of Sports Medicine* investigated the relationship between exercise cadence and heart rate by measuring the heart rates of 20 healthy volunteers at a cadence of 48 cycles per minute (a cycle consisted of two steps).[10] The data are listed here:

87	109	79	80	96	95	90	92	96	98
101	91	78	112	94	98	94	107	81	96

Construct a stem and leaf plot to describe the data. Discuss the characteristics of the data distribution.

1.44 Princeton Review An advertising flyer for the *Princeton Review*, a review course designed for high school students taking the SAT tests, presents the accompanying bar graph to show the average score improvements for students using various study methods.[11]

a. What graphical techniques did the *Princeton Review* use to make their average improvement figures look as dramatic as possible?

b. If you were in charge of promoting a review course at your high school, how would you modify the graph to make the average improvement for students using a school review course look more impressive?

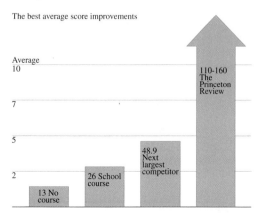

The best average score improvements

(MY) CD **1.45 Ages of Pennies** We collected 50 pen-
EX0145 nies and recorded their ages, by calculating
AGE = CURRENT YEAR − YEAR ON PENNY.

5	1	9	1	2	20	0	25	0	17
1	4	4	3	0	25	3	3	8	28
5	21	19	9	0	5	0	2	1	0
0	1	19	0	2	0	20	16	22	10
19	36	23	0	1	17	6	0	5	0

a. Before drawing any graphs, try to visualize what
   the distribution of penny ages will look like. Will
   it be mound-shaped, symmetric, skewed right, or
   skewed left?

b. Draw a relative frequency histogram to describe
   the distribution of penny ages. How would you describe the shape of the distribution?

(MY) CD **1.46 Ages of Pennies, continued** The data
EX0146 below represent the ages of a different set of
50 pennies, again calculated using AGE = CURRENT
YEAR − YEAR ON PENNY.

41	9	0	4	3	0	3	8	21	3
2	10	4	0	14	0	25	12	24	19
3	1	14	7	2	4	4	5	1	20
14	9	3	5	3	0	8	17	16	0
0	7	3	5	23	7	28	17	9	2

a. Draw a relative frequency histogram to describe
   the distribution of penny ages. Is the shape similar
   to the shape of the relative frequency histogram in
   Exercise 1.41?

b. Draw a stem and leaf plot to describe the penny
   ages. Are there any unusually large or small measurements in the set?

(MY) CD **1.47 Presidential Vetoes** Here is a list of the
EX0147 43 presidents of the United States along with the
number of regular vetoes used by each:[5]

Washington	2		B. Harrison	19
J. Adams	0		Cleveland	42
Jefferson	0		McKinley	6
Madison	5		T. Roosevelt	42
Monroe	1		Taft	30
J.Q. Adams	0		Wilson	33
Jackson	5		Harding	5
Van Buren	0		Coolidge	20
W.H. Harrison	0		Hoover	21
Tyler	6		F.D. Roosevelt	372
Polk	2		Truman	180
Taylor	0		Eisenhower	73
Fillmore	0		Kennedy	12
Pierce	9		L. Johnson	16
Buchanan	4		Nixon	26
Lincoln	2		Ford	48
A. Johnson	21		Carter	13
Grant	45		Reagan	39
Hayes	12		G.H.W. Bush	29
Garfield	0		Clinton	36
Arthur	4		G.W. Bush	0
Cleveland	304			

Source: The World Almanac and Book of Facts, 2004

Use an appropriate graph to describe the number of
vetoes cast by the 43 presidents. Write a summary
paragraph describing this set of data.

(MY) CD **1.48 Windy Cities** Are some cities more
EX0148 windy than others? Does Chicago deserve to be
nicknamed "The Windy City"? These data are the
average wind speeds (in miles per hour) for 45
selected cities in the United States:[5]

8.9	12.4	8.6	11.3	9.2	8.8	35.1	6.2	7.0
7.1	11.8	10.7	7.6	9.1	9.2	8.2	9.0	8.7
9.1	10.9	10.3	9.6	7.8	11.5	9.3	7.9	8.8
8.8	12.7	8.4	7.8	5.7	10.5	10.5	9.6	8.9
10.2	10.3	7.7	10.6	8.3	8.8	9.5	8.8	9.4

Source: The World Almanac and Book of Facts, 2004

a. Construct a relative frequency histogram for the
   data. (HINT: Choose the class boundaries without
   including the value $x = 35.1$ in the range of
   values.)

b. The value $x = 35.1$ was recorded at Mt. Washington, New Hampshire. Does the geography of that
   city explain the observation?

c. The average wind speed in Chicago is recorded as
   10.3 miles per hour. Do you consider this unusually windy?

(MY) CD **1.49 Kentucky Derby** The following data set
EX0149 shows the winning times (in seconds) for the
Kentucky Derby races from 1950 to 2004.[12]

(1950)	121.3	122.3	121.3	122.0	123.0	121.4	123.2	122.1	125.0	122.1
(1960)	122.2	124.0	120.2	121.4	120.0	121.1	122.0	120.3	122.1	121.4
(1970)	123.2	123.1	121.4	119.2[†]	124.0	122.0	121.3	122.1	121.1	122.2
(1980)	122.0	122.0	122.2	121.1	122.2	120.1	122.4	123.2	122.2	125.0
(1990)	122.0	123.0	123.0	122.2	123.3	121.1	121.0	122.4	122.2	123.2
(2000)	121.0	119.97	121.13	121.19	124.06					

[†]Record time set by Secretariat in 1973
Source: www.kentuckyderby.com

a. Do you think there will be a trend in the winning
   times over the years? Draw a line chart to verify
   your answer.

b. Describe the distribution of winning times using an
   appropriate graph. Comment on the shape of the
   distribution and look for any unusual observations.

(MY) CD **1.50 Computer Networks at Home** As
EX0150 Americans become more knowledgeable about
computer hardware and software, as prices drop and
installation becomes easier, home networking of PCs
is expected to penetrate 27 percent of U.S. households
by 2008, with wireless technology leading the way.[13]

U.S. Home Networks (in millions)

Year	Wired	Wireless
2002	6.1	1.7
2003	6.5	4.5
2004	6.2	8.7
2005	5.7	13.7
2006	4.9	19.1
2007	4.1	24.0
2008	3.4	28.2

Source: Jupiter Research

a. What graphical methods could you use to describe the data?

b. Before you draw a graph, look at the predicted number of wired and wireless households in the table. What trends do you expect to see in the graphs?

c. Use a line chart to describe the predicted number of *wired* households for the years 2002 to 2008.

d. Use a bar chart to describe the predicted number of *wireless* households for the years 2002 to 2008.

(MY) CD **1.51 Election Results** The 2000 election was

EX0151 a close race, in which George W. Bush defeated Al Gore, Ralph Nader, and Pat Buchanan by the closest of margins. The popular vote (in thousands) for George W. Bush in each of the 50 states is listed below:[5]

AL	941	HI	138	MA	879	NM	286	SD	191
AK	167	ID	337	MI	1953	NY	2406	TN	1062
AZ	782	IL	2019	MN	1110	NC	1631	TX	3800
AR	473	IN	1246	MS	573	ND	175	UT	515
CA	4567	IA	634	MO	1190	OH	2351	VT	120
CO	884	KS	622	MT	240	OK	744	VA	1437
CT	561	KY	873	NE	434	OR	714	WA	1109
DE	137	LA	928	NV	302	PA	2281	WV	336
FL	2913	ME	287	NH	274	RI	131	WI	1237
GA	1420	MD	814	NJ	1284	SC	787	WY	148

a. By just looking at the table, what shape do you think the data distribution for the popular vote by state will have?

b. Draw a relative frequency histogram to describe the distribution of the popular vote for President Bush in the 50 states.

c. Did the histogram in part b confirm your guess in part a? Are there any outliers? How can you explain them?

(MY) CD **1.52 Election Results, continued** Refer to

EX0152 Exercise 1.51. Listed here is the *percentage* of the popular vote received by President Bush in each of the 50 states:[5]

AL	57	HI	38	MA	33	NM	47	SD	61
AK	60	ID	70	MI	47	NY	36	TN	51
AZ	50	IL	43	MN	46	NC	56	TX	60
AR	52	IN	58	MS	57	ND	61	UT	67
CA	42	IA	48	MO	51	OH	50	VT	41
CO	51	KS	59	MT	59	OK	61	VA	53
CT	39	KY	57	NE	63	OR	49	WA	45
DE	42	LA	53	NV	50	PA	47	WV	52
FL	49	ME	44	NH	48	RI	32	WI	48
GA	56	MD	40	NJ	41	SC	57	WY	70

a. By just looking at the table, what shape do you think the data distribution for the *percentage* of the popular vote by state will have?

b. Draw a relative frequency histogram to describe the distribution. Describe the shape of the distribution and look for outliers. Did the graph confirm your answer to part a?

1.53 Election Results, continued Refer to Exercises 1.51 and 1.52. The accompanying stem and leaf plots were generated using *MINITAB* for the variables named "Popular Vote" and "Percent Vote."

Stem-and-Leaf Display: Popular Vote, Percent Vote

```
Stem-and-leaf of              Stem-and-leaf of
Popular Vote   N = 50         Percent Vote   N = 50
Leaf Unit = 100               Leaf Unit = 1.0

  8    0 11111111              2    3 23
 15    0 2222333              5    3 689
 20    0 44555               12    4 0112234
 (6)   0 667777              22    4 5677788899
 24    0 888899             (10)   5 0001112233
 18    1 0111                18    5 667777899
 14    1 222                  9    6 001113
 11    1 44                   3    6 7
  9    1 6                    2    7 00
  8    1 9
  7    2 0
  6    2 23
  4    2 4

   HI   29, 38, 45
```

a. Describe the shapes of the two distributions. Are there any outliers?

b. Do the stem and leaf plots resemble the relative frequency histograms constructed in Exercises 1.51 and 1.52?

c. Explain why the distribution of the popular vote for President Bush by state is skewed while the percentage of popular votes by state is mound-shaped.

(MY) CD **1.54 Student Heights** The self-reported

EX0154 heights of 105 students in a biostatistics class are described in the relative frequency histogram below.

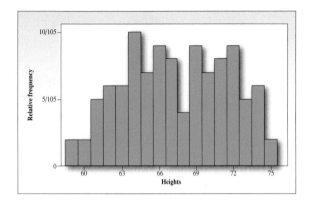

a. Describe the shape of the distribution.

b. Do you see any unusual feature in this histogram?

c. Can you think of an explanation for the two peaks in the histogram? Is there some other factor that is causing the heights to mound up in two separate peaks? What is it?

1.55 Fear of Terrorism One week after the terrorist attack on the World Trade Center on September 11, 2001, the Gallup poll presented a summary of various polls that had asked the question: "How worried are you that you or someone in your family will become a victim of a terrorist attack?" The data are shown in the table below:[14]

Date	Very/Somewhat Worried (%)	Not Too/Not at All Worried (%)
April 1995	41	57
April 1996	35	65
July 1996	39	61
August 1998	32	67
April 2000	24	75
September 11, 2001	58	40
September 14–15, 2001	51	48

a. Draw a line chart to describe the percentages that are very/somewhat worried. Use time as the horizontal axis.

b. Superimpose another line chart on the one drawn in part b to describe the percentages that are not too/not at all worried.

c. Use your line chart to summarize the changes in the polls during the period just after the attack on the World Trade Center.

d. The following line chart was presented on the www.gallup.com Web page. How does it differ from the graph that you drew? What characteristic of the Gallup line chart might cause distortion when the graph is interpreted?

Worried That You or a Family Member Will Become a Victim of a Terrorist Attack?

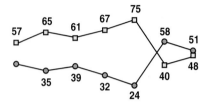

Apr. 21 - 23, 1995	Apr. 9 - 10, 1996	Jul. 20 - 21, 1996	Aug. 20, 1998	Apr. 7 - 9, 2000	Sep. 11, 2001	Sep. 14 - 15, 2001

1.56 Pulse Rates A group of 50 biomedical students recorded their pulse rates by counting the number of beats for 30 seconds and multiplying by 2.

80	70	88	70	84	66	84	82	66	42
52	72	90	70	96	84	96	86	62	78
60	82	88	54	66	66	80	88	56	104
84	84	60	84	88	58	72	84	68	74
84	72	62	90	72	84	72	110	100	58

a. Why are all of the measurements even numbers?

b. Draw a stem and leaf plot to describe the data, splitting each stem in to two lines.

c. Construct a relative frequency histogram for the data.

d. Write a short paragraph describing the distribution of the student pulse rates.

1.57 Getting Up to Speed Small businesses continue to rely on the telephone as their primary communication tool, using the Internet primarily as an information source. A research survey conducted by *Covad/Sprint/Equation Research* indicates that most small businesses are only beginning to adopt broadband Internet access.[15]

Small Business Internet Access Methods

Dial-up/modem	31.7%
DSL/ADSL broadband	30.3%
Cable broadband	26.1%
T1 or higher	9.8%
Other	2.1%

Source: www.clickz.com

a. Do the percentages add up to 100%?

b. Use a pie chart to describe the Internet access methods used by small businesses.

MY CD
EX0158
1.58 How Much Can You Save? An advertisement in a recent *Time* magazine claimed that Geico Insurance will help you save an average of $200 per year on your automobile insurance.[16]

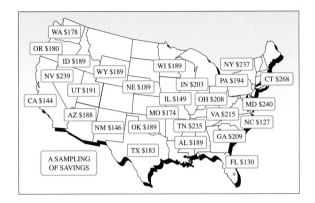

a. Construct a relative frequency histogram to describe the average savings for the 27 states shown on the United States map. Do you see any unusual features in the histogram?

b. Construct a stem and leaf plot for the data provided by Geico Insurance.

c. How do you think that Geico selected the 27 states for inclusion in this advertisement?

MY CD
EX0159
1.59 An Archeological Find An article in *Archaeometry* involved an analysis of 26 samples of Romano-British pottery, found at four different kiln sites in the United Kingdom.[17] The samples were analyzed to determine their chemical composition, and the percentage of aluminum oxide in each of the 26 samples is shown below.

Llanederyn		Caldicot	Island Thorns	Ashley Rails
14.4	11.6	11.8	18.3	17.7
13.8	11.1	11.6	15.8	18.3
14.6	13.4		18.0	16.7
11.5	12.4		18.0	14.8
13.8	13.1		20.8	19.1
10.9	12.7			
10.1	12.5			

a. Construct a relative frequency histogram to describe the aluminum oxide content in the 26 pottery samples.

b. What unusual feature do you see in this graph? Can you think of an explanation for this feature?

c. Draw a dotplot for the data, using a letter (L, C, I, or A) to locate the data point on the horizontal

scale. Does this help explain the unusual feature in part b?

1.60 The Great Calorie Debate Want to lose weight? You can do it by cutting calories, as long as you get enough nutritional value from the foods that you do eat! Below you will see a visual representation of the number of calories in some of America's favorite foods adapted from an article in *The Press-Enterprise*.[18]

a. Comment on the accuracy of the graph shown above. Do the sizes, heights, and volumes of the six items accurately represent the number of calories in the item?

b. Draw an actual bar chart to describe the number of calories in these six food favorites.

MY CD
EX0161
1.61 Laptops and Learning An informal experiment was conducted at McNair Academic High School in Jersey City, New Jersey, to investigate the use of laptop computers as a learning tool in the study of algebra.[19] A freshman class of 20 students was given laptops to use at school and at home, while another freshman class of 27 students was not given laptops; however, many of these students were able to use computers at home. The final exam scores for the two classes are shown below.

Laptops		No Laptops		
98	84	63	83	97
97	93	93	52	74
88	57	83	63	88
100	84	86	81	84
100	81	99	91	49
78	83	80	81	89
68	84	78	29	64
47	93	74	72	89
90	57	67	89	70
94	83			

The histograms below show the distribution of final exam scores for the two groups.

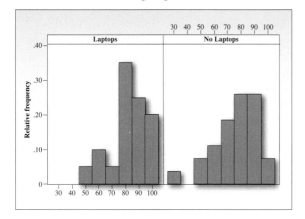

Write a summary paragraph describing and comparing the distribution of final exam scores for the two groups of students.

1.62 Old Faithful The data below are the waiting times between eruptions of the Old Faithful geyser in Yellowstone National Park.[20] Use one of the graphical methods from this chapter to describe the distribution of waiting times. If there are any unusual features in your graph, see if you can think of any practical explanation for them.

56	89	51	79	58	82	52	88	52	78
69	75	77	53	80	54	79	74	65	78
55	87	53	85	61	93	54	76	80	81
59	86	78	71	77	89	45	93	72	71
76	94	75	50	83	82	72	77	75	65
79	72	78	77	79	72	82	74	80	49
75	78	64	80	49	49	88	51	78	85
65	75	77	69	92	91	53	86	49	79
68	87	61	81	55	93	53	84	70	73
93	50	87	77	74	89	87	76	59	80

1.63 Gasoline Tax The following are the 2003 state gasoline tax in cents per gallon for the 50 United States and the District of Columbia.[5]

AL	18.0	HI	16.0	MA	21.0	NM	18.5	SD	22.0
AK	8.0	ID	25.0	MI	19.0	NY	22.0	TN	20.0
AZ	18.0	IL	19.0	MN	20.0	NC	24.1	TX	20.0
AR	21.7	IN	15.0	MS	18.4	ND	21.0	UT	24.5
CA	18.0	IA	20.0	MO	17.0	OH	22.0	VT	20.0
CO	22.0	KS	21.0	MT	27.0	OK	17.0	VA	17.5
CT	25.0	KY	16.4	NE	24.5	OR	24.0	WA	23.0
DE	23.0	LA	20.0	NV	24.75	PA	26.0	WV	25.65
DC	20.0	ME	22.0	NH	19.5	RI	29.0	WI	27.3
FL	13.6	MD	23.5	NJ	10.5	SC	16.0	WY	14.0
GA	7.5								

Source: The World Almanac and Book of Facts, 2004

a. Construct a stem and leaf display for the data.

b. How would you describe the shape of this distribution?

c. Are there states with unusually high or low gasoline taxes? If so, which states are they?

1.64 Hydroelectric Plants The following data represent the planned rated capacities in megawatts (millions of watts) for the world's 23 largest hydroelectric plants.[5]

20,000	6,000	4,000
18,200	5,328	3,600
13,320	5,225	3,409
10,830	5,020	3,300
10,300	4,678	3,200
7,260	4,500	3,000
6,400	4,500	2,715
6,000	4,150	

Source: The World Almanac and Book of Facts, 2004

a. Construct a stem and leaf display for the data.

b. How would you describe the shape of this distribution?

1.65 Car Colors The most popular colors for compact and sports cars in a recent year are given in the table.[5]

Color	Percentage	Color	Percentage
Silver	24.6	Medium Red	5.5
Black	14.3	Light Brown	4.3
Medium/Dark Blue	12.9	Gold	4.1
White	8.8	Dark Red	2.6
Bright Red	6.9	Other	9.3
Medium/Dark Gray	6.7		

Source: The World Almanac and Book of Facts, 2004

Use an appropriate graphical display to describe these data.

1.66 Starbucks The number of Starbucks coffee shops in cities within 25 miles of the University of California, Riverside is shown in the following table.[21]

City	Starbucks	City	Starbucks
Riverside	6	Ontario	6
Grand Terrace	1	Norco	2
Rialto	2	Fontana	1
San Bernardino	3	Perris	1
Redlands	3	Highland	1
Corona	5	Rancho Cucamonga	7
Yucaipa	1	Upland	2
Chino	2	Moreno Valley	0

Source: www.starbucks.com

a. Draw a dotplot to describe the data.

b. Describe the shape of the distribution.

c. Is there another variable that you could measure that might help to explain why some cities have more Starbucks than others? Explain.

1.67 What's Normal? The 98.6 degree standard for human body temperature was derived by a German doctor in 1868. In an attempt to verify his claim, Mackowiak, Wasserman, and Levine[22] took temperatures from 148 healthy people over a three-day period. A data set closely matching the one in Mackowiak's article was derived by Allen Shoemaker, and

appears in the *Journal of Statistics Education*.[23] The body temperatures for these 130 individuals are shown below.

a. Describe the shape of the distribution of temperatures.

b. Are there any unusual observations? Can you think of any explanation for these?

c. Locate the 98.6-degree standard on the horizontal axis of the graph. Does it appear to be near the center of the distribution?

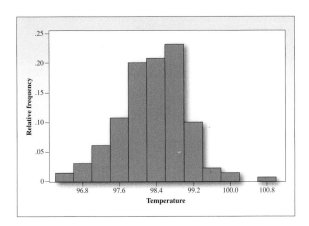

MY APPLET Exercises

1.68 If you have not yet done so, use the first applet in **Building a Dotplot** to create a dotplot for the following data set: 2, 3, 9, 6, 7, 6.

1.69 Cheeseburgers Use the second applet in **Building a Dotplot** to create a dotplot for the number of cheeseburgers consumed in a given week by 10 college students:

4	5	4	2	1
3	3	4	2	7

a. How would you describe the shape of the distribution?

b. What proportion of the students ate more than 4 cheeseburgers that week?

1.70 Social Security Numbers A group of 70 students were asked to record the last digit of their social security number.

1	6	9	1	5	9	0	2	8	4
0	7	3	4	2	3	5	8	4	2
3	2	0	0	2	1	2	7	7	4
0	0	9	9	5	3	8	4	7	4
6	6	9	0	2	6	2	9	5	8
5	1	7	7	7	8	7	5	1	8
3	4	1	9	3	8	6	6	6	6

a. Before graphing the data, use your common sense to guess the shape of the data distribution. Explain your reasoning.

b. Use the second applet in **Building a Dotplot** to create a dotplot to describe the data. Was your intuition correct in part a?

1.71 If you have not yet done so, use the first applet in **Building a Histogram** to create a histogram for the data in Example 1.11, the number of quarts of milk purchased during a particular week.

1.72 The United Fund The following data set
EX0172 records the yearly charitable contributions (in
dollars) to the United Fund for a group of employees
at a public university.

41	81	80	65	47	56	80	69	79	63
28	51	112	71	83	84	82	103	80	70
77	75	59	63	63	80	101	115	99	67
42	78	81	90	103	125	92	79	24	93

Use the second applet in **Building a Histogram** to
construct a relative frequency histogram for the data.
What is the shape of the distribution? Can you see
any obvious outliers?

1.73 Internet Hotspots The most visited site
EX0173 on the Internet is Yahoo!, which boasted
111,271 thousand unique visitors in December 2003.[24]
The number of unique visitors at the top 25 sites are
shown in the table:

a. Look at the data. Can you guess the approximate
shape of the data distribution?

b. Use the second applet in **Building a Histogram** to
construct a relative frequency histogram for the
data. What is the shape of the distribution?

c. Are there any outliers in the set? If so, what sites
boast an unusually large number of unique visitors?

Site	Unique Visitors (Millions)	Site	Unique Visitors (Millions)
Yahoo! Sites	111.3	Weather Channel	23.8
Time Warner Network	110.5	Real.com Network	22.3
MSN-Microsoft Sites	110.0	Verizon Communi-	
Ebay	69.2	cations	22.1
Google Sites	61.5	Wal-Mart	21.4
Terra Lycos	52.1	Shopping.com Sites	21.3
Amazon Sites	45.7	Symantec	19.9
About/Primedia	42.6	AT&T Properties	17.5
Excite Network	25.1	InfoSpace Network	17.3
CNET Networks	25.1	Monster Property	17.3
Walt Disney Internet	25.1	EA Online	16.8
Viacom Online	24.7	SBC Communications	16.5
American Greetings	24.4	Sony Online	16.5

Source: cyberatlas.internet.com

CASE STUDY **How Is Your Blood Pressure?**

 Blood
Pressure

Blood pressure is the pressure that the blood exerts against the walls of the arteries.
When physicians or nurses measure your blood pressure, they take two readings. The
systolic blood pressure is the pressure when the heart is contracting and therefore
pumping. The diastolic blood pressure is the pressure in the arteries when the heart
is relaxing. The diastolic blood pressure is always the lower of the two readings.
Blood pressure varies from one person to another. It will also vary for a single individual from day to day and even within a given day.

   If your blood pressure is too high, it can lead to a stroke or a heart attack. If it is
too low, blood will not get to your extremities and you may feel dizzy. Low blood
pressure is usually not serious.

   So, what should *your* blood pressure be? A systolic blood pressure of 120 would
be considered normal. One of 150 would be high. But since blood pressure varies
with gender and increases with age, a better gauge of the relative standing of your
blood pressure would be obtained by comparing it with the population of blood
pressures of all persons of your gender and age in the United States. Of course, we
cannot supply you with that data set, but we can show you a very large sample selected from it. The CD-ROM provides blood pressure data on 1910 persons, 965 men
and 945 women between the ages of 15 and 20. The data are part of a health survey
conducted by the National Institutes of Health (NIH). Entries for each person include

that person's age and systolic and diastolic blood pressures at the time the blood pressure was recorded.

1. Describe the variables that have been measured in this survey. Are the variables quantitative or qualitative? Discrete or continuous? Are the data univariate, bivariate, or multivariate?

2. What types of graphical methods are available for describing this data set? What types of questions could be answered using various types of graphical techniques?

3. Using the systolic blood pressure data set, construct a relative frequency histogram for the 965 men and another for the 945 women. Use a statistical software package if you have access to one. Compare the two histograms.

4. Consider the 965 men and 945 women as the entire population of interest. Choose a sample of $n = 50$ men and $n = 50$ women, recording their systolic blood pressures and their ages. Draw two relative frequency histograms to graphically display the systolic blood pressures for your two samples. Do the shapes of the histograms resemble the population histograms from part 3?

5. How does your blood pressure compare with that of others of your same gender? Check your systolic blood pressure against the appropriate histogram in part 3 or 4 to determine whether your blood pressure is "normal" or whether it is unusually high or low.

Describing Data with Numerical Measures

© Photodisc/Royalty-Free/Getty Images

GENERAL OBJECTIVES

Graphs are extremely useful for the visual description of a data set. However, they are not always the best tool when you want to make inferences about a population from the information contained in a sample. For this purpose, it is better to use numerical measures to construct a mental picture of the data.

CHAPTER INDEX

MY PERSONAL TRAINER

How Do I Calculate Sample Quartiles?

The Boys of Summer

Are the baseball champions of today better than those of "yesteryear"? Do players in the National League hit better than players in the American League? The case study at the end of this chapter involves the batting averages of major league batting champions. Numerical descriptive measures can be used to answer these and similar questions.

DESCRIBING A SET OF DATA WITH NUMERICAL MEASURES

Graphs can help you describe the basic shape of a data distribution; "a picture is worth a thousand words." There are limitations, however, to the use of graphs. Suppose you need to display your data to a group of people and the bulb on the data projector blows out! Or you might need to describe your data over the telephone—no way to display the graphs! You need to find another way to convey a mental picture of the data to your audience.

A second limitation is that graphs are somewhat imprecise for use in statistical inference. For example, suppose you want to use a sample histogram to make inferences about a population histogram. How can you measure the similarities and differences between the two histograms in some concrete way? If they were identical, you could say "They are the same!" But, if they are different, it is difficult to describe the "degree of difference."

One way to overcome these problems is to use **numerical measures,** which can be calculated for either a sample or a population of measurements. You can use the data to calculate a set of *numbers* that will convey a good mental picture of the frequency distribution. These measures are called **parameters** when associated with the population, and they are called **statistics** when calculated from sample measurements.

Definition   Numerical descriptive measures associated with a population of measurements are called **parameters;** those computed from sample measurements are called **statistics.**

MEASURES OF CENTER

In Chapter 1, we introduced dotplots, stem and leaf plots, and histograms to describe the distribution of a set of measurements on a quantitative variable x. The horizontal axis displays the values of x, and the data are "distributed" along this horizontal line. One of the first important numerical measures is a **measure of center**—a measure along the horizontal axis that locates the center of the distribution.

The GPA data presented in Table 1.9 ranged from a low of 1.9 to a high of 3.4, with the center of the histogram located in the vicinity of 2.6 (see Figure 2.1). Let's consider some rules for locating the center of a distribution of measurements.

FIGURE 2.1

Center of the GPA data

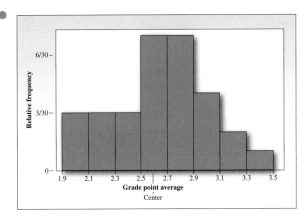

The arithmetic average of a set of measurements is a very common and useful measure of center. This measure is often referred to as the **arithmetic mean,** or simply the **mean,** of a set of measurements. To distinguish between the mean for the sample and the mean for the population, we will use the symbol $\bar{x}$ (x-bar) for a sample mean and the symbol μ (Greek lowercase mu) for the mean of a population.

Definition   The **arithmetic mean** or **average** of a set of n measurements is equal to the sum of the measurements divided by n.

Since statistical formulas often involve adding or "summing" numbers, we use a shorthand symbol to indicate the process of summing. Suppose there are n measurements on the variable x—call them $x_1, x_2, \ldots, x_n$. To add the n measurements together, we use this shorthand notation:

$$\sum_{i=1}^{n} x_i \quad \text{which means } x_1 + x_2 + x_3 + \cdots + x_n$$

The Greek capital sigma (Σ) tells you to add the items that appear to its right, beginning with the number below the sigma ($i = 1$) and ending with the number above ($i = n$). However, since the typical sums in statistical calculations are almost always made on the total set of n measurements, you can use a simpler notation:

Σx_i which means "the sum of all the x measurements"

Using this notation, we write the formula for the sample mean:

NOTATION

Sample mean: $\bar{x} = \dfrac{\Sigma x_i}{n}$

Population mean: μ

EXAMPLE   2.1

Draw a dotplot for the $n = 5$ measurements 2, 9, 11, 5, 6. Find the sample mean and compare its value with what you might consider the "center" of these observations on the dotplot.

Solution   The dotplot in Figure 2.2 seems to be centered between 6 and 8. To find the sample mean, calculate

$$\bar{x} = \frac{\Sigma x_i}{n} = \frac{2 + 9 + 11 + 5 + 6}{5} = 6.6$$

FIGURE 2.2

Dotplot for Example 2.1

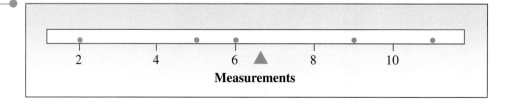

The statistic $\bar{x} = 6.6$ is the balancing point or fulcrum shown on the dotplot. It does seem to mark the center of the data.

TIP

mean = balancing point or fulcrum

Remember that samples are measurements drawn from a larger population that is usually unknown. An important use of the sample mean $\bar{x}$ is as an estimator of the unknown population mean μ. The GPA data in Table 1.9 are a sample from a larger population of GPAs, and the distribution is shown in Figure 2.1. The mean of the 30 GPAs is

$$\bar{x} = \frac{\Sigma x_i}{30} = \frac{77.5}{30} = 2.58$$

shown in Figure 2.1; it marks the balancing point of the distribution. The mean of the entire population of GPAs is unknown, but if you had to guess its value, your best estimate would be 2.58. Although the sample mean $\bar{x}$ changes from sample to sample, the population mean μ stays the same.

A second measure of central tendency is the **median,** which is the value in the middle position in the set of measurements ordered from smallest to largest.

Definition   The **median** m of a set of n measurements is the value of x that falls in the middle position when the measurements are ordered from smallest to largest.

EXAMPLE 2.2

Find the median for the set of measurements 2, 9, 11, 5, 6.

Solution   Rank the $n = 5$ measurements from smallest to largest:

$$2 \quad 5 \quad 6 \quad 9 \quad 11$$
$$\uparrow$$

The middle observation, marked with an arrow, is in the center of the set, or $m = 6$.

EXAMPLE 2.3

Find the median for the set of measurements 2, 9, 11, 5, 6, 27.

Solution   Rank the measurements from smallest to largest:

$$2 \quad 5 \quad \boxed{6 \quad 9} \quad 11 \quad 27$$
$$\uparrow$$

 TIP

Roughly 50% of the measurements are smaller, 50% are larger than the median.

Now there are two "middle" observations, shown in the box. To find the median, choose a value halfway between the two middle observations:

$$m = \frac{6 + 9}{2} = 7.5$$

The value **.5(n + 1)** indicates the **position of the median** in the ordered data set. If the position of the median is a number that ends in the value **.5,** you need to average the two adjacent values.

EXAMPLE 2.4

For the $n = 5$ ordered measurements from Example 2.2, the position of the median is .5(n + 1) =.5(6) = 3, and the median is the *3rd ordered observation,* or $m = 6$. For the $n = 6$ ordered measurements from Example 2.3, the position of the median is .5(n + 1) = .5(7) = 3.5, and the median is the *average of the 3rd and 4th ordered observations,* or $m = (6 + 9)/2 = 7.5$.

Although both the mean and the median are good measures of the center of a distribution, the median is less sensitive to extreme values or *outliers*. For example, the value $x = 27$ in Example 2.3 is much larger than the other five measurements. The median, $m = 7.5$, is not affected by the outlier, whereas the sample average,

$$\bar{x} = \frac{\Sigma x_i}{n} = \frac{60}{6} = 10$$

is affected; its value is not representative of the remaining five observations.

When a data set has extremely small or extremely large observations, the sample mean is drawn toward the direction of the extreme measurements (see Figure 2.3).

FIGURE 2.3

Relative frequency distributions showing the effect of extreme values on the mean and median

If a distribution is skewed to the right, the mean shifts to the right; if a distribution is skewed to the left, the mean shifts to the left. The median is not affected by these extreme values because the numerical values of the measurements are not used in its calculation. When a distribution is symmetric, the mean and the median are equal. If a distribution is strongly skewed by one or more extreme values, you should use the median rather than the mean as a measure of center.

MY APPLET

You can see the effect of extreme values on both the mean and the median using the **How Extreme Values Affect the Mean and Median** applet. The first of three applets (Figure 2.4) shows a dotplot of the data in Example 2.2. Use your mouse to move the largest observation ($x = 11$) even further to the right. How does this larger observation affect the mean? How does it affect the median? We will use this applet again for the MyApplet Exercises at the end of the chapter.

FIGURE 2.4

How Extreme Values Affect the Mean and Median applet

Another way to locate the center of a distribution is to look for the value of x that occurs with the highest frequency. This measure of the center is called the **mode.**

Definition   The **mode** is the category that occurs most frequently, or the most frequently occurring value of x. When measurements on a continuous variable have been grouped as a frequency or relative frequency histogram, the class with the highest peak or frequency is called the **modal class,** and the midpoint of that class is taken to be the mode.

The mode is generally used to describe large data sets, whereas the mean and median are used for both large and small data sets. From the data in Example 1.11, the mode of the distribution of the number of quarts of milk purchased during one particular week is 2. The modal class and the value of x occurring with the highest frequency are the same, as shown in Figure 2.5(a).

For the data in Table 1.9, a GPA of 2.5 occurs five times, and therefore the mode for the distribution of GPAs is 2.5. Using the histogram to find the modal class, you find two classes that occur with equal frequency. Fortunately, these classes are side by side in the tabulation, and the choice for the value of the mode is thus 2.7, the value centered between the fourth and fifth classes. See Figure 2.5(b).

It is possible for a distribution of measurements to have more than one mode. These modes would appear as "local peaks" in the relative frequency distribution. For example, if we were to tabulate the length of fish taken from a lake during one season, we might get a *bimodal distribution,* possibly reflecting a mixture of young and old fish in the population. Sometimes bimodal distributions of sizes or weights reflect a mixture of measurements taken on males and females. In any case, a set or distribution of measurements may have more than one mode.

MY TIP

Remember that there can be several modes or no mode (if each observation occurs only

FIGURE 2.5

Relative frequency histograms for the milk and GPA data

EXERCISES

BASIC TECHNIQUES

2.1 You are given $n = 5$ measurements: 0, 5, 1, 1, 3.

a. Draw a dotplot for the data. (HINT: If two measurements are the same, place one dot above the other.) Guess the approximate "center."

b. Find the mean, median, and mode.

c. Locate the three measures of center on the dotplot in part a. Based on the relative positions of the mean and median, are the measurements symmetric or skewed?

2.2 You are given $n = 8$ measurements: 3, 2, 5, 6, 4, 4, 3, 5.

a. Find $\bar{x}$.

b. Find m.

c. Based on the results of parts a and b, are the measurements symmetric or skewed? Draw a dotplot to confirm your answer.

2.3 You are given $n = 10$ measurements: 3, 5, 4, 6, 10, 5, 6, 9, 2, 8.

a. Calculate $\bar{x}$.

b. Find m.

c. Find the mode.

APPLICATIONS

2.4 Auto Insurance The cost of automobile insurance has become a sore subject in California because insurance rates are dependent on so many different variables, such as the city in which you live, the number of cars you insure, and the company with which you are insured. The website www.insurance.ca.gov reports the 6-month premiums in 2004 for a single male, licensed for 6–8 years, who drives about 15,000 miles per year and has no violations or accidents.[1]

City	Allstate	21st Century
Long Beach	$884	$996
Pomona	730	940
San Bernardino	725	836
Moreno Valley	801	762

Source: www.insurance.ca.gov

a. What is the average premium for Allstate Insurance?

b. What is the average premium for 21st Century Insurance?

c. If you were a consumer, would you be interested in the average premium cost? If not, what would you be interested in?

2.5 DVD Players The DVD player is a common fixture in most American households. In fact, most American households have DVDs, and many have more than one. A sample of 25 households produced the following measurements on x, the number of DVDs in the household:

1	0	2	1	1
1	0	2	1	0
0	1	2	3	2
1	1	1	0	1
3	1	0	1	1

a. Is the distribution of x, the number of DVDs in a household, symmetric or skewed? Explain.

b. Guess the value of the mode, the value of x that occurs most frequently.

c. Calculate the mean, median, and mode for these measurements.

d. Draw a relative frequency histogram for the data set. Locate the mean, median, and mode along the horizontal axis. Are your answers to parts a and b correct?

2.6 Fortune 500 Revenues Ten of the 100 largest businesses in the United States, randomly selected from the *Fortune* 500, are listed below along with their revenues (in millions of dollars):[2]

Company	Revenues	Company	Revenues
General Motors	$186,763.0	Intel	$26,764.0
Home Depot	58,247.0	Viacom	24,522.0
AT&T	46,727.0	FedEx	20,607.0
Johnson & Johnson	36,298.0	Coca-Cola	19,564.0
Safeway	32,399.2	Bristol-Myers Squibb	18,119.0

Source: Time Almanac 2004

a. Draw a stem and leaf plot for the data. Are the data skewed?

b. Calculate the mean revenue for these 10 businesses. Calculate the median revenue.

c. Which of the two measures in part b best describes the center of the data? Explain.

2.7 Birth Order and Personality Does birth order have any effect on a person's personality? A report on a study by an MIT researcher indicates that later-born children are more likely to challenge the establishment, more open to new ideas, and more accepting of change.[3] In fact, the number of later-born children is increasing. During the Depression years of the 1930s, families averaged 2.5 children (59% later born), whereas the parents of baby boomers averaged 3 to 4 children (68% later born). What does the author mean by an average of 2.5 children?

2.8 Tuna Fish An article in *Consumer Reports* gives the price—an estimated average for a 6-ounce can or a 7.06-ounce pouch—for 14 different brands of water-packed light tuna, based on prices paid nationally in supermarkets:[4]

.99	1.92	1.23	.85	.65	.53	1.41
1.12	.63	.67	.69	.60	.60	.66

a. Find the average price for the 14 different brands of tuna.

b. Find the median price for the 14 different brands of tuna.

c. Based on your findings in parts a and b, do you think that the distribution of prices is skewed? Explain.

2.9 Sports Salaries As professional sports teams become a more and more lucrative business for their owners, the salaries paid to the players have also increased. In fact, sports superstars are paid astronomical salaries for their talents. If you were asked by a sports management firm to describe the distribution of players' salaries in several different categories of professional sports, what measure of center would you choose? Why?

2.10 Time on Task In a psychological experiment, the time on task was recorded for 10 subjects under a 5-minute time constraint. These measurements are in seconds:

175	190	250	230	240
200	185	190	225	265

a. Find the average time on task.

b. Find the median time on task.

c. If you were writing a report to describe these data, which measure of central tendency would you use? Explain.

 2.11 Starbucks The number of Starbucks coffee shops in 16 cities within 25 miles of the University of California, Riverside is shown in the following table (www.starbucks.com).[5]

3	6	1	6
5	1	7	2
1	2	2	1
2	3	0	1

a. Find the mean, the median, and the mode.

b. Compare the median and the mean. What can you say about the shape of this distribution?

c. Draw a dotplot for the data. Does this confirm your conclusion about the shape of the distribution from part b?

2.12 LCD TVs The cost of televisions exhibits huge variation—from $100–200 for a standard TV to $8000–10,000 for a large plasma screen TV. *Consumer Reports* gives the prices for the top ten LCD standard definition TVs in the 14- to 20-inch category:[6]

Brand	Price
Sharp LC-20E1U	$1200
Sony KLV-15SR1	800
Panasonic TC-20LA1	1050
Panasonic TC-17LA1	750
Gateway GTW-L18M103	700
Panasonic TC-14LA1	500
Gateway GTW-L17M103	600
Toshiba 14VL43U	670
Toshiba 20VL43U	1200
Sharp LC-15E1U	650

a. What is the average price of these ten LCD televisions?

b. What is the median price of these ten LCD televisions?

c. As a consumer, would you be interested in the average cost of an LCD TV? What other variables would be important to you?

MEASURES OF VARIABILITY

Data sets may have the same center but look different because of the way the numbers *spread out* from the center. Consider the two distributions shown in Figure 2.6. Both distributions are centered at $x = 4$, but there is a big difference in the way the measurements spread out, or *vary*. The measurements in Figure 2.6(a) vary from 3 to 5; in Figure 2.6(b) the measurements vary from 0 to 8.

FIGURE 2.6

Variability or dispersion of data

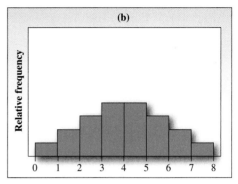

Variability or **dispersion** is a very important characteristic of data. For example, if you were manufacturing bolts, extreme variation in the bolt diameters would cause a high percentage of defective products. On the other hand, if you were trying to discriminate between good and poor accountants, you would have trouble if the examination always produced test grades with little variation, making discrimination very difficult.

Measures of variability can help you create a mental picture of the spread of the data. We will present some of the more important ones. The simplest measure of variation is the **range.**

Definition   The **range, R,** of a set of n measurements is defined as the difference between the largest and smallest measurements.

For the GPA data in Table 1.9, the measurements vary from 1.9 to 3.4. Hence, the range is $(3.4 - 1.9) = 1.5$. The range is easy to calculate, easy to interpret, and is an adequate measure of variation for small sets of data. But, for large data sets, the range is not an adequate measure of variability. For example, the two relative frequency distributions in Figure 2.7 have the same range but very different shapes and variability.

FIGURE 2.7

Distributions with equal range and unequal variability

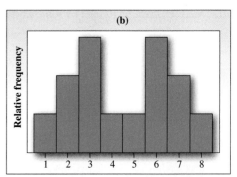

Is there a measure of variability that is more sensitive than the range? Consider, as an example, the sample measurements 5, 7, 1, 2, 4, displayed as a dotplot in Figure 2.8. The mean of these five measurements is

$$\bar{x} = \frac{\Sigma x_i}{n} = \frac{19}{5} = 3.8$$

FIGURE 2.8

Dotplot showing the deviations of points from the mean

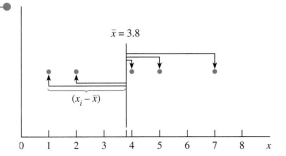

as indicated on the dotplot. The horizontal distances between each dot (measurement) and the mean $\bar{x}$ will help you to measure the variability. If the distances are large, the data are more spread out or *variable* than if the distances are small. If x_i is a particular dot (measurement), then the **deviation** of that measurement from the mean is $(x_i - \bar{x})$. Measurements to the right of the mean produce positive deviations, and those to the left produce negative deviations. The values of x and the deviations for our example are listed in the first and second columns of Table 2.1.

TABLE 2.1 **Computation of $\Sigma(x_i - \bar{x})^2$**

x	$(x_i - \bar{x})$	$(x_i - \bar{x})^2$
5	1.2	1.44
7	3.2	10.24
1	−2.8	7.84
2	−1.8	3.24
4	.2	.04
19	0.0	22.80

Because the deviations in the second column of the table contain information on variability, one way to combine the five deviations into one numerical measure is to average them. Unfortunately, the average will not work because some of the deviations are positive, some are negative, and the sum is always zero (unless round-off errors have been introduced into the calculations). Note that the deviations in the second column of Table 2.1 sum to zero.

Another possibility might be to disregard the signs of the deviations and calculate the average of their absolute values.[†] This method has been used as a measure of variability in exploratory data analysis and in the analysis of time series data. We prefer, however, to overcome the difficulty caused by the signs of the deviations by working

---

[†]The absolute value of a number is its magnitude, ignoring its sign. For example, the absolute value of −2, represented by the symbol $|-2|$, is 2. The absolute value of 2—that is, $|2|$—is 2.

with their sum of squares. From the sum of squared deviations, a single measure called the **variance** is calculated. To distinguish between the variance of a *sample* and the variance of a *population,* we use the symbol s^2 for a sample variance and σ^2 (Greek lowercase sigma) for a population variance. *The variance will be relatively large for highly variable data and relatively small for less variable data.*

Definition   The **variance of a population** of N measurements is the average of the squares of the deviations of the measurements about their mean μ. The population variance is denoted by σ^2 and is given by the formula

$$\sigma^2 = \frac{\Sigma(x_i - \mu)^2}{N}$$

Most often, you will not have all the population measurements available but will need to calculate the *variance of a sample* of n measurements.

Definition   The **variance of a sample** of n measurements is the sum of the squared deviations of the measurements about their mean $\bar{x}$ divided by $(n - 1)$. The sample variance is denoted by s^2 and is given by the formula

$$s^2 = \frac{\Sigma(x_i - \bar{x})^2}{n - 1}$$

For the set of $n = 5$ sample measurements presented in Table 2.1, the square of the deviation of each measurement is recorded in the third column. Adding, we obtain

$$\Sigma(x_i - \bar{x})^2 = 22.80$$

and the sample variance is

$$s^2 = \frac{\Sigma(x_i - \bar{x})^2}{n - 1} = \frac{22.80}{4} = 5.70$$

(MY) TIP

The variance and the standard deviation *cannot* be negative numbers.

The variance is measured in terms of the square of the original units of measurement. If the original measurements are in inches, the variance is expressed in square inches. Taking the square root of the variance, we obtain the **standard deviation,** which returns the measure of variability to the original units of measurement.

Definition   The **standard deviation** of a set of measurements is equal to the positive square root of the variance.

NOTATION

n: number of measurements in the sample

s^2: sample variance

$s = \sqrt{s^2}$: sample standard deviation

N: number of measurements in the population

σ^2: population variance

$\sigma = \sqrt{\sigma^2}$: population standard deviation

TIP

If you are using your calculator, make sure to choose the correct key for the sample standard deviation.

For the set of $n = 5$ sample measurements in Table 2.1, the sample variance is $s^2 = 5.70$, so the sample standard deviation is $s = \sqrt{s^2} = \sqrt{5.70} = 2.39$. The more variable the data set is, the larger the value of s.

For the small set of measurements we used, the calculation of the variance is not too difficult. However, for a larger set, the calculations can become very tedious. Most scientific calculators have built-in programs that will calculate $\bar{x}$ and s or μ and σ, so that your computational work will be minimized. The sample or population mean key is usually marked with $\bar{x}$. The sample standard deviation key is usually marked with s, s_x, or σ_{xn-1}, and the population standard deviation key with σ, σ_x, or σ_{xn}. In using any calculator with these built-in function keys, be sure you know which calculation is being carried out by each key!

If you need to calculate s^2 and s by hand, it is much easier to use the alternative computing formula given next. This computational form is sometimes called the **short-cut method for calculating s^2.**

THE COMPUTING FORMULA FOR CALCULATING s^2

$$s^2 = \frac{\sum x_i^2 - \dfrac{(\sum x_i)^2}{n}}{n - 1}$$

The symbols $(\sum x_i)^2$ and $\sum x_i^2$ in the computing formula are shortcut ways to indicate the arithmetic operation you need to perform. You know from the formula for the sample mean that $\sum x_i$ is the sum of all the measurements. To find $\sum x_i^2$, you square each individual measurement and then add them together.

$\sum x_i^2 =$ Sum of the squares of the individual measurements
$(\sum x_i)^2 =$ Square of the sum of the individual measurements

---

The *sample standard deviation, s,* is the positive square root of s^2.

EXAMPLE **2.5** Calculate the variance and standard deviation for the five measurements in Table 2.2, which are 5, 7, 1, 2, 4. Use the computing formula for s^2 and compare your results with those obtained using the original definition of s^2.

TABLE 2.2 ● **Table for Simplified Calculation of s^2 and s**

x_i	x_i^2
5	25
7	49
1	1
2	4
4	16
19	95

Don't round off partial
results as you go along!

Solution The entries in Table 2.2 are the individual measurements, x_i, and their squares, x_i^2, together with their sums. Using the computing formula for s^2, you have

$$s^2 = \frac{\Sigma x_i^2 - \frac{(\Sigma x_i)^2}{n}}{n-1}$$

$$= \frac{95 - \frac{(19)^2}{5}}{4} = \frac{22.80}{4} = 5.70$$

and $s = \sqrt{s^2} = \sqrt{5.70} = 2.39$, as before.

You may wonder why you need to divide by $(n-1)$ rather than n when computing the sample variance. Just as we used the sample mean $\bar{x}$ to estimate the population mean μ, you may want to use the sample variance s^2 to estimate the population variance σ^2. It turns out that the sample variance s^2 with $(n-1)$ in the denominator provides better estimates of σ^2 than would an estimator calculated with n in the denominator. **For this reason, we always divide by $(n-1)$ when computing the sample variance s^2 and the sample standard deviation s.**

MY APPLET

You can compare the accuracy of estimators of the population variance σ^2 using the **Why Divide by $n-1$?** applet. The applet selects samples from a population with standard deviation $\sigma = 29.2$. It then calculates the standard deviation s using $(n-1)$ in the denominator as well as a standard deviation calculated using n in the denominator. You can choose to compare the estimators for a single new sample, for 10 samples, or for 100 samples. Notice that each of the 10 samples shown in Figure 2.9 has a different sample standard deviation. However, when the 10 standard deviations are averaged at the bottom of the applet, one of the two estimators is closer to the population standard deviation, $\sigma = 29.2$. Which one is it? We will use this applet again for the MyApplet Exercises at the end of the chapter.

FIGURE 2.9
Why Divide by $n-1$?
applet

```
Results displayed for last 10 samples of size3
St. Dev. when dividing by n-1
   38.6 42.8 33.7 13.5 26.0 14.2 9.0 23.3 9.3 32.7
St. Dev. when dividing by n
   31.5 34.9 27.5 11.0 21.2 11.6 7.4 19.0 7.6 26.7
Averages of St. Dev. from all 10 samples:
  when dividing by n-1 : 27.0
  when dividing by n    : 22.0



        New Sample    10 Samples   100 Samples
```

At this point, you have learned how to compute the variance and standard deviation of a set of measurements. Remember these points:

- The value of s is always greater than or equal to zero.
- The larger the value of s^2 or s, the greater the variability of the data set.
- If s^2 or s is equal to zero, all the measurements must have the same value.
- In order to measure the variability in the same units as the original observations, we compute the standard deviation $s = \sqrt{s^2}$.

This information allows you to compare several sets of data with respect to their locations and their variability. How can you use these measures to say something more specific about a single set of data? The theorem and rule presented in the next section will help answer this question.

 2.3 EXERCISES

BASIC TECHNIQUES

2.13 You are given $n = 5$ measurements: 2, 1, 1, 3, 5.

a. Calculate the sample mean, $\bar{x}$.

b. Calculate the sample variance, s^2, using the formula given by the definition.

c. Find the sample standard deviation, s.

d. Find s^2 and s using the computing formula. Compare the results with those found in parts b and c.

2.14 Refer to Exercise 2.13.

a. Use the data entry method in your scientific calculator to enter the five measurements. Recall the proper memories to find the sample mean and standard deviation.

b. Verify that the calculator provides the same values for $\bar{x}$ and s as in Exercise 2.13, parts a and c.

2.15 You are given $n = 8$ measurements: 4, 1, 3, 1, 3, 1, 2, 2.

a. Find the range.

b. Calculate $\bar{x}$.

c. Calculate s^2 and s using the computing formula.

d. Use the data entry method in your calculator to find $\bar{x}$, s, and s^2. Verify that your answers are the same as those in parts b and c.

2.16 You are given $n = 8$ measurements: 3, 1, 5, 6, 4, 4, 3, 5.

a. Calculate the range.

b. Calculate the sample mean.

c. Calculate the sample variance and standard deviation.

d. Compare the range and the standard deviation. The range is approximately how many standard deviations?

APPLICATIONS

2.17 An Archeological Find, again An article in *Archaeometry* involved an analysis of 26 samples of Romano-British pottery found at four different kiln sites in the United Kingdom.[7] The samples were analyzed to determine their chemical composition. The percentage of iron oxide in each of five samples collected at the Island Thorns site was:

  1.28, 2.39, 1.50, 1.88, 1.51

a. Calculate the range.

b. Calculate the sample variance and the standard deviation using the computing formula.

c. Compare the range and the standard deviation. The range is approximately how many standard deviations?

2.18 Utility Bills in Southern California
EX0218   The monthly utility bills for a household in
Riverside, California, were recorded for 12 consecu-
tive months starting in January 2003:

Month	Amount ($)	Month	Amount ($)
January	$204.94	July	$276.70
February	180.00	August	309.70
March	178.23	September	312.40
April	176.43	October	238.66
May	165.12	November	225.47
June	236.72	December	222.23

a. Calculate the range of the utility bills for the year 2003.

b. Calculate the average monthly utility bill for the year 2003.

c. Calculate the standard deviation for the 2003 utility bills.

ON THE PRACTICAL SIGNIFICANCE OF THE STANDARD DEVIATION

2.4

We now introduce a useful theorem developed by the Russian mathematician Tchebysheff. Proof of the theorem is not difficult, but we are more interested in its application than its proof.

Tchebysheff's Theorem   Given a number k greater than or equal to 1 and a set of n measurements, at least $[1 - (1/k^2)]$ of the measurements will lie within k standard deviations of their mean.

Tchebysheff's Theorem applies to *any set of measurements* and can be used to describe either a sample or a population. We will use the notation appropriate for populations, but you should realize that we could just as easily use the mean and the standard deviation for the sample.

The idea involved in Tchebysheff's Theorem is illustrated in Figure 2.10. An interval is constructed by measuring a distance $k\sigma$ on either side of the mean μ. The number k can be any number as long as it is greater than or equal to 1. Then Tchebysheff's Theorem states that at least $1 - (1/k^2)$ of the total number n measurements lies in the constructed interval.

FIGURE 2.10

Illustrating Tchebysheff's Theorem

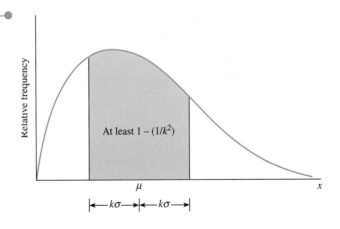

In Table 2.3, we choose a few numerical values for k and compute $[1 - (1/k^2)]$.

TABLE 2.3 ● **Illustrative Values of $[1 - (1/k^2)]$**

k	$1 - (1/k^2)$
1	$1 - 1 = 0$
2	$1 - 1/4 = 3/4$
3	$1 - 1/9 = 8/9$

From the calculations in Table 2.3, the theorem states:

- At least none of the measurements lie in the interval $\mu - \sigma$ to $\mu + \sigma$.
- At least 3/4 of the measurements lie in the interval $\mu - 2\sigma$ to $\mu + 2\sigma$.
- At least 8/9 of the measurements lie in the interval $\mu - 3\sigma$ to $\mu + 3\sigma$.

Although the first statement is not at all helpful, the other two values of k provide valuable information about the proportion of measurements that fall in certain intervals. The values $k = 2$ and $k = 3$ are not the only values of k you can use; for example, the proportion of measurements that fall within $k = 2.5$ standard deviations of the mean is at least $1 - [1/(2.5)^2] = .84$.

EXAMPLE   2.6

The mean and variance of a sample of $n = 25$ measurements are 75 and 100, respectively. Use Tchebysheff's Theorem to describe the distribution of measurements.

Solution   You are given $\bar{x} = 75$ and $s^2 = 100$. The standard deviation is $s = \sqrt{100} = 10$. The distribution of measurements is centered about $\bar{x} = 75$, and Tchebysheff's Theorem states:

- *At least* 3/4 of the 25 measurements lie in the interval $\bar{x} \pm 2s = 75 \pm 2(10)$ —that is, 55 to 95.
- *At least* 8/9 of the measurements lie in the interval $\bar{x} \pm 3s = 75 \pm 3(10)$— that is, 45 to 105.

Since Tchebysheff's Theorem applies to *any* distribution, it is very conservative. This is why we emphasize "at least $1 - (1/k^2)$" in this theorem.

Another rule for describing the variability of a data set does not work for *all* data sets, but it does work very well for data that "pile up" in the familiar mound shape shown in Figure 2.11. The closer your data distribution is to the mound-shaped curve in Figure 2.11, the more accurate the rule will be. Since mound-shaped data distributions occur quite frequently in nature, the rule can often be used in practical applications. For this reason, we call it the **Empirical Rule.**

FIGURE 2.11 ●
Mound-shaped distribution

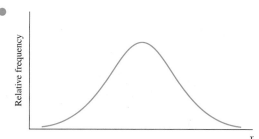

Empirical Rule   Given a distribution of measurements that is approximately mound-shaped:

The interval ($\mu \pm \sigma$) contains approximately 68% of the measurements.

The interval ($\mu \pm 2\sigma$) contains approximately 95% of the measurements.

The interval ($\mu \pm 3\sigma$) contains approximately 99.7% of the measurements.

MY TIP

Remember these three numbers:

68—95—99.7

The mound-shaped distribution shown in Figure 2.11 is commonly known as the **normal distribution** and will be discussed in detail in Chapter 6.

EXAMPLE 2.7

In a time study conducted at a manufacturing plant, the length of time to complete a specified operation is measured for each of $n = 40$ workers. The mean and standard deviation are found to be 12.8 and 1.7, respectively. Describe the sample data using the Empirical Rule.

Solution   To describe the data, calculate these intervals:

$$(\bar{x} \pm s) = 12.8 \pm 1.7 \qquad \text{or} \qquad 11.1 \text{ to } 14.5$$
$$(\bar{x} \pm 2s) = 12.8 \pm 2(1.7) \quad \text{or} \qquad 9.4 \text{ to } 16.2$$
$$(\bar{x} \pm 3s) = 12.8 \pm 3(1.7) \quad \text{or} \qquad 7.7 \text{ to } 17.9$$

According to the Empirical Rule, you expect approximately 68% of the measurements to fall into the interval from 11.1 to 14.5, approximately 95% to fall into the interval from 9.4 to 16.2, and approximately 99.7% to fall into the interval from 7.7 to 17.9.

If you doubt that the distribution of measurements is mound-shaped, or if you wish for some other reason to be conservative, you can apply Tchebysheff's Theorem and be absolutely certain of your statements. Tchebysheff's Theorem tells you that at least 3/4 of the measurements fall into the interval from 9.4 to 16.2 and at least 8/9 into the interval from 7.7 to 17.9.

EXAMPLE 2.8

Student teachers are trained to develop lesson plans, on the assumption that the written plan will help them to perform successfully in the classroom. In a study to assess the relationship between written lesson plans and their implementation in the classroom, 25 lesson plans were scored on a scale of 0 to 34 according to a Lesson Plan Assessment Checklist. The 25 scores are shown in Table 2.4. Use Tchebysheff's Theorem and the Empirical Rule (if applicable) to describe the distribution of these assessment scores.

TABLE 2.4   ● **Lesson Plan Assessment Scores**

26.1	26.0	14.5	29.3	19.7
22.1	21.2	26.6	31.9	25.0
15.9	20.8	20.2	17.8	13.3
25.6	26.5	15.7	22.1	13.8
29.0	21.3	23.5	22.1	10.2

Solution   Use your calculator or the computing formulas to verify that $\bar{x} = 21.6$ and $s = 5.5$. The appropriate intervals are calculated and listed in Table 2.5. We have also referred back to the original 25 measurements and counted the actual number of measurements that fall into each of these intervals. These frequencies and relative frequencies are shown in Table 2.5.

TABLE 2.5

Intervals $\bar{x} \pm ks$ for the Data of Table 2.4

k	Interval $\bar{x} \pm ks$	Frequency in Interval	Relative Frequency
1	16.1–27.1	16	.64
2	10.6–32.6	24	.96
3	5.1–38.1	25	1.00

 TIP

Empirical Rule ⇔ mound-shaped data

Tchebysheff ⇔ any shaped data

Is Tchebysheff's Theorem applicable? Yes, because it can be used for any set of data. According to Tchebysheff's Theorem,

- at least 3/4 of the measurements will fall between 10.6 and 32.6.
- at least 8/9 of the measurements will fall between 5.1 and 38.1.

You can see in Table 2.5 that Tchebysheff's Theorem is true for these data. In fact, the proportions of measurements that fall into the specified intervals exceed the lower bound given by this theorem.

Is the Empirical Rule applicable? You can check for yourself by drawing a graph—either a stem and leaf plot or a histogram. The *MINITAB* histogram in Figure 2.12 shows that the distribution is *relatively* mound-shaped, so the Empirical Rule should work *relatively well.* That is,

- approximately 68% of the measurements will fall between 16.1 and 27.1.
- approximately 95% of the measurements will fall between 10.6 and 32.6.
- approximately 99.7% of the measurements will fall between 5.1 and 38.1.

The relative frequencies in Table 2.5 closely approximate those specified by the Empirical Rule.

FIGURE 2.12

MINITAB histogram for Example 2.8

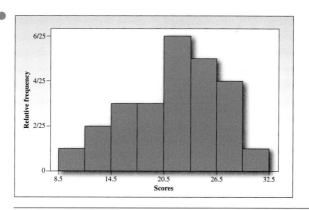

USING TCHEBYSHEFF'S THEOREM AND THE EMPIRICAL RULE

Tchebysheff's Theorem can be proven mathematically. It applies to any set of measurements—sample or population, large or small, mound-shaped or skewed.

Tchebysheff's Theorem gives a *lower bound* to the fraction of measurements to be found in an interval constructed as $\bar{x} \pm ks$. *At least* $1 - (1/k^2)$ of the measurements will fall into this interval, and probably more!

The Empirical Rule is a "rule of thumb" that can be used as a descriptive tool only when the data tend to be roughly mound-shaped (the data tend to pile up near the center of the distribution).

When you use these two tools for describing a set of measurements, Tchebysheff's Theorem will always be satisfied, but it is a very conservative estimate of the fraction of measurements that fall into a particular interval. If it is appropriate to use the Empirical Rule (mound-shaped data), this rule will give you a more accurate estimate of the fraction of measurements that fall into the interval.

A CHECK ON THE CALCULATION OF s

Tchebysheff's Theorem and the Empirical Rule can be used to detect gross errors in the calculation of s. Roughly speaking, these two tools tell you that *most of the time,* measurements lie within *two* standard deviations of their mean. This interval is marked off in Figure 2.13, and it implies that the total range of the measurements, from smallest to largest, should be somewhere around four standard deviations. This is, of course, a very rough approximation, but it can be very useful in checking for large errors in your calculation of s. If the range, R, is about four standard deviations, or $4s$, you can write

$$R \approx 4s \quad \text{or} \quad s \approx \frac{R}{4}$$

The computed value of s using the shortcut formula should be of roughly the same order as the approximation.

FIGURE 2.13

Range approximation to s

EXAMPLE 2.9 Use the range approximation to check the calculation of s for Table 2.2.

Solution The range of the five measurements—5, 7, 1, 2, 4—is

$$R = 7 - 1 = 6$$

Then

$$s \approx \frac{R}{4} = \frac{6}{4} = 1.5$$

This is the same order as the calculated value $s = 2.4$.

$s \approx R/4$ gives only an **approximate** value for s.

The range approximation is *not* intended to provide an accurate value for s. Rather, its purpose is to detect gross errors in calculating, such as the failure to divide the sum of squares of deviations by $(n - 1)$ or the failure to take the square root of s^2. If you make one of these mistakes, your answer will be many times larger than the range approximation of s.

EXAMPLE 2.10　Use the range approximation to determine an approximate value for the standard deviation for the data in Table 2.4.

Solution　The range $R = 31.9 - 10.2 = 21.7$. Then

$$s \approx \frac{R}{4} = \frac{21.7}{4} = 5.4$$

Since the exact value of s is 5.5 for the data in Table 2.4, the approximation is very close.

The range for a sample of n measurements will depend on the sample size, n. For larger values of n, a larger range of the x values is expected. The range for large samples (say, $n = 50$ or more observations) may be as large as $6s$, whereas the range for small samples (say, $n = 5$ or less) may be as small as or smaller than $2.5s$.

The range approximation for s can be improved if it is known that the sample is drawn from a mound-shaped distribution of data. Thus, the calculated s should not differ substantially from the range divided by the appropriate ratio given in Table 2.6.

TABLE 2.6　　● **Divisor for the Range Approximation of s**

Number of Measurements	Expected Ratio of Range to s
5	2.5
10	3
25	4

2.5　EXERCISES

BASIC TECHNIQUES

2.19 A set of $n = 10$ measurements consists of the values 5, 2, 3, 6, 1, 2, 4, 5, 1, 3.

a. Use the range approximation to estimate the value of s for this set. (HINT: Use the table at the end of Section 2.5.)

b. Use your calculator to find the actual value of s. Is the actual value close to your estimate in part a?

c. Draw a dotplot of this data set. Are the data mound-shaped?

d. Can you use Tchebysheff's Theorem to describe this data set? Why or why not?

e. Can you use the Empirical Rule to describe this data set? Why or why not?

2.20 Suppose you want to create a mental picture of the relative frequency histogram for a large data set consisting of 1,000 observations, and you know

that the mean and standard deviation of the data set are 36 and 3, respectively.

a. If you are fairly certain that the relative frequency distribution of the data is mound-shaped, how might you picture the relative frequency distribution? (HINT: Use the Empirical Rule.)

b. If you have no prior information concerning the shape of the relative frequency distribution, what can you say about the relative frequency histogram? (HINT: Construct intervals $\bar{x} \pm ks$ for several choices of k.)

2.21 A distribution of measurements is relatively mound-shaped with mean 50 and standard deviation 10.

a. What proportion of the measurements will fall between 40 and 60?

b. What proportion of the measurements will fall between 30 and 70?

c. What proportion of the measurements will fall between 30 and 60?

d. If a measurement is chosen at random from this distribution, what is the probability that it will be greater than 60?

2.22 A set of data has a mean of 75 and a standard deviation of 5. You know nothing else about the size of the data set or the shape of the data distribution.

a. What can you say about the proportion of measurements that fall between 60 and 90?

b. What can you say about the proportion of measurements that fall between 65 and 85?

c. What can you say about the proportion of measurements that are less than 65?

APPLICATIONS

2.23 Driving Emergencies The length of time required for an automobile driver to respond to a particular emergency situation was recorded for $n = 10$ drivers. The times (in seconds) were .5, .8, 1.1, .7, .6, .9, .7, .8, .7, .8.

a. Scan the data and use the procedure in Section 2.5 to find an approximate value for s. Use this value to check your calculations in part b.

b. Calculate the sample mean $\bar{x}$ and the standard deviation s. Compare with part a.

(MY) (CD) **2.24 Packaging Hamburger Meat** The data
EX0224 listed here are the weights (in pounds) of 27 packages of ground beef in a supermarket meat display:

1.08	.99	.97	1.18	1.41	1.28	.83
1.06	1.14	1.38	.75	.96	1.08	.87
.89	.89	.96	1.12	1.12	.93	1.24
.89	.98	1.14	.92	1.18	1.17	

a. Construct a stem and leaf plot or a relative frequency histogram to display the distribution of weights. Is the distribution relatively mound-shaped?

b. Find the mean and standard deviation of the data set.

c. Find the percentage of measurements in the intervals $\bar{x} \pm s$, $\bar{x} \pm 2s$, and $\bar{x} \pm 3s$.

d. How do the percentages obtained in part c compare with those given by the Empirical Rule? Explain.

e. How many of the packages weigh exactly 1 pound? Can you think of any explanation for this?

2.25 Breathing Rates Is your breathing rate normal? Actually, there is no standard breathing rate for humans. It can vary from as low as 4 breaths per minute to as high as 70 or 75 for a person engaged in strenuous exercise. Suppose that the resting breathing rates for college-age students have a relative frequency distribution that is mound-shaped, with a mean equal to 12 and a standard deviation of 2.3 breaths per minute. What fraction of all students would have breathing rates in the following intervals?

a. 9.7 to 14.3 breaths per minute

b. 7.4 to 16.6 breaths per minute

c. More than 18.9 or less than 5.1 breaths per minute

(MY) (CD) **2.26 Ore Samples** A geologist collected 20
EX0226 different ore samples, all the same weight, and randomly divided them into two groups. She measured the titanium (Ti) content of the samples using two different methods.

Method 1					Method 2				
.011	.013	.013	.015	.014	.011	.016	.013	.012	.015
.013	.010	.013	.011	.012	.012	.017	.013	.014	.015

a. Construct stem and leaf plots for the two data sets. Visually compare their centers and their ranges.

b. Calculate the sample means and standard deviations for the two sets. Do the calculated values confirm your visual conclusions from part a?

2.27 Social Security Numbers The data from Exercise 1.70 (see data set EX0170), reproduced below, show the last digit of the social security number for a group of 70 students.

1	6	9	1	5	9	0	2	8	4
0	7	3	4	2	3	5	8	4	2
3	2	0	0	2	1	2	7	7	4
0	0	9	9	5	3	8	4	7	4
6	6	9	0	2	6	2	9	5	8
5	1	7	7	7	8	7	5	1	8
3	4	1	9	3	8	6	6	6	6

a. You found in Exercise 1.70 that the distribution of this data was relatively "flat," with each different value from 0 to 9 occurring with nearly equal frequency. Using this fact, what would be your best estimate for the mean of the data set?

b. Use the range approximation to guess the value of s for this set.

c. Use your calculator to find the actual values of $\bar{x}$ and s. Compare with your estimates in parts a and b.

2.28 Social Security Numbers, continued Refer to the data set in Exercise 2.27.

a. Find the percentage of measurements in the intervals $\bar{x} \pm s$, $\bar{x} \pm 2s$, and $\bar{x} \pm 3s$.

b. How do the percentages obtained in part a compare with those given by the Empirical Rule? Should they be approximately the same? Explain.

2.29 Survival Times A group of experimental animals is infected with a particular form of bacteria, and their survival time is found to average 32 days, with a standard deviation of 36 days.

a. Visualize the distribution of survival times. Do you think that the distribution is relatively mound-shaped, skewed right, or skewed left? Explain.

b. Within what limits would you expect at least 3/4 of the measurements to lie?

2.30 Survival Times, continued Refer to Exercise 2.29. You can use the Empirical Rule to see why the distribution of survival times could not be mound-shaped.

a. Find the value of x that is exactly one standard deviation below the mean.

b. If the distribution is in fact mound-shaped, approximately what percentage of the measurements should be less than the value of x found in part a?

c. Since the variable being measured is time, is it possible to find any measurements that are more than one standard deviation below the mean?

d. Use your answers to part b and c to explain why the data distribution cannot be mound-shaped.

2.31 Timber Tracts To estimate the amount of lumber in a tract of timber, an owner decided to count the number of trees with diameters exceeding 12 inches in randomly selected 50-by-50-foot squares. Seventy 50-by-50-foot squares were chosen, and the selected trees were counted in each tract. The data are listed here:

7	8	7	10	4	8	6	8	9	10
9	6	4	9	10	9	8	8	7	9
3	9	5	9	9	8	7	5	8	8
10	2	7	4	8	5	10	7	7	7
9	6	8	8	8	7	8	9	6	8
6	11	9	11	7	7	11	7	9	13
10	8	8	5	9	9	8	5	9	8

a. Construct a relative frequency histogram to describe the data.

b. Calculate the sample mean $\bar{x}$ as an estimate of μ, the mean number of timber trees for all 50-by-50-foot squares in the tract.

c. Calculate s for the data. Construct the intervals $\bar{x} \pm s$, $\bar{x} \pm 2s$, and $\bar{x} \pm 3s$. Calculate the percentage of squares falling into each of the three intervals, and compare with the corresponding percentages given by the Empirical Rule and Tchebysheff's Theorem.

2.32 Tuna Fish, again Refer to Exercise 2.8 and data set EX0208. The prices of a 6-ounce can or a 7.06 pouch for 14 different brands of water-packed light tuna, based on prices paid nationally in supermarkets are reproduced here.[4]

| .99 | 1.92 | 1.23 | .85 | .65 | .53 | 1.41 |
| 1.12 | .63 | .67 | .69 | .60 | .60 | .66 |

a. Use the range approximation to find an estimate of s.

b. How does it compare to the computed value of s?

2.33 Old Faithful The data below are 30 waiting times between eruptions of the Old Faithful geyser in Yellowstone National Park.[8]

| 56 | 89 | 51 | 79 | 58 | 82 | 52 | 88 | 52 | 78 | 69 | 75 | 77 | 72 | 71 |
| 55 | 87 | 53 | 85 | 61 | 93 | 54 | 76 | 80 | 81 | 59 | 86 | 78 | 71 | 77 |

a. Calculate the range.

b. Use the range approximation to approximate the standard deviation of these 30 measurements.

c. Calculate the sample standard deviation s.

d. What proportion of the measurements lie within two standard deviations of the mean? Within three standard deviations of the mean? Do these proportions agree with the proportions given in Tchebysheff's Theorem?

2.34 The President's Kids The table below shows the names of the 42 presidents of the United States along with the number of children in their family.[2]

Washington	0	Van Buren	4	Buchanan	0
Adams	5	W.H. Harrison	10	Lincoln	4
Jefferson	6	Tyler*	14	A. Johnson	5
Madison	0	Polk	0	Grant	4
Monroe	2	Taylor	6	Hayes	8
J.Q. Adams	4	Fillmore*	2	Garfield	7
Jackson	0	Pierce	3	Arthur	3
Cleveland	5	Coolidge	2	Nixon	2
B. Harrison*	3	Hoover	2	Ford	4
McKinley	2	F.D. Roosevelt	6	Carter	4
T. Roosevelt*	6	Truman	1	Reagan*	4
Taft	3	Eisenhower	2	G.H.W. Bush	6
Wilson*	3	Kennedy	3	Clinton	1
Harding	0	L.B. Johnson	2	G.W. Bush	2

*Married twice   Source: Time Almanac 2004

a. Construct a relative frequency histogram to describe the data. How would you describe the shape of this distribution?

b. Calculate the mean and the standard deviation for the data set.

c. Construct the intervals $\bar{x} \pm s$, $\bar{x} \pm 2s$, and $\bar{x} \pm 3s$. Find the percentage of measurements falling into these three intervals and compare with the corresponding percentages given by Tchebysheff's Theorem and the Empirical Rule.

2.35 An Archeological Find, again Refer to Exercise 2.17. The percentage of iron oxide in each of five pottery samples collected at the Island Thorns site was:

1.28 2.39 1.50 1.88 1.51

a. Use the range approximation to find an estimate of s, using an appropriate divisor from Table 2.6.

b. Calculate the standard deviation s. How close did your estimate come to the actual value of s?

(MY)(CD) 2.36 Brett Favre The number of passes com-
EX0236 pleted by Bret Favre, quarterback for the Green Bay Packers, was recorded at each of the 16 regular season and 2 postseason games in the fall of 2003 (ESPN.com):[9]

25 15 23 21 19 25
23 18 14 13 10 23
22 23 22 12 26 15

a. Draw a stem and leaf plot to describe the data.

b. Calculate the mean and standard deviation for Brett Favre's per-game pass completions.

c. What proportion of the measurements lie within two standard deviations of the mean?

CALCULATING THE MEAN AND STANDARD DEVIATION FOR GROUPED DATA (OPTIONAL)

2.37 Suppose that some measurements occur more than once and that the data $x_1, x_2, \ldots, x_k$ are arranged in a frequency table as shown here:

Observations	Frequency f_i
x_1	f_1
x_2	f_2
.	.
.	.
.	.
x_k	f_k

The formulas for the mean and variance for grouped data are

$$\bar{x} = \frac{\Sigma x_i f_i}{n}, \qquad \text{where } n = \Sigma f_i$$

and

$$s^2 = \frac{\Sigma x_i^2 f_i - \dfrac{(\Sigma x_i f_i)^2}{n}}{n - 1}$$

Notice that if each value occurs once, these formulas reduce to those given in the text. Although these formulas for grouped data are primarily of value when you have a large number of measurements, demonstrate their use for the sample 1, 0, 0, 1, 3, 1, 3, 2, 3, 0, 0, 1, 1, 3, 2.

a. Calculate $\bar{x}$ and s^2 directly, using the formulas for ungrouped data.

b. The frequency table for the $n = 15$ measurements is as follows:

x	f
0	4
1	5
2	2
3	4

Calculate $\bar{x}$ and s^2 using the formulas for grouped data. Compare with your answers to part a.

2.38 International Baccalaureate The International Baccalaureate (IB) program is an accelerated academic program offered at a growing number of high schools throughout the country. Students enrolled in this program are placed in accelerated or advanced courses and must take IB examinations in each of six subject areas at the end of their junior or senior year. Students are scored on a scale of 1–7, with 1–2 being poor, 3 mediocre, 4 average, and 5–7 excellent. During its first year of operation at John W. North High School in Riverside, California, 17 juniors attempted the IB economics exam, with these results:

Exam Grade	Number of Students
7	1
6	4
5	4
4	4
3	4

Calculate the mean and standard deviation for these scores.

2.39 A Skewed Distribution To illustrate the utility of the Empirical Rule, consider a distribution that is heavily skewed to the right, as shown in the accompanying figure.

a. Calculate $\bar{x}$ and s for the data shown. (NOTE: There are 10 zeros, 5 ones, and so on.)

b. Construct the intervals $\bar{x} \pm s$, $\bar{x} \pm 2s$, and $\bar{x} \pm 3s$ and locate them on the frequency distribution.

c. Calculate the proportion of the $n = 25$ measurements that fall into each of the three intervals. Compare with Tchebysheff's Theorem and the Empirical Rule. Note that, although the proportion that falls into the interval $\bar{x} \pm s$ does not agree closely with the Empirical Rule, the proportions

that fall into the intervals $\bar{x} \pm 2s$ and $\bar{x} \pm 3s$ agree very well. Many times this is true, even for non-mound-shaped distributions of data.

Distribution for Exercise 2.39

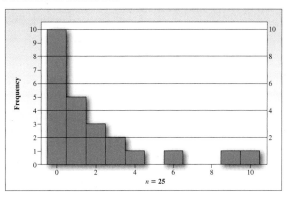

$n = 25$

MEASURES OF RELATIVE STANDING

2.6

Sometimes you need to know the position of one observation relative to others in a set of data. For example, if you took an examination with a total of 35 points, you might want to know how your score of 30 compared to the scores of the other students in the class. The mean and standard deviation of the scores can be used to calculate a **z-score**, which measures the relative standing of a measurement in a data set.

Definition The **sample z-score** is a measure of relative standing defined by

 TIP

Positive z-score ⇔ *x* is above the mean.

Negative z-score ⇔ *x* is below the mean.

$$z\text{-score} = \frac{x - \bar{x}}{s}$$

A z-score measures the distance between an observation and the mean, measured in units of standard deviation. For example, suppose that the mean and standard deviation of the test scores (based on a total of 35 points) are 25 and 4, respectively. The z-score for your score of 30 is calculated as follows:

$$z\text{-score} = \frac{x - \bar{x}}{s} = \frac{30 - 25}{4} = 1.25$$

Your score of 30 lies 1.25 standard deviations above the mean ($30 = \bar{x} + 1.25s$).

The *z*-score is a valuable tool for determining whether a particular observation is likely to occur quite frequently or whether it is unlikely and might be considered an **outlier.**

According to Tchebysheff's Theorem and the Empirical Rule,

- at least 75% and more likely 95% of the observations lie within two standard deviations of their mean: their *z*-scores are between -2 and $+2$. *Observations with z-scores exceeding 2 in absolute value happen less than 5% of the time and are considered somewhat unlikely.*
- at least 89% and more likely 99.7% of the observations lie within three standard deviations of their mean: their *z*-scores are between -3 and $+3$. *Observations with z-scores exceeding 3 in absolute value happen less than 1% of the time and are considered very unlikely.*

MY TIP

z-scores above 3 in absolute value are very unusual.

You should look carefully at any observation that has a *z*-score exceeding 3 in absolute value. Perhaps the measurement was recorded incorrectly or does not belong to the population being sampled. Perhaps it is just a highly unlikely observation, but a valid one nonetheless!

EXAMPLE 2.11

Consider this sample of $n = 10$ measurements:

$$1, 1, 0, 15, 2, 3, 4, 0, 1, 3$$

The measurement $x = 15$ appears to be unusually large. Calculate the *z*-score for this observation and state your conclusions.

Solution Calculate $\bar{x} = 3.0$ and $s = 4.42$ for the $n = 10$ measurements. Then the *z*-score for the suspected outlier, $x = 15$, is calculated as

$$z\text{-score} = \frac{x - \bar{x}}{s} = \frac{15 - 3}{4.42} = 2.71$$

Hence, the measurement $x = 15$ lies 2.71 standard deviations above the sample mean, $\bar{x} = 3.0$. Although the *z*-score does not exceed 3, it is close enough so that you might suspect that $x = 15$ is an outlier. You should examine the sampling procedure to see whether $x = 15$ is a faulty observation.

A **percentile** is another measure of relative standing and is most often used for large data sets. (Percentiles are not very useful for small data sets.)

Definition A set of n measurements on the variable x has been arranged in order of magnitude. The **pth percentile** is the value of x that is greater than $p\%$ of the measurements and is less than the remaining $(100 - p)\%$.

EXAMPLE 2.12

Suppose you have been notified that your score of 610 on the Verbal Graduate Record Examination placed you at the 60th percentile in the distribution of scores. Where does your score of 610 stand in relation to the scores of others who took the examination?

Solution Scoring at the 60th percentile means that 60% of all the examination scores were lower than your score and 40% were higher.

In general, the 60th percentile for the variable x is a point on the *horizontal axis* of the data distribution that is greater than 60% of the measurements and less than the others. That is, 60% of the measurements are less than the 60th percentile and 40% are greater (see Figure 2.14). Since the total area under the distribution is 100%, 60% of the area is to the left and 40% of the area is to the right of the 60th percentile. Remember that the median, m, of a set of data is the middle measurement; that is, 50% of the measurements are smaller and 50% are larger than the median. Thus, the *median is the same as the 50th percentile!*

FIGURE 2.14

The 60th percentile shown on the relative frequency histogram for a data set

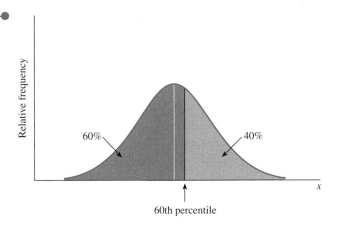

The 25th and 75th percentiles, called the **lower** and **upper quartiles,** along with the median (the 50th percentile), locate points that divide the data into four sets, each containing an equal number of measurements. Twenty-five percent of the measurements will be less than the lower (first) quartile, 50% will be less than the median (the second quartile), and 75% will be less than the upper (third) quartile. Thus, the median and the lower and upper quartiles are located at points on the x-axis so that the area under the relative frequency histogram for the data is partitioned into four equal areas, as shown in Figure 2.15.

FIGURE 2.15

Location of quartiles

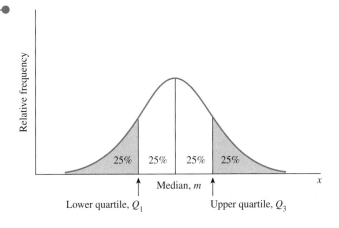

Definition A set of n measurements on the variable x has been arranged in order of magnitude. The **lower quartile (first quartile), Q_1,** is the value of x that is greater than one-fourth of the measurements and is less than the remaining three-fourths. The **second quartile** is the median. The **upper quartile (third quartile), Q_3,** is the value of x that is greater than three-fourths of the measurements and is less than the remaining one-fourth.

For small data sets, it is often impossible to divide the set into four groups, each of which contains exactly 25% of the measurements. For example, when $n = 10$, you would need to have $2\frac{1}{2}$ measurements in each group! Even when you can perform this task (for example, if $n = 12$), there are many numbers that would satisfy the preceding definition, and could therefore be considered "quartiles." To avoid this ambiguity, we use the following rule to locate sample quartiles.

CALCULATING SAMPLE QUARTILES

- When the measurements are arranged in order of magnitude, the **lower quartile, Q_1,** is the value of x in position $.25(n + 1)$, and the **upper quartile, Q_3,** is the value of x in position $.75(n + 1)$.
- When $.25(n + 1)$ and $.75(n + 1)$ are not integers, the quartiles are found by interpolation, using the values in the two adjacent positions.[†]

EXAMPLE 2.13 Find the lower and upper quartiles for this set of measurements:

16, 25, 4, 18, 11, 13, 20, 8, 11, 9

Solution Rank the $n = 10$ measurements from smallest to largest:

4, 8, 9, 11, 11, 13, 16, 18, 20, 25

Calculate

Position of $Q_1 = .25(n + 1) = .25(10 + 1) = 2.75$
Position of $Q_3 = .75(n + 1) = .75(10 + 1) = 8.25$

Since these positions are not integers, the lower quartile is taken to be the value 3/4 of the distance between the second and third ordered measurements, and the upper quartile is taken to be the value 1/4 of the distance between the eighth and ninth ordered measurements. Therefore,

$$Q_1 = 8 + .75(9 - 8) = 8 + .75 = 8.75$$

and

$$Q_3 = 18 + .25(20 - 18) = 18 + .5 = 18.5$$

Because the median and the quartiles divide the data distribution into four parts, each containing approximately 25% of the measurements, Q_1 and Q_3 are the upper and lower boundaries for the middle 50% of the distribution. We can measure the range of this "middle 50%" of the distribution using a numerical measure called the **interquartile range.**

[†]This definition of quartiles is consistent with the one used in the *MINITAB* package. Some textbooks use ordinary rounding when finding quartile positions, whereas other compute sample quartiles as the medians of the upper and lower halves of the data set.

Definition The **interquartile range (IQR)** for a set of measurements is the difference between the upper and lower quartiles; that is, $IQR = Q_3 - Q_1$.

For the data in Example 2.13, $IQR = Q_3 - Q_1 = 18.50 - 8.75 = 9.75$. We will use the IQR along with the quartiles and the median in the next section to construct another graph for describing data sets.

(MY) PERSONAL TRAINER

How Do I Calculate Sample Quartiles?

1. Arrange the data set in order of magnitude from smallest to largest.
2. Calculate the quartile positions:
   - Position of Q_1: $.25(n + 1)$
   - Position of Q_3: $.75(n + 1)$
3. If the positions are integers, than Q_1 and Q_3 are the values in the ordered data set found in those positions.
4. If the positions in step 2 are not integers, find the two measurements in positions just above and just below the calculated position. Calculate the quartile by finding a value either one-fourth, one-half, or three-fourths of the way between these two measurements.

Exercise Reps

A. Below you will find two practice data sets. Fill in the blanks to find the necessary quartiles.

Data Set	Sorted	n	Position of Q_1	Position of Q_3	Lower Quartile, Q_1	Upper Quartile, Q_3
2, 5, 7, 1, 1, 2, 8						
5, 0, 1, 3, 1, 5						
5, 2, 4, 4, 1						

B. Below you will find three data sets that have been already been sorted. The positions of the upper and lower quartiles are shown in the table. Fill in the blanks to find the upper and lower quartiles.

Sorted Data Set	Position of Q_1	Above and Below	Q_1	Position of Q_3	Above and Below	Q_3
0, 1, 4, 4, 5, 9	1.75	_____ and _____		5.25	_____ and _____	
0, 1, 3, 3, 4, 7, 7, 8	2.25	_____ and _____		6.75	_____ and _____	
1, 1, 2, 5, 6, 6, 7, 9, 9	2.5	_____ and _____		7.5	_____ and _____	

Progress Report

- Still having trouble? Try again using the Exercise Reps at the end of this section.
- Mastered sample quartiles? You can skip the Exercise Reps at the end of this section!

Answers are located on the perforated card at the back of this book.

Many of the numerical measures that you have learned are easily found using computer programs or even graphics calculators. The *MINITAB* command **Stat → Basic Statistics → Display Descriptive Statistics** (see the section "My *MINITAB*" at the end of this chapter) produces output containing the mean, the standard deviation, the median, and the lower and upper quartiles, as well as the values of some other statistics that we have not discussed yet. The data from Example 2.13 produced the *MINITAB* output shown in Figure 2.16. Notice that the quartiles are identical to the hand-calculated values in that example.

FIGURE 2.16

MINITAB output for the data in Example 2.13

Descriptive Statistics: x

Variable	N	N*	Mean	SE Mean	StDev	Minimum	Q1	Median	Q3	Maximum
X	10	0	13.50	1.98	6.28	4.00	8.75	12.00	18.50	25.00

2.7 THE FIVE-NUMBER SUMMARY AND THE BOX PLOT

The median and the upper and lower quartiles shown in Figure 2.15 divide the data into four sets, each containing an equal number of measurements. If we add the largest number (Max) and the smallest number (Min) in the data set to this group, we will have a set of numbers that provide a quick and rough summary of the data distribution.

> The **five-number summary** consists of the smallest number, the lower quartile, the median, the upper quartile, and the largest number, presented in order from smallest to largest:
>
> **Min Q_1 Median Q_3 Max**
>
> By definition, one-fourth of the measurements in the data set lie between each of the four adjacent pairs of numbers.

The five-number summary can be used to create a simple graph called a **box plot** to visually describe the data distribution. From the box plot, you can quickly detect any skewness in the shape of the distribution and see whether there are any outliers in the data set.

An outlier may result from transposing digits when recording a measurement, from incorrectly reading an instrument dial, from a malfunctioning piece of equipment, or from other problems. Even when there are no recording or observational errors, a data set may contain one or more valid measurements that, for one reason or another, differ markedly from the others in the set. These outliers can cause a marked distortion in commonly used numerical measures such as $\bar{x}$ and s. In fact, outliers may themselves contain important information not shared with the other measurements in the set. Therefore, isolating outliers, if they are present, is an important step in any preliminary analysis of a data set. The box plot is designed expressly for this purpose.

TO CONSTRUCT A BOX PLOT

- Calculate the median, the upper and lower quartiles, and the IQR for the data set.
- Draw a horizontal line representing the scale of measurement. Form a box just above the horizontal line with the right and left ends at Q_1 and Q_3. Draw a vertical line through the box at the location of the median.

A box plot is shown in Figure 2.17.

FIGURE 2.17

Box plot

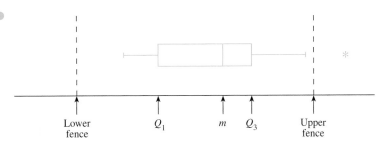

In Section 2.6, the z-score provided boundaries for finding unusually large or small measurements. You looked for z-scores greater than 2 or 3 in absolute value. The box plot uses the IQR to create imaginary "fences" to separate outliers from the rest of the data set:

DETECTING OUTLIERS—OBSERVATIONS THAT ARE BEYOND:

- Lower fence: $Q_1 - 1.5(IQR)$
- Upper fence: $Q_3 + 1.5(IQR)$

The upper and lower fences are shown with broken lines in Figure 2.17, but they are not usually drawn on the box plot. Any measurement beyond the upper or lower fence is an **outlier;** the rest of the measurements, inside the fences, are not unusual. Finally, the box plot marks the range of the data set using "whiskers" to connect the smallest and largest measurements (*excluding outliers*) to the box.

TO FINISH THE BOX PLOT

- Mark any **outliers** with an asterisk (*) on the graph.
- Extend horizontal lines called "whiskers" from the ends of the box to the smallest and largest observations that are *not* outliers.

EXAMPLE 2.14

As American consumers become more careful about the foods they eat, food processors try to stay competitive by avoiding excessive amounts of fat, cholesterol, and sodium in the foods they sell. The following data are the amounts of sodium per slice (in milligrams) for each of eight brands of regular American cheese. Construct a box plot for the data and look for outliers.

340, 300, 520, 340, 320, 290, 260, 330

Solution The $n = 8$ measurements are first ranked from smallest to largest:

260, 290, 300, 320, 330, 340, 340, 520

The positions of the median, Q_1, and Q_3 are:

$.5(n + 1) = .5(9) = 4.5$
$.25(n + 1) = .25(9) = 2.25$
$.75(n + 1) = .75(9) = 6.75$

so that $m = (320 + 330)/2 = 325$, $Q_1 = 290 + .25(10) = 292.5$, and $Q_3 = 340$. The interquartile range is calculated as

$$IQR = Q_3 - Q_1 = 340 - 292.5 = 47.5$$

Calculate the upper and lower fences:

Lower fence: $292.5 - 1.5(47.5) = 221.25$
Upper fence: $340 + 1.5(47.5) = 411.25$

The value $x = 520$, a brand of cheese containing 520 milligrams of sodium, is the only *outlier,* lying beyond the upper fence.

The box plot for the data is shown in Figure 2.18. The outlier is marked with an asterisk (*). Once the outlier is excluded, we find (from the ranked data set) that the smallest and largest measurements are $x = 260$ and $x = 340$. These are the two values that form the whiskers. Since the value $x = 340$ is the same as Q_3, there is no whisker on the right side of the box.

FIGURE 2.18

Box plot for Example 2.14

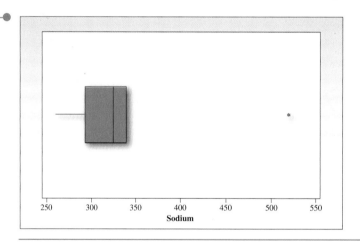

APPLET

Now would be a good time to try the **Building a Box Plot** applet. The applet in Figure 2.19 shows a dotplot of the data in Example 2.14. Using the `Next Step` button, you will see a step-by-step description explaining how the box plot is constructed. We will use this applet again for the MyApplet Exercises at the end of the chapter.

FIGURE 2.19

Building a Box Plot

applet

You can use the box plot to describe the shape of a data distribution by looking at the position of the median line compared to Q_1 and Q_3, the left and right ends of the box. If the median is close to the middle of the box, the distribution is fairly symmetric, providing equal-sized intervals to contain the two middle quarters of the data. If the median line is to the left of center, the distribution is skewed to the right; if the median is to the right of center, the distribution is skewed to the left. Also, for most skewed distributions, the whisker on the skewed side of the box tends to be longer than the whisker on the other side.

We used the *MINITAB* command **Graph → Boxplot** to draw two box plots, one for the sodium contents of the eight brands of cheese in Example 2.14, and another for five brands of fat-free cheese with these sodium contents:

$$300, \quad 300, \quad 320, \quad 290, \quad 180$$

The two box plots are shown together in Figure 2.20. Look at the long whisker on the left side of both box plots and the position of the median lines. Both distributions are skewed to the left; that is, there are a few unusually small measurements. The regular cheese data, however, also show one brand ($x = 520$) with an unusually large amount of sodium. In general, it appears that the sodium content of the fat-free brands is lower than that of the regular brands, but the variability of the sodium content for regular cheese (excluding the outlier) is less than that of the fat-free brands.

FIGURE 2.20

MINITAB output for regular and fat-free cheese

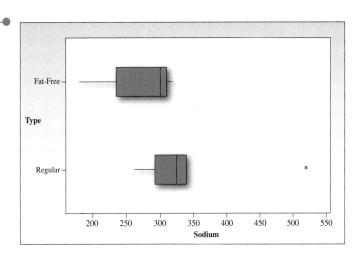

2.7 **EXERCISES**

EXERCISE REPS

2.40 Below you will find two practice data sets. Fill in the blanks to find the necessary quartiles.

Data Set	Sorted	n	Position of Q_1	Position of Q_3	Lower Quartile, Q_1	Upper Quartile, Q_3
.13, .76, .34, .88, .21, .16, .28						
2.3, 1.0, 2.1, 6.5, 2.8, 8.8, 1.7, 2.9, 4.4, 5.1, 2.0						

2.41 Below you will find three data sets that have already been sorted. Fill in the blanks to find the upper and lower quartiles.

Sorted Data Set	Position of Q_1	Above and Below	Q_1	Position of Q_3	Above and Below	Q_3
1, 1.5, 2, 2, 2.2		_____ and _____			_____ and _____	
0, 1.7, 1.8, 3.1, 3.2, 7, 8, 8.8, 8.9, 9, 10		_____ and _____			_____ and _____	
.23, .30, .35, .41, .56, .58, .76, .80		_____ and _____			_____ and _____	

BASIC TECHNIQUES

2.42 Given the following data set: 8, 7, 1, 4, 6, 6, 4, 5, 7, 6, 3, 0

a. Find the five-number summary and the IQR.

b. Calculate $\bar{x}$ and s.

c. Calculate the z-score for the smallest and largest observations. Is either of these observations unusually large or unusually small?

2.43 Find the five-number summary and the IQR for these data:

19, 12, 16, 0, 14, 9, 6, 1, 12, 13, 10, 19, 7, 5, 8

2.44 Construct a box plot for these data and identify any outliers:

25, 22, 26, 23, 27, 26, 28, 18, 25, 24, 12

2.45 Construct a box plot for these data and identify any outliers:

3, 9, 10, 2, 6, 7, 5, 8, 6, 6, 4, 9, 22

APPLICATIONS

2.46 If you scored at the 69th percentile on a placement test, how does your score compare with others?

 2.47 Mercury Concentration in Dolphins
EX0247 Environmental scientists are increasingly concerned with the accumulation of toxic elements in marine mammals and the transfer of such elements to the animals' offspring. The striped dolphin (*Stenella coeruleoalba*), considered to be the top predator in the marine food chain, was the subject of one such study. The mercury concentrations (micrograms/gram) in the livers of 28 male striped dolphins were as follows:

1.70	183.00	221.00	286.00
1.72	168.00	406.00	315.00
8.80	218.00	252.00	241.00
5.90	180.00	329.00	397.00
101.00	264.00	316.00	209.00
85.40	481.00	445.00	314.00
118.00	485.00	278.00	318.00

a. Calculate the five-number summary for the data.

b. Construct a box plot for the data.

c. Are there any outliers?

d. If you knew that the first four dolphins were all less than 3 years old, while all the others were more than 8 years old, would this information help explain the difference in the magnitude of those four observations? Explain.

2.48 Hamburger Meat The weights (in pounds) of the 27 packages of ground beef from Exercise 2.24 (see data set EX0224) are listed here in order from smallest to largest:

.75	.83	.87	.89	.89	.89	.92
.93	.96	.96	.97	.98	.99	1.06
1.08	1.08	1.12	1.12	1.14	1.14	1.17
1.18	1.18	1.24	1.28	1.38	1.41	

a. Confirm the values of the mean and standard deviation, calculated in Exercise 2.24 as $\bar{x} = 1.05$ and $s = .17$.

b. The two largest packages of meat weigh 1.38 and 1.41 pounds. Are these two packages unusually heavy? Explain.

c. Construct a box plot for the package weights. What does the position of the median line and the length of the whiskers tell you about the shape of the distribution?

 2.49 Comparing NFL Quarterbacks How does Brett Favre, quarterback for the Green Bay Packers, compare to Donovan McNabb, quarterback for the Philadelphia Eagles. The table below shows the number of completed passes for each athlete during the 2003 NFL football season (ESPN.com):[9]

Brett Favre			Donovan McNabb		
25	23	22	19	17	18
15	18	23	18	21	15
23	14	22	18	15	27
21	13	12	16	24	23
19	10	26	11	16	21
25	23	15	9	18	10

a. Calculative five-number summaries for the number of passes completed by both Brett Favre and Donovan McNabb.

b. Construct box plots for the two sets of data. Are there any outliners? What do the box plots tell you about the shapes of the two distributions?

c. Write a short paragraph comparing the number of pass completions for the two quarterbacks.

2.50 Presidential Vetoes The set of presidential vetoes in Exercise 1.47 and data set EX0147 is listed here, along with a box plot generated by *MINITAB*. Use the box plot to describe the shape of the distribution and identify any outliers.

Washington	2	B. Harrison	19
J. Adams	0	Cleveland	42
Jefferson	0	McKinley	6
Madison	5	T. Roosevelt	42
Monroe	1	Taft	30
J. Q. Adams	0	Wilson	33
Jackson	5	Harding	5
Van Buren	0	Coolidge	20
W. H. Harrison	0	Hoover	21
Tyler	6	F. D. Roosevelt	372
Polk	2	Truman	180
Taylor	0	Eisenhower	73
Fillmore	0	Kennedy	12
Pierce	9	L. Johnson	16
Buchanan	4	Nixon	26
Lincoln	2	Ford	48
A. Johnson	21	Carter	13
Grant	45	Reagan	39
Hayes	12	G. H. W. Bush	29
Garfield	0	Clinton	36
Arthur	4	G. W. Bush	0
Cleveland	304		

Source: The World Almanac and Book of Facts, 2004

Box plot for Exercise 2.50

2.51 Internet Hotspots The most visited site on the Internet is Yahoo!, which boasted 111,271 thousand unique visitors in December 2003 (Exercise 1.73 and data set EX0173). The number of unique visitors at the top 25 sites are shown in the table:

Site	Unique Visitors (Millions)	Site	Unique Visitors (Millions)
Yahoo! Sites	111.3	Weather Channel	23.8
Time Warner Network	110.5	Real.com Network	22.3
MSN-Microsoft Sites	110.0	Verizon Communi- cations	22.1
Ebay	69.2	Wal-Mart	21.4
Google Sites	61.5	Shopping.com Sites	21.3
Terra Lycos	52.1	Symantec	19.9
Amazon Sites	45.7	AT&T Properties	17.5
About/Primedia	42.6	InfoSpace Network	17.3
Excite Network	25.1	Monster Property	17.3
CNET Networks	25.1	EA Online	16.8
Walt Disney Internet	25.1	SBC Communications	16.5
Viacom Online	24.7	Sony Online	16.5
American Greetings	24.4		

a. Can you tell by looking at the data whether it is roughly symmetric? Or is it skewed?

b. Calculate the mean and the median. Use these measures to decide whether or not the data are symmetric or skewed.

c. Draw a box plot to describe the data. Explain why the box plot confirms your conclusions in part b.

2.52 Utility Bills in Southern California, again The monthly utility bills for a household in Riverside, California, were recorded for 12 consecutive months starting in January 2003:

Month	Amount ($)	Month	Amount ($)
January	$204.94	July	$276.70
February	180.00	August	309.70
March	178.23	September	312.40
April	176.43	October	238.66
May	165.12	November	225.47
June	236.72	December	222.23

a. Construct a box plot for the monthly utility costs.

b. What does the box plot tell you about the distribution of utility costs for this household?

2.53 What's Normal?, again Refer to Exercise 1.67 and data set EX0167. In addition to the normal body temperature in degrees Fahrenheit for the 130 individuals, the data record the gender of the individuals. Box plots for the two groups, male and female, are shown below:[10]

Box plots for Exercise 2.53

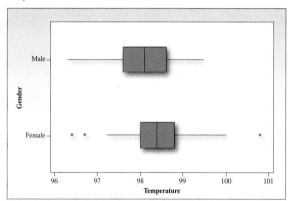

How would you describe the similarities and differences between male and female temperatures in this data set?

CHAPTER REVIEW

Key Concepts and Formulas

I. Measures of the Center of a Data Distribution

1. Arithmetic mean (mean) or average
   a. Population: μ
   b. Sample of n measurements: $\bar{x} = \dfrac{\Sigma x_i}{n}$

2. Median; **position** of the median $= .5(n + 1)$

3. Mode

4. The median may be preferred to the mean if the data are highly skewed.

II. Measures of Variability

1. Range: $R = \text{largest} - \text{smallest}$

2. Variance
   a. Population of N measurements:
   $$\sigma^2 = \frac{\Sigma(x_i - \mu)^2}{N}$$

   b. Sample of n measurements:
   $$s^2 = \frac{\Sigma(x_i - \bar{x})^2}{n - 1} = \frac{\Sigma x_i^2 - \dfrac{(\Sigma x_i)^2}{n}}{n - 1}$$

3. Standard deviation
   a. Population: $\sigma = \sqrt{\sigma^2}$
   b. Sample: $s = \sqrt{s^2}$

4. A rough approximation for s can be calculated as $s \approx R/4$. The divisor can be adjusted depending on the sample size.

III. Tchebysheff's Theorem and the Empirical Rule

1. Use Tchebysheff's Theorem for any data set, regardless of its shape or size.
   a. At least $1 - (1/k^2)$ of the measurements lie within k standard deviations of the mean.
   b. This is only a lower bound; there may be more measurements in the interval.

2. The Empirical Rule can be used only for relatively mound-shaped data sets. Approximately 68%, 95%, and 99.7% of the measurements are within one, two, and three standard deviations of the mean, respectively.

IV. Measures of Relative Standing

1. Sample z-score: $z = \dfrac{x - \bar{x}}{s}$

2. pth percentile; $p\%$ of the measurements are smaller, and $(100 - p)\%$ are larger.

3. Lower quartile, Q_1; **position** of $Q_1 = .25 (n + 1)$

4. Upper quartile, Q_3; **position** of $Q_3 = .75 (n + 1)$

5. Interquartile range: $\text{IQR} = Q_3 - Q_1$

V. The Five-Number Summary and Box Plots

1. The **five-number summary:**

   Min Q_1 Median Q_3 Max

   One-fourth of the measurements in the data set lie between each of the four adjacent pairs of numbers.

2. Box plots are used for detecting outliers and shapes of distributions.

3. Q_1 and Q_3 form the ends of the box. The median line is in the interior of the box.

4. Upper and lower fences are used to find outliers, observations that lie outside these fences.
   a. **Lower fence:** $Q_1 - 1.5(\text{IQR})$
   b. **Upper fence:** $Q_3 + 1.5(\text{IQR})$

5. **Outliers** are marked on the box plot with an asterisk (*).

6. **Whiskers** are connected to the box from the smallest and largest observations that are *not* outliers.

7. Skewed distributions usually have a long whisker *in the direction of the skewness,* and the median line is drawn *away from the direction of the skewness.*

MINITAB

Numerical Descriptive Measures

MINITAB provides most of the basic descriptive statistics presented in Chapter 2 using a single command in the drop-down menus. Once you are on the Windows desktop, double-click on the *MINITAB* icon or use the Start button to start *MINITAB.*

Practice entering some data into the Data window, naming the columns appropriately in the gray cell just below the column number. When you have finished entering your data, you will have created a *MINITAB* **worksheet,** which can be saved either singly or as a *MINITAB* **project** for future use. Click on **File → Save Current Worksheet** or **File → Save Project.** You will need to name the worksheet (or project)—perhaps "test data"—so that you can retrieve it later.

The following data are the floor lengths (in inches) behind the second and third seats in nine different minivans:[11]

Second seat:　62.0, 62.0, 64.5, 48.5, 57.5, 61.0, 45.5, 47.0, 33.0
Third seat:　　27.0, 27.0, 24.0, 16.5, 25.0, 27.5, 14.0, 18.5, 17.0

Since the data involve two variables, we enter the two rows of numbers into columns C1 and C2 in the *MINITAB* worksheet and name them "2nd Seat" and "3rd Seat," respectively. Using the drop-down menus, click on **Stat → Basic Statistics → Display Descriptive Statistics.** The Dialog box is shown in Figure 2.21.

FIGURE 2.21

Now click on the Variables box and **select** both columns from the list on the left. (You can click on the **Graphs** option and choose one of several graphs if you like. You may also click on the **Statistics** option to select the statistics you would like to see displayed.) Click **OK.** A display of descriptive statistics for both columns will appear in the Session window (see Figure 2.22). You may print this output using **File → Print Session Window** if you choose.

To examine the distribution of the two variables and look for outliers, you can create box plots using the command **Graph → Boxplot → One Y → Simple.** Click **OK.** Select the appropriate column of measurements in the Dialog box (see Figure 2.23). You can change the appearance of the box plot in several ways. **Scale → Axes and Ticks** will allow you to transpose the axes and orient the box plot horizontally.

Multiple Graphs provides printing options for multiple box plots. **Labels** will let you annotate the graph with titles and footnotes. If you have entered data into the worksheet as a frequency distribution (values in one column, frequencies in another), the **Data Options** will allow the data to be read in that format. The box plot for the third seat lengths is shown in Figure 2.24.

You can use the *MINITAB* commands from Chapter 1 to display stem and leaf plots or histograms for the two variables. How would you describe the similarities and differences in the two data sets? Save this worksheet in a file called "Minivans" before exiting *MINITAB*. We will use it again in Chapter 3.

FIGURE 2.22

Session

Descriptive Statistics: 2nd Seat, 3rd Seat

Variable	N	N*	Mean	SE Mean	StDev	Minimum	Q1	Median	Q3	Maximum
2nd Seat	9	0	53.44	3.51	10.54	33.00	46.25	57.50	62.00	64.50
3rd Seat	9	0	21.83	1.77	5.30	14.00	16.75	24.00	27.00	27.50

FIGURE 2.23

FIGURE 2.24

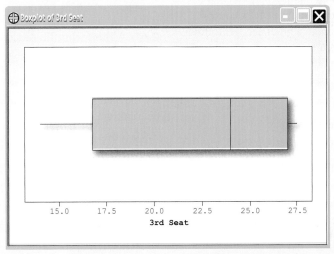

Supplementary Exercises

(MY) CD **2.54 Raisins** The number of raisins in each of
EX0254 14 miniboxes (1/2-ounce size) was counted for a
generic brand and for Sunmaid brand raisins. The two
data sets are shown here:

Generic Brand				Sunmaid			
25	26	25	28	25	29	24	24
26	28	28	27	28	24	28	22
26	27	24	25	25	28	30	27
26	26			28	24		

a. What are the mean and standard deviation for the
generic brand?

b. What are the mean and standard deviation for the
Sunmaid brand?

c. Compare the centers and variabilities of the two
brands using the results of parts a and b.

2.55 Raisins, continued Refer to Exercise 2.54.

a. Find the median, the upper and lower quartiles,
and the IQR for each of the two data sets.

b. Construct two box plots on the same horizontal
scale to compare the two sets of data.

c. Draw two stem and leaf plots to depict the shapes
of the two data sets. Do the box plots in part b verify these results?

d. If we can assume that none of the boxes of raisins
are being underfilled (that is, they all weigh approximately 1/2 ounce), what do your results say
about the average number of raisins for the two
brands?

(MY) CD **2.56 TV Viewers** The number of television
EX0256 viewing hours per household and the prime
viewing times are two factors that affect television
advertising income. A random sample of 25 households in a particular viewing area produced the following estimates of viewing hours per household:

3.0	6.0	7.5	15.0	12.0
6.5	8.0	4.0	5.5	6.0
5.0	12.0	1.0	3.5	3.0
7.5	5.0	10.0	8.0	3.5
9.0	2.0	6.5	1.0	5.0

a. Scan the data and use the range to find an approximate value for s. Use this value to check your calculations in part b.

b. Calculate the sample mean $\bar{x}$ and the sample standard deviation s. Compare s with the approximate
value obtained in part a.

c. Find the percentage of the viewing hours per
household that falls into the interval $\bar{x} \pm 2s$. Compare with the corresponding percentage given by
the Empirical Rule.

2.57 A Recurring Illness Refer to Exercise 1.26 and
data set EX0126. The lengths of time (in months)
between the onset of a particular illness and its recurrence were recorded:

2.1	4.4	2.7	32.3	9.9
9.0	2.0	6.6	3.9	1.6
14.7	9.6	16.7	7.4	8.2
19.2	6.9	4.3	3.3	1.2
4.1	18.4	.2	6.1	13.5
7.4	.2	8.3	.3	1.3
14.1	1.0	2.4	2.4	18.0
8.7	24.0	1.4	8.2	5.8
1.6	3.5	11.4	18.0	26.7
3.7	12.6	23.1	5.6	.4

a. Find the range.

b. Use the range approximation to find an approximate value for s.

c. Compute s for the data and compare it with your
approximation from part b.

2.58 A Recurring Illness, continued Refer to
Exercise 2.57.

a. Examine the data and count the number of obser-
vations that fall into the intervals $\bar{x} \pm s$, $\bar{x} \pm 2s$,
and $\bar{x} \pm 3s$.

b. Do the percentages that fall into these intervals
agree with Tchebysheff's Theorem? With the Empirical Rule?

c. Why might the Empirical Rule be unsuitable for
describing these data?

2.59 A Recurring Illness, again Find the median
and the lower and upper quartiles for the data on times
until recurrence of an illness in Exercise 2.57. Use
these descriptive measures to construct a box plot
for the data. Use the box plot to describe the data
distribution.

2.60 Tuna Fish, again Refer to Exercise 2.8. The
prices of a 6-ounce can or a 7.06-ounce pouch for 14
different brands of water-packed light tuna, based on
prices paid nationally in supermarkets are reproduced
here.[4]

.99	1.92	1.23	.85	.65	.53	1.41
1.12	.63	.67	.69	.60	.60	.66

a. Calculate the five-number summary.

b. Construct a box plot for the data. Are there any outliers?

c. The value $x = 1.92$ looks large in comparison to the other prices. Use a z-score to decide whether this is an unusually expensive brand of tuna.

2.61 Electrolysis An analytical chemist wanted to use electrolysis to determine the number of moles of cupric ions in a given volume of solution. The solution was partitioned into $n = 30$ portions of .2 milliliter each, and each of the portions was tested. The average number of moles of cupric ions for the $n = 30$ portions was found to be .17 mole; the standard deviation was .01 mole.

a. Describe the distribution of the measurements for the $n = 30$ portions of the solution using Tchebysheff's Theorem.

b. Describe the distribution of the measurements for the $n = 30$ portions of the solution using the Empirical Rule. (Do you expect the Empirical Rule to be suitable for describing these data?)

c. Suppose the chemist had used only $n = 4$ portions of the solution for the experiment and obtained the readings .15, .19, .17, and .15. Would the Empirical Rule be suitable for describing the $n = 4$ measurements? Why?

2.62 Chloroform According to the EPA, chloroform, which in its gaseous form is suspected of being a cancer-causing agent, is present in small quantities in all of the country's 240,000 public water sources. If the mean and standard deviation of the amounts of chloroform present in the water sources are 34 and 53 micrograms per liter, respectively, describe the distribution for the population of all public water sources.

2.63 Aptitude Tests In contrast to aptitude tests, which are predictive measures of what one can accomplish with training, achievement tests tell what an individual can do at the time of the test. Mathematics achievement test scores for 400 students were found to have a mean and a variance equal to 600 and 4900, respectively. If the distribution of test scores was mound-shaped, approximately how many of the scores would fall into the interval 530 to 670? Approximately how many scores would be expected to fall into the interval 460 to 740?

2.64 Sleep and the College Student How much sleep do you get on a typical school night? A group of 10 college students were asked to report the number of

hours that they slept on the previous night with the following results:

$$7, \quad 6, \quad 7.25, \quad 7, \quad 8.5, \quad 5, \quad 8, \quad 7, \quad 6.75, \quad 6$$

a. Find the mean and the standard deviation of the number of hours of sleep for these 10 students.

b. Calculate the z-score for the largest value $(x = 8.5)$. Is this an unusually sleepy college student?

c. What is the most frequently reported measurement? What is the name for this measure of center?

d. Construct a box plot for the data. Does the box plot confirm your results in part b? [HINT: Since the z-score and the box plot are two unrelated methods for detecting outliers, and use different types of statistics, they do not necessarily have to (but usually do) produce the same results.]

2.65 Gas Mileage The miles per gallon (mpg) for each of 20 medium-sized cars selected from a production line during the month of March.

EX0265

23.1	21.3	23.6	23.7
20.2	24.4	25.3	27.0
24.7	22.7	26.2	23.2
25.9	24.7	24.4	24.2
24.9	22.2	22.9	24.6

a. What are the maximum and minimum miles per gallon? What is the range?

b. Construct a relative frequency histogram for these data. How would you describe the shape of the distribution?

c. Find the mean and the standard deviation.

d. Arrange the data from smallest to largest. Find the z-scores for the largest and smallest observations. Would you consider them to be outliers? Why or why not?

e. What is the median?

f. Find the lower and upper quartiles.

2.66 Gas Mileage, continued Refer to Exercise 2.65. Construct a box plot for the data. Are there any outliers? Does this conclusion agree with your results in Exercise 2.65?

2.67 Polluted Seawater Petroleum pollution in seas and oceans stimulates the growth of some types of bacteria. A count of petroleumlytic micro-organisms (bacteria per 100 milliliters) in ten portions of seawater gave these readings:

$$49, \quad 70, \quad 54, \quad 67, \quad 59, \quad 40, \quad 61, \quad 69, \quad 71, \quad 52$$

a. Guess the value for *s* using the range approximation.

b. Calculate $\bar{x}$ and *s* and compare with the range approximation of part a.

c. Construct a box plot for the data and use it to describe the data distribution.

2.68 Basketball Attendances at a high school's basketball games were recorded and found to have a sample mean and variance of 420 and 25, respectively. Calculate $\bar{x} \pm s$, $\bar{x} \pm 2s$, and $\bar{x} \pm 3s$ and then state the approximate fractions of measurements you would expect to fall into these intervals according to the Empirical Rule.

2.69 SAT Tests The College Board's verbal and mathematics scholastic aptitude tests are scored on a scale of 200 to 800. Although the tests were originally designed to produce mean scores of approximately 500, the mean verbal and math scores in recent years have been as low as 463 and 493, respectively, and have been trending downward. It seems reasonable to assume that a distribution of all test scores, either verbal or math, is mound-shaped. If σ is the standard deviation of one of these distributions, what is the largest value (approximately) that σ might assume? Explain.

2.70 Summer Camping A favorite summer pastime for many Americans is camping. In fact, camping has become so popular at the California beaches that reservations must sometimes be made months in advance! Data from a *USA Today* Snapshot is shown below.[12]

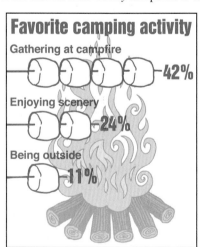

Favorite camping activity

Gathering at campfire —42%

Enjoying scenery —24%

Being outside —11%

The Snapshot also reports that men go camping 2.9 times a year, women go 1.7 times a year; and men are more likely than women to want to camp more often. What does the magazine mean when they talk about 2.9 or 1.7 times a year?

2.71 Long-Stemmed Roses A strain of long-stemmed roses has an approximate normal distribution with a mean stem length of 15 inches and standard deviation of 2.5 inches.

a. If one accepts as "long-stemmed roses" only those roses with a stem length greater than 12.5 inches, what percentage of such roses would be unacceptable?

b. What percentage of these roses would have a stem length between 12.5 and 20 inches?

2.72 Drugs for Hypertension A pharmaceutical company wishes to know whether an experimental drug being tested in its laboratories has any effect on systolic blood pressure. Fifteen randomly selected subjects were given the drug, and their systolic blood pressures (in millimeters) are recorded.

172	148	123
140	108	152
123	129	133
130	137	128
115	161	142

a. Guess the value of *s* using the range approximation.

b. Calculate $\bar{x}$ and *s* for the 15 blood pressures.

c. Find two values, *a* and *b*, such that at least 75% of the measurements fall between *a* and *b*.

2.73 Lumber Rights A company interested in lumbering rights for a certain tract of slash pine trees is told that the mean diameter of these trees is 14 inches with a standard deviation of 2.8 inches. Assume the distribution of diameters is roughly mound-shaped.

a. What fraction of the trees will have diameters between 8.4 and 22.4 inches?

b. What fraction of the trees will have diameters greater than 16.8 inches?

2.74 Social Ambivalence The following data represent the social ambivalence scores for 15 people as measured by a psychological test. (The higher the score, the stronger the ambivalence.)

9	13	12
14	15	11
10	4	10
8	19	13
11	17	9

a. Guess the value of *s* using the range approximation.

b. Calculate $\bar{x}$ and *s* for the 15 social ambivalence scores.

c. What fraction of the scores actually lie in the interval $\bar{x} \pm 2s$?

2.75 TV Commercials The mean duration of television commercials on a given network is 75 seconds, with a standard deviation of 20 seconds. Assume that durations are approximately normally distributed.

a. What is the approximate probability that a commercial will last less than 35 seconds?

b. What is the approximate probability that a commercial will last longer than 55 seconds?

2.76 Parasites in Foxes A random sample of 100 foxes was examined by a team of veterinarians to determine the prevalence of a particular type of parasite. Counting the number of parasites per fox, the veterinarians found that 69 foxes had no parasites, 17 had one parasite, and so on. A frequency tabulation of the data is given here:

Number of Parasites, x	0	1	2	3	4	5	6	7	8
Number of Foxes, f	69	17	6	3	1	2	1	0	1

a. Construct a relative frequency histogram for x, the number of parasites per fox.

b. Calculate $\bar{x}$ and s for the sample.

c. What fraction of the parasite counts fall within two standard deviations of the mean? Within three standard deviations? Do these results agree with Tchebysheff's Theorem? With the Empirical Rule?

2.77 College Teachers Consider a population consisting of the number of teachers per college at small 2-year colleges. Suppose that the number of teachers per college has an average $\mu = 175$ and a standard deviation $\sigma = 15$.

a. Use Tchebysheff's Theorem to make a statement about the percentage of colleges that have between 145 and 205 teachers.

b. Assume that the population is normally distributed. What fraction of colleges have more than 190 teachers?

MY CD 2.78 Is It Accurate? From the following data,
EX0278 a student calculated s to be .263. On what grounds might we doubt his accuracy? What is the correct value (to the nearest hundredth)?

17.2 17.1 17.0 17.1 16.9 17.0 17.1 17.0 17.3 17.2
17.1 17.0 17.1 16.9 17.0 17.1 17.3 17.2 17.4 17.1

MY CD 2.79 Homerun Kings In the summer of 2001,
EX0279 Barry Bonds began his quest to break Mark

McGwire's record of 70 home runs hit in a single season. At the end of the 2003 major league baseball season, the number of home runs hit per season by each of four major league superstars over each player's career were recorded and shown in the box plots below:[13]

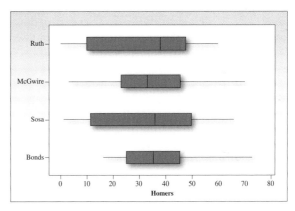

Write a short paragraph comparing the home run hitting patterns of these four players.

MY CD 2.80 Barry Bonds At the end of the 2002 and
EX0280 2003 seasons Barry Bonds hit 46 and 45 homers (www.mlb.com).[13] Two box plots, one of Bond's homers through 2001 and a second including the years 2002–2003, follow.

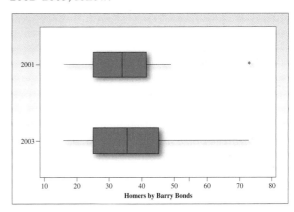

The statistics used to construct these box plots are given in the table.

Years	Min	Q_1	Median	Q_3	IQR	Max	n
2001	16	25.00	34.00	41.50	16.5	73	16
2003	16	25.00	35.50	45.25	20.25	73	18

a. Calculate the upper fences for both of these box plots.

b. Can you explain why the record number of homers is an outlier in the 2001 box plot, but not in the 2003 box plot?

2.81 Ages of Pennies Here are the ages of 50 pennies from Exercise 1.45 and data set EX0145. The data have been sorted from smallest to largest.

0	0	0	0	0	0	0	0	0	0
0	0	1	1	1	1	1	1	2	2
2	3	3	3	4	4	5	5	5	5
6	8	9	9	10	16	17	17	19	19
19	20	20	21	22	23	25	25	28	36

a. What is the average age of the pennies?

b. What is the median age of the pennies?

c. Based on the results of parts a and b, how would you describe the age distribution of these 50 pennies?

d. Construct a box plot for the data set. Are there any outliers? Does the box plot confirm your description of the distribution's shape?

2.82 Snapshots Here are a few facts reported as snapshots in *USA Today*.

- Fifty-two percent of Americans believe that the ideal family size is 2 or fewer children.[14]

- Seventy percent of Americans reheat leftovers in their microwave at least two times a week.[15]

- Fifty percent of Americans typically wait 15 minutes or less to have a prescription filled.[16]

Identify the variable x being measured, and any percentiles you can determine from this information.

(MY) (CD) 2.83 Breathing Patterns Research psychologists are interested in finding out whether a person's breathing patterns are affected by a particular experimental treatment. To determine the general respiratory patterns of the $n = 30$ people in the study, the researchers collected some baseline measurements—the total ventilation in liters of air per minute adjusted for body size—for each person before the treatment. The data are shown here, along with some descriptive tools generated by *MINITAB*.

5.23	4.79	5.83	5.37	4.35	5.54	6.04	5.48	6.58	4.82
5.92	5.38	6.34	5.12	5.14	4.72	5.17	4.99	4.51	5.70
4.67	5.77	5.84	6.19	5.58	5.72	5.16	5.32	4.96	5.63

Descriptive Statistics: Liters

Variable	N	N*	Mean	SE Mean	StDev
Liters	30	0	5.3953	0.0997	0.5462

Minimum	Q1	Median	Q3	Variable	Maximum
4.3500	4.9825	5.3750	5.7850	Liters	6.5800

Stem-and-Leaf Display: Liters

```
Stem-and-leaf of Liters   N = 30
Leaf Unit =  0.10

 1     4  3
 2     4  5
 5     4  677
 8     4  899
12     5  1111
(4)    5  2333
14     5  455
11     5  6777
 7     5  889
 4     6  01
 2     6  3
 1     6  5
```

a. Summarize the characteristics of the data distribution using the *MINITAB* output.

b. Does the Empirical Rule provide a good description of the proportion of measurements that fall within two or three standard deviations of the mean? Explain.

c. How large or small does a ventilation measurement have to be before it is considered unusual?

(MY) (CD) 2.84 Arranging Objects The following data are the response times in seconds for $n = 25$ first graders to arrange three objects by size.

5.2	3.8	5.7	3.9	3.7
4.2	4.1	4.3	4.7	4.3
3.1	2.5	3.0	4.4	4.8
3.6	3.9	4.8	5.3	4.2
4.7	3.3	4.2	3.8	5.4

a. Find the mean and the standard deviation for these 25 response times.

b. Order the data from smallest to largest.

c. Find the z-scores for the smallest and largest response times. Is there any reason to believe that these times are unusually large or small? Explain.

2.85 Arranging Objects, continued Refer to Exercise 2.84.

a. Find the five-number summary for this data set.

b. Construct a box plot for the data.

c. Are there any unusually large or small response times identified by the box plot?

d. Construct a stem and leaf display for the response times. How would you describe the shape of the distribution? Does the shape of the box plot confirm this result?

2.86 Refer to Data Set #1 in the **How Extreme Values Affect the Mean and Median** applet. This applet loads with a dotplot for the following $n = 5$ observations: 2, 5, 6, 9, 11.

a. What are the mean and median for this data set?

b. Use your mouse to change the value $x = 11$ (the moveable green dot) to $x = 13$. What are the mean and median for the new data set?

c. Use your mouse to move the green dot to $x = 33$. When the largest value is extremely large compared to the other observations, which is larger, the mean or the median?

d. What effect does an extremely large value have on the mean? What effect does it have on the median?

2.87 Refer to Data Set #2 in the **How Extreme Values Affect the Mean and Median** applet. This applet loads with a dotplot for the following $n = 5$ observations: 2, 5, 10, 11, 12.

a. Use your mouse to move the value $x = 12$ to the left until it is smaller than the value $x = 11$.

b. As the value of x gets smaller, what happens to the sample mean?

c. As the value of x gets smaller, at what point does the value of the median finally change?

d. As you move the green dot, what are the largest and smallest possible values for the median?

2.88 Refer to Data Set #3 in the **How Extreme Values Affect the Mean and Median** applet. This applet loads with a dotplot for the following $n = 5$ observations: 27, 28, 32, 34, 37.

a. What are the mean and median for this data set?

b. Use your mouse to change the value $x = 27$ (the moveable green dot) to $x = 25$. What are the mean and median for the new data set?

c. Use your mouse to move the green dot to $x = 5$. When the smallest value is extremely small compared to the other observations, which is larger, the mean or the median?

d. At what value of x does the mean equal the median?

e. What are the smallest and largest possible values for the median?

f. What effect does an extremely small value have on the mean? What effect does it have on the median?

2.89 Refer to the **Why Divide by $n - 1$** applet. The first applet on the page randomly selects sample of

$n = 3$ from a population in which the standard deviation is $\sigma = 29.2$.

a. Click New Sample. A sample consisting of $n = 3$ observations will appear. Use your calculator to verify the values of the standard deviation when dividing by $n - 1$ and n as shown in the applet.

b. Click New Sample again. Calculate the average of the two standard deviations (dividing by $n - 1$) from parts a and b. Repeat the process for the two standard deviations (dividing by n). Compare your results to those shown in red on the applet.

c. You can look at how the two estimators in part a behave "in the long run" by clicking 10 Samples or 100 Samples a number of times, until the average of all the standard deviations begins to stabilize. Which of the two methods gives a standard deviation closer to $\sigma = 29.2$?

d. In the long run, how far off is the standard deviation when dividing by n?

2.90 Refer to **Why Divide by $n - 1$** applet. The second applet on the page randomly selects sample of $n = 10$ from the same population in which the standard deviation is $\sigma = 29.2$.

a. Repeat the instructions in part c and d of Exercise 2.89.

b. Based on your simulation, when the sample size is larger, does it make as much difference whether you divide by n or $n - 1$ when computing the sample standard deviation?

2.91 If you have not yet done so, use the first **Building a Box Plot** applet to construct a box plot for the data in Example 2.14.

a. Compare the finished box plot to the plot shown in Figure 2.18.

b. How would you describe the shape of the data distribution?

c. Are there any outliers? If so, what is the value of the unusual observation?

2.92 Use the second **Building a Box Plot** applet to construct a box plot for the data in Example 2.13.

a. How would you describe the shape of the data distribution?

b. Use the box plot to approximate the values of the median, the lower quartile and the upper quartile. Compare your results to the actual values calculated in Example 2.13.

CASE STUDY

 Batting

The Boys of Summer

Which baseball league has had the best hitters? Many of us have heard of baseball greats like Stan Musial, Hank Aaron, Roberto Clemente, and Pete Rose of the National League and Ty Cobb, Babe Ruth, Ted Williams, Rod Carew, and Wade Boggs of the American League. But have you ever heard of Willie Keeler, who batted .432 for the Baltimore Orioles, or Nap Lajoie, who batted .422 for the Philadelphia A's? The batting averages for the batting champions of the National and American Leagues are given on the CD-ROM. The batting averages for the National League begin in 1876 with Roscoe Barnes, whose batting average was .403 when he played with the Chicago Cubs.

The last entry for the National League is for the year 2003, when Albert Pujols of the St. Louis Cardinals averaged .359. The American League records begin in 1901 with Nap Lajoie of the Philadelphia A's, who batted .422, and end in 2003 with Bill Mueller of the Boston Red Sox who batted .326.[17] How can we summarize the information in this data set?

1. Use *MINITAB* or another statistical software package to describe the batting averages for the American and National League batting champions. Generate any graphics that may help you in interpreting these data sets.

2. Does one league appear to have a higher percentage of hits than the other? Do the batting averages of one league appear to be more variable than the other?

3. Are there any outliers in either league?

4. Summarize your comparison of the two baseball leagues.

3

Describing Bivariate Data

© Photodisc/Royalty-Free/Getty Images

Do You *Think* Your Dishes Are *Really* Clean?

Does the price of an appliance, such as a dishwasher, convey something about its quality? In the case study at the end of this chapter, we rank 20 different brands of dishwashers according to their prices, and then we rate them on various characteristics, such as how the dishwasher performs, how much noise it makes, its cost for either gas or electricity, its cycle time, and its water use. The techniques presented in this chapter will help to answer our question.

BIVARIATE DATA

Very often researchers are interested in more than just one variable that can be measured during their investigation. For example, an auto insurance company might be interested in the number of vehicles owned by a policyholder as well as the number of drivers in the household. An economist might need to measure the amount spent per week on groceries in a household and also the number of people in that household. A real estate agent might measure the selling price of a residential property and the square footage of the living area.

 TIP

"Bi" means "two." Bivariate data generate pairs of measurements.

When two variables are measured on a single experimental unit, the resulting data are called **bivariate data.** How should you display these data? Not only are both variables important when studied separately, but you also may want to explore the *relationship between the two variables.* Methods for graphing bivariate data, whether the variables are qualitative or quantitative, allow you to study the two variables together. As with *univariate data,* you use different graphs depending on the type of variables you are measuring.

GRAPHS FOR QUALITATIVE VARIABLES

When at least one of the two variables is *qualitative,* you can use either simple or more intricate pie charts, line charts, and bar charts to display and describe the data. Sometimes you will have one qualitative and one quantitative variable that have been measured in two different populations or groups. In this case, you can use two **side-by-side pie charts** or a bar chart in which the bars for the two populations are placed side by side. Another option is to use a **stacked bar chart,** in which the bars for each category are stacked on top of each other.

 EXAMPLE 3.1

Are professors in private colleges paid more than professors at public colleges? The data in Table 3.1 were collected from a sample of 400 college professors whose rank, type of college, and salary were recorded. The number in each cell is the average salary (in thousands of dollars) for all professors who fell into that category. Use a graph to answer the question posed for this sample.

TABLE 3.1 ● **Salaries of Professors by Rank and Type of College**

	Full Professor	Associate Professor	Assistant Professor
Public	55.8	42.2	35.2
Private	61.6	43.3	35.5

Solution   To display the average salaries of these 400 professors, you can use a side-by-side bar chart, as shown in Figure 3.1. The height of the bars is the average salary, with each pair of bars along the horizontal axis representing a different professorial rank. Salaries are substantially higher for full professors in private colleges, but there is very little difference at the lower two ranks.

FIGURE 3.1

Comparative bar charts for
Example 3.1

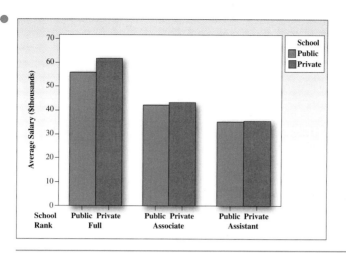

EXAMPLE (3.2) Along with the salaries for the 400 college professors in Example 3.1, the researcher recorded two qualitative variables for each professor: rank and type of college. Table 3.2 shows the number of professors in each of the $2 \times 3 = 6$ categories. Use comparative charts to describe the data. Do the private colleges employ as many high-ranking professors as the public colleges do?

TABLE 3.2   ●   **Number of Professors by Rank and Type of College**

	Full Professor	Associate Professor	Assistant Professor	Total
Public	24	57	69	150
Private	60	78	112	250

Solution   The numbers in the table are not quantitative measurements on a single experimental unit (the professor). They are *frequencies,* or counts, of the number of professors who fall into each category. To compare the numbers of professors at public and private colleges, you might draw two pie charts and display them side by side, as in Figure 3.2.

FIGURE 3.2

Comparative pie charts for
Example 3.2

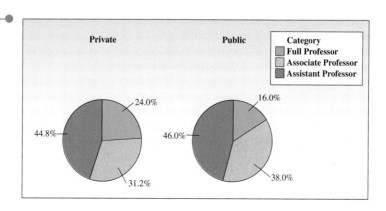

Alternatively, you could draw either a stacked or a side-by-side bar chart. The stacked bar chart is shown in Figure 3.3.

FIGURE 3.3

Stacked bar chart for Example 3.2

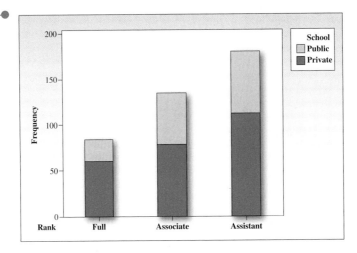

Although the graphs are not strikingly different, you can see that public colleges have fewer full professors and more associate professors than private colleges. The reason for these differences is not clear, but you might speculate that private colleges, with their higher salaries, are able to attract more full professors. Or perhaps public colleges are not as willing to promote professors to the higher-paying ranks. In any case, the graphs provide a means for comparing the two sets of data.

You can also compare the distributions for public versus private colleges by creating *conditional data distributions.* These conditional distributions are shown in Table 3.3. One distribution shows the proportion of professors in each of the three ranks under the *condition* that the college is public, and the other shows the proportions under the *condition* that the college is private. These *relative frequencies* are easier to compare than the *actual frequencies* and lead to the same conclusions:

- That the proportion of assistant professors is roughly the same for both public and private colleges
- That public colleges have a smaller proportion of full professors and a larger proportion of associate professors.

TABLE 3.3

Proportions of Professors by Rank for Public and Private Colleges

	Full Professor	Associate Professor	Assistant Professor	Total
Public	$\frac{24}{150} = .16$	$\frac{57}{150} = .38$	$\frac{69}{150} = .46$	1.00
Private	$\frac{60}{250} = .24$	$\frac{78}{250} = .31$	$\frac{112}{250} = .45$	1.00

BASIC TECHNIQUES

3.1 Gender Differences Male and female respondents to a questionnaire about gender differences are categorized into three groups according to their answers on the first question:

	Group 1	Group 2	Group 3
Men	37	49	72
Women	7	50	31

a. Create side-by-side pie charts to describe these data.

b. Create a side-by-side bar chart to describe these data.

c. Draw a stacked bar chart to describe these data.

d. Which of the three charts best depicts the difference or similarity of the responses of men and women?

3.2 State-by-State A group of items are categorized according to a certain attribute—X, Y, Z—and according to the state in which they are produced:

	X	Y	Z
New York	20	5	5
California	10	10	5

a. Create a comparative (side-by-side) bar chart to compare the numbers of items of each type made in California and New York.

b. Create a stacked bar chart to compare the numbers of items of each type made in the two states.

c. Which of the two types of presentation in parts a and b is more easily understood? Explain.

d. What other graphical methods could you use to describe the data?

3.3 Consumer Spending The table below shows the average amounts spent per week by men and women in each of four spending categories:

	A	B	C	D
Men	$54	$27	$105	$22
Women	21	85	100	75

a. What possible graphical methods could you use to compare the spending patterns of women and men?

b. Choose two different methods of graphing and display the data in graphical form.

c. What can you say about the similarities or differences in the spending patterns for men and women?

d. Which of the two methods used in part b provides a better descriptive graph?

APPLICATIONS

3.4 M&Ms The color distributions for two snack-size bags of M&M® candies, one plain and one peanut, are displayed in the table. Choose an appropriate graphical method and compare the distributions.

	Brown	Yellow	Red	Orange	Green	Blue
Plain	15	14	12	4	5	6
Peanut	6	2	2	3	3	5

3.5 How Much Free Time? When you were growing up, did you feel that you did not have enough free time? Parents and children have differing opinions on this subject. A research group surveyed 198 parents and 200 children and recorded their responses to the question, "How much free time does your child have?" or "How much free time do you have?" The responses are shown in the table below:[1]

	Just the Right Amount	Not Enough	Too Much	Don't Know
Parents	138	14	40	6
Children	130	48	16	6

a. Define the sample and the population of interest to the researchers.

b. Describe the variables that have been measured in this survey. Are the variables qualitative or quantitative? Are the data univariate or bivariate?

c. What do the entries in the cells represent?

d. Use comparative pie charts to compare the responses for parents and children.

e. What other graphical techniques could be used to describe the data? Would any of these techniques be more informative than the pie charts constructed in part d?

3.6 Consumer Price Index The price of living in the United States has increased dramatically in the past decade, as demonstrated by the consumer price indexes (CPIs) for housing and transportation. These CPIs are listed in the table for the years 1994 to 2003:[2]

EX0306

Year	1994	1995	1996	1997	1998
Housing	144.8	148.5	152.8	156.8	160.4
Transportation	134.3	139.1	143.0	144.3	141.6
Year	1999	2000	2001	2002	2003
Housing	163.9	169.6	176.4	180.3	184.8
Transportation	144.4	153.3	154.3	152.9	157.6

Source: www.bls.gov

a. Create side-by-side comparative bar charts to describe the CPIs over time.

b. Draw two line charts on the same set of axes to describe the CPIs over time.

c. What conclusions can you draw using the two graphs in parts a and b? Which is the most effective?

(MY) CD **3.7 How Big Is the Household?** A local
EX0307 chamber of commerce surveyed 126 households in their city and recorded the type of residence and the number of family members in each of the households. The data are shown in the table.

	Type of Residence		
Family Members	Apartment	Duplex	Single Residence
1	8	10	2
2	15	4	14
3	9	5	24
4 or more	6	1	28

a. Use a side-by-side bar chart to compare the number of family members living in the three types of residences.

b. Use a stacked bar chart to compare the number of family members living in the three types of residences.

c. What conclusions can you draw using the graphs in parts a and b?

(MY) CD **3.8 Charitable Contributions** Charitable
EX0308 organizations count on support from both private donations and other sources. Here are the sources of income in a recent year for several well-known charitable organizations in the United States:[3]

	Amounts ($ millions)		
Organization	Private	Other	Total
Salvation Army	$1372.0	$775.4	$2147.4
YMCA	713.9	3557.8	4271.7
American Red Cross	1736.4	2351.0	4087.4
American Cancer Society	777.4	12.0	789.4
American Heart Association	437.5	88.2	525.7
Total	$5037.2	$6784.4	$11821.6

Source: The World Almanac and Book of Facts, 2004

a. Construct a stacked bar chart to display the sources of income given in the table.

b. Construct two comparative pie charts to display the sources of income given in the table.

c. Write a short paragraph summarizing the information that can be gained by looking at these graphs. Which of the two types of comparative graphs is more effective?

SCATTERPLOTS FOR TWO QUANTITATIVE VARIABLES

3.3

When both variables to be displayed on a graph are *quantitative,* one variable is plotted along the horizontal axis and the second along the vertical axis. The first variable is often called x and the second is called y, so that the graph takes the form of a plot on the (x, y) axes, which is familiar to most of you. Each pair of data values is plotted as a point on this two-dimensional graph, called a **scatterplot.** It is the two-dimensional extension of the dotplot we used to graph one quantitative variable in Section 1.4.

You can describe the relationship between two variables, x and y, using the patterns shown in the scatterplot.

- **What type of pattern do you see?** Is there a constant upward or downward trend that follows a straight-line pattern? Is there a curved pattern? Is there no pattern at all, but just a random scattering of points?

- **How strong is the pattern?** Do all of the points follow the pattern exactly, or is the relationship only weakly visible?

- **Are there any unusual observations?** An outlier is a point that is far from the cluster of the remaining points. Do the points cluster into groups? If so, is there an explanation for the observed groupings?

EXAMPLE 3.3

The number of household members, x, and the amount spent on groceries per week, y, are measured for six households in a local area. Draw a scatterplot of these six data points.

x	2	2	3	4	1	5
y	$45.75	$60.19	$68.33	$100.92	$35.86	$130.62

Solution Label the horizontal axis x and the vertical axis y. Plot the points using the coordinates (x, y) for each of the six pairs. The scatterplot in Figure 3.4 shows the six pairs marked as dots. You can see a pattern even with only six data pairs. The cost of weekly groceries increases with the number of household members in an apparent straight-line relationship.

Suppose you found that a seventh household with two members spent $115 on groceries. This observation is shown as an X in Figure 3.4. It does not fit the linear pattern of the other six observations and is classified as an outlier. Possibly these two people were having a party the week of the survey!

FIGURE 3.4

Scatterplot for
Example 3.3

EXAMPLE 3.4

A distributor of table wines conducted a study of the relationship between price and demand using a type of wine that ordinarily sells for $10.00 per bottle. He sold this wine in ten different marketing areas over a 12-month period, using five different price levels—from $10 to $14. The data are given in Table 3.4. Construct a scatterplot for the data, and use the graph to describe the relationship between price and demand.

TABLE 3.4

Cases of Wine Sold at Five Price Levels

Cases Sold per 10,000 Population	Price per Bottle
23, 21	$10
19, 18	11
15, 17	12
19, 20	13
25, 24	14

Solution The 10 data points are plotted in Figure 3.5. As the price increases from $10 to $12 the demand decreases. However, as the price continues to increase, from $12 to $14, the demand begins to *increase*. The data show a curved pattern, with the relationship changing as the price changes. How do you explain this relationship? Possibly, the increased price is a signal of increased quality for the consumer, which causes the increase in demand once the cost exceeds $12. You might be able to think of other reasons, or perhaps some other variable, such as the income of people in the marketing areas, that may be causing the change.

FIGURE 3.5

Scatterplot for
Example 3.4

MY APPLET

Now would be a good time for you to try creating a scatterplot on your own. Use the applets in **Building a Scatterplot** to create the scatterplots that you see in Figures 3.5 and 3.7. You will find step-by-step instructions on the left-hand side of the applet (Figure 3.6), and you will be corrected if you make a mistake!

FIGURE 3.6

Building a Scatterplot
applet

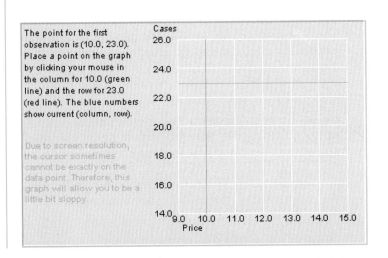

The point for the first observation is (10.0, 23.0). Place a point on the graph by clicking your mouse in the column for 10.0 (green line) and the row for 23.0 (red line). The blue numbers show current (column, row).

Due to screen resolution, the cursor sometimes cannot be exactly on the data point. Therefore, this graph will allow you to be a little bit sloppy.

NUMERICAL MEASURES FOR QUANTITATIVE BIVARIATE DATA

3.4

A constant rate of increase or decrease is perhaps the most common pattern found in bivariate scatterplots. The scatterplot in Figure 3.4 exhibits this *linear* pattern—that is, a straight line with the data points lying both above and below the line and within a fixed distance from the line. When this is the case, we say that the two variables exhibit a *linear relationship.*

EXAMPLE (3.5)

The data in Table 3.5 are the size of the living area (in square feet), x, and the selling price, y, of 12 residential properties. The *MINITAB* scatterplot in Figure 3.7 shows a linear pattern in the data.

TABLE 3.5

Living Area and Selling Price of 12 Properties

Residence	x (sq. ft.)	y (in thousands)
1	1360	$178.5
2	1940	275.7
3	1750	239.5
4	1550	229.8
5	1790	195.6
6	1750	210.3
7	2230	360.5
8	1600	205.2
9	1450	188.6
10	1870	265.7
11	2210	325.3
12	1480	168.8

FIGURE 3.7

Scatterplot of x versus y for Example 3.5

For the data in Example 3.5, you could describe each variable, x and y, individually using descriptive measures such as the means ($\bar{x}$ and $\bar{y}$) or the standard deviations (s_x and s_y). However, these measures do not describe the relationship between x and y for a particular residence—that is, how the size of the living space affects the selling price of the home. A simple measure that serves this purpose is called the **correlation coefficient,** denoted by r, and is defined as

$$r = \frac{s_{xy}}{s_x s_y}$$

The quantities s_x and s_y are the standard deviations for the variables x and y, respectively, which can be found by using the statistics function on your calculator or the computing formula in Section 2.3. The new quantity s_{xy} is called the **covariance** between x and y and is defined as

$$s_{xy} = \frac{\Sigma(x_i - \bar{x})(y_i - \bar{y})}{n - 1}$$

There is also a computing formula for the covariance:

$$s_{xy} = \frac{\Sigma x_i y_i - \dfrac{(\Sigma x_i)(\Sigma y_i)}{n}}{n - 1}$$

where $\Sigma x_i y_i$ is the sum of the products $x_i y_i$ for each of the n pairs of measurements. How does this quantity detect and measure a linear pattern in the data?

Look at the signs of the cross-products $(x_i - \bar{x})(y_i - \bar{y})$ in the numerator of r, or s_{xy}. When a data point (x, y) is in either area I or III in the scatterplot shown in Figure 3.8, the cross-product will be positive; when a data point is in area II or IV, the cross-product will be negative. We can draw these conclusions:

- If most of the points are in areas I and III (forming a positive pattern), s_{xy} and r will be positive.

- If most of the points are in areas II and IV (forming a negative pattern), s_{xy} and r will be negative.

- If the points are scattered across all four areas (forming *no* pattern), s_{xy} and r will be close to 0.

FIGURE 3.8

The signs of the cross-products $(x_i - \bar{x})(y_i - \bar{y})$ in the covariance formula

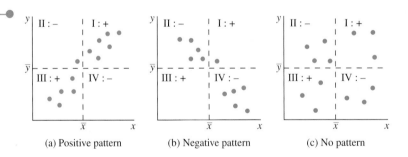

(a) Positive pattern   (b) Negative pattern   (c) No pattern

(MY) APPLET

The applet called **Exploring Correlation** will help you to visualize how the pattern of points affects the correlation coefficient. Use your mouse to move the slider at the bottom of the scatterplot (Figure 3.9). You will see the value of r change as the pattern of the points changes. Notice that a positive pattern (a) results in a positive value of r; no pattern (c) gives a value of r close to zero; and a negative pattern (b) results in a negative value of r. What pattern do you see when $r = 1$? When $r = -1$? You will use this applet again for the MyApplet Exercises section at the end of the chapter.

FIGURE 3.9

Exploring Correlation applet

 TIP

$r > 0 \Leftrightarrow$ positive *linear* relationship

$r < 0 \Leftrightarrow$ negative *linear* relationship

$r \approx 0 \Leftrightarrow$ no relationship

Most scientific and graphics calculators can compute the correlation coefficient, r, when the data are entered in the proper way. Check your calculator manual for the proper sequence of entry commands. Computer programs such as *MINITAB* are also programmed to perform these calculations. The *MINITAB* output in Figure 3.10 shows the covariance and correlation coefficient for x and y in Example 3.5. In the covariance table, you will find these values:

$$s_{xy} = 15,545.20 \qquad s_x^2 = 79,233.33 \qquad s_y^2 = 3571.16$$

and in the correlation output, you find $r = .924$.

However you decide to calculate the correlation coefficient, it can be shown that the value of r always lies between -1 and 1. When r is positive, x increases when y increases, and vice versa. When r is negative, x decreases when y increases, or x increases when y decreases. When r takes the value 1 or -1, all the points lie exactly on a straight line. If $r = 0$, then there is no apparent linear relationship between the two variables. The closer the value of r is to 1 or -1, the stronger the linear relationship between the two variables.

FIGURE 3.10

MINITAB output of covariance and correlation for Example 3.5

Covariances: x, y

	x	y
x	79233.33	
y	15545.20	3571.16

Correlations: x, y

Pearson correlation of x and y = 0.924
P-Value = 0.000

EXAMPLE 3.6

Find the correlation coefficient for the number of square feet of living area and the selling price of a home for the data in Example 3.5.

Solution Three quantities are needed to calculate the correlation coefficient. The standard deviations of the x and y variables are found using a calculator with a statistical function. You can verify that $s_x = 281.4842$ and $s_y = 59.7592$. Finally,

$$s_{xy} = \frac{\Sigma x_i y_i - \dfrac{(\Sigma x_i)(\Sigma y_i)}{n}}{n-1}$$

$$= \frac{5{,}142{,}383 - \dfrac{(20{,}980)(2843.5)}{12}}{11} = 15{,}545.19697$$

This agrees with the value given in the *MINITAB* printout in Figure 3.10. Then

$$r = \frac{s_{xy}}{s_x s_y} = \frac{15{,}545.19697}{(281.4842)(59.7592)} = .9241$$

which also agrees with the value of the correlation coefficient given in Figure 3.10. (You may wish to verify the value of r using your calculator.) This value of r is fairly close to 1, which indicates that the linear relationship between these two variables is very strong. Additional information about the correlation coefficient and its role in analyzing linear relationships, along with alternative calculation formulas, can be found in Chapter 12.

TIP

x "explains" *y* or *y* "depends on" *x*.

x is the **explan**atory or independent variable.

y is the response or **depend**ent variable.

Sometimes the two variables, *x* and *y*, are related in a particular way. It may be that the value of *y* depends on the value of *x;* that is, the value of *x* in some way explains the value of *y*. For example, the cost of a home (*y*) may *depend* on its amount of floor space (*x*); a student's grade point average (*x*) may *explain* her score on an achievement test (*y*). In these situations, we call *y* the **dependent variable,** while *x* is called the **independent variable.**

If one of the two variables can be classified as the dependent variable *y* and the other as *x*, and if the data exhibit a straight-line pattern, it is possible to describe the relationship relating *y* to *x* using a straight line given by the equation

$$y = a + bx$$

as shown in Figure 3.11.

FIGURE 3.11
The graph of a straight line

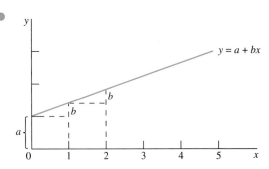

As you can see, *a* is where the line crosses or intersects the *y*-axis: *a* is called the *y-intercept.* You can also see that for every one-unit increase in *x*, *y* increases by an amount *b*. The quantity *b* determines whether the line is increasing ($b > 0$), decreasing ($b < 0$), or horizontal ($b = 0$) and is appropriately called the **slope** of the line.

MY **APPLET**

You can see the effect of changing the slope and the y-intercept of a line using the applet called **How a Line Works.** Use your mouse to move the slider on the right side of the scatterplot. As you move the slider, the slope of the line, shown as the vertical side of the green triangle (light gray in Figure 3.12), will change. Moving the slider on the left side of the applet causes the y-intercept, shown in red (teal in Figure 3.12), to change. What is the slope and y-intercept for the line shown in the applet in Figure 3.12? You will use this applet again for the MyApplet Exercises section at the end of the chapter.

FIGURE 3.12

How a Line Works applet

Our points (x, y) do not all fall on a straight line, but they do show a trend that could be described as a linear pattern. We can describe this trend by fitting a line as best we can through the points. This best-fitting line relating y to x, often called the **regression** or **least-squares line,** is found by minimizing the sum of the squared differences between the data points and the line itself, as shown in Figure 3.13. The formulas for computing b and a, which are derived mathematically, are shown below.

COMPUTING FORMULAS FOR THE LEAST-SQUARES REGRESSION LINE

$$b = r\left(\frac{s_y}{s_x}\right) \quad \text{and} \quad a = \bar{y} - b\bar{x}$$

and the least-squares regression line is: $y = a + bx$

FIGURE 3.13

The best-fitting line

TIP

Remember that r and b have the same sign!

Since s_x and s_y are both positive, b and r have the same sign, so that:

- When r is positive, so is b, and the line is increasing with x.
- When r is negative, so is b, and the line is decreasing with x.
- When r is close to 0, then b is close to 0.

EXAMPLE 3.7

Find the best-fitting line relating y to x for the following data. Plot the line and the data points on the same graph.

x	2	3	4	5	6	7
y	3.0	5.0	5.5	6.0	8.0	9.5

Solution Use the data entry method for your calculator to find these descriptive statistics for the bivariate data set:

$$\bar{x} = 4.5 \qquad \bar{y} = 6.167 \qquad s_x = 1.871 \qquad s_y = 2.295 \qquad r = .978$$

TIP

Use the regression line to predict y for a given value of x.

Then

$$b = r\left(\frac{s_y}{s_x}\right) = .978\left(\frac{2.295}{1.871}\right) = 1.1996311 \cong 1.200$$

and

$$a = \bar{y} - b\bar{x} = 6.167 - 1.200(4.5) = 6.167 - 5.4 = .767$$

Therefore, the best-fitting line is $y = .767 + 1.200x$. The plot of the regression line and the actual data points are shown in Figure 3.14.

The best-fitting line can be used to estimate or predict the value of the variable y when the value of x is known. For example, if the value $x = 3$ was observed at some time in the future, what would you predict for the value of y? From the best-fitting line in Figure 3.14, the best estimate would be

$$y = a + bx = .767 + 1.200(3) = 4.367$$

FIGURE 3.14

Fitted line and data points for Example 3.7

PERSONAL TRAINER

How Do I Calculate the Correlation Coefficient?

1. First, create a table or use your calculator to find Σx, Σy, and Σxy.
2. Calculate the covariance, s_{xy}.
3. Use your calculator or the computing formula from Chapter 2 to calculate s_x and s_y.
4. Calculate $r = \dfrac{s_{xy}}{s_x s_y}$.

How Do I Calculate the Regression Line?

1. First, calculate $\bar{y}$ and $\bar{x}$. Then, calculate $r = \dfrac{s_{xy}}{s_x s_y}$.

2. Find the slope, $b = r\left(\dfrac{s_y}{s_x}\right)$ and the y-intercept, $a = \bar{y} - b\bar{x}$.

3. Write the regression line by substituting the values for a and b into the equation: $y = a + bx$.

Exercise Reps

A. Below you will find a simple set of bivariate data. Fill in the blanks to find the correlation coefficient.

x	y	xy	Calculate:	Covariance
0	1		$n =$ _____	
				$s_{xy} = \dfrac{\Sigma xy - \dfrac{(\Sigma x)(\Sigma y)}{n}}{n - 1} =$ _____
2	5		$s_x =$ _____	
4	2		$s_y =$ _____	**Correlation Coefficient**
$\Sigma x =$ _____	$\Sigma y =$ _____	$\Sigma xy =$ _____		$r = \dfrac{s_{xy}}{s_x s_y} =$ _____

B. Use the information from part A and find the regression line.

From Part A	From Part A	Calculate:	Slope	y-intercept
$\Sigma x =$ _____	$s_x =$ _____	$\bar{x} =$ _____	$b = r\left(\dfrac{s_y}{s_x}\right) =$ _____	$a = \bar{y} - b\bar{x} =$ _____
$\Sigma y =$ _____	$s_y =$ _____	$\bar{y} =$ _____		
	$r =$ _____		**Regression Line: $y =$** _____	

Answers are located on the perforated card at the back of this book.

When should you describe the linear relationship between x and y using the correlation coefficient r, and when should you use the regression line $y = a + bx$? The regression approach is used when the values of x are set in advance and then the corresponding value of y is measured. The correlation approach is used when an experimental unit is selected at random and then measurements are made on both variables x and y. This technical point will be taken up in Chapter 12 on regression analysis.

Most data analysts begin any data-based investigation by examining plots of the variables involved. If the relationship between two variables is of interest, bivariate plots are also explored in conjunction with numerical measures of location, dispersion, and correlation. Graphs and numerical descriptive measures are only the first of many statistical tools you will soon have at your disposal.

3.4 EXERCISES

EXERCISE REPS

3.9 Below you will find a simple set of bivariate data. Fill in the blanks to find the correlation coefficient.

x	y	xy	Calculate:	Covariance	
1	6		$n =$ ____	$s_{xy} = \dfrac{\Sigma xy - \dfrac{(\Sigma x)(\Sigma y)}{n}}{n-1} =$ ____	
3	2		$s_x =$ ____		
2	4		$s_y =$ ____	**Correlation Coefficient**	
$\Sigma x =$ ____	$\Sigma y =$ ____	$\Sigma xy =$ ____		$r = \dfrac{s_{xy}}{s_x s_y} =$ ____	

3.10 Use the information from Exercise 3.9 and find the regression line.

From Part A	From Part A	Calculate:	Slope	y-intercept
$\Sigma x =$ ____	$s_x =$ ____	$\bar{x} =$ ____	$b = r\left(\dfrac{s_y}{s_x}\right) =$ ____	
$\Sigma y =$ ____	$s_y =$ ____	$\bar{y} =$ ____		$a = \bar{y} - b\bar{x} =$ ____
	$r =$ ____		**Regression Line: $y =$** _____	

BASIC TECHNIQUES

 3.11 A set of bivariate data consists of these
EX0311   measurements on two variables, x and y:

(3, 6) (5, 8) (2, 6) (1, 4) (4, 7) (4, 6)

a. Draw a scatterplot to describe the data.

b. Does there appear to be a relationship between x and y? If so, how do you describe it?

c. Calculate the correlation coefficient, r, using the computing formula given in this section.

d. Find the best-fitting line using the computing formulas. Graph the line on the scatterplot from part a. Does the line pass through the middle of the points?

3.12 Refer to Exercise 3.11.

a. Use the data entry method in your scientific calculator to enter the six pairs of measurements. Recall the proper memories to find the correlation coefficient, r, the y-intercept, a, and the slope, b, of the line.

b. Verify that the calculator provides the same values for r, a, and b as in Exercise 3.11.

 3.13 Consider this set of bivariate data:
EX0313

x	1	2	3	4	5	6
y	5.6	4.6	4.5	3.7	3.2	2.7

a. Draw a scatterplot to describe the data.

b. Does there appear to be a relationship between x and y? If so, how do you describe it?

c. Calculate the correlation coefficient, r. Does the value of r confirm your conclusions in part b? Explain.

MY CD **3.14** The value of a quantitative variable is
EX0314 measured once a year for a 10-year period:

Year	Measurement	Year	Measurement
1	61.5	6	58.2
2	62.3	7	57.5
3	60.7	8	57.5
4	59.8	9	56.1
5	58.0	10	56.0

a. Draw a scatterplot to describe the variable as it changes over time.

b. Describe the measurements using the graph constructed in part a.

c. Use this *MINITAB* output to calculate the correlation coefficient, r:

MINITAB output for Exercise 3.14

```
Covariances
             x          y
x        9.16667
y       -6.42222    4.84933
```

d. Find the best-fitting line using the results of part c. Verify your answer using the data entry method in your calculator.

e. Plot the best-fitting line on your scatterplot from part a. Describe the fit of the line.

APPLICATIONS

MY CD **3.15** **Grocery Costs** These data relating the
EX0315 amount spent on groceries per week and the number of household members are from Example 3.3:

x	2	2	3	4	1	5
y	$45.75	$60.19	$68.33	$100.92	$35.86	$130.62

a. Find the best-fitting line for these data.

b. Plot the points and the best-fitting line on the same graph. Does the line summarize the information in the data points?

c. What would you estimate a household of six to spend on groceries per week? Should you use the fitted line to estimate this amount? Why or why not?

MY CD **3.16** **Real Estate Prices** The data relating the
EX0316 square feet of living space and the selling price

of 12 residential properties given in Example 3.5 are reproduced here. First, find the best-fitting line that describes these data, and then plot the line and the data points on the same graph. Comment on the goodness of the fitted line in describing the selling price of a residential property as a linear function of the square feet of living area.

x (sq. ft.)	y (in thousands)
1360	$178.5
1940	275.7
1750	239.5
1550	229.8
1790	195.6
1750	210.3
2230	360.5
1600	205.2
1450	188.6
1870	265.7
2210	325.3
1480	168.8

MY CD **3.17** **Disabled Students** A social skills train-
EX0317 ing program, reported in *Psychology in the Schools,* was implemented for seven students with mild handicaps in a study to determine whether the program caused improvement in pre/post measures and behavior ratings.[4] For one such test, these are the pretest and posttest scores for the seven students:

Student	Pretest	Posttest
Earl	101	113
Ned	89	89
Jasper	112	121
Charlie	105	99
Tom	90	104
Susie	91	94
Lori	89	99

a. Draw a scatterplot relating the posttest score to the pretest score.

b. Describe the relationship between pretest and posttest scores using the graph in part a. Do you see any trend?

c. Calculate the correlation coefficient and interpret its value. Does it reinforce any relationship that was apparent from the scatterplot? Explain.

MY CD **3.18** **Lexus, Inc.** The makers of the *Lexus*
EX0318 automobile have steadily increased their sales since their U.S. launch in 1989. However, the rate of increase changed in 1996 when *Lexus* introduced a line of trucks. The sales of *Lexus* from 1996 to 2003 are shown in the table.[5]

Year	1996	1997	1998	1999	2000	2001	2002	2003
Sales ($000)	80	100	155	180	210	225	230	260

Source: Adapted from: *Automotive News*, 26 January 2004.

a. Plot the data using a scatterplot. How would you describe the relationship between year and sales of *Lexus*?

b. Find the least-squares regression line relating the sales of *Lexus* to the year being measured.

c. If you were to predict the sales of *Lexus* in the year 2015, what problems might arise with your prediction?

(MY) CD **3.19** LCD TVs, again In Exercise 2.12, *Con-*
EX0319 *sumer Reports* gave the prices for the top ten LCD standard definition TVs in the 14- to 20-inch category. Does the price of an LCD TV depend on the size of the screen? The table below shows the ten costs again, along with the screen size in inches.[6]

Brand	Price	Size
Sharp LC-20E1U	$1200	20
Sony KLV-15SR1	800	15
Panasonic TC-20LA1	1050	20
Panasonic TC-17LA1	750	17
Gateway GTW-L18M103	700	18
Panasonic TC-14LA1	500	14
Gateway GTW-L17M103	600	17
Toshiba 14VL43U	670	14
Toshiba 20VL43U	1200	20
Sharp LC-15E1U	650	15

a. Which of the two variables (price and size) is the independent variable, and which is the dependent variable?

b. Construct a scatterplot for the data. Does the relationship appear to be linear?

3.20 LCD TVs, continued Refer to Exercise 3.19. Suppose we assume that the relationship between x and y is linear.

a. Find the correlation coefficient, r. What does this value tell you about the strength and direction of the relationship between size and price?

b. What is the equation of the regression line used to predict the price of the TV based on the size of the screen?

c. The Sony Corporation is introducing a new 18″ LCD TV. What would you predict its price to be?

d. Would it be reasonable to try to predict the price of a 30″ LCD TV? Explain.

CHAPTER REVIEW

Key Concepts

I. Bivariate Data

1. Both qualitative and quantitative variables
2. Describing each variable separately
3. Describing the relationship between the two variables

II. Describing Two Qualitative Variables

1. Side-by-side pie charts
2. Comparative line charts

3. Comparative bar charts
   a. Side-by-side
   b. Stacked
4. Relative frequencies to describe the relationship between the two variables

III. Describing Two Quantitative Variables

1. Scatterplots
   a. Linear or nonlinear pattern

b. Strength of relationship

c. Unusual observations: clusters and outliers

2. Covariance and correlation coefficient

3. The best-fitting regression line

a. Calculating the slope and y-intercept

b. Graphing the line

c. Using the line for prediction

 MINITAB

Describing Bivariate Data

MINITAB provides different graphical techniques for *qualitative* and *quantitative* bivariate data, as well as commands for obtaining bivariate descriptive measures when the data are quantitative. To explore both types of bivariate procedures, you need to enter two different sets of bivariate data into a *MINITAB* worksheet. Once you are on the Windows desktop, double-click on the *MINITAB* icon or use the Start button to start *MINITAB.*

Start a new project using **File → New → Minitab Project.** Then open the existing project called "Chapter 1." We will use the college student data, which should be in Worksheet 1. Suppose that the 105 students already tabulated were from the University of California, Riverside, and that another 100 students from an introductory statistics class at UC Berkeley were also interviewed. Table 3.6 shows the status distribution for both sets of students. Create another variable in C3 of the worksheet called "College" and enter UCR for the first five rows. Now enter the UCB data in columns C1–C3. You can use the familiar Windows cut-and-paste icons if you like.

TABLE 3.6

	Freshman	Sophomore	Junior	Senior	Grad Student
Frequency (UCR)	5	23	32	35	10
Frequency (UCB)	10	35	24	25	6

The other worksheet in "Chapter 1" is not needed and can be deleted by clicking on the X in the top right corner of the worksheet. We *will* use the worksheet called "Minivans" from Chapter 2, which you should open using **File → Open Worksheet** and selecting "Minivans.mtw." Now **save** this new **project** as "Chapter 3."

To graphically describe the UCR/UCB student data, you can use comparative pie charts—one for each school (see Chapter 1). Alternatively, you can use either stacked or side-by-side bar charts. Use **Graph → Bar Chart.** In the "Bar Charts" Dialog box (Figure 3.15), select **Values from a Table** in the drop-down list and click either **Stack** or **Cluster** in the row marked "One Column of Values." Click **OK.** In the next Dialog box (Figure 3.16), select "Frequency" for the **Graph variables** box and "Status" and "College" for the **Categorical variable for grouping** box. Click **OK.** Once the bar chart is displayed (Figure 3.17), you can *right-click* on various items in the bar chart to edit. If you *right-click* on the bars and select **Update Graph Automatically,** the bar chart will automatically update when you change the data in the Minitab worksheet.

FIGURE 3.15

FIGURE 3.16

FIGURE 3.17

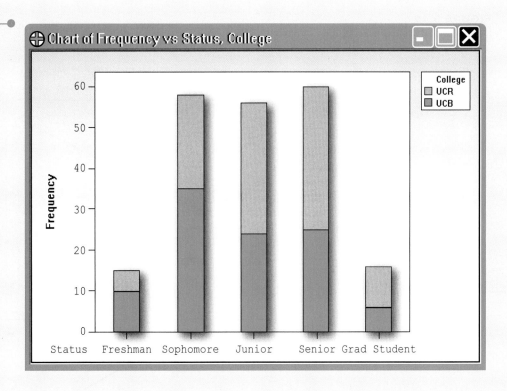

Turn to Worksheet 2, in which the bivariate minivan data from Chapter 2 are located. To examine the relationship between the second and third car seat lengths, you can plot the data and numerically describe the relationship with the correlation coefficient and the best-fitting line. Use **Stat → Regression → Fitted Line Plot,** and select "2nd Seat" and "3rd Seat" for **Y** and **X,** respectively (see Figure 3.18). Make sure that the dot next to **Linear** is selected, and click **OK.** The plot of the nine data points and the best-fitting line will be generated as in Figure 3.19.

FIGURE 3.18

FIGURE 3.19

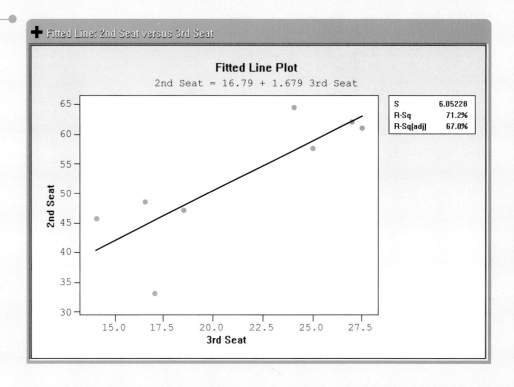

To calculate the correlation coefficient, use **Stat → Basic Statistics → Correlation,** selecting "2nd Seat" and "3rd Seat" for the Variables box. To select both variables at once, hold the **Shift** key down as you highlight the variables and then click **Select.** Click **OK,** and the correlation coefficient will appear in the Session window (see Figure 3.20). Notice the relatively strong positive correlation and the positive slope of the regression line, indicating that a minivan with a long floor length behind the second seat will also tend to have a long floor length behind the third seat.

Save "Chapter 3" before you exit *MINITAB*!

FIGURE 3.20

Supplementary Exercises

3.21 Professor Asimov Professor Isaac Asimov was one of the most prolific writers of all time. He wrote nearly 500 books during a 40-year career prior to his death in 1992. In fact, as his career progressed, he became even more productive in terms of the number of books written within a given period of time.[7] These data are the times (in months) required to write his books, in increments of 100:

Number of Books	100	200	300	400	490
Time (in months)	237	350	419	465	507

a. Plot the accumulated number of books as a function of time using a scatterplot.

b. Describe the productivity of Professor Asimov in light of the data set graphed in part a. Does the relationship between the two variables seem to be linear?

(MY) CD 3.22 Cheese, Please! Health-conscious
EX0322 Americans often consult the nutritional information on food packages in an attempt to avoid foods with large amounts of fat, sodium, or cholesterol. The following information was taken from eight different brands of American cheese slices:

Brand	Fat (g)	Saturated Fat (g)	Cholesterol (mg)	Sodium (mg)	Calories
Kraft Deluxe American	7	4.5	20	340	80
Kraft Velveeta Slices	5	3.5	15	300	70
Private Selection	8	5.0	25	520	100
Ralphs Singles	4	2.5	15	340	60
Kraft 2% Milk Singles	3	2.0	10	320	50
Kraft Singles American	5	3.5	15	290	70
Borden Singles	5	3.0	15	260	60
Lake to Lake American	5	3.5	15	330	70

a. Which pairs of variables do you expect to be strongly related?

b. Draw a scatterplot for fat and saturated fat. Describe the relationship.

c. Draw a scatterplot for fat and calories. Compare the pattern to that found in part b.

d. Draw a scatterplot for fat versus sodium and another for cholesterol versus sodium. Compare the patterns. Are there any clusters or outliers?

e. For the pairs of variables that appear to be linearly related, calculate the correlation coefficients.

f. Write a paragraph to summarize the relationships you can see in these data. Use the correlations and the patterns in the four scatterplots to verify your conclusions.

3.23 Army versus Marine Corps Who are the men and women who serve in our armed forces? Are they male or female, officers or enlisted? What is their ethnic origin and their average age? An article in *Time* magazine provided some insight into the demographics of the U.S. armed forces.[8] Two of the bar charts are shown below.

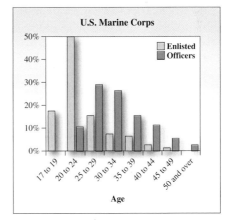

a. What variables have been measured in this study? Are the variables qualitative or quantitative?

b. Describe the population of interest. Do these data represent a population or a sample drawn from the population?

c. What type of graphical presentation has been used? What other type could have been used?

d. How would you describe the similarities and differences in the age distributions of enlisted persons and officers?

e. How would you describe the similarities and differences in the age distributions of personnel in the U.S. Army and the Marine Corps?

3.24 Cheese, again! The demand for healthy foods that are low in fats and calories has resulted in a large number of "low-fat" and "fat-free" products at the supermarket. The table shows the numbers of calories and the amounts of sodium (in milligrams) per slice for five different brands of fat-free American cheese.

Brand	Sodium (mg)	Calories
Kraft Fat Free Singles	300	30
Ralphs Fat Free Singles	300	30
Borden Fat Free	320	30
Healthy Choice Fat Free	290	30
Smart Beat American	180	25

a. Draw a scatterplot to describe the relationship between the amount of sodium and the number of calories.

b. Describe the plot in part a. Do you see any outliers? Do the rest of the points seem to form a pattern?

c. Based *only* on the relationship between sodium and calories, can you make a clear decision about which of the five brands to buy? Is it reasonable to base your choice on only these two variables? What other variables should you consider?

MY CD 3.25 Peak Current Using a chemical procedure called *differential pulse polarography*, a chemist measured the peak current generated (in microamperes) when a solution containing a given amount of nickel (in parts per billion) is added to a buffer. The data are shown here:

x = Ni (ppb)	y = Peak Current (μA)
19.1	.095
38.2	.174
57.3	.256
76.2	.348
95	.429
114	.500
131	.580
150	.651
170	.722

Use a graph to describe the relationship between x and y. Add any numerical descriptive measures that are appropriate. Write a paragraph summarizing your results.

MY CD 3.26 Movie Money Does the opening weekend adequately predict the success or failure of a

new movie? In a recent summer, 36 movies were investigated in *Entertainment Weekly*, and the following variables were recorded.[9]

• The movie's first weekend's gross earnings (in millions)

• The movie's total gross earnings in the United States (in millions)

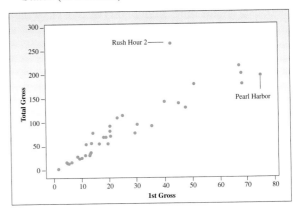

a. How would you describe the relationship between the first weekend's gross and the total gross?

b. Are there any outliers? If so, explain how they do not fit the pattern of the other movies.

c. Which dot represents the movie with the best opening weekend? Did it also have the highest total gross?

d. The film *Pearl Harbor* opened on a 3-day weekend (Memorial Day). Does that help to explain its position in relation to the other data points?

3.27 Movie Money, continued The data from Exercise 3.26 were entered into a *MINITAB* worksheet, and the following output was obtained.

Covariances: 1st Gross, Total Gross

```
                1st Gross      Total Gross
1st Gross       412.528
Total Gross     1232.231       4437.109
```

a. Use the *MINITAB* output or the original data to find the correlation between first weekend and total gross.

b. Which of the two variables would you classify as the independent variable? The dependent variable?

c. If the average first weekend gross is 25.66 million dollars and the average total gross is 86.71 million dollars, find the regression line for predicting total gross as a function of the first weekend's gross.

d. If another film was released and grossed $30 million on the first weekend, what would you predict that its total gross earnings will be?

3.28 Heights and Gender Refer to Exercise 1.54 and data set EX0154. When the heights of these 105 students were recorded, their gender was also recorded.

a. What variables have been measured in this experiment? Are they qualitative or quantitative?

b. Look at the histogram from Exercise 1.54 along with the comparative box plots shown below. Do the box plots help to explain the two local peaks in the histogram? Explain.

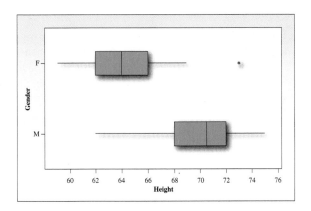

State	Sites	Area	State	Sites	Area
AL	15	52	MT	15	147
AK	6	663	NE	11	77
AZ	9	114	NV	1	111
AR	12	53	NH	20	9
CA	99	164	NJ	116	9
CO	18	104	NM	12	122
CT	16	6	NY	91	55
DE	15	2	NC	29	54
DC	1	0	ND	0	71
FL	52	66	OH	35	45
GA	15	59	OK	11	70
HI	3	11	OR	12	98
ID	9	84	PA	96	46
IL	45	58	RI	12	2
IN	29	36	SC	25	32
IA	14	56	SD	2	77
KS	12	82	TN	13	42
KY	14	40	TX	45	269
LA	15	52	UT	19	85
ME	12	35	VT	9	10
MD	19	12	VA	30	43
MA	32	11	WA	47	71
MI	69	97	WV	9	24
MN	24	87	WI	40	65
MS	4	48	WY	2	98
MO	25	70			

MINITAB printout for Exercise 3.29

Covariances: Sites, Area

```
                  Sites            Area
Sites          712.603
Area           -83.530         9346.603
```

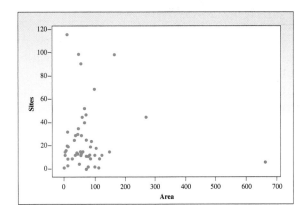

3.29 Hazardous Waste The data in Exercise 1.37 gave the number of hazardous waste sites in each of the 50 states and the District of Columbia in 2003.[3] Suspecting that there might be a relationship between the number of waste sites and the size of the state (in thousands of square miles), researchers recorded both variables and generated a scatterplot with *MINITAB*.

a. Is there any clear pattern in the scatterplot? Describe the relationship between number of waste sites and the size of the state.

b. Use the *MINITAB* output to calculate the correlation coefficient. Does this confirm your answer to part a?

c. Are there any outliers or clusters in the data? If so, can you explain them?

d. What other variables could you consider in trying to understand the distribution of hazardous waste sites in the United States?

3.30 Brett Favre, again The number of passes completed and the total number of passing yards were recorded for Brett Favre for each of the 16 regular season and 2 postseason games in the fall of 2003:[10]

Week	Completions	Total Yards
1	25	248
2	15	132
3	23	245
4	21	179
5	19	185
6	25	272
7	23	268
8	18	194
9	14	109
10	13	92
11	10	138
12	23	296
13	22	210
14	23	278
15	22	399
16	12	116
17	26	319
18	15	180

Source: ESPN.com

a. Draw a scatterplot to describe the relationship between number of completions and total passing yards for Brett Favre.

b. Describe the plot in part a. Do you see any outliers? Do the rest of the points seem to form a pattern?

c. Calculate the correlation coefficient, r, between the number of completions and total passing yards.

d. What is the regression line for predicting total number of passing yards y based on the total number of completions x?

e. If Brett Favre had 20 pass completions in his next game, what would you predict his total number of passing yards to be?

3.31 Pottery, continued In Exercise 1.59, we analyzed the percentage of aluminum oxide in 26 samples of Romano-British pottery found at four different kiln sites in the United Kingdom.[11] Since one of the sites only provided two measurements, that site is eliminated, and comparative box plots of aluminum oxide at the other three sites are shown.

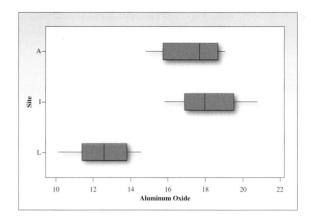

a. What two variables have been measured in this experiment? Are they qualitative or quantitative?

b. How would you compare the amount of aluminum oxide in the samples at the three sites?

3.32 Pottery, continued Here is the percentage of aluminum oxide, the percentage of iron oxide, and the percentage of magnesium oxide in five samples collected at Ashley Rails in the United Kingdom.

Sample	Al	Fe	Mg
1	17.7	1.12	0.56
2	18.3	1.14	0.67
3	16.7	0.92	0.53
4	14.8	2.74	0.67
5	19.1	1.64	0.60

a. Find the correlation coefficients describing the relationships between aluminum and iron oxide content, between iron oxide and magnesium oxide, and between aluminum oxide and magnesium oxide.

b. Write a sentence describing the relationships between these three chemicals in the pottery samples.

3.33 Computer Networks at Home The table below (Exercise 1.50) shows the predicted rise of home networking of PCs in the next few years.[12]

U.S Home Networks (in millions)

Year	Wired	Wireless
2002	6.1	1.7
2003	6.5	4.5
2004	6.2	8.7
2005	5.7	13.7
2006	4.9	19.1
2007	4.1	24.0
2008	3.4	28.2

Source: Jupiter Research

a. What variables have been measured in this experiment? Are they qualitative or quantitative?

b. Use one of the graphical methods given in this chapter to describe the data.

c. Write a sentence describing the relationships between wired and wireless technology in the next few years.

3.34 Politics and Religion A survey was conducted prior to the 2004 presidential election to explore the relationship between a person's religious fervor and their choice of a political candidate. Voters were asked how often they attended church and which of the two major presidential candidates (George W. Bush or his democratic opponent) they would favor in the 2004 election.[13] The results are shown below.

Church Attendance	G. W. Bush	Democratic Candidate
More than once a week	63%	37%
Once a week	56%	44%
Once or twice a month	52%	48%
Once or twice a year	46%	54%
Seldom/never	38%	62%

Source: Press-Enterprise

a. What variables have been measured in this survey? Are they qualitative or quantitative?

b. Draw side-by-side comparative bar charts to describe the percentages favoring the two candidates, categorized by church attendance.

c. Draw two line charts on the same set of axes to describe the same percentages for the two candidates.

d. What conclusions can you draw using the two graphs in parts b and c? Which is more effective?

(MY)(CD) **3.35 Armspan and Height** Leonardo
EX0335   DaVinci (1452–1519) drew a sketch of a man, indicating that a persons armspan (measuring across the back with arms outstretched to make a "T") is roughly equal to the person's height. To test this claim, we measured eight people with the following results:

Person	1	2	3	4
Armspan (inches)	68	62.25	65	69.5
Height (inches)	69	62	65	70

Person	5	6	7	8
Armspan (inches)	68	69	62	60.25
Height (inches)	67	67	63	62

a. Draw a scatterplot for armspan and height. Use the same scale on both the horizontal and vertical axes. Describe the relationship between the two variables.

b. Calculate the correlation coefficient relating armspan and height.

c. If you were to calculate the regression line for predicting height based on a person's armspan, how would you estimate the slope of this line?

d. Find the regression line relating armspan to a person's height.

e. If a person has an armspan of 62 inches, what would you predict the person's height to be?

(MY)(CD) **3.36 Airline Revenues** The number of pas-
EX0336   sengers x (in millions) and the revenue y (in billions of dollars) for the top nine U.S. airlines in a recent year are given in the table.[3]

	American	United	Delta	Northwest	Continental
x	94.0	68.4	90.8	51.7	39.5
y	17.3	14.3	13.3	9.5	8.4

	U.S. Air	Southwest	Alaska	American West
x	47.2	72.4	14.1	19.4
y	7.0	5.5	2.2	2.0

Source: The World Almanac and Book of Facts, 2004

a. Construct a scatterplot for the data.

b. Describe the form, direction, and strength of the pattern in the scatterplot.

3.37 Test Interviews Of two personnel evalu- EX0337 ation techniques available, the first requires a two-hour test-interview while the second can be completed in less than an hour. The scores for each of the eight individuals who took both tests are given in the next table.

Applicant	Test 1 (x)	Test 2 (y)
1	75	38
2	89	56
3	60	35
4	71	45
5	92	59
6	105	70
7	55	31
8	87	52

a. Construct a scatterplot for the data.

b. Describe the form, direction, and strength of the pattern in the scatterplot.

3.38 Test Interviews, continued Refer to Exercise 3.37.

a. Find the correlation coefficient, r, to describe the relationship between the two tests.

b. Would you be willing to use the second and quicker test rather than the longer test-interview to evaluate personnel? Explain.

 3.39 Happy in the Air? continued Of the EX0339 6,229 consumer complaints against major U.S. airlines in a recent year, the distribution by airline (Exercise 1.29) is shown in the table, along with the systemwide passenger boardings (millions of passengers).[14]

Airline	Number of Complaints	Passengers (millions)
United	1172	68.6
American West	318	19.5
Northwest	765	52.8
Continental	563	40.0
Delta	1231	89.9
American	1212	94.1
U.S. Airways	532	47.2
Alaska	129	14.2
American Eagle	71	11.8
Southwest	236	72.5

Source: Time Almanac 2004

a. Construct a scatterplot for the data.

b. Describe the form, direction, and strength of the pattern in the scatterplot.

c. Are there any outliers in the scatterplot? If so, which airline does this outlier represent?

d. Does the outlier from part c indicate that this airline is doing better or worse than the other airlines with respect to customer satisfaction.

MY APPLET Exercises

3.40 If you have not yet done so, use the first applet in **Building a Scatterplot** to create a scatterplot for the data in Example 3.4.

3.41 If you have not yet done so, use the second applet in **Building a Scatterplot** to create a scatterplot for the data in Example 3.5.

3.42 CD Players The table below shows the EX0342 prices of eight portable CD players along with their overall score (on a scale of 0–100) in a consumer rating survey presented by *Consumer Reports*.[15]

Brand and Model	Price	Overall Score
Sony D-EJ611	$80	70
Panasonic SL-SX280	50	66
Aiwa XP-V713	70	60
Aiwa XP-SP911	80	65
Panasonic SL-CT470	100	59
Phillips AZ9213	80	60
GPX C3948B1	60	47
RCA RP-2360FM	65	42

a. Calculate the correlation coefficient r between price and overall score. How would you describe the relationship between price and overall score?

b. Use the applet called **Correlation and the Scatterplot** to plot the eight data points. What is the correlation coefficient shown on the applet? Compare with the value you calculated in part a.

c. Describe the pattern that you see in the scatterplot. Are there any clusters or outliers? If so, how would you explain them?

3.43 SAT Scores Is there a correlation EX0343 between Math and Verbal SAT test scores? That is, do students who do well on the Math portion typically do well on the Verbal portion of the test? The data below (score − 400) show the average scores on standardized math and verbal tests for seven high schools in Southern California.[16]

School	Verbal	Math
Centennial	64	84
Norco	74	95
Moreno Valley	27	45
Valley View	75	71
Ramona	20	50
San Bernardino	38	27
Canyon Springs	68	85
North	85	98

a. Calculate the correlation coefficient r between verbal and math scores. How would you describe the relationship between verbal and math scores?

b. Use the applet called **Correlation and the Scatterplot** to plot the eight data points. What is the correlation coefficient shown on the applet? Compare with the value you calculated in part a.

c. Describe the pattern that you see in the scatterplot. Are there any clusters or outliers? If so, how would you explain them?

3.44 Acess the applet called **Exploring Correlation.**

a. Move the slider in the first applet so that $r \approx .75$. Now switch the sign using the [Switch Sign] button at the bottom of the applet. Describe the change in the pattern of the points.

b. Move the slider in the first applet so that $r \approx 0$. Describe the pattern of points on the scatterplot.

c. Refer to part b. In the second applet labeled "Correlation and the Quadrants," with $r \approx 0$, count the number of points falling in each of the four quadrants of the scatterplot. Is the distribution of points in the quadrants relatively uniform, or do more points fall into certain quadrants than others?

d. Use the second applet labeled "Correlation and the Quadrants" and change the correlation coefficient to $r \approx -0.9$. Is the distribution of points in the quadrants relatively uniform, or do more points fall into certain quadrants than others? What happens if $r \approx 0.9$?

e. Use the third applet labeled "Correlation and the Regression Line." Move the slider to see the relationship between the correlation coefficient r, the slope of the regression line and the direction of the relationship between x and y. Describe the relationship.

3.45 Suppose that the relationship between two variables x and y can be described by the regression line $y = 2.0 + 0.5x$. Use the applet in **How a Line Works** to answer the following questions:

a. What is the change in y for a one-unit change in x?

b. Do the values of y increase or decrease as x increases?

c. At what point does the line cross the y-axis? What is the name given to this value?

d. If $x = 2.5$, use the least squares equation to predict the value of y. What value would you predict for y if $x = 4.0$?

3.46 Access the applet in **How a Line Works.**

a. Use the slider to change the *y-intercept* of the line, but do not change the *slope*. Describe the changes that you see in the line.

b. Use the slider to change the *slope* of the line, but do not change the *y-intercept*. Describe the changes that you see in the line.

CASE STUDY

 Dishwashers

Do You *Think* Your Dishes Are *Really* Clean?

Does the price of an appliance convey something about its quality? Twenty different dishwashers, beginning with a Kenmore model 16941 that sold for $565 and ranging down to another Kenmore, this time model 16541 that sold for $319, were ranked on characteristics including an overall satisfaction score, the yearly cost of gas or electricity, the cycle time (in minutes), and the water used per wash (in gallons).[17] The information shown in the table can also be found on the CD-ROM. Use a statistical computer package to explore the relationships between various pairs of variables in the table.

Brand and Model	Price ($)	Overall Score	Electricity Cost/Year ($)	Gas Cost/Year ($)	Time (minutes)	Water Used (gallons)
Kenmore (Sears) 16941	565	84	61	38	85	7.5
Kenmore (Sears) 16779 A BEST BUY	400	83	66	43	95	7.5
KitchenAid KUDA23SBWH0	840	78	56	32	80	7.5
Kenmore (Sears) 16649 A BEST BUY	369	77	69	47	100	7.5
Maytag DWU9921AAE	550	76	66	35	90	9.5
KitchenAid KUDP230B0	510	75	58	35	75	7.5
Maytag DWU9200AAX	460	74	65	35	95	9.5
Magic Chef DU5JV	330	73	60	30	90	9.5
Maytag DWU7400AAE	400	71	66	35	95	9.5
Jenn-Air DW960W	470	70	57	30	95	9.0
General Electric GSD1230TWW	400	66	71	42	110	9.0
KitchenAid KUDB230B0	380	64	59	35	80	7.5
White-Westinghouse WDB632RS0	350	61	56	32	70	7.5
Caloric CDU600CWW	375	60	70	40	105	9.5
General Electric GSD1930TWW	450	60	69	41	110	9.0
Hotpoint HDA430VWW	290	56	70	40	110	9.5
Tappan TDB668RBS0	379	55	71	45	95	8.0
Frigidaire FDB878RS0	280	54	73	45	95	9.0
Frigidaire FDB663RBS0	320	53	70	45	95	8.0
Kenmore (Sears) 16541	319	52	70	40	105	9.5

Source: © 1995 by Consumers Union of U.S., Inc., Yonkers, NY 10703-1057, a nonprofit organization. Reprinted with permission from the August 1995 issues of CONSUMER REPORTS® for educational purposes only. No commercial use or photocopying permitted. www.ConsumerReports.org.

1. Look at each variable individually. What can you say about symmetry? About outliers?

2. Look at the variables in pairs. Which pairs of variables are positively correlated? Which are negatively correlated? Do any pairs exhibit little or no correlation? Are some of these results counterintuitive? Can you offer an explanation for these cases?

3. Answer the questions: Does the price of an appliance, specifically a dishwasher, convey something about its quality? Which variables did you use in arriving at your answer?

Probability and Probability Distributions

© Getty Images

GENERAL OBJECTIVES

Now that you have learned to describe a data set, how can you use sample data to draw conclusions about the sampled populations? The technique involves a statistical tool called *probability*. To use this tool correctly, you must first understand how it works. The first part of this chapter will teach you the new language of probability, presenting the basic concepts with simple examples.

The variables that we measured in Chapters 1 and 2 can now be redefined as random variables, whose values depend on the chance selection of the elements in the sample. Using probability as a tool, you can create probability distributions that serve as models for discrete random variables, and you can describe these random variables using a mean and standard deviation similar to those in Chapter 2.

CHAPTER INDEX

Probability and Decision Making in the Congo

In his exciting novel *Congo*, author Michael Crichton describes an expedition racing to find boron-coated blue diamonds in the rain forests of eastern Zaire. Can probability help the heroine Karen Ross in her search for the Lost City of Zinj? The case study at the end of this chapter involves Ross's use of probability in decision-making situations.

PERSONAL TRAINER

What's the Difference between Mutually Exclusive and Independent Events?

THE ROLE OF PROBABILITY IN STATISTICS

4.1

Probability and statistics are related in an important way. Probability is used as a *tool;* it allows you to evaluate the reliability of your conclusions about the population when you have only sample information. Consider these situations:

- When you toss a single coin, you will see either a head (H) or a tail (T). If you toss the coin repeatedly, you will generate an infinitely large number of Hs and Ts—the entire population. What does this population look like? If the coin is fair, then the population should contain 50% Hs and 50% Ts. Now toss the coin one more time. What is the chance of getting a head? Most people would say that the "probability" or chance is 1/2.

- Now suppose you are not sure whether the coin is fair; that is, you are not sure whether the makeup of the population is 50–50. You decide to perform a simple experiment. You toss the coin $n = 10$ times and observe 10 heads in a row. Can you conclude that the coin is fair? Probably not, because if the coin were fair, observing 10 heads in a row would be very *unlikely*; that is, the "probability" would be very small. It is more *likely* that the coin is biased.

As in the coin-tossing example, statisticians use probability in two ways. When the population is *known,* probability is used to describe the likelihood of observing a particular sample outcome. When the population is *unknown* and only a sample from that population is available, probability is used in making statements about the makeup of the population—that is, in making statistical inferences.

In Chapters 4–7, you will learn many different ways to calculate probabilities. You will assume that the population is *known* and calculate the probability of observing various sample outcomes. Once you begin to use probability for statistical inference in Chapter 8, the population will be *unknown* and you will use your knowledge of probability to make reliable inferences from sample information. We begin with some simple examples to help you grasp the basic concepts of probability.

EVENTS AND THE SAMPLE SPACE

4.2

Data are obtained by observing either uncontrolled events in nature or controlled situations in a laboratory. We use the term **experiment** to describe either method of data collection.

Definition An **experiment** is the process by which an observation (or measurement) is obtained.

The observation or measurement generated by an experiment may or may not produce a numerical value. Here are some examples of experiments:

- Recording a test grade
- Measuring daily rainfall
- Interviewing a householder to obtain his or her opinion on a greenbelt zoning ordinance

- Testing a printed circuit board to determine whether it is a defective product or an acceptable product
- Tossing a coin and observing the face that appears

When an experiment is performed, what we observe is an outcome called a **simple event,** often denoted by the capital *E* with a subscript.

Definition A **simple event** is the outcome that is observed on a single repetition of the experiment.

EXAMPLE 4.1

Experiment: Toss a die and observe the number that appears on the upper face. List the simple events in the experiment.

Solution When the die is tossed once, there are six possible outcomes. There are the simple events, listed below.

Event E_1: Observe a 1 Event E_4: Observe a 4
Event E_2: Observe a 2 Event E_5: Observe a 5
Event E_3: Observe a 3 Event E_6: Observe a 6

We can now define an **event** as a collection of simple events, often denoted by a capital letter.

Definition An **event** is a collection of simple events.

EXAMPLE
continued 4.1

We can define the events *A* and *B* for the die tossing experiment:

A: Observe an odd number
B: Observe a number less than 4

Since event *A* occurs if the upper face is 1, 3, or 5, it is a collection of three simple events and we write $A = \{E_1, E_3, E_5\}$. Similarly, the event *B* occurs if the upper face is 1, 2, or 3 and is defined as a collection or set of these three simple events: $B = \{E_1, E_2, E_3\}$.

Sometimes when one event occurs, it means that another event cannot.

Definition Two events are **mutually exclusive** if, when one event occurs, the others cannot, and vice versa.

In the die-tossing experiment, events *A* and *B* are *not* mutually exclusive because they have two outcomes in common—if the number on the upper face of the die is a 1 or a 3. Both events *A* and *B* will occur if either E_1 or E_3 is observed when the experiment is performed. In contrast, the six simple events $E_1, E_2, \ldots, E_6$ form a set of all mutually exclusive outcomes of the experiment. When the experiment is performed once, one and only one of these simple events can occur.

Definition The set of all simple events is called the sample space, *S*.

Sometimes it helps to visualize an experiment using a picture called a **Venn diagram,** shown in Figure 4.1. The outer box represents the *sample space,* which contains all of the *simple events,* represented by labeled points. Since an event is a collection of one or more simple events, the appropriate points are circled and labeled with the event letter. For the die-tossing experiment, the sample space is $S = \{E_1, E_2, E_3, E_4, E_5, E_6\}$ or, more simply, $S = \{1, 2, 3, 4, 5, 6\}$. The events $A = \{1, 3, 5\}$ and $B = \{1, 2, 3\}$ are circled in the Venn diagram.

FIGURE 4.1

Venn diagram for die tossing

EXAMPLE 4.2

Experiment: Toss a single coin and observe the result. These are the simple events:

E_1: Observe a head (H)
E_2: Observe a tail (T)

The sample space is $S = \{E_1, E_2\}$ or more simply, $S = \{H, T\}$.

EXAMPLE 4.3

Experiment: Record a person's blood type. The four mutually exclusive possible outcomes are these simple events:

E_1: Blood type A
E_2: Blood type B
E_3: Blood type AB
E_4: Blood type O

The sample space is $S = \{E_1, E_2, E_3, E_4\}$, or $S = \{A, B, AB, O\}$.

Some experiments can be generated in stages, and the sample space can be displayed in a **tree diagram.** Each successive level of branching on the tree corresponds to a step required to generate the final outcome.

EXAMPLE 4.4

A medical technician records a person's blood type and Rh factor. List the simple events in the experiment.

Solution   For each person, a two-stage procedure is needed to record the two variables of interest. The tree diagram is shown in Figure 4.2. The eight simple events in the tree diagram form the sample space, $S = \{A+, A-, B+, B-, AB+, AB-, O+, O-\}$.

An alternative way to display the simple events is to use a **probability table,** as shown in Table 4.1. The rows and columns show the possible outcomes at the first and second stages, respectively, and the simple events are shown in the cells of the table.

FIGURE 4.2

Tree Diagram for Example 4.4

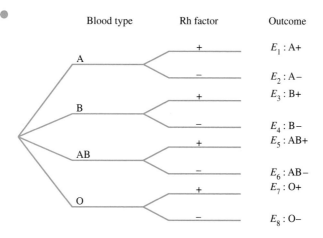

TABLE 4.1

Probability Table for Example 4.4

	Blood Type			
Rh Factor	A	B	AB	O
Negative	A−	B−	AB−	O−
Positive	A+	B+	AB+	O+

CALCULATING PROBABILITIES USING SIMPLE EVENTS

4.3

The probability of an event A is a measure of our belief that the event A will occur. One practical way to interpret this measure is with the concept of *relative frequency.* Recall from Chapter 1 that if an experiment is performed n times, then the relative frequency of a particular occurrence—say, A—is

$$\text{Relative frequency} = \frac{\text{Frequency}}{n}$$

where the frequency is the number of times the event A occurred. If you let n, the number of repetitions of the experiment, become larger and larger ($n \to \infty$), you will eventually generate the entire population. In this population, the relative frequency of the event A is defined as the **probability of event A;** that is,

$$P(A) = \lim_{n \to \infty} \frac{\text{Frequency}}{n}$$

Since $P(A)$ behaves like a relative frequency, $P(A)$ must be a proportion lying between 0 and 1; $P(A) = 0$ if the event A never occurs, and $P(A) = 1$ if the event A always occurs. The closer $P(A)$ is to 1, the more likely it is that A will occur.

For example, if you tossed a balanced, six-sided die an infinite number of times, you would expect the relative frequency for any of the six values, $x = 1, 2, 3, 4, 5, 6$, to be 1/6. Needless to say, it would be very time-consuming, if not impossible, to repeat an experiment an infinite number of times. For this reason, there are alternative methods for calculating probabilities that make use of the relative frequency concept.

An important consequence of the relative frequency definition of probability involves the simple events. Since the simple events are mutually exclusive, their probabilities must satisfy two conditions.

REQUIREMENTS FOR SIMPLE-EVENT PROBABILITIES

- Each probability must lie between 0 and 1.
- The sum of the probabilities for all simple events in S equals 1.

When it is possible to write down the simple events associated with an experiment and to determine their respective probabilities, we can find the probability of an event A by summing the probabilities for all the simple events contained in the event A.

Definition The **probability of an event A** is equal to the sum of the probabilities of the simple events contained in A.

EXAMPLE 4.5 — Toss two fair coins and record the outcome. Find the probability of observing exactly one head in the two tosses.

Solution To list the simple events in the sample space, you can use a tree diagram as shown in Figure 4.3. The letters H and T mean that you observed a head or a tail, respectively, on a particular toss. To assign probabilities to each of the four simple events, you need to remember that the coins are fair. Therefore, any of the four simple events is as likely as any other. Since the sum of the four simple events must be 1, each must have probability $P(E_i) = 1/4$. The simple events in the sample space are shown in Table 4.2, along with their *equally likely probabilities*. To find $P(A) = P$(observe exactly one head), you need to find all the simple events that result in event A—namely, E_2 and E_3:

$$P(A) = P(E_2) + P(E_3)$$

$$= \frac{1}{4} + \frac{1}{4} = \frac{1}{2}$$

 MY TIP

Probabilities must lie between 0 and 1.

 MY TIP

The probabilities of all the simple events must add to 1.

FIGURE 4.3 ●

Tree Diagram for Example 4.5

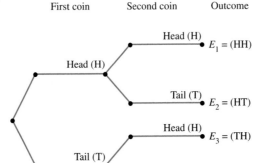

First coin   Second coin   Outcome

Head (H) — E_1 = (HH)

Head (H)

Tail (T) — E_2 = (HT)

Head (H) — E_3 = (TH)

Tail (T)

Tail (T) — E_4 = (TT)

TABLE 4.2 ● **Simple Events and Their Probabilities**

Event	First Coin	Second Coin	$P(E_i)$
E_1	H	H	1/4
E_2	H	T	1/4
E_3	T	H	1/4
E_4	T	T	1/4

EXAMPLE 4.6

The proportions of blood phenotypes A, B, AB, and O in the population of all Caucasians in the United States are reported as .41, .10, .04, and .45, respectively.[1] If a single Caucasian is chosen randomly from the population, what is the probability that he or she will have either type A or type AB blood?

Solution   The four simple events, A, B, AB, and O, do *not* have equally likely probabilities. Their probabilities are found using the relative frequency concept as

$$P(A) = .41 \qquad P(B) = .10 \qquad P(AB) = .04 \qquad P(O) = .45$$

The event of interest consists of two simple events, so

$$P(\text{person is either type A or type AB}) = P(A) + P(AB)$$
$$= .41 + .04 = .45$$

EXAMPLE 4.7

A candy dish contains one yellow and two red candies. You close your eyes, choose two candies one at a time from the dish, and record their colors. What is the probability that both candies are red?

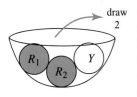

draw 2

Solution   Since no probabilities are given, you must list the simple events in the sample space. The two-stage selection of the candies suggests a tree diagram, shown in Figure 4.4. There are two red candies in the dish, so you can use the letters R_1, R_2, and Y to indicate that you have selected the first red, the second red, or the yellow candy, respectively. Since you closed your eyes when you chose the candies, all six choices should be *equally likely* and are assigned probability 1/6. If A is the event that both candies are red, then

$$A = \{R_1R_2, R_2R_1\}$$

Thus,

$$P(A) = P(R_1R_2) + P(R_2R_1)$$
$$= \frac{1}{6} + \frac{1}{6} = \frac{1}{3}$$

FIGURE 4.4 ●

Tree Diagram for Example 4.7

 TIP

A tree diagram helps to find simple events.

Branch = step toward outcome

Following branches ⇒ list of simple events.

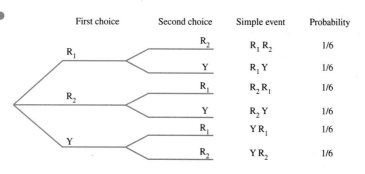

First choice	Second choice	Simple event	Probability
R_1	R_2	$R_1 R_2$	1/6
	Y	$R_1 Y$	1/6
R_2	R_1	$R_2 R_1$	1/6
	Y	$R_2 Y$	1/6
Y	R_1	$Y R_1$	1/6
	R_2	$Y R_2$	1/6

CALCULATING THE PROBABILITY OF AN EVENT

1. List all the simple events in the sample space.
2. Assign an appropriate probability to each simple event.
3. Determine which simple events result in the event of interest.
4. Sum the probabilities of the simple events that result in the event of interest.

In your calculation, you must always be careful that you satisfy these two conditions:

- Include all simple events in the sample space.
- Assign realistic probabilities to the simple events.

When the sample space is large, it is easy to unintentionally omit some of the simple events. If this happens, or if your assigned probabilities are wrong, your answers will not be useful in practice.

One way to determine the required number of simple events is to use the counting rules presented in the next optional section. These rules can be used to solve more complex problems, which generally involve a large number of simple events. If you need to master only the basic concepts of probability, you may choose to skip the next section.

4.3 EXERCISES

BASIC TECHNIQUES

4.1 Tossing a Die An experiment involves tossing a single die. These are some events:

A: Observe a 2

B: Observe an even number

C: Observe a number greater than 2

D: Observe both A and B

E: Observe A or B or both

F: Observe both A and C

a. List the simple events in the sample space.
b. List the simple events in each of the events A through F.
c. What probabilities should you assign to the simple events?
d. Calculate the probabilities of the six events A through F by adding the appropriate simple-event probabilities.

4.2 A sample space S consists of five simple events with these probabilities:

$$P(E_1) = P(E_2) = .15 \qquad P(E_3) = .4$$
$$P(E_4) = 2P(E_5)$$

a. Find the probabilities for simple events E_4 and E_5.
b. Find the probabilities for these two events:

$$A = \{E_1, E_3, E_4\}$$
$$B = \{E_2, E_3\}$$

c. List the simple events that are either in event A or event B or both.
d. List the simple events that are in both event A and event B.

4.3 A sample space contains 10 simple events: E_1, E_2, ..., E_{10}. If $P(E_1) = 3P(E_2) = .45$ and the remaining simple events are equiprobable, find the probabilities of these remaining simple events.

4.4 Free Throws A particular basketball player hits 70% of her free throws. When she tosses a pair of free throws, the four possible simple events and three of their associated probabilities are as given in the table:

Simple Event	Outcome of First Free Throw	Outcome of Second Free Throw	Probability
1	Hit	Hit	.49
2	Hit	Miss	?
3	Miss	Hit	.21
4	Miss	Miss	.09

a. Find the probability that the player will hit on the first throw and miss on the second.

b. Find the probability that the player will hit on at least one of the two free throws.

4.5 Four Coins A jar contains four coins: a nickel, a dime, a quarter, and a half-dollar. Three coins are randomly selected from the jar.

a. List the simple events in S.

b. What is the probability that the selection will contain the half-dollar?

c. What is the probability that the total amount drawn will equal 60¢ or more?

4.6 Preschool or Not? On the first day of kindergarten, the teacher randomly selects 1 of his 25 students and records the student's gender, as well as whether or not that student had gone to preschool.

a. How would you describe the experiment?

b. Construct a tree diagram for this experiment. How many simple events are there?

c. The table below shows the distribution of the 25 students according to gender and preschool experience. Use the table to assign probabilities to the simple events in part b.

	Male	Female
Preschool	8	9
No preschool	6	2

d. What is the probability that the randomly selected student is male? What is the probability that the student is a female and did not go to preschool?

4.7 The Urn Problem A bowl contains three red and two yellow balls. Two balls are randomly selected and their colors recorded. Use a tree diagram to list the 20 simple events in the experiment, keeping in mind the order in which the balls are drawn.

4.8 The Urn Problem, continued Refer to Exercise 4.7. A ball is randomly selected from the bowl containing three red and two yellow balls. Its color is noted, and the ball is returned to the bowl before a second ball is selected. List the additional five simple events that must be added to the sample space in Exercise 4.7.

APPLICATIONS

4.9 Need Eyeglasses? A survey classified a large number of adults according to whether they were judged to need eyeglasses to correct their reading vision and whether they used eyeglasses when reading. The proportions falling into the four categories are shown in the table. (Note that a small proportion, .02, of adults used eyeglasses when in fact they were judged not to need them.)

Judged to Need Eyeglasses	Used Eyeglasses for Reading	
	Yes	No
Yes	.44	.14
No	.02	.40

If a single adult is selected from this large group, find the probability of each event:

a. The adult is judged to need eyeglasses.

b. The adult needs eyeglasses for reading but does not use them.

c. The adult uses eyeglasses for reading whether he or she needs them or not.

4.10 Roulette The game of roulette uses a wheel containing 38 pockets. Thirty-six pockets are numbered 1, 2, . . . , 36, and the remaining two are marked 0 and 00. The wheel is spun, and a pocket is identified as the "winner." Assume that the observance of any one pocket is just as likely as any other.

a. Identify the simple events in a single spin of the roulette wheel.

b. Assign probabilities to the simple events.

c. Let A be the event that you observe either a 0 or a 00. List the simple events in the event A and find $P(A)$.

d. Suppose you placed bets on the numbers 1 through 18. What is the probability that one of your numbers is the winner?

4.11 Jury Duty Three people are randomly selected from voter registration and driving records to report for jury duty. The gender of each person is noted by the county clerk.

a. Define the experiment.

b. List the simple events in S.

c. If each person is just as likely to be a man as a woman, what probability do you assign to each simple event?

d. What is the probability that only one of the three is a man?

e. What is the probability that all three are women?

4.12 Jury Duty II Refer to Exercise 4.11. Suppose that there are six prospective jurors, four men and two women, who might be impaneled to sit on the jury in a criminal case. Two jurors are randomly selected from these six to fill the two remaining jury seats.

a. List the simple events in the experiment (HINT: There are 15 simple events if you ignore the order of selection of the two jurors.)

b. What is the probability that both impaneled jurors are women?

4.13 Tea Tasters A food company plans to conduct an experiment to compare its brand of tea with that of two competitors. A single person is hired to taste and rank each of three brands of tea, which are unmarked except for identifying symbols A, B, and C.

a. Define the experiment.

b. List the simple events in S.

c. If the taster has no ability to distinguish a difference in taste among teas, what is the probability that the taster will rank tea type A as the most desirable? As the least desirable?

4.14 100-Meter Run Four equally qualified runners, John, Bill, Ed, and Dave, run a 100-meter sprint, and the order of finish is recorded.

a. How many simple events are in the sample space?

b. If the runners are equally qualified, what probability should you assign to each simple event?

c. What is the probability that Dave wins the race?

d. What is the probability that Dave wins and John places second?

e. What is the probability that Ed finishes last?

4.15 Fruit Flies In a genetics experiment, the researcher mated two *Drosophila* fruit flies and observed the traits of 300 offspring. The results are shown in the table.

Eye Color	Wing Size	
	Normal	Miniature
Normal	140	6
Vermillion	3	151

One of these offspring is randomly selected and observed for the two genetic traits.

a. What is the probability that the fly has normal eye color and normal wing size?

b. What is the probability that the fly has vermillion eyes?

c. What is the probability that the fly has either vermillion eyes or miniature wings, or both?

4.16 The Eyes of the World About a year after the war in Iraq began, Americans worry about how our country is viewed by people in other nations. When asked "In general, how do you think the United States rates in the eyes of the world?" The proportions of Americans with varying opinions are approximately as shown in the following table:[2]

Opinion	Proportion
Very favorably	.10
Somewhat favorably	.44
Somewhat unfavorably	.34
Very unfavorably	.11
No opinion	.01

Source: Adapted from www.pollingreport.com

Suppose that one person is randomly selected and his or her opinion on this question is recorded.

a. What are the simple events in the experiment?

b. Are the simple events in part a equally likely? If not, what are the probabilities?

c. What is the probability that the person feels our country is rated at least somewhat favorably by people in other nations?

d. What is the probability that the person feels the United States is rated very unfavorably by people in other nations?

USEFUL COUNTING RULES (OPTIONAL)

Suppose that an experiment involves a large number *N* of simple events and you know that all the simple events are *equally likely*. Then each simple event has probability $1/N$, and the probability of an event *A* can be calculated as

$$P(A) = \frac{n_A}{N}$$

where n_A is the number of simple events that result in the event *A*. In this section, we present three simple rules that can be used to count either *N*, the number of simple events in the sample space, or n_A, the number of simple events in event *A*. Once you have obtained these counts, you can find $P(A)$ without actually listing all the simple events.

THE *mn* RULE

Consider an experiment that is performed in two stages. If the first stage can be accomplished in *m* ways and for each of these ways, the second stage can be accomplished in *n* ways, then there are *mn* ways to accomplish the experiment.

For example, suppose that you can order a car in one of three styles and in one of four paint colors. To find out how many options are available, you can think of first picking one of the $m = 3$ styles and then selecting one of the $n = 4$ paint colors. Using the *mn* Rule, as shown in Figure 4.5, you have $mn = (3)(4) = 12$ possible options.

FIGURE 4.5

Style–color combinations

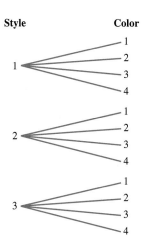

EXAMPLE 4.8 — Two dice are tossed. How many simple events are in the sample space *S*?

Solution The first die can fall in one of $m = 6$ ways, and the second die can fall in one of $n = 6$ ways. Since the experiment involves two stages, forming the pairs of numbers shown on the two faces, the total number of simple events in *S* is

$$mn = (6)(6) = 36$$

APPLET

The Java applet called **Tossing Dice** gives a visual display of the 36 simple events described in Example 4.8. You can use this applet to find probabilities for any event involving the tossing of two fair dice. By clicking on the appropriate dice combinations, we have found the probability of observing a sum of 3 on the upper faces to be 2/36 = .056. What is the probability that the sum equals 4? You will use this applet for the MyApplet Exercises at the end of the chapter.

FIGURE 4.6

Tossing Dice applet

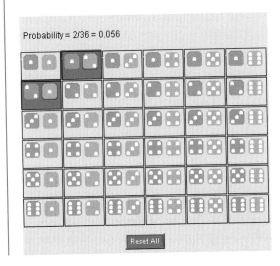

We can extend the *mn* Rule for an experiment that is performed in more than two stages.

EXAMPLE 4.9 — A candy dish contains one yellow and two red candies. Two candies are selected one at a time from the dish, and their colors are recorded. How many simple events are in the sample space *S*?

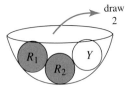

draw 2

Solution The first candy can be chosen in *m* = 3 ways. Since one candy is now gone, the second candy can be chosen in *n* = 2 ways. The total number of simple events is

$$mn = (3)(2) = 6$$

These six simple events were listed in Example 4.7.

We can extend the *mn* Rule for an experiment that is performed in more than two stages.

THE EXTENDED *mn* RULE

If an experiment is performed in *k* stages, with n_1 ways to accomplish the first stage, n_2 ways to accomplish the second stage, . . . , and n_k ways to accomplish the *k*th stage, then the number of ways to accomplish the experiment is

$$n_1 n_2 n_3 \cdots n_k$$

EXAMPLE   4.10

How many simple events are in the sample space when three coins are tossed?

Solution   Each coin can land in one of two ways. Hence, the number of simple events is

$$(2)(2)(2) = 8$$

EXAMPLE 4.11

A truck driver can take three routes from city A to city B, four from city B to city C, and three from city C to city D. If, when traveling from A to D, the driver must drive from A to B to C to D, how many possible A-to-D routes are available?

Solution   Let

$$n_1 = \text{Number of routes from } A \text{ to } B = 3$$
$$n_2 = \text{Number of routes from } B \text{ to } C = 4$$
$$n_3 = \text{Number of routes from } C \text{ to } D = 3$$

Then the total number of ways to construct a complete route, taking one subroute from each of the three groups, (A to B), (B to C), and (C to D), is

$$n_1 n_2 n_3 = (3)(4)(3) = 36$$

A second useful counting rule follows from the *mn* Rule and involves **orderings** or **permutations.** For example, suppose you have three books, A, B, and C, but you have room for only two on your bookshelf. In how many ways can you select and arrange the two books? There are three choices for the two books—A and B, A and C, or B and C—but each of the pairs can be arranged in two ways on the shelf. All the permutations of the two books, chosen from three, are listed in Table 4.3. The *mn* Rule implies that there are 6 ways, because the first book can be chosen in $m = 3$ ways and the second in $n = 2$ ways, so the result is $mn = 6$.

TABLE 4.3 ● **Permutations of Two Books Chosen from Three**

Combinations of Two	Reordering of Combinations
AB	BA
AC	CA
BC	CB

In how many ways can you arrange all three books on your bookshelf? These are the six permutations:

ABC   ACB   BAC
BCA   CAB   CBA

Since the first book can be chosen in $n_1 = 3$ ways, the second in $n_2 = 2$ ways, and the third in $n_3 = 1$ way, the total number of orderings is $n_1 n_2 n_3 = (3)(2)(1) = 6$.

Rather than applying the *mn* Rule each time, you can find the number of orderings using a general formula involving *factorial notation*.

A COUNTING RULE FOR PERMUTATIONS

The number of ways we can arrange n distinct objects, taking them r at a time, is

$$P_r^n = \frac{n!}{(n-r)!}$$

where $n! = n(n-1)(n-2) \cdots (3)(2)(1)$ and $0! = 1$.

Since r objects are chosen, this is an *r-stage* experiment. The first object can be chosen in n ways, the second in $(n-1)$ ways, the third in $(n-2)$ ways, and the rth in $(n-r+1)$ ways. We can simplify this awkward notation using the counting rule for permutations because

$$\frac{n!}{(n-r)!} = \frac{n(n-1)(n-2) \cdots (n-r+1)(n-r) \cdots (2)(1)}{(n-r) \cdots (2)(1)}$$

$$= n(n-1) \cdots (n-r+1)$$

A SPECIAL CASE: ARRANGING n ITEMS

The number of ways to arrange an entire set of n distinct items is $P_n^n = n!$

EXAMPLE 4.12 Three lottery tickets are drawn from a total of 50. If the tickets will be distributed to each of three employees in the order in which they are drawn, the order will be important. How many simple events are associated with the experiment?

Solution The total number of simple events is

$$P_3^{50} = \frac{50!}{47!} = 50(49)(48) = 117,600$$

EXAMPLE 4.13 A piece of equipment is composed of five parts that can be assembled in any order. A test is to be conducted to determine the time necessary for each order of assembly. If each order is to be tested once, how many tests must be conducted?

Solution The total number of tests equals

$$P_5^5 = \frac{5!}{0!} = 5(4)(3)(2)(1) = 120$$

When we counted the number of permutations of the two books chosen for your bookshelf, we used a systematic approach:

- First we counted the number of *combinations* or pairs of books to be chosen.
- Then we counted the number of ways to arrange the two chosen books on the shelf.

Sometimes the ordering or arrangement of the objects is not important, but only the objects that are chosen. In this case, you can use a counting rule for **combinations.** For example, you may not care in what order the books are placed on the shelf, but only which books you are able to shelve. When a five-person committee is chosen from a group of 12 students, the order of choice is unimportant because all five students will be equal members of the committee.

A COUNTING RULE FOR COMBINATIONS

The number of distinct combinations of n distinct objects that can be formed, taking them r at a time, is

$$C_r^n = \frac{n!}{r!(n - r)!}$$

The number of *combinations* and the number of *permutations* are related:

$$C_r^n = \frac{P_r^n}{r!}$$

You can see that C_r^n results when you divide the number of permutations by $r!$, the number of ways of rearranging each distinct group of r objects chosen from the total n.

EXAMPLE 4.14 — A printed circuit board may be purchased from five suppliers. In how many ways can three suppliers be chosen from the five?

Solution Since it is important to know only which three have been chosen, not the order of selection, the number of ways is

$$C_3^5 = \frac{5!}{3!2!} = \frac{(5)(4)}{2} = 10$$

The next example illustrates the use of counting rules to solve a probability problem.

EXAMPLE 4.15 — Five manufacturers produce a certain electronic device, whose quality varies from manufacturer to manufacturer. If you were to select three manufacturers at random, what is the chance that the selection would contain exactly two of the best three?

Solution The simple events in this experiment consist of all possible combinations of three manufacturers, chosen from a group of five. Of these five, three have been designated as "best" and two as "not best." You can think of a candy dish containing three red and two yellow candies, from which you will select three, as illustrated in Figure 4.7. The total number of simple events N can be counted as the number of ways to choose three of the five manufacturers, or

$$N = C_3^5 = \frac{5!}{3!2!} = 10$$

FIGURE 4.7

Illustration for
Example 4.15

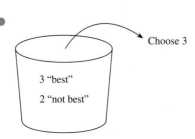

Choose 3

3 "best"

2 "not best"

Since the manufacturers are selected at random, any of these 10 simple events will be *equally likely,* with probability 1/10. But how many of these simple events result in the event

A: Exactly two of the "best" three

You can count n_A, the number of events in A, in two steps because event A will occur when you select two of the "best" three and one of the two "not best." There are

$$C_2^3 = \frac{3!}{2!1!} = 3$$

ways to accomplish the first stage and

$$C_1^2 = \frac{2!}{1!1!} = 2$$

ways to accomplish the second stage. Applying the *mn* Rule, we find there are $n_A = (3)(2) = 6$ of the 10 simple events in event A and $P(A) = n_A/N = 6/10$.

Many other counting rules are available in addition to the three presented in this section. If you are interested in this topic, you should consult one of the many textbooks on combinatorial mathematics.

4.4 EXERCISES

BASIC TECHNIQUES

4.17 You have *two* groups of distinctly different items, 10 in the first group and 8 in the second. If you select one item from each group, how many different pairs can you form?

4.18 You have *three* groups of distinctly different items, four in the first group, seven in the second, and three in the third. If you select one item from each group, how many different triplets can you form?

4.19 Permutations Evaluate the following *permutations.* (HINT: Your scientific calculator may have a function that allows you to calculate permutations and combinations quite easily.)

a. P_3^5 **b.** P_9^{10} **c.** P_6^6 **d.** P_1^{20}

4.20 Combinations Evaluate these *combinations:*

a. C_3^5 **b.** C_9^{10} **c.** C_6^6 **d.** C_1^{20}

4.21 Choosing People In how many ways can you select five people from a group of eight if the order of selection is important?

4.22 Choosing People, again In how many ways can you select two people from a group of 20 if the order of selection is not important?

4.23 Dice Three dice are tossed. How many simple events are in the sample space?

4.24 Coins Four coins are tossed. How many simple events are in the sample space?

4.25 The Urn Problem, again Three balls are selected from a box containing 10 balls. The order of selection is not important. How many simple events are in the sample space?

APPLICATIONS

4.26 What to Wear? You own 4 pairs of jeans, 12 clean T-shirts, and 4 wearable pairs of sneakers. How many outfits (jeans, T-shirt, and sneakers) can you create?

4.27 Itineraries A businessman in New York is preparing an itinerary for a visit to six major cities. The distance traveled, and hence the cost of the trip, will depend on the order in which he plans his route. How many different itineraries (and trip costs) are possible?

4.28 Vacation Plans Your family vacation involves a cross-country air flight, a rental car, and a hotel stay in Boston. If you can choose from four major air carriers, five car rental agencies, and three major hotel chains, how many options are available for your vacation accommodations?

4.29 A Card Game Three students are playing a card game. They decide to choose the first person to play by each selecting a card from the 52-card deck and looking for the highest card in value and suit. They rank the suits from lowest to highest: clubs, diamonds, hearts, and spades.

a. If the card is replaced in the deck after each student chooses, how many possible configurations of the three choices are possible?

b. How many configurations are there in which each student picks a different card?

c. What is the probability that all three students pick exactly the same card?

d. What is the probability that all three students pick different cards?

4.30 Dinner at Gerards A French restaurant in Riverside, California, offers a special summer menu in which, for a fixed dinner cost, you can choose from one of two salads, one of two entrees, and one of two desserts. How many different dinners are available?

4.31 Playing Poker Five cards are selected from a 52-card deck for a poker hand.

a. How many simple events are in the sample space?

b. A *royal flush* is a hand that contains the A, K, Q, J, and 10, all in the same suit. How many ways are there to get a royal flush?

c. What is the probability of being dealt a royal flush?

4.32 Poker II Refer to Exercise 4.31. You have a poker hand containing four of a kind.

a. How many possible poker hands can be dealt?

b. In how many ways can you receive four cards of the same face value *and* one card from the other 48 available cards?

c. What is the probability of being dealt four of a kind?

4.33 A Hospital Survey A study is to be conducted in a hospital to determine the attitudes of nurses toward various administrative procedures. If a sample of 10 nurses is to be selected from a total of 90, how many different samples can be selected? (HINT: Is order important in determining the makeup of the sample to be selected for the survey?)

4.34 Traffic Problems Two city council members are to be selected from a total of five to form a subcommittee to study the city's traffic problems.

a. How many different subcommittees are possible?

b. If all possible council members have an equal chance of being selected, what is the probability that members Smith and Jones are both selected?

4.35 The WNBA Professional basketball is now a reality for women basketball players in the United States. There are two conferences in the WNBA, one with six and one with seven teams, as shown in the table:

Western Conference	Eastern Conference
Houston Comets	Indiana Fever
Minnesota Lynx	New York Liberty
Phoenix Mercury	Washington Mystics
Sacramento Monarchs	Detroit Shock
Los Angeles Sparks	Charlotte Sting
Seattle Storm	Connecticut Sun
San Antonio Silver Stars	

Two teams, one from each conference, are randomly selected to play an exhibition game.

a. How many pairs of teams can be chosen?

b. What is the probability that the two teams are Los Angeles and New York?

c. What is the probability that the Western Conference team is from California?

4.36 100-Meter Run, again Refer to Exercise 4.14, in which a 100-meter sprint is run by John, Bill, Ed, and Dave. Assume that all of the runners are equally qualified, so that any order of finish is equally likely. Use the *mn* Rule or permutations to answer these questions:

a. How many orders of finish are possible?

b. What is the probability that Dave wins the sprint?

c. What is the probability that Dave wins and John places second?

d. What is the probability that Ed finishes last?

4.37 Gender Bias? The following case occurred in Gainesville, Florida. The eight-member Human Relations Advisory Board considered the complaint of a woman who claimed discrimination, based on her gender, on the part of a local surveying company. The board, composed of five women and three men, voted 5–3 in favor of the plaintiff, the five women voting for the plaintiff and the three men against. The attorney representing the company appealed the board's decision by claiming gender bias on the part of the board members. If the vote in favor of the plaintiff was 5–3 and the board members were not biased by gender, what is the probability that the vote would split along gender lines (five women for, three men against)?

4.38 Cramming A student prepares for an exam by studying a list of 10 problems. She can solve 6 of them. For the exam, the instructor selects 5 questions at random from the list of 10. What is the probability that the student can solve all 5 problems on the exam?

4.39 Monkey Business A monkey is given 12 blocks: 3 shaped like squares, 3 like rectangles, 3 like triangles, and 3 like circles. If it draws three of each kind in order—say, 3 triangles, then 3 squares, and so on—would you suspect that the monkey associates identically shaped figures? Calculate the probability of this event.

EVENT RELATIONS AND PROBABILITY RULES

4.5

Sometimes the event of interest can be formed as a combination of several other events. Let A and B be two events defined on the sample space S. Here are three important relationships between events.

Definition   The **union** of events A and B, denoted by $A \cup B$, is the event that either A or B or both occur.

Definition   The **intersection** of events A and B, denoted by $A \cap B$, is the event that both A and B occur.[†]

Definition   The **complement** of an event A, denoted by A^c, is the event that A *does not* occur.

Figures 4.8, 4.9, and 4.10 show Venn diagram representations of $A \cup B$, $A \cap B$, and A^c, respectively. Any simple event in the shaded area is a possible outcome resulting in the appropriate event. One way to find the probabilities of the union, the intersection, or the complement is to sum the probabilities of all the associated simple events.

[†]Some authors use the notation AB.

FIGURE 4.8 ●

Venn Diagram of $A \cup B$

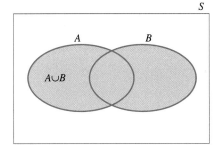

FIGURE 4.9 ●

Venn Diagram $A \cap B$

 TIP

Intersection ⇔ "both . . . and" or just "and"

Union ⇔ "either . . . or . . . or both" or just "or"

FIGURE 4.10 ●

The Complement of an Event

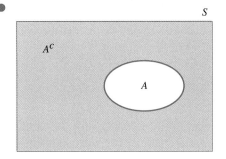

EXAMPLE 4.16

Two fair coins are tossed, and the outcome is recorded. These are the events of interest:

  A: Observe at least one head

  B: Observe at least one tail

Define the events A, B, $A \cap B$, $A \cup B$, and A^c as collections of simple events, and find their probabilities.

Solution Recall from Example 4.5 that the simple events for this experiment are:

  E_1: HH (head on first coin, head on second)

  E_2: HT

  E_3: TH

  E_4: TT

and that each simple event has probability 1/4. Event A, at least one head, occurs if E_1, E_2, or E_3 occurs, so that

$$A = \{E_1, E_2, E_3\} \qquad P(A) = \frac{3}{4}$$

and

$$A^c = \{E_4\} \qquad P(A^c) = \frac{1}{4}$$

Similarly,

$$B = \{E_2, E_3, E_4\} \qquad\qquad P(B) = \frac{3}{4}$$

$$A \cap B = \{E_2, E_3\} \qquad\qquad P(A \cap B) = \frac{1}{2}$$

$$A \cup B = \{E_1, E_2, E_3, E_4\} \qquad P(A \cup B) = \frac{4}{4} = 1$$

Note that $(A \cup B) = S$, the sample space, and is thus certain to occur.

The concept of unions and intersections can be extended to more than two events. For example, the union of three events A, B, and C, which is written as $A \cup B \cup C$, is the set of simple events that are in A or B or C or in any combination of those events. Similarly, the intersection of three events A, B, and C, which is written as $A \cap B \cap C$, is the collection of simple events that are common to the three events A, B, and C.

Calculating Probabilities for Unions and Complements

When we can write the event of interest in the form of a union, a complement, or an intersection, there are special probability rules that can simplify our calculations. The first rule deals with *unions* of events.

THE ADDITION RULE

Given two events, A and B, the probability of their union, $A \cup B$, is equal to

$$P(A \cup B) = P(A) + P(B) - P(A \cap B)$$

Notice in the Venn diagram in Figure 4.11 that the sum $P(A) + P(B)$ double counts the simple events that are common to both A and B. Subtracting $P(A \cap B)$ gives the correct result.

FIGURE 4.11

The Addition Rule

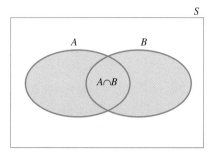

When two events A and B are **mutually exclusive** or **disjoint,** it means that when A occurs, B cannot, and vice versa. This means that the probability that they both occur, $P(A \cap B)$, must be zero. Figure 4.12 is a Venn diagram representation of two such events with no simple events in common.

FIGURE 4.12

Two Disjoint Events

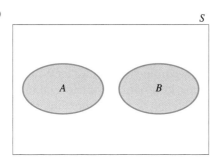

When two events A and B are **mutually exclusive,** then $P(A \cap B) = 0$ and the Addition Rule simplifies to

$$P(A \cup B) = P(A) + P(B)$$

 MY TIP

Remember, mutually exclusive $\Leftrightarrow P(A \cap B)$ $= 0$

The second rule deals with *complements* of events. You can see from the Venn diagram in Figure 4.10 that A and A^c are mutually exclusive and that $A \cup A^c = S$, the entire sample space. It follows that

$$P(A) + P(A^c) = 1 \text{ and } P(A^c) = 1 - P(A)$$

RULE FOR COMPLEMENTS

$$P(A^c) = 1 - P(A)$$

EXAMPLE 4.17   An oil-prospecting firm plans to drill two exploratory wells. Past evidence is used to assess the possible outcomes listed in Table 4.4.

TABLE 4.4   **Outcomes for Oil-Drilling Experiment**

Event	Description	Probability
A	Neither well produces oil or gas	.80
B	Exactly one well produces oil or gas	.18
C	Both wells produce oil or gas	.02

Find $P(A \cup B)$ and $P(B \cup C)$.

Solution   By their definition, events A, B, and C are jointly mutually exclusive because the occurrence of one event precludes the occurrence of either of the other two. Therefore,

$$P(A \cup B) = P(A) + P(B) = .80 + .18 = .98$$

and

$$P(B \cup C) = P(B) + P(C) = .18 + .02 = .20$$

The event $A \cup B$ can be described as the event that *at most* one well produces oil or gas, and $B \cup C$ describes the event that *at least* one well produces gas or oil.

EXAMPLE 4.18   In a telephone survey of 1000 adults, respondents were asked about the expense of a college education and the relative necessity of some form of financial assistance. The respondents were classified according to whether they currently had a child in college and whether they thought the loan burden for most college students is too high, the right amount, or too little. The proportions responding in each category are shown in the **probability table** in Table 4.5. Suppose one respondent is chosen at random from this group.

TABLE 4.5   ●   **Probability Table**

	Too High (A)	Right Amount (B)	Too Little (C)
Child in College (D)	.35	.08	.01
No Child in College (E)	.25	.20	.11

1. What is the probability that the respondent has a child in college?
2. What is the probability that the respondent does not have a child in college?
3. What is the probability that the respondent has a child in college or thinks that the loan burden is too high?

Solution   Table 4.5 gives the probabilities for the six simple events in the cells of the table. For example, the entry in the top left corner of the table is the probability that a respondent has a child in college *and* thinks the loan burden is too high $(A \cap D)$.

1. The event that a respondent has a child in college will occur regardless of his or her response to the question about loan burden. That is, event D consists of the simple events in the first row:

$$P(D) = .35 + .08 + .01 = .44$$

   In general, the probabilities of *marginal* events such as D and A are found by summing the probabilities in the appropriate row or column.

2. The event that the respondent does not have a child in college is the complement of the event D denoted by D^c. The probability of D^c is found as

$$P(D^c) = 1 - P(D)$$

   Using the result of part 1, we have

$$P(D^c) = 1 - .44 = .56$$

3. The event of interest is $P(A \cup D)$. Using the addition rule

$$P(A \cup D) = P(A) + P(D) - P(A \cap D)$$
$$= .60 + .44 - .35$$
$$= .69$$

INDEPENDENCE, CONDITIONAL PROBABILITY, AND THE MULTIPLICATION RULE

4.6

There is a probability rule that can be used to calculate the probability of the intersection of several events. However, this rule depends on the important statistical concept of **independent** or **dependent events.**

Definition Two events, *A* and *B*, are said to be **independent** if and only if the probability of event *B* is not influenced or changed by the occurrence of event *A*, or vice versa.

Colorblindness Suppose a researcher notes a person's gender and whether or not the person is colorblind to red and green. Does the probability that a person is colorblind change depending on whether the person is male or not? Define two events:

   A: person is a male

   B: person is colorblind

In this case, since colorblindness is a male sex-linked characteristic, the probability that a man is colorblind will be greater than the probability that a person chosen from the general population will be colorblind. The probability of event *B*, that a person is colorblind, depends on whether or not event *A*, that the person is a male, has occurred. We say that *A* and *B* are *dependent events.*

Tossing Dice On the other hand, consider tossing a single die two times, and define two events:

   A: observe a 2 on the first toss

   B: observe a 2 on the second toss

If the die is fair, the probability of event *A* is $P(A) = 1/6$. Consider the probability of event *B*. Regardless of whether event *A* has or has not occurred, the probability of observing a 2 on the second toss is still 1/6. We could write:

   P(*B* given that *A* occurred) = 1/6
   P*B* given that *A* did not occur) = 1/6

Since the probability of event *B* is not changed by the occurrence of event *A*, we say that *A* and *B* are *independent events.*

   The probability of an event *A*, given that the event *B* has occurred, is called the **conditional probability of A, given that B has occurred,** denoted by $P(A|B)$. The vertical bar is read "given" and the events appearing to the right of the bar are those that you know have occurred. We will use these probabilities to calculate the probability that *both A and B* occur when the experiment is performed.

THE GENERAL MULTIPLICATION RULE

The probability that *both A and B* occur when the experiment is performed is

   $$P(A \cap B) = P(A)P(B|A)$$

or

   $$P(A \cap B) = P(B)P(A|B)$$

EXAMPLE (4.19)

In a color preference experiment, eight toys are placed in a container. The toys are identical except for color—two are red, and six are green. A child is asked to choose two toys *at random.* What is the probability that the child chooses the two red toys?

Solution   You can visualize the experiment using a tree diagram as shown in Figure 4.13. Define the following events:

  R: Red toy is chosen

  G: Green toy is chosen

FIGURE 4.13

Tree Diagram for
Example 4.19

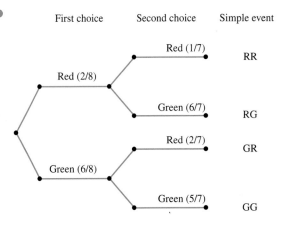

The event *A* (both toys are red) can be constructed as the intersection of two events:

  A = (R on first choice) ∩ (R on second choice)

Since there are only two red toys in the container, the probability of choosing red on the first choise is 2/8. However, once this red toy has been chosen, the probability of red on the second choice is *dependent* on the outcome of the first choice (see Figure 4.13). If the first choice was red, the probability of choosing a second red toy is only 1/7 because there is only one red toy among the seven remaining. If the first choice was green, the probability of choosing red on the second choice is 2/7 because there are two red toys among the seven remaining. Using this information and the Multiplication Rule, you can find the probability of event *A.*

$$P(A) = P(\text{R on first choice} \cap \text{R on second choice})$$
$$= P(\text{R on first choice}) \, P(\text{R on second choice})|\text{R on first})$$
$$= \left(\frac{2}{8}\right)\left(\frac{1}{7}\right) = \frac{2}{56} = \frac{1}{28}$$

Sometimes you may need to use the Multiplication Rule in a slightly different form, so that you can calculate the **conditional probability, *P(A|B).*** Just rearrange the terms in the Multiplication Rule.

CONDITIONAL PROBABILITIES

The conditional probability of event A, given that event B has occurred is

$$P(A|B) = \frac{P(A \cap B)}{P(B)} \quad \text{if} \quad P(B) \neq 0$$

The conditional probability of event B, given that event A has occurred is

$$P(B|A) = \frac{P(A \cap B)}{P(A)} \quad \text{if} \quad P(A) \neq 0$$

Colorblindness, continued   Suppose that in the general population, there are 51% men and 49% women, and that the proportions of colorblind men and women are shown in the probability table below:

	Men(B)	Women (B^C)	Total
Colorblind (A)	.04	.002	.042
Not Colorblind (A^C)	.47	.488	.958
Total	.51	.49	1.00

If a person is drawn at random from this population and is found to be a man (event B), what is the probability that the man is colorblind (event A)? If we know that the event B has occurred, we must restrict our focus to only the 51% of the population that is male. The probability of being colorblind, given that the person is male, is 4% of the 51% or

$$P(A|B) = \frac{P(A \cap B)}{P(B)} = \frac{.04}{.51} = .078$$

What is the probability of being colorblind, given that the person is female? Now we are restricted to only the 49% of the population is female, and

$$P(A|B^C) = \frac{P(A \cap B^C)}{P(B^C)} = \frac{.002}{.49} = .004$$

Notice that the probability of event A changed, depending on whether event B occurred. This indicates that these two events are *dependent*.

When two events are **independent**—that is, if the probability of event B is the same, whether or not event A has occurred, then event A does not affect event B and

$$P(B|A) = P(B)$$

The Multiplication Rule can now be simplified.

THE MULTIPLICATION RULE FOR INDEPENDENT EVENTS

If two events A and B are independent, the probability that *both A and B* occur is

$$P(A \cap B) = P(A)P(B)$$

Similarly, if A, B, and C are mutually independent events (all pairs of events are independent), then the probability that A, B, and C all occur is

$$P(A \cap B \cap C) = P(A)P(B)P(C)$$

Coin Tosses at Football Games A football team is involved in two overtime periods during a given game, so that there are three coins tosses. If the coin is fair, what is the probability that they lose all three tosses?

Solution If the coin is fair, the event can be described in three steps:

A: lose the first toss

B: lose the second toss

C: lose the third toss

Since the tosses are independent, and since P(win) = P(lose) = .5 for any of the three tosses,

$$P(A \cap B \cap C) = P(A)P(B)P(C) = (.5)(.5)(.5) = .125$$

How can you check to see if two events are independent or dependent? The easiest solution is to redefine the concept of **independence** in a more formal way.

CHECKING FOR INDEPENDENCE

Two events A and B are said to be **independent** if and only if either

$$P(A \cap B) = P(A)P(B)$$

or

$$P(B|A) = P(B)$$

Otherwise, the events are said to be **dependent.**

EXAMPLE 4.20

Toss two coins and observe the outcome. Define these events:

A: Head on the first coin

B: Tail on the second coin

Are events A and B independent?

Solution From previous examples, you know that S = {HH, HT, TH, TT}. Use these four simple events to find

MY TIP

Remember,
independence ⇔
P(A ∩ B) = P(A)P(B)

$$P(A) = \frac{1}{2}, P(B) = \frac{1}{2}, \text{ and } P(A \cap B) = \frac{1}{4}.$$

Since $P(A)P(B) = \left(\frac{1}{2}\right)\left(\frac{1}{2}\right) = \frac{1}{4}$ and $P(A \cap B) = \frac{1}{4}$, we have $P(A)P(B) = P(A \cap B)$

and the two events must be independent.

EXAMPLE 4.21

Refer to the probability table in Example 4.18, which is reproduced below.

	Too High (A)	Right Amount (B)	Too Little (C)
Child in College (D)	.35	.08	.01
No Child in College (E)	.25	.20	.11

Are events D and A independent? Explain.

Solution

1. Use the probability table to find $P(A \cap D) = .35$, $P(A) = .60$, and $P(D) = .44$. Then

   $$P(A)P(D) = (.60)(.44) = .264 \text{ and } P(A \cap D) = .35$$

   Since these two probabilities are not the same, events A and D are *dependent*.

2. Alternately, calculate

   $$P(A|D) = \frac{P(A \cap D)}{P(D)} = \frac{.35}{.44} = .80$$

   Since $P(A|D) = .80$ and $P(A) = .60$, we are again led to the conclusion that events A and D are *dependent*.

 PERSONAL TRAINER

What's the Difference between Mutually Exclusive and Independent Events?

Many students find it hard to tell the difference between *mutually exclusive* and *independent* events.

- When two events are *mutually exclusive* or *disjoint*, they cannot both happen when the experiment is performed. Once the event B has occurred, event A cannot occur, so that $P(A|B) = 0$, or vice versa. The occurrence of event B certainly affects the probability that event A can occur.

- Therefore, mutually exclusive events must be *dependent*.

- When two events are *mutually exclusive* or *disjoint*, $P(A \cap B) = 0$ and $P(A \cup B) = P(A) + P(B)$.

- When two events are *independent*, $P(A \cap B) = P(A)P(B)$, and $P(A \cup B) = P(A) + P(B) - P(A)P(B)$.

Exercise Reps

Use the relationships above to find probabilities in the table below.

P(A)	P(B)	Conditions for events A and B	P(A ∩ B)	P(A ∪ B)	P(A|B)
.3	.4	Mutually exclusive			
.3	.4	Independent			
.1	.5			.6	
.2	.5		.10		

Answers are located on the perforated card at the back of this book.

Using probability rules to calculate the probability of an event requires some experience and ingenuity. You need to express the event of interest as a union or intersection (or the combination of both) of two or more events whose probabilities are known or easily calculated. Often you can do this in different ways; the key is to find the right combination.

EXAMPLE 4.22 Two cards are drawn from a deck of 52 cards. Calculate the probability that the draw includes an ace and a ten.

Solution Consider the event of interest:

$$A\text{: Draw an ace and a ten}$$

Then $A = B \cup C$, where

B: Draw the ace on the first draw and the ten on the second
C: Draw the ten on the first draw and the ace on the second

Events B and C were chosen to be mutually exclusive and also to be intersections of events with known probabilities; that is,

$$B = B_1 \cap B_2 \text{ and } C = C_1 \cap C_2$$

where

B_1: Draw an ace on the first draw
B_2: Draw a ten on the second draw
C_1: Draw a ten on the first draw
C_2: Draw an ace on the second draw

Applying the Multiplication Rule, you get

$$P(B_1 \cap B_2) = P(B_1)P(B_2|B_1)$$

$$= \left(\frac{4}{52}\right)\left(\frac{4}{51}\right)$$

and

$$P(C_1 \cap C_2) = \left(\frac{4}{52}\right)\left(\frac{4}{51}\right)$$

Then, applying the Addition Rule,

$$P(A) = P(B) + P(C)$$

$$= \left(\frac{4}{52}\right)\left(\frac{4}{51}\right) + \left(\frac{4}{52}\right)\left(\frac{4}{51}\right) = \frac{8}{663}$$

Check each composition carefully to be certain that it is actually equal to the event of interest.

4.6 EXERCISES

EXERCISE REPS

4.40 Use the relationships above to find probabilities in the table below.

P(A)	P(B)	Conditions for events A and B	P(A ∩ B)	P(A ∪ B)	P(A|B)
.3	.4		.12		
.3	.4			.7	
.1	.5	Mutually exclusive			
.2	.5	Independent			

4.41 Use the relationships above to find probabilities in the table below.

$P(A)$	$P(B)$	Conditions for events A and B	$P(A \cap B)$	$P(A \cup B)$	$P(A\|B)$
.3	.4	Mutually exclusive			
.3	.4	Independent			
.1	.5				.1
.2	.5		0		

BASIC TECHNIQUES

4.42 An experiment can result in one of five equally likely simple events, $E_1, E_2, \ldots, E_5$. Events A, B, and C are defined as follows:

$A: E_1, E_3$ $P(A) = .4$

$B: E_1, E_2, E_4, E_5$ $P(B) = .8$

$C: E_3, E_4$ $P(C) = .4$

Find the probabilities associated with these compound events by listing the simple events in each.

a. A^c **b.** $A \cap B$ **c.** $B \cap C$

d. $A \cup B$ **e.** $B\|C$ **f.** $A\|B$

g. $A \cup B \cup C$ **h.** $(A \cap B)^c$

4.43 Refer to Exercise 4.42. Use the definition of a complementary event to find these probabilities:

a. $P(A^c)$ **b.** $P((A \cap B)^c)$

Do the results agree with those obtained in Exercise 4.42?

4.44 Refer to Exercise 4.42. Use the definition of conditional probability to find these probabilities:

a. $P(A\|B)$ **b.** $P(B\|C)$

Do the results agree with those obtained in Exercise 4.42?

4.45 Refer to Exercise 4.42. Use the Addition and Multiplication Rules to find these probabilities:

a. $P(A \cup B)$ **b.** $P(A \cap B)$ **c.** $P(B \cap C)$

Do the results agree with those obtained in Exercise 4.42?

4.46 Refer to Exercise 4.42.

a. Are events A and B independent?

b. Are events A and B mutually exclusive?

4.47 Dice An experiment consists of tossing a single die and observing the number of dots that show on the upper face. Events A, B, and C are defined as follows:

A: Observe a number less than 4

B: Observe a number less than or equal to 2

C: Observe a number greater than 3

Find the probabilities associated with the events below using either the simple event approach or the rules and definitions from this section.

a. S **b.** $A\|B$ **c.** B

d. $A \cap B \cap C$ **e.** $A \cap B$ **f.** $A \cap C$

g. $B \cap C$ **h.** $A \cup C$ **i.** $B \cup C$

4.48 Refer to Exercise 4.47.

a. Are events A and B independent? Mutually exclusive?

b. Are events A and C independent? Mutually exclusive?

4.49 Suppose that $P(A) = .4$ and $P(B) = .2$. If events A and B are independent, find these probabilities:

a. $P(A \cap B)$ **b.** $P(A \cup B)$

4.50 Suppose that $P(A) = .3$ and $P(B) = .5$. If events A and B are mutually exclusive, find these probabilities:

a. $P(A \cap B)$ **b.** $P(A \cup B)$

4.51 Suppose that $P(A) = .4$ and $P(A \cap B) = .12$.

a. Find $P(B\|A)$.

b. Are events A and B mutually exclusive?

c. If $P(B) = .3$, are events A and B independent?

4.52 An experiment can result in one or both of events A and B with the probabilities shown in this probability table:

	A	A^c
B	.34	.46
B^c	.15	.05

Find the following probabilities:

a. $P(A)$ **b.** $P(B)$ **c.** $P(A \cap B)$

d. $P(A \cup B)$ **e.** $P(A\|B)$ **f.** $P(B\|A)$

4.53 Refer to Exercise 4.52.

a. Are events A and B mutually exclusive? Explain.

b. Are events A and B independent? Explain.

APPLICATIONS

4.54 Drug Testing Many companies are testing prospective employees for drug use, with the intent of improving efficiency and reducing absenteeism, accidents, and theft. Opponents claim that this procedure is creating a class of unhirables and that some persons may be placed in this class because the tests themselves are not 100% reliable. Suppose a company uses a test that is 98% accurate—that is, it correctly identifies a person as a drug user or nonuser with probability .98—and to reduce the chance of error, each job applicant is required to take two tests. If the outcomes of the two tests on the same person are independent events, what are the probabilities of these events?

a. A nonuser fails both tests.

b. A drug user is detected (i.e., he or she fails at least one test).

c. A drug user passes both tests.

4.55 Grant Funding Whether a grant proposal is funded quite often depends on the reviewers. Suppose a group of research proposals was evaluated by a group of experts as to whether the proposals were worthy of funding. When these same proposals were submitted to a second independent group of experts, the decision to fund was reversed in 30% of the cases. If the probability that a proposal is judged worthy of funding by the first peer review group is .2, what are the probabilities of these events?

a. A worthy proposal is approved by both groups.

b. A worthy proposal is disapproved by both groups.

c. A worthy proposal is approved by one group.

4.56 Drug Offenders A study of the behavior of a large number of drug offenders after treatment for drug abuse suggests that the likelihood of conviction within a 2-year period after treatment may depend on the offender's education. The proportions of the total number of cases that fall into four education/conviction categories are shown in the table below:

	Status Within 2 Years After Treatment		
Education	Convicted	Not Convicted	Totals
10 Years or More	.10	.30	.40
9 Years or Less	.27	.33	.60
Totals	.37	.63	1.00

Suppose a single offender is selected from the treatment program. Here are the events of interest:

A: The offender has 10 or more years of education

B: The offender is convicted within 2 years after completion of treatment

Find the appropriate probabilities for these events:

a. A **b.** B **c.** $A \cap B$

d. $A \cup B$ **e.** A^c **f.** $(A \cup B)^c$

g. $(A \cap B)^c$ **h.** A given that B has occurred

i. B given that A has occurred

4.57 Use the probabilities of Exercise 4.56 to show that these equalities are true:

a. $P(A \cap B) = P(A)P(B|A)$

b. $P(A \cap B) = P(B)P(A|B)$

c. $P(A \cup B) = P(A) + P(B) - P(A \cap B)$

4.58 The Birthday Problem Two people enter a room and their birthdays (ignoring years) are recorded.

a. Identify the nature of the simple events in S.

b. What is the probability that the two people have a specific pair of birthdates?

c. Identify the simple events in event A: Both people have the same birthday.

d. Find $P(A)$.   **e.** Find $P(A^c)$.

4.59 The Birthday Problem, continued If n people enter a room, find these probabilities:

A: None of the people have the same birthday

B: At least two of the people have the same birthday

Solve for

a. $n = 3$ **b.** $n = 4$

[NOTE: Surprisingly, $P(B)$ increases rapidly as n increases. For example, for $n = 20$, $P(B) = .411$; for $n = 40$, $P(B) = .891$.]

4.60 Starbucks or Peetes? A college student frequents one of two coffee houses on campus, choosing Starbucks 70% of the time and Peetes 30% of the time. Regardless of where she goes, she buys a cafe mocha on 60% of her visits.

a. The next time she goes into a coffee house on campus, what is the probability that she goes to Starbucks and orders a cafe mocha?

b. Are the two events in part a independent? Explain.

c. If she goes into a coffee house and orders a cafe mocha, what is the probability that she is at Peetes?

d. What is the probability that she goes to Starbucks or orders a cafe mocha or both?

4.61 Inspection Lines A certain manufactured item is visually inspected by two different inspectors. When a defective item comes through the line, the probability that it gets by the first inspector is .1. Of those that get past the first inspector, the second inspector will "miss" 5 out of 10. What fraction of the defective items get by both inspectors?

4.62 Smoking and Cancer A survey of people in a given region showed that 20% were smokers. The probability of death due to lung cancer, given that a person smoked, was roughly 10 times the probability of death due to lung cancer, given that a person did not smoke. If the probability of death due to lung cancer in the region is .006, what is the probability of death due to lung cancer given that a person is a smoker?

4.63 Smoke Detectors A smoke-detector system uses two devices, A and B. If smoke is present, the probability that it will be detected by device A is .95; by device B, .98; and by both devices, .94.

a. If smoke is present, find the probability that the smoke will be detected by device A or device B or both devices.

b. Find the probability that the smoke will not be detected.

4.64 Plant Genetics Gregor Mendel was a monk who suggested in 1865 a theory of inheritance based on the science of genetics. He identified heterozygous individuals for flower color that had two alleles (one r = recessive white color allele and one R = dominant red color allele). When these individuals were mated, 3/4 of the offspring were observed to have red flowers and 1/4 had white flowers. The table summarizes this mating; each parent gives one of its alleles to form the gene of the offspring.

	Parent 2	
Parent 1	r	R
r	rr	rR
R	Rr	RR

We assume that each parent is equally likely to give either of the two alleles and that, if either one or two of the alleles in a pair is dominant (R), the offspring will have red flowers.

a. What is the probability that an offspring in this mating has at least one dominant allele?

b. What is the probability that an offspring has at least one recessive allele?

c. What is the probability that an offspring has one recessive allele, given that the offspring has red flowers?

4.65 Soccer Injuries During the inaugural season of Major League Soccer in the United States, the medical teams documented 256 injuries that caused a loss of participation time to the player. The results of this investigation, reported in *The American Journal of Sports Medicine,* is shown in the table.[3]

Severity	Practice (P)	Game (G)	Total
Minor (A)	66	88	154
Moderate (B)	23	44	67
Major (C)	12	23	35
Total	101	155	256

If one individual is drawn at random from this group of 256 soccer players, find the following probabilities:

a. $P(A)$ **b.** $P(G)$ **c.** $P(A \cap G)$
d. $P(G|A)$ **e.** $P(G|B)$ **f.** $P(G|C)$
g. $P(C|P)$ **h.** $P(B^c)$

4.66 Choosing a Mate Men and women often disagree on how they think about selecting a mate. Suppose that a poll of 1000 individuals in their twenties gave the following responses to the question of whether it is more important for their future mate to be able to communicate their feelings (F) than it is for that person to make a good living (G).

	Feelings (F)	Good Living (G)	Totals
Men (M)	.35	.20	.55
Women (W)	.36	.09	.45
Totals	.71	.29	1.00

If an individual is selected at random from this group of 1000 individuals, calculate the following probabilities:

a. $P(F)$ **b.** $P(G)$ **c.** $P(F|M)$
d. $P(F|W)$ **e.** $P(M|F)$ **f.** $P(W|G)$

4.67 Kobe and Shaq The two stars of the 2003–04 Los Angeles Lakers are very different when it comes to making free throws. ESPN.com reports that Kobe Bryant makes about 85% of his free throws, while Shaquille O'Neal makes only 50% of his free throws.[4] Assume that the free throws are independent, and that each player takes two free throws during a particular game.

a. What is the probability that Kobe makes both of his free throws?

b. What is the probability that Shaq makes exactly one of his two free throws?

c. What is the probability that Shaq makes both of his free throws, and Kobe makes neither of his?

4.68 Golfing Player A has entered a golf tournament but it is not certain whether player B will enter.

Player A has probability 1/6 of winning the tournament if player B enters and probability 3/4 of winning if player B does not enter the tournament. If the probability that player B enters is 1/3, find the probability that player A wins the tournament.

BAYES' RULE (OPTIONAL)

Colorblindness Let us reconsider the experiment involving colorblindness from Section 4.6. Notice that the two events:

B: the person selected is a man

B^C: the person selected is a woman

taken together make up the sample space S, consisting of both men and women. Since colorblind people can be either male or female, the event A, which is that a person is colorblind, consists of both those simple events that are in A **and** B and those simple events that are in A **and** B^C. Since these two *intersections* are *mutually exclusive*, you can write the event A as

$$A = (A \cap B) \cup (A \cap B^c)$$

and

$$P(A) = P(A \cap B) + P(A \cap B^c)$$
$$= .04 + .002 = .042$$

Suppose now that the sample space can be partitioned into k subpopulations, $S_1, S_2, S_3, \ldots, S_k$, that, as in the colorblindness example, are **mutually exclusive and exhaustive;** that is, taken together they make up the entire sample space. In a similar way, you can express an event A as

$$A = (A \cap S_1) \cup (A \cap S_2) \cup (A \cap S_3) \cup \cdots \cup (A \cap S_k)$$

Then

$$P(A) = P(A \cap S_1) + P(A \cap S_2) + P(A \cap S_3) + \cdots + P(A \cap S_k)$$

This is illustrated for $k = 3$ in Figure 4.14.

FIGURE 4.14

Decomposition of event A

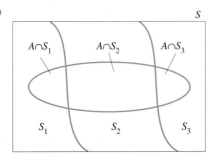

You can go one step further and use the Multiplication Rule to write $P(A \cap S_i)$ as $P(S_i)P(A|S_i)$, for $i = 1, 2, \ldots, k$. The result is known as the **Law of Total Probability.**

LAW OF TOTAL PROBABILITY

Given a set of events $S_1, S_2, S_3, \ldots, S_k$ that are mutually exclusive and exhaustive and an event A, the probability of the event A can be expressed as

$$P(A) = P(S_1)P(A|S_1) + P(S_2)P(A|S_2) + P(S_3)P(A|S_3) + \cdots + P(S_k)P(A|S_k)$$

EXAMPLE 4.23

Sneakers are no longer just for the young. In fact, most adults own multiple pairs of sneakers. Table 4.6 gives the fraction of U.S. adults 20 years of age and older who own five or more pairs of wearable sneakers, along with the fraction of the U.S. adult population 20 years or older in each of five age groups.[5] Use the Law of Total Probability to determine the unconditional probability of an adult 20 years and older owning five or more pairs of wearable sneakers.

TABLE 4.6

Probability Table

	Groups and Ages				
	G_1 20–24	G_2 25–34	G_3 35–49	G_4 50–64	G_5 ≥ 65
Fraction with ≥ 5 Pairs	.26	.20	.13	.18	.14
Fraction of U.S. Adults 20 and Older	.09	.20	.31	.23	.17

Solution Let A be the event that a person chosen at random from the U.S. adult population 18 years of age and older owns five or more pairs of wearable sneakers. Let $G_1, G_2, \ldots, G_5$ represent the event that the person selected belongs to each of the five age groups, respectively. Since the five groups are *exhaustive,* you can write the event A as

$$A = (A \cap G_1) \cup (A \cap G_2) \cup (A \cap G_3) \cup (A \cap G_4) \cup (A \cap G_5)$$

Using the Law of Total Probability, you can find the probability of A as

$$P(A) = P(A \cap G_1) + P(A \cap G_2) + P(A \cap G_3) + P(A \cap G_4) + P(A \cap G_5)$$
$$= P(G_1)P(A|G_1) + P(G_2)P(A|G_2) + P(G_3)P(A|G_3)$$
$$+ P(G_4)P(A|G_4) + P(G_5)P(A|G_5)$$

From the probabilities in Table 4.6,

$$P(A) = (.09)(.26) + (.20)(.20) + (.31)(.13) + (.23)(.18) + (.17)(.14)$$
$$= .0234 + .0400 + .0403 + .0414 + .0238 = .1689$$

The *unconditional probability* that a person selected at random from the population of U.S. adults 20 years of age and older owns at least five pairs of wearable sneakers is about .17. Notice that the Law of Total Probability is a weighted average of the probabilities within each group, with weights .09, .20, .31, .23, and .17, which reflect the relative sizes of the groups.

Often you need to find the conditional probability of an event B, given that an event A has occurred. One such situation occurs in screening tests, which used to be associated primarily with medical diagnostic tests but are now finding applications in a variety of fields. Automatic test equipment is routinely used to inspect parts in high-volume production processes. Steroid testing of athletes, home pregnancy tests, and AIDS testing are some other applications. Screening tests are evaluated on the probability of a false negative or a false positive, and both of these are *conditional probabilities.*

A **false positive** is the event that the test is positive for a given condition, given that the person does not have the condition. A **false negative** is the event that the test is negative for a given condition, given that the person has the condition. You can evaluate these conditional probabilities using a formula derived by the probabilist Thomas Bayes.

The experiment involves selecting a sample from one of k subpopulations that are mutually exclusive and exhaustive. Each of these subpopulations, denoted by $S_1, S_2, \ldots, S_k$, has a selection probability $P(S_1), P(S_2), P(S_3), \ldots, P(S_k)$, called *prior probabilities.* An event A is observed in the selection. What is the probability that the sample came from subpopulation S_i, given that A has occurred?

You know from Section 4.6 that $P(S_i|A) = [P(A \cap S_i)]/P(A)$, which can be rewritten as $P(S_i|A) = [P(S_i)P(A|S_i)]/P(A)$. Using the Law of Total Probability to rewrite $P(A)$, you have

$$P(S_i|A) = \frac{P(S_i)P(A|S_i)}{P(S_1)P(A|S_1) + P(S_2)P(A|S_2) + P(S_3)P(A|S_3) + \cdots + P(S_k)P(A|S_k)}$$

These new probabilities are often referred to as *posterior probabilities*—that is, probabilities of the subpopulations (also called *states of nature*) that have been updated after observing the sample information contained in the event A. Bayes suggested that if the prior probabilities are unknown, they can be taken to be $1/k$, which implies that each of the events S_1 through S_k is equally likely.

BAYES' RULE

Let $S_1, S_2, \ldots, S_k$ represent k mutually exclusive and exhaustive subpopulations with prior probabilities $P(S_1), P(S_2), \ldots, P(S_k)$. If an event A occurs, the posterior probability of S_i given A is the conditional probability

$$P(S_i|A) = \frac{P(S_i)P(A|S_i)}{\sum\limits_{j=1}^{k} P(S_j)P(A|S_j)}$$

for $i = 1, 2, \ldots, k$.

EXAMPLE 4.24 — Refer to Example 4.23. Find the probability that the person selected was 65 years of age or older, given that the person owned at least five pairs of wearable sneakers.

Solution   You need to find the conditional probability given by

$$P(G_5|A) = \frac{P(A \cap G_5)}{P(A)}$$

You have already calculated $P(A) = .1689$ using the Law of Total Probability. Therefore,

$P(G_5|A) =$

$$\frac{P(G_5)P(A|G_5)}{P(G_1)P(A|G_1) + P(G_2)P(A|G_2) + P(G_3)P(A|G_3) + P(G_4P(A|G_4) + P(G_5)P(A|G_5)}$$

$$= \frac{(.17)(.14)}{(.09)(.26) + (.20)(.20) + (.31)(.13) + (.23)(.18) + (.17)(.14)}$$

$$= \frac{.0238}{.1689} = .1409$$

In this case, the posterior probability of .14 is somewhat less than the prior probability of .17 (from Table 4.6). This group *a priori* was the second smallest, and only a small proportion of this segment had five or more pairs of wearable sneakers.

What is the posterior probability for those aged 35 to 49? For this group of adults, we have

$$P(G_3|A) = \frac{(.31)(.13)}{(.09)(.26) + (.20)(.20) + (.31)(.13) + (.23)(.18) + (.17)(.14)} = .2386$$

This posterior probability of .24 is substantially less than the prior probability of .31. In effect, this group was *a priori* the largest segment of the population sampled, but at the same time, the proportion of individuals in this group who had at least five pairs of wearable sneakers was the smallest of any of the groups. These two facts taken together cause a downward adjustment of almost a third in the *a priori* probability of .31.

4.7 EXERCISES

BASIC TECHNIQUES

4.69 Bayes' Rule A sample is selected from one of two populations, S_1 and S_2, with probabilities $P(S_1) = .7$ and $P(S_2) = .3$. If the sample has been selected from S_1, the probability of observing an event A is $P(A|S_1) = .2$. Similarly, if the sample has been selected from S_2, the probability of observing A is $P(A|S_2) = .3$.

a. If a sample is randomly selected from one of the two populations, what is the probability that event A occurs?

b. If the sample is randomly selected and event A is observed, what is the probability that the sample was selected from population S_1? From population S_2?

4.70 Bayes' Rule II If an experiment is conducted, one and only one of three mutually exclusive events S_1, S_2, and S_3 can occur, with these probabilities:

$$P(S_1) = .2 \qquad P(S_2) = .5 \qquad P(S_3) = .3$$

The probabilities of a fourth event A occurring, given that event S_1, S_2, or S_3 occurs, are

$$P(A|S_1) = .2 \quad P(A|S_2) = .1 \quad P(A|S_3) = .3$$

If event A is observed, find $P(S_1|A)$, $P(S_2|A)$, and $P(S_3|A)$.

4.71 Law of Total Probability A population can be divided into two subgroups that occur with probabilities 60% and 40%, respectively. An event A occurs 30% of the time in the first subgroup and 50% of the time in the second subgroup. What is the unconditional probability of the event A, regardless of which subgroup it comes from?

APPLICATIONS

4.72 Violent Crime City crime records show that 20% of all crimes are violent and 80% are nonviolent, involving theft, forgery, and so on. Ninety percent of violent crimes are reported versus 70% of nonviolent crimes.

a. What is the overall reporting rate for crimes in the city?

b. If a crime in progress is reported to the police, what is the probability that the crime is violent? What is the probability that it is nonviolent?

c. Refer to part b. If a crime in progress is reported to the police, why is it more likely that it is a nonviolent crime? Wouldn't violent crimes be more likely to be reported? Can you explain these results?

4.73 Worker Error A worker-operated machine produces a defective item with probability .01 if the worker follows the machine's operating instructions exactly, and with probability .03 if he does not. If the worker follows the instructions 90% of the time, what proportion of all items produced by the machine will be defective?

4.74 Airport Security Suppose that, in a particular city, airport A handles 50% of all airline traffic, and airports B and C handle 30% and 20%, respectively. The detection rates for weapons at the three airports are .9, .5, and .4, respectively. If a passenger at one of the airports is found to be carrying a weapon through the boarding gate, what is the probability that the passenger is using airport A? Airport C?

4.75 Football Strategies A particular football team is known to run 30% of its plays to the left and 70% to the right. A linebacker on an opposing team notes that the right guard shifts his stance most of the time (80%) when plays go to the right and that he uses a balanced stance the remainder of the time. When plays go to the left, the guard takes a balanced stance 90% of the time and the shift stance the remaining 10%. On a particular play, the linebacker notes that the guard takes a balanced stance.

a. What is the probability that the play will go to the left?

b. What is the probability that the play will go to the right?

c. If you were the linebacker, which direction would you prepare to defend if you saw the balanced stance?

4.76 No Pass, No Play Many public schools are implementing a "no pass, no play" rule for athletes. Under this system, a student who fails a course is disqualified from participating in extracurricular activities during the next grading period. Suppose the probability that an athlete who has not previously been disqualified will be disqualified is .15 and the probability that an athlete who has been disqualified will be disqualified again in the next time period is .5. If 30% of the athletes have been disqualified before, what is the unconditional probability that an athlete will be disqualified during the next grading period?

4.77 Medical Diagnostics Medical case histories indicate that different illnesses may produce identical symptoms. Suppose a particular set of symptoms, which we will denote as event H, occurs only when any one of three illnesses—A, B, or C—occurs. (For the sake of simplicity, we will assume that illnesses A, B, and C are mutually exclusive.) Studies show these probabilities of getting the three illnesses:

$$P(A) = .01$$
$$P(B) = .005$$
$$P(C) = .02$$

The probabilities of developing the symptoms H, given a specific illness, are

$$P(H|A) = .90$$
$$P(H|B) = .95$$
$$P(H|C) = .75$$

Assuming that an ill person shows the symptoms H, what is the probability that the person has illness A?

4.78 Cheating on Your Taxes? Suppose 5% of all people filing the long income tax form seek deductions that they know are illegal, and an additional 2% incorrectly list deductions because they are unfamiliar with income tax regulations. Of the 5% who are guilty of cheating, 80% will deny knowledge of the error if confronted by an investigator. If the filer of the long form is confronted with an unwarranted deduction and he or she denies the knowledge of the error, what is the probability that he or she is guilty?

4.79 Screening Tests Suppose that a certain disease is present in 10% of the population, and that there is a screening test designed to detect this disease if present. The test does not always work perfectly. Sometimes the test is negative when the disease is present, and sometimes it is positive when the disease is absent. The table below shows the proportion of times that the test produces various results:

	Test is Positive (P)	Test is Negative (N)
Disease present (D)	.08	.22
Disease absent (Dᶜ)	.05	.85

a. Find the following probabilities from the table: $P(D)$, $P(D^c)$, $P(N|D^c)$, $P(N|D)$.

b. Use Bayes' Rule and the results of part a to find $P(D|N)$.

c. Use the definition of conditional probability to find $P(D|N)$. (Your answer should be the same as the answer to part b.)

d. Find the probability of a false positive, that the test is positive, given that the person is disease-free.

e. Find the probability of a false negative, that the test is negative, given that the person has the disease.

f. Are either of the probabilities in parts d or e large enough that you would be concerned about the reliability of this screening method? Explain.

4.8 DISCRETE RANDOM VARIABLES AND THEIR PROBABILITY DISTRIBUTIONS

In Chapter 1, *variables* were defined as characteristics that change or vary over time and/or for different individuals or objects under consideration. *Quantitative variables* generate numerical data, whereas *qualitative variables* generate categorical data. However, even qualitative variables can generate numerical data if the categories are numerically coded to form a scale. For example, if you toss a single coin, the qualitative outcome could be recorded as "0" if a head and "1" if a tail.

Random Variables

A numerically valued variable x will vary or change depending on the particular outcome of the experiment being measured. For example, suppose you toss a die and measure x, the number observed on the upper face. The variable x can take on any of six values—1, 2, 3, 4, 5, 6—depending on the *random* outcome of the experiment. For this reason, we refer to the variable x as a **random variable.**

Definition   A variable x is a **random variable** if the value that it assumes, corresponding to the outcome of an experiment, is a chance or random event.

You can think of many examples of random variables:

- x = Number of defects on a *randomly selected* piece of furniture
- x = SAT score for a *randomly selected* college applicant
- x = Number of telephone calls received by a crisis intervention hotline during a *randomly selected* time period

As in Chapter 1, quantitative random variables are classified as either *discrete* or *continuous,* according to the values that x can assume. It is important to distinguish between discrete and continuous random variables because different techniques are used to describe their distributions. We focus on discrete random variables in the remainder of this chapter; continuous random variables are the subject of Chapter 6.

Probability Distributions

In Chapters 1 and 2, you learned how to construct the *relative frequency distribution* for a set of numerical measurements on a variable x. The distribution gave this information about x:

- What values of x occurred
- How often each value of x occurred

You also learned how to use the mean and standard deviation to measure the center and variability of this data set.

In this chapter, we defined *probability* as the limiting value of the relative frequency as the experiment is repeated over and over again. Now we define the **probability distribution** for a random variable x as the *relative frequency distribution* constructed for the entire population of measurements.

Definition   The **probability distribution** for a discrete random variable is a formula, table, or graph that gives the possible values of x, and the probability $p(x)$ associated with each value of x.

The values of x represent mutually exclusive numerical events. Summing $p(x)$ over all values of x is equivalent to adding the probabilities of all simple events and therefore equals 1.

REQUIREMENTS FOR A DISCRETE PROBABILITY DISTRIBUTION

- $0 \leq p(x) \leq 1$
- $\Sigma \, p(x) = 1$

EXAMPLE 4.25

Toss two fair coins and let x equal the number of heads observed. Find the probability distribution for x.

Solution   The simple events for this experiment with their respective probabilities are listed in Table 4.7. Since $E_1 = $ HH results in two heads, this simple event results in the value $x = 2$. Similarly, the value $x = 1$ is assigned to E_2, and so on. For each

TABLE 4.7 ● **Simple Events and Probabilities in Tossing Two Coins**

Simple Event	Coin 1	Coin 2	$P(E_i)$	x
E_1	H	H	1/4	2
E_2	H	T	1/4	1
E_3	T	H	1/4	1
E_4	T	T	1/4	0

value of x, you can calculate $p(x)$ by adding the probabilities of the simple events in that event. For example, when $x = 0$,

$$p(0) = P(E_4) = \frac{1}{4}$$

and when $x = 1$,

$$p(1) = P(E_2) + P(E_3) = \frac{1}{2}$$

The values of x and their respective probabilities, $p(x)$, are listed in Table 4.8. Notice that the probabilities add to 1.

TABLE 4.8 ● **Probability Distribution for x (x = Number of Heads)**

x	Simple Events in x	$p(x)$
0	E_4	1/4
1	E_2, E_3	1/2
2	E_1	1/4
	$\Sigma\, p(x) = 1$	

The probability distribution in Table 4.8 can be graphed using the methods of Section 1.5 to form the **probability histogram** in Figure 4.15.[†] The three values of the random variable x are located on the horizontal axis, and the probabilities $p(x)$ are located on the vertical axis (replacing the relative frequencies used in Chapter 1). Since the width of each bar is 1, the area under the bar is the probability of observing the particular value of x and the total area equals 1.

FIGURE 4.15 ●
Probability histogram for
Example 4.25

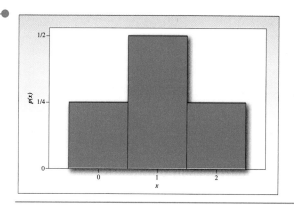

MY APPLET

There are two Java applets that will allow you to approximate discrete probability distributions using *simulation methods*. That is, even though the probabilities $p(x)$ can only be found as the long-run relative frequencies when the experiment is repeated an *infinite* number of times, we can get close to these probabilities if we repeat the experiment a *large* number of times. The applets called **Flipping Fair Coins** and **Flipping Weighted Coins** are two such simulations. The fastest way to generate the approximate probability distribution for x, the number of heads in n tosses of the coin is to repeat the experiment "100 at a Time," using the `100 at a Time` button at the bottom of the applet. The probability distribution will build up rather quickly. You can approximate the values of $p(x)$ and compare to the actual values calculated using probability rules. We will use these applets for the MyApplet Exercises at the end of the chapter.

[†]The probability distribution in Table 4.8 can also be presented using a formula, which is given in Section 5.2.

FIGURE 4.16

Flipping Fair Coins
applet

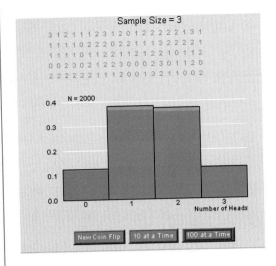

FIGURE 4.17

Flipping Weighted Coins
applet

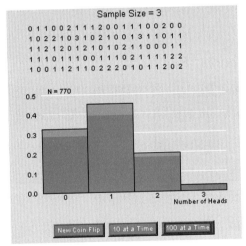

The Mean and Standard Deviation
for a Discrete Random Variable

The probability distribution for a discrete random variable looks very similar to the relative frequency distribution discussed in Chapter 1. The difference is that the relative frequency distribution describes a *sample* of n measurements, whereas the probability distribution is constructed as a model for the *entire population* of measurements. Just as the mean $\bar{x}$ and the standard deviation s measured the center and spread of the sample data, you can calculate similar measures to describe the center and spread of the population.

The population mean, which measures the average value of x in the population, is also called the **expected value** of the random variable x. It is the value that you would *expect* to observe on *average* if the experiment is repeated over and over again. The formula for calculating the population mean is easier to understand by example. Toss those two fair coins again, and let x be the number of heads observed. We constructed this probability distribution for x:

x	0	1	2
$p(x)$	1/4	1/2	1/4

Suppose the experiment is repeated a large number of times—say, $n = 4,000,000$ times. Intuitively, you would expect to observe approximately 1 million zeros, 2 million ones, and 1 million twos. Then the average value of x would equal

$$\frac{\text{Sum of measurements}}{n} = \frac{1,000,000(0) + 2,000,000(1) + 1,000,000(2)}{4,000,000}$$

$$= \left(\frac{1}{4}\right)(0) + \left(\frac{1}{2}\right)(1) + \left(\frac{1}{4}\right)(2)$$

Note that the first term in this sum is $(0)p(0)$, the second is equal to $(1)p(1)$, and the third is $(2)p(2)$. The average value of x, then, is

$$\Sigma xp(x) = 0 + \frac{1}{2} + \frac{2}{4} = 1$$

This result provides some intuitive justification for the definition of the expected value of a discrete random variable x.

Definition Let x be a discrete random variable with probability distribution $p(x)$. The mean or **expected value of x** is given as

$$\mu = E(x) = \Sigma\, xp(x)$$

where the elements are summed over all values of the random variable x.

We could use a similar argument to justify the formulas for the **population variance** σ^2 and the **population standard deviation** σ. These numerical measures describe the spread or variability of the random variable using the "average" or "expected value" of the squared deviations of the x-values from their mean μ.

Definition Let x be a discrete random variable with probability distribution $p(x)$ and mean μ. The **variance of x** is

$$\sigma^2 = E[(x - \mu)^2] = \Sigma(x - \mu)^2 p(x)$$

where the summation is over all values of the random variable x.[†]

Definition The **standard deviation σ of a random variable x** is equal to the positive square root of its variance.

EXAMPLE 4.26

An electronics store sells a particular model of computer notebook. There are only four notebooks in stock, and the manager wonders what today's demand for this particular model will be. She learns from the marketing department that the probability distribution for x, the daily demand for the laptop, is as shown in the table. Find the

---

[†]It can be shown (proof omitted) that

$$\sigma^2 = \Sigma(x - \mu)^2 p(x) = \Sigma x^2 p(x) - \mu^2$$

This result is analogous to the computing formula for the sum of squares of deviations given in Chapter 2.

mean, variance, and standard deviation of x. Is it likely that five or more customers will want to buy a laptop today?

x	0	1	2	3	4	5
p(x)	.10	.40	.20	.15	.10	.05

Solution Table 4.9 shows the values of x and $p(x)$, along with the individual terms used in the formulas for μ and σ^2. The sum of the values in the third column is

$$\mu = \Sigma\, xp(x) = (0)(.10) + (1)(.40) + \cdots + (5)(.05) = 1.90$$

while the sum of the values in the fifth column is

$$\sigma^2 = \Sigma(x - \mu)^2 p(x)$$
$$= (0 - 1.9)^2(.10) + (1 - 1.9)^2(.40) + \cdots + (5 - 1.9)^2(.05) = 1.79$$

and

$$\sigma = \sqrt{\sigma^2} = \sqrt{1.79} = 1.34$$

TABLE 4.9 ● **Calculations for Example 4.26**

x	p(x)	xp(x)	$(x - \mu)^2$	$(x - \mu)^2\, p(x)$
0	.10	.00	3.61	.361
1	.40	.40	.81	.324
2	.20	.40	.01	.002
3	.15	.45	1.21	.1815
4	.10	.40	4.41	.441
5	.05	.25	9.61	.4805
Totals	1.00	$\mu = 1.90$		$\sigma^2 = 1.79$

The graph of the probability distribution is shown in Figure 4.18. Since the distribution is approximately mound-shaped, approximately 95% of all measurements should lie within *two* standard deviations of the mean—that is,

$$\mu \pm 2\sigma \Rightarrow 1.90 \pm 2(1.34) \qquad \text{or} \ -.78 \text{ to } 4.58$$

Since $x = 5$ lies outside this interval, you can say it is unlikely that five or more customers will want to buy a laptop today. In fact, $P(x \geqslant 5)$ is exactly .05, or 1 time in 20.

FIGURE 4.18 ●

Probability Distribution for
Example 4.26

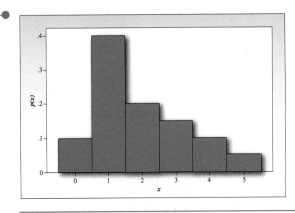

EXAMPLE 4.27

In a lottery conducted to benefit the local fire company, 8000 tickets are to be sold at $5 each. The prize is a $12,000 automobile. If you purchase two tickets, what is your expected gain?

Solution   Your gain x may take one of two values. You will either lose $10 (i.e., your "gain" will be $-$10) or win $11,990, with probabilities 7998/8000 and 2/8000, respectively. The probability distribution for the gain x is shown in the table:

x	$p(x)$
$-$10	7998/8000
$11,990	2/8000

The expected gain will be

$$\mu = E(x) = \Sigma \, xp(x)$$

$$= (-\$10)\left(\frac{7998}{8000}\right) + (\$11,900)\left(\frac{2}{8000}\right) = -\$7$$

Recall that the expected value of x is the average of the theoretical population that would result if the lottery were repeated an infinitely large number of times. If this were done, your average or expected gain per lottery ticket would be a loss of $7.

EXAMPLE 4.28

Determine the yearly premium for a $1000 insurance policy covering an event that, over a long period of time, has occurred at the rate of 2 times in 100. Let x equal the yearly financial gain to the insurance company resulting from the sale of the policy, and let C equal the unknown yearly premium. Calculate the value of C such that the expected gain $E(x)$ will equal zero. Then C is the premium required to break even. To this, the company would add administrative costs and profit.

Solution   The first step in the solution is to determine the values that the gain x may take and then to determine $p(x)$. If the event does not occur during the year, the insurance company will gain the premium of $x = C$ dollars. If the event does occur, the gain will be negative; that is, the company will lose $1000 less the premium of C dollars already collected. Then $x = -(1000 - C)$ dollars. The probabilities associated with these two values of x are 98/100 and 2/100, respectively. The probability distribution for the gain is shown in the table:

$x = $ Gain	$p(x)$
C	98/100
$-(1000 - C)$	2/100

Since the company wants the insurance premium C such that, in the long run (for many similar policies), the mean gain will equal zero, you can set the expected value of x equal to zero and solve for C. Then

$$\mu = E(x) = \Sigma xp(x)$$

$$= C\left(\frac{98}{100}\right) + [-1000 - C)]\left(\frac{2}{100}\right) = 0$$

or

$$\frac{98}{100}C + \frac{2}{100}C - 20 = 0$$

Solving this equation for C, you obtain $C = \$20$. Therefore, if the insurance company charged a yearly premium of $20, the average gain calculated for a large number of similar policies would equal zero. The actual premium would equal $20 plus administrative costs and profit.

The method for calculating the expected value of x for a continuous random variable is similar to what you have done, but in practice it involves the use of calculus. Nevertheless, the basic results concerning expectations are the same for continuous and discrete random variables. For example, regardless of whether x is continuous or discrete, $\mu = E(x)$ and $\sigma^2 = E[(x - \mu)^2]$.

4.8 EXERCISES

BASIC TECHNIQUES

4.80 Discrete or Continuous? Identify the following as discrete or continuous random variables:

a. Total number of points scored in a football game

b. Shelf life of a particular drug

c. Height of the ocean's tide at a given location

d. Length of a 2-year-old black bass

e. Number of aircraft near-collisions in a year

4.81 Discrete or Continuous II Identify the following as discrete or continuous random variables:

a. Increase in length of life attained by a cancer patient as a result of surgery

b. Tensile breaking strength (in pounds per square inch) of 1-inch-diameter steel cable

c. Number of deer killed per year in a state wildlife preserve

d. Number of overdue accounts in a department store at a particular time

e. Your blood pressure

4.82 Probability Distribution I A random variable x has this probability distribution:

x	0	1	2	3	4	5
$p(x)$	.1	.3	.4	.1	?	.05

a. Find $p(4)$.

b. Construct a probability histogram to describe $p(x)$.

c. Find μ, σ^2, and σ.

d. Locate the interval $\mu \pm 2\sigma$ on the x-axis of the histogram. What is the probability that x will fall into this interval?

e. If you were to select a very large number of values of x from the population, would most fall into the interval $\mu \pm 2\sigma$? Explain.

4.83 Probability Distribution II A random variable x can assume five values: 0, 1, 2, 3, 4. A portion of the probability distribution is shown here:

x	0	1	2	3	4
$p(x)$	.1	.3	.3	?	.1

a. Find $p(3)$.

b. Construct a probability histogram for $p(x)$.

c. Calculate the population mean, variance, and standard deviation.

d. What is the probability that x is greater than 2?

e. What is the probability that x is 3 or less?

4.84 Dice Let x equal the number observed on the throw of a single balanced die.

a. Find and graph the probability distribution for x.

b. What is the average or expected value of x?

c. What is the standard deviation of x?

d. Locate the interval $\mu \pm 2\sigma$ on the x-axis of the graph in part a. What proportion of all the measurements would fall into this range?

4.85 Grocery Visits Let x represent the number of times a customer visits a grocery store in a 1-week period. Assume this is the probability distribution of x:

x	0	1	2	3
$p(x)$	.1	.4	.4	.1

Find the expected value of x, the average number of times a customer visits the store.

APPLICATIONS

4.86 Letterman or Leno? Who is the king of late night TV? An Internet survey estimates that, when given a choice between David Letterman and Jay Leno, 52% of the population prefers to watch Jay

Leno. Suppose that you randomly select three late night TV watchers and ask them which of the two talk show hosts they prefer.

a. Find the probability distribution for x, the number of people in the sample of three who would prefer Jay Leno.

b. Construct the probability histogram for $p(x)$.

c. What is the probability that exactly one of the three would prefer Jay Leno?

d. What are the population mean and standard deviation for the random variable x?

4.87 Which Key Fits? A key ring contains four office keys that are identical in appearance, but only one will open your office door. Suppose you randomly select one key and try it. If it does not fit, you randomly select one of the three remaining keys. If it does not fit, you randomly select one of the last two. Each different sequence that could occur in selecting the keys represents one of a set of equiprobable simple events.

a. List the simple events in S and assign probabilities to the simple events.

b. Let x equal the number of keys that you try before you find the one that opens the door ($x = 1, 2, 3, 4$). Then assign the appropriate value of x to each simple event.

c. Calculate the values of $p(x)$ and display them in a table.

d. Construct a probability histogram for $p(x)$.

4.88 Roulette Exercise 4.10 described the game of roulette. Suppose you bet $5 on a single number—say, the number 18. The payoff on this type of bet is usually 35 to 1. What is your expected gain?

4.89 Gender Bias? A company has five applicants for two positions: two women and three men. Suppose that the five applicants are equally qualified and that no preference is given for choosing either gender. Let x equal the number of women chosen to fill the two positions.

a. Find $p(x)$.

b. Construct a probability histogram for x.

4.90 Defective Equipment A piece of electronic equipment contains six computer chips, two of which are defective. Three chips are selected at random, removed from the piece of equipment, and inspected. Let x equal the number of defectives observed, where $x = 0, 1,$ or 2. Find the probability distribution for x.

Express the results graphically as a probability histogram.

4.91 Drilling Oil Wells Past experience has shown that, on the average, only 1 in 10 wells drilled hits oil. Let x be the number of drillings until the first success (oil is struck). Assume that the drillings represent independent events.

a. Find $p(1)$, $p(2)$, and $p(3)$.

b. Give a formula for $p(x)$.

c. Graph $p(x)$.

4.92 Tennis, Anyone? Two tennis professionals, A and B, are scheduled to play a match; the winner is the first player to win three sets in a total that cannot exceed five sets. The event that A wins any one set is independent of the event that A wins any other, and the probability that A wins any one set is equal to .6. Let x equal the total number of sets in the match; that is, $x = 3, 4,$ or 5. Find $p(x)$.

4.93 Tennis, again The probability that a tennis player A can win a set from tennis player B is one measure of the comparative abilities of the two players. In Exercise 4.92 you found the probability distribution for x, the number of sets required to play a best-of-five-sets match, given that the probability that A wins any one set—call this $P(A)$—is .6.

a. Find the expected number of sets required to complete the match for $P(A) = .6$.

b. Find the expected number of sets required to complete the match when the players are of equal ability—that is, $P(A) = .5$.

c. Find the expected number of sets required to complete the match when the players differ greatly in ability—that is, say, $P(A) = .9$.

4.94 The PGA One professional golfer plays best on short-distance holes. Experience has shown that the numbers x of shots required for 3-, 4-, and 5-par holes have the probability distributions shown in the table:

Par-3 Holes		Par-4 Holes		Par-5 Holes	
x	$p(x)$	x	$p(x)$	x	$p(x)$
2	.12	3	.14	4	.04
3	.80	4	.80	5	.80
4	.06	5	.04	6	.12
5	.02	6	.02	7	.04

What is the golfer's expected score on these holes?

a. A par-3 hole

b. A par-4 hole

c. A par-5 hole

4.95 Insuring Your Diamonds You can insure a $50,000 diamond for its total value by paying a premium of D dollars. If the probability of theft in a given year is estimated to be .01, what premium should the insurance company charge if it wants the expected gain to equal $1000?

4.96 FDA Testing The maximum patent life for a new drug is 17 years. Subtracting the length of time required by the FDA for testing and approval of the drug provides the actual patent life of the drug—that is, the length of time that a company has to recover research and development costs and make a profit. Suppose the distribution of the lengths of patent life for new drugs is as shown here:

Years, x	3	4	5	6	7	8
$p(x)$	.03	.05	.07	.10	.14	.20

Years, x	9	10	11	12	13
$p(x)$	.18	.12	.07	.03	.01

a. Find the expected number of years of patent life for a new drug.
b. Find the standard deviation of x.
c. Find the probability that x falls into the interval $\mu \pm 2\sigma$.

4.97 Heaven Do you believe in heaven? In a survey conducted for *Time* magazine, 81% of the adult Americans surveyed expressed a belief in "heaven, where people live forever with God after they die."[6] Suppose you had conducted your own telephone survey at the same time. You randomly called people and asked them whether they believe in heaven. Assume that the percentage given in the *Time* survey can be taken to approximate the percentage of all adult Americans who believe in heaven.

a. Find the probability distribution for x, the number of calls until you find the first person who *does not* believe in heaven.
b. What problems might arise as you randomly call people and ask them to take part in your survey? How would this affect the reliability of the probabilities calculated in part a?

4.98 Shipping Charges From experience, a shipping company knows that the cost of delivering a small package within 24 hours is $14.80. The company charges $15.50 for shipment but guarantees to refund the charge if delivery is not made within 24 hours. If the company fails to deliver only 2% of its packages within the 24-hour period, what is the expected gain per package?

4.99 Actuaries A manufacturing representative is considering taking out an insurance policy to cover possible losses incurred by marketing a new product. If the product is a complete failure, the representative feels that a loss of $80,000 would be incurred; if it is only moderately successful, a loss of $25,000 would be incurred. Insurance actuaries have determined from market surveys and other available information that the probabilities that the product will be a failure or only moderately successful are .01 and .05, respectively. Assuming that the manufacturing representative is willing to ignore all other possible losses, what premium should the insurance company charge for a policy in order to break even?

CHAPTER REVIEW

Key Concepts and Formulas

I. Experiments and the Sample Space

1. Experiments, events, mutually exclusive events, simple events
2. The sample space
3. Venn diagrams, tree diagrams, probability tables

II. Probabilities

1. Relative frequency definition of probability

2. Properties of probabilities
   a. Each probability lies between 0 and 1.
   b. Sum of all simple-event probabilities equals 1.
3. $P(A)$, the sum of the probabilities for all simple events in A

III. Counting Rules

1. *mn* Rule; extended *mn* Rule

2. Permutations: $P_r^n = \dfrac{n!}{(n-r)!}$

3. Combinations: $C_r^n = \dfrac{n!}{r!(n-r)!}$

6. Multiplication Rule: $P(A \cap B) = P(A)P(B|A)$

7. Law of Total Probability

8. Bayes' Rule

IV. Event Relations

1. Unions and intersections

2. Events

   a. Disjoint or mutually exclusive:
$P(A \cap B) = 0$

   b. Complementary: $P(A) = 1 - P(A^c)$

3. Conditional probability: $P(A|B) = \dfrac{P(A \cap B)}{P(B)}$

4. Independent and dependent events

5. Addition Rule:
$P(A \cup B) = P(A) + P(B) - P(A \cap B)$

V. Discrete Random Variables and Probability Distributions

1. Random variables, discrete and continuous

2. Properties of probability distributions

   a. $0 \leq p(x) \leq 1$

   b. $\Sigma p(x) = 1$

3. Mean or expected value of a discrete random variable: $\mu = \Sigma x p(x)$

4. Variance and standard deviation of a discrete random variable: $\sigma^2 = \Sigma(x - \mu)^2 p(x)$ and $\sigma = \sqrt{\sigma^2}$

 MINITAB

Discrete Probability Distributions

Although *MINITAB* cannot help you solve the types of general probability problems presented in this chapter, it is useful for graphing the probability distribution $p(x)$ for a general discrete random variable x when the probabilities are known, and for calculating the mean, variance, and standard deviation of the random variable x. In Chapters 5 and 6, we will use *MINITAB* to calculate exact probabilities for three special cases: the binomial, the Poisson, and the normal random variables.

Suppose you have this general probability distribution:

x	0	1	3	5
$p(x)$	.25	.35	.25	.15

Enter the values of x and $p(x)$ into columns C1 and C2 of a new *MINITAB* worksheet. You can now use the **Calc → Calculator** command to calculate μ, σ^2, and σ and to store the results in columns C3–C5 (named "Mean," "Variance," and "Std Dev") of the worksheet. Use the same approach for the three parameters. In the Calculator dialog box, **select** "Mean" as the column in which to store μ. In the Expression box, use the Functions list, the calculator keys, and the variables list on the left to highlight, **select,** and create the expression for the mean (see Figure 4.19):

SUM('x'*'p(x)')

FIGURE 4.19

MINITAB will multiply each row element in C1 times the corresponding row element in C2, sum the resulting products, and store the result in C3! You can check the result by hand if you like. The formulas for the variance and standard deviation are selected in a similar way:

$$\text{Variance:} \quad \text{SUM}((\text{'x'} - \text{'Mean'})**2*\text{'p(x)'})$$
$$\text{Std Dev: SQRT('Variance')}$$

To see the tabular form of the probability distribution and the three parameters, use **Data → Display Data** and select all five columns. Click **OK** and the results will be displayed in the Session window, as shown in Figure 4.20.

The probability histogram can be plotted using the *MINITAB* command **Graph → Scatterplot → Simple → OK.** In the Scatterplot dialog box (Figure 4.21), select 'p(x)' for **Y variables** and 'x' for **X variables.** To display the discrete probability bars, click on **Data View,** uncheck the box marked "Symbols," and check the box marked "Project Lines." Click **OK** twice to see the plot. You will see a single straight line projected at each of the four values of *x.* If you want the plot to look more like the discrete probability histograms in Section 4.8, position your cursor on one of the lines, right-click the mouse and choose "Edit Project Lines." Under the "Attributes" tab, select **Custom** and change the line size to **75.** Click **OK.** If the bar width is not satisfactory, you can readjust the line size. Finally, right-click on the X-axis, choose "Edit X Scale" and select − **.5** and **5.5** for the minimum and maximum **Scale Ranges.** Click **OK.** The probability histogram is shown in Figure 4.22.

Locate the mean on the graph. Is it at the center of the distribution? If you mark off two standard deviations on either side of the mean, do most of the possible values of *x* fall into this interval?

FIGURE 4.20

FIGURE 4.21

FIGURE 4.22

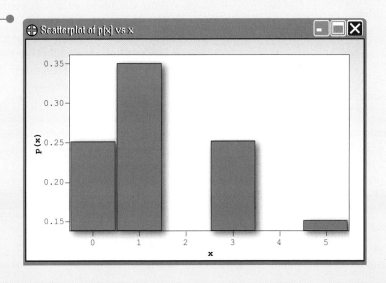

Supplementary Exercises

Starred (*) exercises are optional.

4.100 Playing the Slots A slot machine has three slots; each will show a cherry, a lemon, a star, or a bar when spun. The player wins if all three slots show the same three items. If each of the four items is equally likely to appear on a given spin, what is your probability of winning?

4.101 Whistle Blowers "Whistle blowers" is the name given to employees who report corporate fraud, theft, and other unethical and perhaps criminal activities by fellow employees or by their employer. Although there is legal protection for whistle blowers, it has been reported that approximately 23% of those who reported fraud suffered reprisals such as demotion or poor performance ratings. Suppose the probability that an employee will fail to report a case of fraud is .69. Find the probability that a worker who observes a case of fraud will report it and will subsequently suffer some form of reprisal.

4.102 Aspirin Two cold tablets are accidentally placed in a box containing two aspirin tablets. The four tablets are identical in appearance. One tablet is selected at random from the box and is swallowed by the first patient. A tablet is then selected at random from the three remaining tablets and is swallowed by the second patient. Define the following events as specific collections of simple events:

a. The sample space S

b. The event A that the first patient obtained a cold tablet

c. The event B that exactly one of the two patients obtained a cold tablet

d. The event C that neither patient obtained a cold tablet

4.103 Refer to Exercise 4.102. By summing the probabilities of simple events, find $P(A)$, $P(B)$, $P(A \cap B)$, $P(A \cup B)$, $P(C)$, $P(A \cap C)$, and $P(A \cup C)$.

4.104 DVRs A retailer sells two styles of high-priced digital video recorders (DVR) that experience indicates are in equal demand. (Fifty percent of all potential customers prefer style 1, and 50% favor style 2.) If the retailer stocks four of each, what is the probability that the first four customers seeking a DVR all purchase the same style?

4.105 Boxcars A boxcar contains seven complex electronic systems. Unknown to the purchaser, three are defective. Two of the seven are selected for thorough testing and are then classified as defective or nondefective. What is the probability that no defectives are found?

4.106 Heavy Equipment A heavy-equipment salesman can contact either one or two customers per day with probability 1/3 and 2/3, respectively. Each contact will result in either no sale or a $50,000 sale with probability 9/10 and 1/10, respectively. What is the expected value of his daily sales?

4.107 Fire Insurance A county containing a large number of rural homes is thought to have 60% of those homes insured against fire. Four rural homeowners are chosen at random from the entire population, and x are found to be insured against fire. Find the probability distribution for x. What is the probability that at least three of the four will be insured?

4.108 Fire Alarms A fire-detection device uses three temperature-sensitive cells acting independently of one another in such a manner that any one or more can activate the alarm. Each cell has a probability $p = .8$ of activating the alarm when the temperature reaches 100°F or higher. Let x equal the number of cells activating the alarm when the temperature reaches 100°F.

a. Find the probability distribution of x.

b. Find the probability that the alarm will function when the temperature reaches 100°F.

c. Find the expected value and the variance for the random variable x.

4.109 Catching a Cold Is your chance of getting a cold influenced by the number of social contacts you have? A study by Sheldon Cohen, a psychology professor at Carnegie Mellon University, seems to show that the more social relationships you have, the *less susceptible* you are to colds. A group of 276 healthy men and women were grouped according to their number of relationships (such as parent, friend, church member, neighbor). They were then exposed to a virus that causes colds. An adaptation of the results is shown in the table:[7]

	Number of Relationships		
	Three or Fewer	Four or Five	Six or More
Cold	49	43	34
No Cold	31	57	62
Total	80	100	96

a. If one person is selected at random from the 276 people in the study, what is the probability that the person got a cold?

b. If two people are randomly selected, what is the probability that one has four or five relationships and the other has six or more relationships?

c. If a single person is randomly selected and has a cold, what is the probability that he or she has three or fewer relationships?

4.110 Plant Genetics Refer to the experiment conducted by Gregor Mendel in Exercise 4.64. Suppose you are interested in following two independent traits in snap peas—seed texture (S = smooth, s = wrinkled) and seed color (Y = yellow, y = green)—in a second-generation cross of heterozygous parents. Remember that the capital letter represents the dominant trait. Complete the table with the gene pairs for both traits. All possible pairings are equally likely.

Seed Texture	Seed Color			
	yy	yY	Yy	YY
ss	(ss yy)	(ss yY)		
sS				
Ss				
SS				

a. What proportion of the offspring from this cross will have smooth yellow peas?

b. What proportion of the offspring will have smooth green peas?

c. What proportion of the offspring will have wrinkled yellow peas?

d. What proportion of the offspring will have wrinkled green peas?

e. Given that an offspring has smooth yellow peas, what is the probability that this offspring carries one s allele? One s allele *and* one y allele?

4.111 Profitable Stocks An investor has the option of investing in three of five recommended stocks. Unknown to her, only two will show a substantial profit within the next 5 years. If she selects the three stocks at random (giving every combination of three stocks an equal chance of selection), what is the probability that she selects the two profitable stocks? What is the probability that she selects only one of the two profitable stocks?

4.112 Racial Bias? Four union men, two from a minority group, are assigned to four distinctly different one-man jobs, which can be ranked in order of desirability.

a. Define the experiment.

b. List the simple events in *S*.

c. If the assignment to the jobs is unbiased—that is, if any one ordering of assignments is as probable as any other—what is the probability that the two men from the minority group are assigned to the least desirable jobs?

4.113 A Reticent Salesman A salesperson figures that the probability of her consummating a sale during the first contact with a client is .4 but improves to .55 on the second contact if the client did not buy during the first contact. Suppose this salesperson makes one and only one callback to any client. If she contacts a client, calculate the probabilities for these events:

a. The client will buy.

b. The client will not buy.

4.114 Bus or Subway A man takes either a bus or the subway to work with probabilities .3 and .7, respectively. When he takes the bus, he is late 30% of the days. When he takes the subway, he is late 20% of the days. If the man is late for work on a particular day, what is the probability that he took the bus?

4.115 Guided Missiles The failure rate for a guided missile control system is 1 in 1000. Suppose that a duplicate, but completely independent, control system is installed in each missile so that, if the first fails, the second can take over. The reliability of a missile is the probability that it does not fail. What is the reliability of the modified missile?

4.116 Rental Trucks A rental truck agency services its vehicles on a regular basis, routinely checking for mechanical problems. Suppose that the agency has six moving vans, two of which need to have new brakes. During a routine check, the vans are tested one at a time.

a. What is the probability that the last van with brake problems is the fourth van tested?

b. What is the probability that no more than four vans need to be tested before both brake problems are detected?

c. Given that one van with bad brakes is detected in the first two tests, what is the probability that the remaining van is found on the third or fourth test?

4.117 Pennsylvania Lottery Probability played a role in the rigging of the April 24, 1980, Pennsylvania state lottery. To determine each digit of the three-digit winning number, each of the numbers 0, 1, 2, . . . , 9 is written on a Ping-Pong ball, the ten balls are blown into a compartment, and the number selected for the digit is the one on the ball that floats to the top of the machine. To alter the odds, the conspirators injected a liquid into all balls used in the game except those numbered 4 and 6, making it almost certain that the lighter balls would be selected and determine the digits in the winning number. They then proceeded to buy lottery tickets bearing the potential winning numbers. How many potential winning numbers were there (666 was the eventual winner)?

*4.118 Lottery, continued Refer to Exercise 4.117. Hours after the rigging of the Pennsylvania state lottery was announced on September 19, 1980, Connecticut state lottery officials were stunned to learn that *their* winning number for the day was 666.

a. All evidence indicates that the Connecticut selection of 666 was pure chance. What is the probability that a 666 would be drawn in Connecticut, given that a 666 had been selected in the April 24, 1980, Pennsylvania lottery?

b. What is the probability of drawing a 666 in the April 24, 1980, Pennsylvania lottery (remember, this drawing was rigged) *and* a 666 on the September 19, 1980, Connecticut lottery?

*4.119 ACL/MCL Tears *The American Journal of Sports Medicine* published a study of 810 women collegiate rugby players who have a history of knee injuries. For these athletes, the two common knee injuries investigated were medial cruciate ligament (MCL) sprains and anterior cruciate ligament (ACL) tears.[8] For backfield players, it was found that 39% had MCL sprains and 61% had ACL tears. For forwards, it was found that 33% had MCL sprains and 67% had ACL tears. Since a rugby team consists of eight forwards and seven backs, you can assume that 47% of the players with knee injuries are backs and 53% are forwards.

a. Find the unconditional probability that a rugby player selected at random from this group of players has experienced an MCL sprain.

b. Given that you have selected a player who has an MCL sprain, what is the probability that the player is a forward?

c. Given that you have selected a player who has an ACL tear, what is the probability that the player is a back?

4.120 MRIs Magnetic resonance imaging (MRI) is an accepted noninvasive test to evaluate changes in the cartilage in joints. An article in *The American Journal of Sports Medicine* compared the results of MRI evaluation with arthroscopic surgical evaluation of cartilage tears at two sites in the knees of 35 patients. The $2 \times 35 = 70$ examinations produced the classifications shown in the table.[9] Actual tears were confirmed by arthroscopic surgical examination.

	Tears	No Tears	Total
MRI Positive	27	0	27
MRI Negative	4	39	43
Total	31	39	70

a. What is the probability that a site selected at random has a tear and has been identified as a tear by MRI?

b. What is the probability that a site selected at random has no tear and has been identified as having a tear?

c. What is the probability that a site selected at random has a tear and has not been identified by MRI?

d. What is the probability of a positive MRI, given that there is a tear?

e. What is the probability of a false negative—that is, a negative MRI, given that there is a tear?

4.121 The Match Game Two men each toss a coin. They obtain a "match" if either both coins are heads or both are tails. Suppose the tossing is repeated three times.

a. What is the probability of three matches?

b. What is the probability that all six tosses (three for each man) result in tails?

c. Coin tossing provides a model for many practical experiments. Suppose that the coin tosses represent the answers given by two students for three specific true–false questions on an examination. If the two students gave three matches for answers, would the low probability found in part a suggest collusion?

4.122 Contract Negotiations Experience has shown that, 50% of the time, a particular union–management contract negotiation led to a contract settlement within a 2-week period, 60% of the time the union strike fund was adequate to support a strike, and 30% of the time both conditions were satisfied. What is the probability of a contract settlement given that the union strike fund is adequate to support a strike? Is settlement of a contract within a 2-week period dependent on whether the union strike fund is adequate to support a strike?

4.123 Work Tenure Suppose the probability of remaining with a particular company 10 years or longer is 1/6. A man and a woman start work at the company on the same day.

a. What is the probability that the man will work there less than 10 years?

b. What is the probability that both the man and the woman will work there less than 10 years? (Assume they are unrelated and their lengths of service are independent of each other.)

c. What is the probability that one or the other or both will work 10 years or longer?

4.124 Accident Insurance Accident records collected by an automobile insurance company give the following information: The probability that an insured driver has an automobile accident is .15; if an accident has occurred, the damage to the vehicle amounts to 20% of its market value with probability .80, 60% of its market value with probability .12, and a total loss with probability .08. What premium should the company charge on a $22,000 car so that the expected gain by the company is zero?

4.125 Waiting Times Suppose that at a particular supermarket the probability of waiting 5 minutes or longer for checkout at the cashier's counter is .2. On a given day, a man and his wife decide to shop individually at the market, each checking out at different cashier counters. They both reach cashier counters at the same time.

a. What is the probability that the man will wait less than 5 minutes for checkout?

b. What is probability that both the man and his wife will be checked out in less than 5 minutes? (Assume that the checkout times for the two are independent events.)

c. What is the probability that one or the other or both will wait 5 minutes or longer?

4.126 Quality Control A quality-control plan calls for accepting a large lot of crankshaft bearings if a

sample of seven is drawn and none are defective. What is the probability of accepting the lot if none in the lot are defective? If 1/10 are defective? If 1/2 are defective?

4.127 Mass Transit Only 40% of all people in a community favor the development of a mass transit system. If four citizens are selected at random from the community, what is the probability that all four favor the mass transit system? That none favors the mass transit system?

4.128 Blood Pressure Meds A research physician compared the effectiveness of two blood pressure drugs A and B by administering the two drugs to each of four pairs of identical twins. Drug A was given to one member of a pair; drug B to the other. If, in fact, there is no difference in the effects of the drugs, what is the probability that the drop in the blood pressure reading for drug A exceeds the corresponding drop in the reading for drug B for all four pairs of twins? Suppose drug B created a greater drop in blood pressure than drug A for each of the four pairs of twins. Do you think this provides sufficient evidence to indicate that drug B is more effective in lowering blood pressure than drug A?

4.129 Blood Tests To reduce the cost of detecting a disease, blood tests are conducted on a pooled sample of blood collected from a group of n people. If no indication of the disease is present in the pooled blood sample (as is usually the case), none have the disease. If analysis of the pooled blood sample indicates that the disease is present, each individual must submit to a blood test. The individual tests are conducted in sequence. If, among a group of five people, one person has the disease, what is the probability that six blood tests (including the pooled test) are required to detect the single diseased person? If two people have the disease, what is the probability that six tests are required to locate both diseased people?

4.130 Tossing a Coin How many times should a coin be tossed to obtain a probability equal to or greater than .9 of observing at least one head?

4.131 Flextime The number of companies offering flexible work schedules has increased as companies try to help employees cope with the demands of home and work. One flextime schedule is to work four 10-hour shifts. However, a big obstacle to flextime schedules for workers paid hourly is state legislation on overtime. A survey provided the following information for 220 firms located in two cities in California.

Flextime Schedule

City	Available	Not Available	Total
A	39	75	114
B	25	81	106
Totals	64	156	220

A company is selected at random from this pool of 220 companies.

a. What is the probability that the company is located in city *A*?

b. What is the probability that the company is located in city *B* and offers flextime work schedules?

c. What is the probability that the company does not have flextime schedules?

d. What is the probability that the company is located in city *B,* given that the company has flextime schedules available?

4.132 A Color Recognition Experiment An experiment is run as follows—the colors red, yellow, and blue are each flashed on a screen for a short period of time. A subject views the colors and is asked to choose the one he feels was flashed for the longest time. The experiment is repeated three times with the same subject.

a. If all the colors were flashed for the same length of time, find the probability distribution for *x,* the number of times that the subject chose the color red. Assume that his three choices are independent.

b. Construct the probability histogram for the random variable *x.*

4.133 Pepsi or Coke? A taste-testing experiment is conducted at a local supermarket, where passing shoppers are asked to taste two soft-drink samples—one Pepsi and one Coke—and state their preference. Suppose that four shoppers are chosen at random and asked to participate in the experiment, and that there is actually no difference in the taste of the two brands.

a. What is the probability that all four shoppers choose Pepsi?

b. What is the probability that exactly one of the four shoppers chooses Pepsi?

4.134 Viruses A certain virus afflicted the families in three adjacent houses in a row of 12 houses. If three houses were randomly chosen from a row of 12 houses, what is the probability that the three houses would be adjacent? Is there reason to believe that this virus is contagious?

4.135 Orchestra Politics The board of directors of a major symphony orchestra has voted to create a players' committee for the purpose of handling employee complaints. The council will consist of the president and vice president of the symphony board and two orchestra representatives. The two orchestra representatives will be randomly selected from a list of six volunteers, consisting of four men and two women.

a. Find the probability distribution for *x,* the number of women chosen to be orchestra representatives.

b. Find the mean and variance for the random variable *x.*

c. What is the probability that both orchestra representatives will be women?

(MY) **APPLET Exercises**

4.136 Two fair dice are tossed. Use the **Tossing Dice** applet to answer the following questions.

a. What is the probability that the sum of the number of dots shown on the upper faces is equal to 7? To 11?

b. What is the probability that you roll "doubles"— that is, both dice have the same number on the upper face?

c. What is the probability that both dice show an odd number?

4.137 If you toss a pair of dice, the sum *T* of the number of dots appearing on the upper faces of the dice can assume the value of an integer in the interval $2 \leq T \leq 12$.

a. Use the **Tossing Dice** applet to find the probability distribution for *T.* Display this probability distribution in a table.

b. Construct a probability histogram for *p(T).* How would you describe the shape of this distribution?

4.138 Access the **Flipping Fair Coins** applet. The experiment consists of tossing three fair coins and recording x, the number of heads.

a. Use the laws of probability to write down the simple events in this experiment.

b. Find the probability distribution for x. Display the distribution in a table and in a probability histogram.

c. Use the **Flipping Fair Coins** applet to simulate the probability distribution—that is, repeat the coin-tossing experiment a large number of times until the relative frequency histogram is very close to the actual probability distribution. Start by performing the experiment once (click New Coin Flip) to see what is happening. Then speed up the process by clicking 100 at a Time. Generate at least 2000 values of x. Sketch the histogram that you have generated.

d. Compare the histograms in parts b and c. Does the simulation confirm your answer from part b?

4.139 Refer to Exercise 4.138.

a. If you were to toss only one coin, what would the probability distribution for x look like?

b. Perform a simulation using the **Flipping Fair Coins** applet with $n = 1$, and compare your results with part a.

4.140 Refer to Exercise 4.138. Access the **Flipping Weighted Coins** applet. The experiment consists of tossing three coins that are *not fair*, and recording x, the number of heads.

a. Perform a simulation of the experiment using the **Flipping Weighted Coins** applet. Is the distribution symmetric or skewed? Which is more likely, heads or tails?

b. Suppose that we do not know the probability of getting a head, $P(H)$. Write a formula for calculating the probability of no heads in three tosses.

c. Use the approximate probability $P(x = 0)$ from your simulation and the results of part b to approximate the value of $P(T)$. What is the probability of getting a head?

CASE STUDY ## Probability and Decision Making in the Congo

In his exciting novel *Congo,* Michael Crichton describes a search by Earth Resources Technology Service (ERTS), a geological survey company, for deposits of boron-coated blue diamonds, diamonds that ERTS believes to be the key to a new generation of optical computers.[10] In the novel, ERTS is racing against an international consortium to find the Lost City of Zinj, a city that thrived on diamond mining and existed several thousand years ago (according to African fable), deep in the rain forests of eastern Zaire.

After the mysterious destruction of its first expedition, ERTS launches a second expedition under the leadership of Karen Ross, a 24-year-old computer genius who is accompanied by Professor Peter Elliot, an anthropologist; Amy, a talking gorilla; and the famed mercenary and expedition leader, "Captain" Charles Munro. Ross's efforts to find the city are blocked by the consortium's offensive actions, by the deadly rain forest, and by hordes of "talking" killer gorillas whose perceived mission is to defend the diamond mines. Ross overcomes these obstacles by using space-age computers to evaluate the probabilities of success for all possible circumstances and all possible actions that the expedition might take. At each stage of the expedition, she is able to quickly evaluate the chances of success.

At one stage in the expedition, Ross is informed by her Houston headquarters that their computers estimate that she is 18 hours and 20 minutes behind the competing Euro-Japanese team, instead of 40 hours ahead. She changes plans and decides to have the 12 members of her team—Ross, Elliot, Munro, Amy, and eight native porters—parachute into a volcanic region near the estimated location of Zinj. As Crichton relates, "Ross had double-checked outcome probabilities from the Houston computer, and the results were unequivocal. The probability of a successful jump was .7980, meaning that there was approximately one chance in five that someone would be badly hurt. However, given a successful jump, the probability of expedition success was .9943, making it virtually certain that they would beat the consortium to the site."

Keeping in mind that this is an excerpt from a novel, let us examine the probability, .7980, of a successful jump. If you were one of the 12-member team, what is the probability that you would successfully complete your jump? In other words, if the probability of a successful jump by all 12 team members is .7980, what is the probability that a single member could successfully complete the jump?

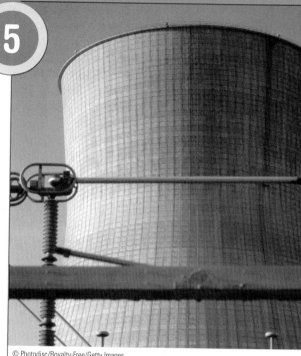

© Photodisc/Royalty-Free/Getty Images

5

Several Useful Discrete Distributions

GENERAL OBJECTIVES

Discrete random variables are used in many practical applications. Three important discrete random variables—the binomial, the Poisson, and the hypergeometric—are presented in this chapter. These random variables are often used to describe the number of occurrences of a specified event in a fixed number of trials or a fixed unit of time or space.

CHAPTER INDEX

 PERSONAL TRAINER

How Do I Use Table 1 to Calculate Binomial Probabilities?
How Do I Calculate Poisson Probabilities Using the Formula?
How Do I Use Table 2 to Calculate Poisson Probabilities?

A Mystery: Cancers Near a Reactor

Is the Pilgrim I nuclear reactor responsible for an increase in cancer cases in the surrounding area? A political controversy was set off when the Massachusetts Department of Public Health found an unusually large number of cases in a 4-mile-wide coastal strip just north of the nuclear reactor in Plymouth, Massachusetts. The case study at the end of this chapter examines how this question can be answered using one of the discrete probability distributions presented here.

INTRODUCTION

Examples of *discrete random variables* can be found in a variety of everyday situations and across most academic disciplines. However, there are three discrete probability distributions that serve as *models* for a large number of these applications. In this chapter we study the binomial, the Poisson, and the hypergeometric probability distributions and discuss their usefulness in different physical situations.

THE BINOMIAL PROBABILITY DISTRIBUTION

A coin-tossing experiment is a simple example of an important discrete random variable called the **binomial random variable.** Many practical experiments result in data similar to the head or tail outcomes of the coin toss. For example, consider the political polls used to predict voter preferences in elections. Each sampled voter can be compared to a coin because the voter may be in favor of our candidate—a "head"— or not—a "tail." In most cases, the proportion of voters who favor our candidate does not equal 1/2; that is, the coin is not fair. In fact, the proportion of voters who favor our candidate is exactly what the poll is designed to measure!

Here are some other situations that are similar to the coin-tossing experiment:

- A sociologist is interested in the proportion of elementary school teachers who are men.
- A soft-drink marketer is interested in the proportion of cola drinkers who prefer her brand.
- A geneticist is interested in the proportion of the population who possess a gene linked to Alzheimer's disease.

Each sampled person is analogous to tossing a coin, but the probability of a "head" is not necessarily equal to 1/2. Although these situations have different practical objectives, they all exhibit the common characteristics of the **binomial experiment.**

Definition A **binomial experiment** is one that has these five characteristics:

1. The experiment consists of n identical trials.
2. Each trial results in one of two outcomes. For lack of a better name, the one outcome is called a success, *S,* and the other a failure, *F.*
3. The probability of success on a single trial is equal to p and remains the same from trial to trial. The probability of failure is equal to $(1 - p) = q$.
4. The trials are independent.
5. We are interested in x, the number of successes observed during the n trials, for $x = 0, 1, 2, \ldots, n$.

EXAMPLE 5.1

Suppose there are approximately 1,000,000 adults in a county and an unknown proportion p favor term limits for politicians. A sample of 1000 adults will be chosen in such a way that every one of the 1,000,000 adults has an equal chance of being selected, and each adult is asked whether he or she favors term limits. (The ultimate objective of this survey is to estimate the unknown proportion p, a problem that we will discuss in Chapter 8.) Is this a binomial experiment?

Solution   Does the experiment have the five binomial characteristics?

1. A "trial" is the choice of a single adult from the 1,000,000 adults in the county. This sample consists of $n = 1000$ identical trials.

2. Since each adult will either favor or not favor term limits, there are two outcomes that represent the "successes" and "failures" in the binomial experiment.[†]

3. The probability of success, p, is the probability that an adult favors term limits. Does this probability remain the same for each adult in the sample? For all practical purposes, the answer is *yes*. For example, if 500,000 adults in the population favor term limits, then the probability of a "success" when the first adult is chosen is 500,000/1,000,000 = 1/2. When the second adult is chosen, the probability p changes slightly, depending on the first choice. That is, there will be either 499,999 or 500,000 successes left among the 999,999 adults. In either case, p is still approximately equal to 1/2.

4. The independence of the trials is guaranteed because of the large group of adults from which the sample is chosen. The probability of an adult favoring term limits does not change depending on the responses of previously chosen people.

5. The random variable x is the number of adults in the sample who favor term limits.

Because the survey satisfies the five characteristics reasonably well, for all practical purposes it can be viewed as a binomial experiment.

EXAMPLE 5.2

A purchaser who has received a shipment containing 20 personal computers (PCs) wants to sample three of the PCs to see whether they are in working order before accepting the shipment. The nearest three PCs are selected for testing and, afterward, are declared either defective or nondefective. Unknown to the purchaser, two of the PCs in the shipment of 20 are defective. Is this a binomial experiment?

Solution   Again, check the sampling procedure for the characteristics of a binomial experiment.

1. A "trial" is the selecting and testing of one PC from the total of 20. This experiment consists of $n = 3$ identical trials.

2. Each trial results in one of two outcomes. Either a PC is defective (call this a "success") or it is not (a "failure").

3. Suppose the PCs were randomly loaded into a boxcar so that any one of the 20 PCs could have been placed near the boxcar door. Then the unconditional probability of drawing a defective PC on a given trial would be 2/20.

4. The condition of independence between trials is *not* satisfied because the probability of drawing a defective PC on the second and third trials is dependent on the outcome of the first trial. For example, if the first trial results in a defective PC, then there is only one defective left among the remaining 19 in the shipment. Therefore,

$$P(\text{defective on trial 2}|\text{defective on trial 1}) = 1/19$$

[†]Although it is traditional to call the two possible outcomes of a trial "success" and "failure," they could have been called "head" and "tail," "red" and "white," or any other pair of words. Consequently, the outcome called a "success" does not need to be viewed as a success in the ordinary use of the word.

If the first trial *does not* result in a defective, then there are still two defective PCs in the shipment and the probability of a "success" (a defective) changes to

$$P(\text{defective on trial 2}|\text{nondefective on trial 1}) = 2/19$$

Therefore, the trials are dependent and the sampling does not represent a binomial experiment.

Think about the difference between these two examples. When the sample (the *n* identical trials) came from a large population, the probability of success *p* stayed about the same from trial to trial. When the population size *N* was small, the probability of success *p* changed quite dramatically from trial to trial, and the experiment *was not* binomial.

RULE OF THUMB

If the sample size is large relative to the population size—in particular, if $n/N \geq .05$—then the resulting experiment is not binomial.

In Chapter 4, we tossed two fair coins and constructed the probability distribution for *x*, the number of heads—a binomial experiment with $n = 2$ and $p = .5$. The general binomial probability distribution is constructed in the same way, but the procedure gets complicated as *n* gets large. Fortunately, the probabilities $p(x)$ follow a general pattern. This allows us to use a single formula to find $p(x)$ for any given value of *x*.

THE BINOMIAL PROBABILITY DISTRIBUTION

A binomial experiment consists of *n* identical trials with probability of success *p* on each trial. The probability of *k* successes in *n* trials is

$$P(x = k) = C_k^n p^k q^{n-k} = \frac{n!}{k!(n-k)!} p^k q^{n-k}$$

for values of $k = 0, 1, 2, \ldots, n$. The symbol

$$C_k^n = \frac{n!}{k!(n-k)!} \qquad \text{where } n! = n(n-1)(n-2) \cdots (2)(1) \text{ and } 0! \equiv 1$$

The general formulas for μ, σ^2, and σ given in Chapter 4 can be used to derive the following simpler formulas for the binomial mean and standard deviation.

MEAN AND STANDARD DEVIATION FOR THE BINOMIAL RANDOM VARIABLE

The random variable *x*, the number of successes in *n* trials, has a probability distribution with this center and spread:

$$\text{Mean:} \quad \mu = np$$
$$\text{Variance:} \quad \sigma^2 = npq$$
$$\text{Standard deviation:} \quad \sigma = \sqrt{npq}$$

EXAMPLE 5.3

Find $P(x = 2)$ for a binomial random variable with $n = 10$ and $p = .1$.

Solution $P(x = 2)$ is the probability of observing two successes and eight failures in a sequence of 10 trials. You might observe the 2 successes first, followed by 8 consecutive failures:

S, S, F, F, F, F, F, F, F, F

Since p is the probability of success and q is the probability of failure, this particular sequence has probability

$$ppqqqqqqqq = p^2q^8$$

MY TIP

$n! =$
$n(n - 1)(n - 2) \dots (2)(1)$

For example,
$5! = 5(4)(3)(2)(1) = 120$
and $0! \equiv 1$.

However, many *other* sequences also result in $x = 2$ successes. The binomial formula uses C_2^{10} to count the number of sequences and gives the exact probability when you use the binomial formula with $k = 2$:

$$P(x = 2) = C_2^{10}(.1)^2(.9)^{10-2}$$

$$= \frac{10!}{2!(10 - 2)!}(.1)^2(.9)^8 = \frac{10(9)}{2(1)}(.01)(.430467) = .1937$$

You could repeat the procedure in Example 5.3 for each value of x—0, 1, 2, . . . , 10—and find all the values of $p(x)$ necessary to construct a probability histogram for x. This would be a long and tedious job, but the resulting graph would look like Figure 5.1(a). You can check the height of the bar for $x = 2$ and find $p(2) = P(x = 2) = .1937$. The graph is skewed right; that is, most of the time you will observe small values of x. The mean or "balancing point" is around $x = 1$; in fact, you can use the formula to find the exact mean:

$$\mu = np = 10(.1) = 1$$

Figures 5.1(b) and 5.1(c) show two other binomial distributions with $n = 10$ but with different values of p. Look at the shapes of these distributions. When $p = .5$, the distribution is exactly symmetric about the mean, $\mu = np = 10(.5) = 5$. When $p = .9$, the distribution is the "mirror image" of the distribution for $p = .1$ and is skewed to the left.

FIGURE 5.1

Binomial probability distributions

(a)

(b)

(c)

EXAMPLE 5.4

Over a long period of time it has been observed that a given marksman can hit a target on a single trial with probability equal to .8. Suppose he fires four shots at the target.

1. What is the probability that he will hit the target exactly two times?
2. What is the probability that he will hit the target at least once?

Solution   A "trial" is a single shot at the target, and you can define a "success" as a hit and a "failure" as a miss, so that $n = 4$ and $p = .8$. If you assume that the marksman's chance of hitting the target does not change from shot to shot, then the number x of times he hits the target is a *binomial random variable.*

1. $P(x = 2) = p(2) = C_2^4(.8)^2(.2)^{4-2}$

$$= \frac{4!}{2!2!}(.64)(.04) = \frac{4(3)(2)(1)}{2(1)(2)(1)}(.64)(.04) = .1536$$

The probability is .1536 that he will hit the target exactly two times.

2. $P(\text{at least once}) = P(x \geq 1) = p(1) + p(2) + p(3) + p(4)$
$$= 1 - p(0)$$
$$= 1 - C_0^4(.8)^0(.2)^4$$
$$= 1 - .0016 = .9984$$

Although you could calculate $P(x = 1)$, $P(x = 2)$, $P(x = 3)$, and $P(x = 4)$ to find this probability, using the complement of the event makes your job easier; that is,

$$P(x \geq 1) = 1 - P(x < 1) = 1 - P(x = 0)$$

Can you think of any reason your assumption of independent trials might be wrong? If the marksman learns from his previous shots (that is, he notices the location of his previous shot and adjusts his aim), then his probability p of hitting the target may increase from shot to shot. The trials would *not* be independent, and the experiment would *not* be binomial.

 MY TIP

Use Table 1 rather than the binomial formula whenever possible. *This is an easier way!*

Calculating binomial probabilities can become tedious even for relatively small values of n. As n gets larger, it becomes almost impossible without the help of a calculator or computer. Fortunately, both of these tools are available to us. Computer-generated tables of **cumulative binomial probabilities** are given in Table 1 of Appendix I for values of n ranging from 2 to 25 and for selected values of p. These probabilities can also be generated using *MINITAB* or the Java applets on your CD-Rom.

Cumulative binomial probabilities differ from the *individual* binomial probabilities that you calculated with the binomial formula. Once you find the column of probabilities for the correct values of n and p in Table 1, the row marked k gives the sum of all the binomial probabilities from $x = 0$ to $x = k$. Table 5.1 shows part of Table 1 for $n = 5$ and $p = .6$. If you look in the row marked $k = 3$, you will find

$$P(x \leq 3) = p(0) + p(1) + p(2) + p(3) = .663$$

TABLE 5.1 ● **Portion of Table 1 in Appendix I for *n* = 5**

k	.01	.05	.10	.20	.30	.40	.50	.60	.70	.80	.90	.95	.99	*k*
0	—	—	—	—	—	—	—	.010	—	—	—	—	—	0
1	—	—	—	—	—	—	—	.087	—	—	—	—	—	1
2	—	—	—	—	—	—	—	.317	—	—	—	—	—	2
3	—	—	—	—	—	—	—	.663	—	—	—	—	—	3
4	—	—	—	—	—	—	—	.922	—	—	—	—	—	4
5	—	—	—	—	—	—	—	1.000	—	—	—	—	—	5

If the probability you need to calculate is not in this form, you will need to think of a way to rewrite your probability to make use of the tables!

EXAMPLE 5.5 Use the cumulative binomial table for $n = 5$ and $p = .6$ to find the probabilities of these events:

1. Exactly three successes
2. Three or more successes

Solution

1. If you find $k = 3$ in Table 5.1, the tabled value is

$$P(x \le 3) = p(0) + p(1) + p(2) + p(3)$$

Since you want only $P(x = 3) = p(3)$, you must subtract out the unwanted probability:

$$P(x \le 2) = p(0) + p(1) + p(2)$$

which is found in Table 5.1 with $k = 2$. Then

$$P(x = 3) = P(x \le 3) - P(x \le 2)$$
$$= .663 - .317 = .346$$

2. To find $P(\text{three or more successes}) = P(x \ge 3)$ using Table 5.1, you must use the complement of the event of interest. Write

$$P(x \ge 3) = 1 - P(x < 3) = 1 - P(x \le 2)$$

You can find $P(x \le 2)$ in Table 5.1 with $k = 2$. Then

$$P(x \ge 3) = 1 - P(x \le 2)$$
$$= 1 - .317 = .683$$

PERSONAL TRAINER

How Do I Use Table 1 to Calculate Binomial Probabilities?

1. Find the necessary values of n and p. Isolate the appropriate column in Table. 1.

2. Table 1 gives $P(x \leq k)$ in the row marked k. Rewrite the probability you need so that it is in this form.

   • List the values of x in your event.

   • From the list, write the event as either the difference of two probabilities:

   $$P(x \leq a) - P(x \leq b) \quad \text{for } a > b$$

   or

   $$1 - P(x \leq a)$$

Exercise Reps

A. Consider a binomial random variable with $n = 5$ and $p = .6$. Isolate the appropriate column in Table 1 and fill in the probabilities below.

k	0	1	2	3	4	5
$P(x \leq k)$						

B. Fill in the blanks in the table below.

The Problem	List the Values of x	Write the Probability	Rewrite the Probability (if needed)	Find the Probability
Four or less				
Four or more				
More than four				
Fewer than four				
Between 2 and 4 (inclusive)				
Exactly four				

Progress Report

• Still having trouble? Try again using the Exercise Reps at the end of this section.

• Mastered binomial probabilities? You can skip the Exercise Reps at the end of this section!

Answers are located on the perforated card at the back of this book.

 APPLET

The Java applet called **Calculating Binomial Probabilities** gives a visual display of the binomial distribution for values of $n \leq 100$ and any p that you choose. You can use this applet to calculate binomial probabilities for any value of x or for any interval $a \leq x \leq b$. To reproduce the results of Example 5.5, enter **5** in the box labeled "n:" and **0.6** in the box labeled "p," pressing the "Enter" key after each entry. Next enter the beginning and ending values for x (if you need to calculate an individual probability, both entries will be the same). The probability will be calculated and shaded in red on your monitor (light teal in Figure 5.2) when you press "Enter." What is the probability of three or more successes from Figure 5.2? Does this confirm our answer in Example 5.5? You will use this applet again for the MyApplet Exercises section at the end of the chapter.

FIGURE 5.2

Calculating Binomial Probabilities applet

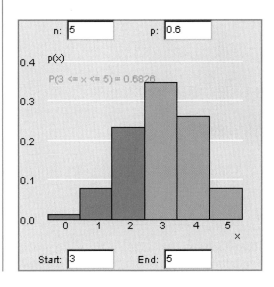

EXAMPLE 5.6

A regimen consisting of a daily dose of vitamin C was tested to determine its effectiveness in preventing the common cold. Ten people who were following the prescribed regimen were observed for a period of 1 year. Eight survived the winter without a cold. Suppose the probability of surviving the winter without a cold is .5 when the vitamin C regimen is not followed. What is the probability of observing eight or more survivors, given that the regimen is ineffective in increasing resistance to colds?

Solution If you assume that the vitamin C regimen is ineffective, then the probability p of surviving the winter without a cold is .5. The probability distribution for x, the number of survivors, is

$$p(x) = C_x^{10}(.5)^x(.5)^{10-x}$$

You have learned four ways to find P(8 or more survivors) = $P(x \geq 8)$. You will get the same results with any of the four; choose the most convenient method for your particular problem.

1. *The binomial formula:*

$$P(8 \text{ or more}) = p(8) + p(9) + p(10)$$
$$= C_8^{10}(.5)^{10} + C_9^{10}(.5)^{10} + C_{10}^{10}(.5)^{10}$$
$$= .055$$

2. *The cumulative binomial tables:* Find the column corresponding to p = .5 in the table for n = 10:

$$P(8 \text{ or more}) = P(x \geq 8) = 1 - P(x \leq 7)$$
$$= 1 - .945 = .055$$

3. The ***Calculating Binomial Probabilities*** *applet:* Enter $n = 10$, $p = .5$ and calculate the probability that x is between 8 and 10. The probability, $P(x \geq 8) = .0547$, is shaded in red on your monitor (light teal in Figure 5.3).

FIGURE 5.3

Java applet for Example 5.6

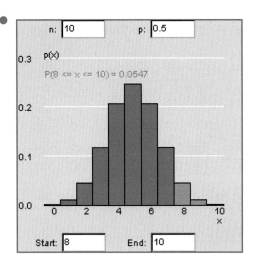

4. *Output from MINITAB:* The output shown in Figure 5.4 gives the **cumulative distribution function,** which gives the same probabilities you found in the cumulative binomial tables. The **probability density function** gives the individual binomial probabilities, which you found using the binomial formula.

FIGURE 5.4

MINITAB output for Example 5.6

Cumulative Distribution Function

Binomial with n = 10 and p = 0.5

x	P(X <= x)
0	0.00098
1	0.01074
2	0.05469
3	0.17187
4	0.37695
5	0.62305
6	0.82813
7	0.94531
8	0.98926
9	0.99902
10	1.00000

Probability Density Function

Binomial with n = 10 and p = 0.5

x	P(X = x)
0	0.000977
1	0.009766
2	0.043945
3	0.117188
4	0.205078
5	0.246094
6	0.205078
7	0.117188
8	0.043945
9	0.009766
10	0.000977

Using the cumulative distribution function, calculate

$$P(x \geq 8) = 1 - P(x \leq 7)$$
$$= 1 - .94531 = .05469$$

Or, using the probability density function, calculate

$$P(x \geq 8) = p(8) + p(9) + p(10)$$
$$= .043945 + .009766 + .000977 = .05469$$

EXAMPLE 5.7

Would you rather take a multiple-choice or a full recall test? If you have absolutely no knowledge of the material, you will score zero on a full recall test. However, if you are given five choices for each question, you have at least one chance in five of guessing correctly! If a multiple-choice exam contains 100 questions, each with five possible answers, what is the expected score for a student who is guessing on each question? Within what limits will the "no-knowledge" scores fall?

Solution If x is the number of correct answers on the 100-question exam, the probability of a correct answer, p, is one in five, so that $p = .2$. Since the student is randomly selecting answers, the $n = 100$ answers are independent, and the expected score for this binomial random variable is

$$\mu = np = 100(.2) = 20 \quad \text{correct answers}$$

To evaluate the spread or variability of the scores, you can calculate

$$\sigma = \sqrt{npq} = \sqrt{100(.2)(.8)} = 4$$

Then, using your knowledge of variation from Tchebysheff's Theorem and the Empirical Rule, you can make these statements:

- A large proportion of the scores will lie within two standard deviations of the mean, or from $20 - 8 = 12$ to $20 + 8 = 28$.

- Almost all the scores will lie within three standard deviations of the mean, or from $20 - 12 = 8$ to $20 + 12 = 32$.

The "guessing" option gives the student a better score than the zero score on the full recall test, but the student still will not pass the exam. What other options does the student have?

5.2 EXERCISES

EXERCISE REPS

5.1 Consider a binomial random variable with $n = 8$ and $p = .7$. Isolate the appropriate column in Table 1 and fill in the probabilities below.

k	0	1	2	3	4	5	6	7	8
$P(x \le k)$									

Fill in the blanks in the table below.

The Problem	List the Values of x	Write the Probability	Rewrite the Probability (if needed)	Find the Probability
Three or less		$P(x \le \_\_)$		
Three or more		$P(x \ge \_\_)$	$1 - P(x \le \_\_)$	
More than three		$P(x > \_\_)$	$1 - P(x \le \_\_)$	
Fewer than three		$P(x < \_\_)$	$P(x \le \_\_) - P(x \le \_\_)$	
Between 3 and 5 (inclusive)		$P(\_\_ \le x \le \_\_)$	$P(x \le \_\_)$	
Exactly three		$P(x = \_\_)$	$P(x \le \_\_) - P(x \le \_\_)$	

5.2 Consider a binomial random variable with $n = 9$ and $p = .3$. Isolate the appropriate column in Table 1 and fill in the probabilities below.

k	0	1	2	3	4	5	6	7	8	9
$P(x \le k)$										

Fill in the blanks in the table below.

The Problem	List the Values of x	Write the Probability	Rewrite the Probability	Find the Probability
Exactly two				
More than two				
Two or more				
Less than two				
Between 2 and 4 (inclusive)				
Two or less				

BASIC TECHNIQUES

5.3 The Urn Problem A jar contains five balls: three red and two white. Two balls are randomly selected without replacement from the jar, and the number x of red balls is recorded. Explain why x is or is not a binomial random variable. (HINT: Compare the characteristics of this experiment with the characteristics of a binomial experiment given in this section.) If the experiment is binomial, give the values of n and p.

5.4 The Urn Problem, continued Refer to Exercise 5.3. Assume that the sampling was conducted with replacement. That is, assume that the first ball was selected from the jar, observed, and then replaced, and that the balls were then mixed before the second ball was selected. Explain why x, the number of red balls observed, is or is not a binomial random variable. If the experiment is binomial, give the values of n and p.

5.5 Evaluate these binomial probabilities:

a. $C_2^8(.3)^2(.7)^6$ **b.** $C_0^4(.05)^0(.95)^4$
c. $C_3^{10}(.5)^3(.5)^7$ **d.** $C_1^7(.2)^1(.8)^6$

5.6 Evaluate these binomial probabilities:

a. $C_0^8(.2)^0(.8)^8$ **b.** $C_1^8(.2)^1(.8)^7$ **c.** $C_2^8(.2)^2(.8)^6$
d. $P(x \le 1)$ when $n = 8, p = .2$
e. P(two or fewer successes)

5.7 Let x be a binomial random variable with $n = 7$, $p = .3$. Find these values:

a. $P(x = 4)$ **b.** $P(x \le 1)$ **c.** $P(x > 1)$
d. $\mu = np$ **e.** $\sigma = \sqrt{npq}$

5.8 Use the formula for the binomial probability distribution to calculate the values of $p(x)$, and construct the probability histogram for x when $n = 6$ and $p = .2$. [HINT: Calculate $P(x = k)$ for seven different values of k.]

5.9 Refer to Exercise 5.8. Construct the probability histogram for a binomial random variable x with $n = 6$ and $p = .8$. Use the results of Exercise 5.8; do not recalculate all the probabilities.

5.10 If x has a binomial distribution with $p = .5$, will the shape of the probability distribution be symmetric, skewed to the left, or skewed to the right?

5.11 Let x be a binomial random variable with $n = 10$ and $p = .4$. Find these values:

a. $P(x = 4)$ **b.** $P(x \ge 4)$ **c.** $P(x > 4)$
d. $P(x \le 4)$ **e.** $\mu = np$ **f.** $\sigma = \sqrt{npq}$

5.12 Use Table 1 in Appendix I to find the sum of the binomial probabilities from $x = 0$ to $x = k$ for these cases:

a. $n = 10, p = .1, k = 3$
b. $n = 15, p = .6, k = 7$
c. $n = 25, p = .5, k = 14$

5.13 Use Table 1 in Appendix I to evaluate the following probabilities for $n = 6$ and $p = .8$:

a. $P(x \ge 4)$ **b.** $P(x = 2)$
c. $P(x < 2)$ **d.** $P(x > 1)$

Verify these answers using the values of $p(x)$ calculated in Exercise 5.9.

5.14 $P(x \leq k)$ in each case:

a. $n = 20, p = .05, k = 2$

b. $n = 15, p = .7, k = 8$

c. $n = 10, p = .9, k = 9$

5.15 Use Table 1 in Appendix I to find the following:

a. $P(x < 12)$ for $n = 20, p = .5$

b. $P(x \leq 6)$ for $n = 15, p = .4$

c. $P(x > 4)$ for $n = 10, p = .4$

d. $P(x \geq 6)$ for $n = 15, p = .6$

e. $P(3 < x < 7)$ for $n = 10, p = .5$

5.16 Find the mean and standard deviation for a binomial distribution with these values:

a. $n = 1000, p = .3$ **b.** $n = 400, p = .01$

c. $n = 500, p = .5$ **d.** $n = 1600, p = .8$

5.17 Find the mean and standard deviation for a binomial distribution with $n = 100$ and these values of p:

a. $p = .01$ **b.** $p = .9$ **c.** $p = .3$

d. $p = .7$ **e.** $p = .5$

5.18 In Exercise 5.17, the mean and standard deviation for a binomial random variable were calculated for a fixed sample size, $n = 100$, and for different values of p. Graph the values of the standard deviation for the five values of p given in Exercise 5.17. For what value of p does the standard deviation seem to be a maximum?

5.19 Let x be a binomial random variable with $n = 20$ and $p = .1$.

a. Calculate $P(x \leq 4)$ using the binomial formula.

b. Calculate $P(x \leq 4)$ using Table 1 in Appendix I.

c. Use the MINITAB output at the top of the page to calculate $P(x \leq 4)$. Compare the results of parts a, b, and c.

d. Calculate the mean and standard deviation of the random variable x.

e. Use the results of part d to calculate the intervals $\mu \pm \sigma$, $\mu \pm 2\sigma$, and $\mu \pm 3\sigma$. Find the probability that an observation will fall into each of these intervals.

f. Are the results of part e consistent with Tchebysheff's Theorem? With the Empirical Rule? Why or why not?

MINITAB Output for Exercise 5.19

Probability Density Function

Binomial with n = 20 and p = 0.1

x	P(X = x)
0	0.121577
1	0.270170
2	0.285180
3	0.190120
4	0.089779
5	0.031921
6	0.008867
7	0.001970
8	0.000356
9	0.000053
10	0.000006
11	0.000001
12	0.000000
13	0.000000
14	0.000000
15	0.000000
16	0.000000
17	0.000000
18	0.000000
19	0.000000
20	0.000000

APPLICATIONS

5.20 Chicago Weather A meteorologist in Chicago recorded the number of days of rain during a 30-day period. If the random variable x is defined as the number of days of rain, does x have a binomial distribution? If not, why not? If so, are both values of n and p known?

5.21 Telemarketers A market research firm hires operators to conduct telephone surveys. The computer randomly dials a telephone number, and the operator asks the respondent whether or not he has time to answer some questions. Let x be the number of telephone calls made until the first respondent is willing to answer the operator's questions. Is this a binomial experiment? Explain.

5.22 SAT Scores In 2003 the average combined SAT score (math + verbal) for college-bound students in the United States was 1026. Suppose that approximately 45% of all high school graduates took this test, and that 100 high school graduates are randomly selected from throughout the United States.[1] Which of the following random variables has an approximate binomial distribution? If possible, give the values for n and p.

a. The number of students who took the SAT

b. The scores of the 100 students on the SAT

c. The number of students who scored above average on the SAT

d. The amount of time it took each student to complete the SAT

5.23 Security Systems A home security system is designed to have a 99% reliability rate. Suppose that nine homes equipped with this system experience an attempted burglary. Find the probabilities of these events:

a. At least one of the alarms is triggered.

b. More than seven of the alarms are triggered.

c. Eight or fewer alarms are triggered.

5.24 Blood Types In a certain population, 85% of the people have Rh-positive blood. Suppose that two people from this population get married. What is the probability that they are both Rh-negative, thus making it inevitable that their children will be Rh-negative?

5.25 Car Colors Car color preferences change over the years and according to the particular model that the customer selects. In a recent year, 10% of all luxury cars sold were black. If 25 cars of that year and type are randomly selected, find the following probabilities:

a. At least five cars are black.

b. At most six cars are black.

c. More than four cars are black.

d. Exactly four cars are black.

e. Between three and five cars (inclusive) are black.

f. More than 20 cars are not black.

5.26 Harry Potter Of all the Harry Potter books purchased in a recent year, about 60% were purchased for readers 14 or older.[2] If 12 Harry Potter fans who bought books that year are surveyed, find the following probabilities.

Who's reading Harry Potter

59% 14 and older

41% Under 14

a. At least five of them are 14 or older.

b. Exactly nine of them are 14 or older.

c. Less than three of them are 14 or older.

5.27 Medical Bills Records show that 30% of all patients admitted to a medical clinic fail to pay their bills and that eventually the bills are forgiven. Suppose $n = 4$ new patients represent a random selection from the large set of prospective patients served by the clinic. Find these probabilities:

a. All the patients' bills will eventually have to be forgiven.

b. One will have to be forgiven.

c. None will have to be forgiven.

5.28 Medical Bills II Consider the medical payment problem in Exercise 5.27 in a more realistic setting. Of all patients admitted to a medical clinic, 30% fail to pay their bills and the debts are eventually forgiven. If the clinic treats 2000 different patients over a period of 1 year, what is the mean (expected) number of debts that have to be forgiven? If x is the number of forgiven debts in the group of 2000 patients, find the variance and standard deviation of x. What can you say about the probability that x will exceed 700? (HINT: Use the values of μ and σ, along with Tchebysheff's Theorem, to answer this question.)

5.29 Whitefly Infestation Suppose that 10% of the fields in a given agricultural area are infested with the sweet potato whitefly. One hundred fields in this area are randomly selected and checked for whitefly.

a. What is the average number of fields sampled that are infested with whitefly?

b. Within what limits would you expect to find the number of infested fields, with probability approximately 95%?

c. What might you conclude if you found that $x = 25$ fields were infested? Is it possible that one of the characteristics of a binomial experiment is not satisfied in this experiment? Explain.

5.30 Color Preferences in Mice In a psychology experiment, the researcher plans to test the color preference of mice under certain experimental conditions. She designs a maze in which the mouse must choose one of two paths, colored either red or blue, at each of 10 intersections. At the end of the maze, the mouse is given a food reward. The researcher counts the number of times the mouse chooses the red path. If you were the researcher, how would you use this count to decide whether the mouse has any preference for color?

5.31 Pet Peeves Across the board, 22% of car leisure travelers rank "traffic and other drivers" as their pet peeve while traveling. Of car leisure travelers in the densely populated Northeast, 33% list this as their pet peeve.[3] A random sample of $n = 8$ such travelers in the Northeast were asked to state their pet peeve while traveling. The *MINITAB* printout shows the *cumulative* and *individual* probabilities.

MINITAB Output for Exercise 5.31

Cumulative Distribution Function		Probability Density Function	
Binomial with n = 8 and p = 0.33		Binomial with n = 8 and p = .033	
x	P(X <= x)	x	P(X = x)
0	0.04061	0	0.040607
1	0.20061	1	0.160003
2	0.47644	2	0.275826
3	0.74814	3	0.271709
4	0.91543	4	0.167283
5	0.98134	5	0.065915
6	0.99758	6	0.016233
7	0.99986	7	0.002284
8	1.00000	8	0.000141

a. Use the binomial formula to find the probability that all eight give "traffic and other drivers" as their pet peeve.

b. Confirm the results of part a using the *MINITAB* printout.

c. What is the probability that at most seven give "traffic and other drivers" as their pet peeve?

5.32 Fast Food and Gas Stations Forty percent of all Americans who travel by car look for gas stations and food outlets that are close to or visible from the highway. Suppose a random sample of $n = 25$ Americans who travel by car are asked how they determine where to stop for food and gas. Let x be the number in the sample who respond that they look for gas stations and food outlets that are close to or visible from the highway.

a. What are the mean and variance of x?

b. Calculate the interval $\mu \pm 2\sigma$. What values of the binomial random variable x fall into this interval?

c. Find $P(6 \le x \le 14)$. How does this compare with the fraction in the interval $\mu \pm 2\sigma$ for any distribution? For mound-shaped distributions?

5.33 Taste Test for PTC The taste test for PTC (phenylthiocarbamide) is a favorite exercise for every human genetics class. It has been established that a single gene determines the characteristic, and that 70% of Americans are "tasters," while 30% are "nontasters." Suppose that 20 Americans are randomly chosen and are tested for PTC.

a. What is the probability that 17 or more are "tasters"?

b. What is the probability that 15 or fewer are "tasters"?

5.34 Man's Best Friend According to the *Humane Society of the United States,* there are approximately 68 million owned dogs in the United States, and 40% of all U.S. households own at least one dog.[4] Suppose that the 40% figure is correct and that 15 households are randomly selected for a pet ownership survey.

a. What is the probability that exactly eight of the households have at least one dog?

b. What is the probability that at most four of the households have at least one dog?

c. What is the probability that more than ten households have at least one dog?

THE POISSON PROBABILITY DISTRIBUTION

5.3

Another discrete random variable that has numerous practical applications is the **Poisson random variable**. Its probability distribution provides a good model for data that represent the number of occurrences of a specified event in a given unit of time or space. Here are some examples of experiments for which the random variable x can be modeled by the Poisson random variable:

- The number of calls received by a switchboard during a given period of time

- The number of bacteria per small volume of fluid

- The number of customer arrivals at a checkout counter during a given minute

- The number of machine breakdowns during a given day
- The number of traffic accidents at a given intersection during a given time period

In each example, *x* **represents the number of events that occur in a period of time or space during which an average of μ such events can be expected to occur.** The only assumptions needed when one uses the Poisson distribution to model experiments such as these are that the counts or events occur **randomly and independently** of one another. The formula for the Poisson probability distribution, as well as its mean and variance, are given next.

THE POISSON PROBABILITY DISTRIBUTION

Let μ be the average number of times that an event occurs in a certain period of time or space. The probability of k occurrences of this event is

$$P(x = k) = \frac{\mu^k e^{-\mu}}{k!}$$

for values of $k = 0, 1, 2, 3, \ldots$. The mean and standard deviation of the Poisson random variable x are

Mean: μ

Standard deviation: $\sigma = \sqrt{\mu}$

The symbol $e = 2.71828 \ldots$ is evaluated using your scientific calculator, which should have a function such as e^x. For each value of k, you can obtain the individual probabilities for the Poisson random variable, just as you did for the binomial random variable.

Alternatively, you can use **cumulative Poisson tables** (Table 2 in Appendix I) or the cumulative or individual probabilities generated by *MINITAB*. Both of these options are usually more convenient than hand calculation. The procedures are similar to those used for the binomial random variable.

MY TIP

Use either the Poisson formula or Table 2 to calculate Poisson probabilities.

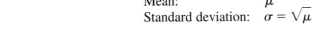

MY PERSONAL TRAINER

How Do I Calculate Poisson Probabilities Using the Formula?

1. Find the necessary value of μ.
2. List the values of x in your event.
3. For each value of x, substitute $x = k$ in the formula, $P(x = k) = \frac{\mu^k e^{-\mu}}{k!}$.
4. Add the individual probabilities in (3) to find the probability of interest.

Exercise Reps

A. Consider a Poisson random variable with $\mu = 1.5$. Calculate the following probabilities using the table below:

Probability	Formula	Calculated Value
$P(x = 0)$	$\dfrac{\mu^k e^{-\mu}}{k!} =$ _____	
$P(x = 1)$	$\dfrac{\mu^k e^{-\mu}}{k!} =$ _____	
$P(1 \text{ or fewer successes})$	$P(x =$ ____$) + P(x =$ ____$)$	

How Do I Use Table 2 to Calculate Poisson Probabilities?

1. Find the necessary value of μ. Isolate the appropriate column in Table 2.
2. Table 2 gives $P(x \leq k)$ in the row marked k. Rewrite the probability you need so that it is in this form.

   - List the values of x in your event.

   - From the list, write the event as either the difference of two probabilities:

   $$P(x \leq a) - P(x \leq b) \quad \text{for } a > b$$

   or

   $$1 - P(x \leq a)$$

Exercise Reps

B. Consider a Poisson random variable with $\mu = 1.5$. Isolate the appropriate column in Table 2 and fill in the probabilities below.

k	0	1	2	3	4	5	6	7
$P(x \leq k)$								

C. Fill in the blanks in the table below.

The Problem	List the Values of x	Write the Probability	Rewrite the Probability (if needed)	Find the Probability
Three or less				
Three or more				
More than three				
Fewer than three				
Between 2 and 4 (inclusive)				
Exactly three				

Progress Report

- Still having trouble? Try again using the Exercise Reps at the end of this section.

- Mastered Poisson probabilities? You can skip the Exercises Reps at the end of this section.

Answers are located on the perforated card at the back of this book.

Once the values for $p(x)$ have been calculated, you can use them to construct a probability histogram for the random variable x. Graphs of the Poisson probability distribution for $\mu = .5$, 1, and 4 are shown in Figure 5.5.

FIGURE 5.5

Poisson probability distributions for $\mu = .5$, 1, and 4

EXAMPLE 5.8 The average number of traffic accidents on a certain section of highway is two per week. Assume that the number of accidents follows a Poisson distribution with $\mu = 2$.

1. Find the probability of no accidents on this section of highway during a 1-week period.
2. Find the probability of at most three accidents on this section of highway during a 2-week period.

Solution

1. The average number of accidents per week is $\mu = 2$. Therefore, the probability of no accidents on this section of highway during a given week is

$$P(x = 0) = p(0) = \frac{2^0 e^{-2}}{0!} = e^{-2} = .135335$$

2. During a 2-week period, the average number of accidents on this section of highway is $2(2) = 4$. The probability of at most three accidents during a 2-week period is

$$P(x \le 3) = p(0) + p(1) + p(2) + p(3)$$

where

$$p(0) = \frac{4^0 e^{-4}}{0!} = .018316 \qquad p(2) = \frac{4^2 e^{-4}}{2!} = .146525$$

$$p(1) = \frac{4^1 e^{-4}}{1!} = .073263 \qquad p(3) = \frac{4^3 e^{-4}}{3!} = .195367$$

Therefore,

$$P(x \le 3) = .018316 + .073263 + .146525 + .195367 = .433471$$

This value could be read directly from Table 2 in Appendix I, indexing $\mu = 4$ and $k = 3$, as $P(x \le 3) = .433$.

In Section 5.2, we used the cumulative binomial tables to simplify the calculation of binomial probabilities. Unfortunately, in practical situations, n is often large and no tables are available.

MY TIP

You can estimate binomial probabilities with the Poisson when n is large and p is small.

THE POISSON APPROXIMATION TO THE BINOMIAL DISTRIBUTION

The Poisson probability distribution provides a simple, easy-to-compute, and accurate approximation to binomial probabilities when n is large and $\mu = np$ is small, preferably with $np < 7$. An approximation suitable for larger values of $\mu = np$ will be given in Chapter 6.

EXAMPLE 5.9

Suppose a life insurance company insures the lives of 5000 men aged 42. If actuarial studies show the probability that any 42-year-old man will die in a given year to be .001, find the exact probability that the company will have to pay $x = 4$ claims during a given year.

Solution The exact probability is given by the binomial distribution as

$$P(x = 4) = p(4) = \frac{5000!}{4!4996!}(.001)^4(.999)^{4996}$$

for which binomial tables are not available. To compute $P(x = 4)$ without the aid of a computer would be very time-consuming, but the Poisson distribution can be used to provide a good approximation to $P(x = 4)$. Computing $\mu = np = (5000)(.001) = 5$ and substituting into the formula for the Poisson probability distribution, we have

$$p(4) \approx \frac{\mu^4 e^{-\mu}}{4!} = \frac{5^4 e^{-5}}{4!} = \frac{(625)(.006738)}{24} = .175$$

The value of $p(4)$ could also be obtained using Table 2 in Appendix I with $\mu = 5$ as

$$p(4) = P(x \le 4) - P(x \le 3) = .440 - .265 = .175$$

EXAMPLE 5.10

A manufacturer of power lawn mowers buys 1-horsepower, two-cycle engines in lots of 1000 from a supplier. She then equips each of the mowers produced by her plant with one of the engines. History shows that the probability of any one engine from that supplier proving unsatisfactory is .001. In a shipment of 1000 engines, what is the probability that none is defective? Three are? Four are?

Solution This is a binomial experiment with $n = 1000$ and $p = .001$. The expected number of defectives in a shipment of $n = 1000$ engines is $\mu = np = (1000)(.001) = 1$. Since this is a binomial experiment with $np < 7$, the probability of x defective engines in the shipment may be approximated by

$$P(x = k) = p(k) = \frac{\mu^k e^{-\mu}}{k!} = \frac{1^k e^{-1}}{k!} = \frac{e^{-1}}{k!}$$

Therefore,

$$p(0) \approx \frac{e^{-1}}{0!} = \frac{.368}{1} = .368$$

$$p(3) \approx \frac{e^{-1}}{3!} = \frac{.368}{6} = .061$$

$$p(4) \approx \frac{e^{-1}}{4!} = \frac{.368}{24} = .015$$

The individual Poisson probabilities for $\mu = 1$ along with the individual binomial probabilities for $n = 1000$ and $p = .001$ were generated by *MINITAB* and are shown in Figure 5.6. The individual probabilities, even though they are computed with totally different formulas, are almost the same. The exact binomial probabilities are in the left section of Figure 5.6, and the Poisson approximations are on the right. Notice that *MINITAB* stops computing probabilities once the value is equal to zero within a preassigned accuracy level.

FIGURE 5.6

MINITAB output of binomial and Poisson probabilities

Probability Density Function

```
Binomial with n = 1000 and p = 0.001

 x    P( X = x )
 0      0.367695
 1      0.368063
 2      0.184032
 3      0.061283
 4      0.015290
 5      0.003049
 6      0.000506
 7      0.000072
 8      0.000009
 9      0.000001
10      0.000000
```

Probability Density Function

```
Poisson with mean = 1

 x    P( X = x )
 0      0.367879
 1      0.367879
 2      0.183940
 3      0.061313
 4      0.015328
 5      0.003066
 6      0.000511
 7      0.000073
 8      0.000009
 9      0.000001
10      0.000000
```

5.3 EXERCISES

EXERCISE REPS

5.35 Consider a Poisson random variable with $\mu = 2.5$. Calculate the following probabilities using the table below.

Probability	Formula	Calculated Value
$P(x = 0)$	$\dfrac{\mu^k e^{-\mu}}{k!} =$ _____	
$P(x = 1)$	$\dfrac{\mu^k e^{-\mu}}{k!} =$ _____	
$P(x = 2)$	$\dfrac{\mu^k e^{-\mu}}{k!} =$ _____	
$P(2$ or fewer successes$)$	$P(x =$ ___$) + P(x =$ ___$) + P(x =$ ___$)$	

5.36 Consider a Poisson random variable with $\mu = 3$. Calculate the following probabilities using the table below.

Probability	Formula	Calculated Value
$P(x = 0)$	$\dfrac{\mu^k e^{-\mu}}{k!} =$ _____	
$P(x = 1)$	$\dfrac{\mu^k e^{-\mu}}{k!} =$ _____	
$P($more than one success$)$	$1 - [P(x =$ ___$) + P(x =$ ___$)]$	

5.37 Consider a Poisson random variable with $\mu = 3$. Isolate the appropriate column in Table 2 and fill in the probabilities below.

k	0	1	2	3	4	5	6	7	8	9	10
$P(x \le k)$											

Fill in the blanks in the table below.

The Problem	List the Values of x	Write the Probability	Rewrite the Probability (if needed)	Find the Probability
Three or less		$P(x \le$ ___$)$		
Three or more		$P(x \ge$ ___$)$	$1 - P(x \le$ ___$)$	
More than three		$P(x >$ ___$)$	$1 - P(x \le$ ___$)$	
Fewer than three		$P(x <$ ___$)$	$P(x \le$ ___$)$	
Between 3 and 5 (inclusive)		$P($___$\le x \le$ ___$)$	$P(x \le$ ___$) - P(x \le$ ___$)$	
Exactly three		$P(x =$ ___$)$	$P(x \le$ ___$) - P(x \le$ ___$)$	

5.38 Consider a Poisson random variable with $\mu = 0.8$. Isolate the appropriate column in Table 2 and fill in the probabilities below.

k	0	1	2	3	4	5
$P(x \le k)$						

Fill in the blanks in the table below.

The Problem	List the Values of x	Write the Probability	Rewrite the Probability (if needed)	Find the Probability
Exactly two				
More than two				
Two or more				
Less than two				
Between 2 and 4 (inclusive)				
Two or less				

BASIC TECHNIQUES

5.39 Let x be a Poisson random variable with mean $\mu = 2$. Calculate these probabilities:

a. $P(x = 0)$ **b.** $P(x = 1)$
c. $P(x > 1)$ **d.** $P(x = 5)$

5.40 Let x be a Poisson random variable with mean $\mu = 2.5$. Use Table 2 in Appendix I to calculate these probabilities:

a. $P(x \geq 5)$ **b.** $P(x < 6)$
c. $P(x = 2)$ **d.** $P(1 \leq x \leq 4)$

5.41 Poisson vs. Binomial Let x be a binomial random variable with $n = 20$ and $p = .1$.

a. Calculate $P(x \leq 2)$ using Table 1 in Appendix I to obtain the exact binomial probability.

b. Use the Poisson approximation to calculate $P(x \leq 2)$.

c. Compare the results of parts a and b. Is the approximation accurate?

5.42 Poisson vs. Binomial II To illustrate how well the Poisson probability distribution approximates the binomial probability distribution, calculate the Poisson approximate values for $p(0)$ and $p(1)$ for a binomial probability distribution with $n = 25$ and $p = .05$. Compare the answers with the exact values obtained from Table 1 in Appendix I.

APPLICATIONS

5.43 Airport Safety The increased number of small commuter planes in major airports has heightened concern over air safety. An eastern airport has recorded a monthly average of five near-misses on landings and takeoffs in the past 5 years.

a. Find the probability that during a given month there are no near-misses on landings and takeoffs at the airport.

b. Find the probability that during a given month there are five near-misses.

c. Find the probability that there are at least five near-misses during a particular month.

5.44 Intensive Care The number x of people entering the intensive care unit at a particular hospital on any one day has a Poisson probability distribution with mean equal to five persons per day.

a. What is the probability that the number of people entering the intensive care unit on a particular day is two? Less than or equal to two?

b. Is it likely that x will exceed 10? Explain.

5.45 Accident Prone Parents who are concerned that their children are "accident prone" can be reassured, according to a study conducted by the Department of Pediatrics at the University of California, San Francisco. Children who are injured two or more times tend to sustain these injuries during a relatively limited time, usually 1 year or less. If the average number of injuries per year for school-age children is two, what are the probabilities of these events?

a. A child will sustain two injuries during the year.

b. A child will sustain two or more injuries during the year.

c. A child will sustain at most one injury during the year.

5.46 Accident Prone, continued Refer to Exercise 5.45.

a. Calculate the mean and standard deviation for x, the number of injuries per year sustained by a school-age child.

b. Within what limits would you expect the number of injuries per year to fall?

5.47 Bacteria in Water Samples If a drop of water is placed on a slide and examined under a microscope, the number x of a particular type of bacteria present has been found to have a Poisson probability distribution. Suppose the maximum permissible count per water specimen for this type of bacteria is five. If the mean count for your water supply is two and you test a single specimen, is it likely that the count will exceed the maximum permissible count? Explain.

5.48 E. coli Outbreak Increased research and discussion have focused on the number of illnesses involving the organism *Escherichia coli* (01257:H7), which causes a breakdown of red blood cells and intestinal hemorrhages in its victims.[4] Sporadic out-

breaks of *E. coli* have occurred in Colorado at a rate of 2.5 per 100,000 for a period of 2 years. Let us suppose that this rate has not changed.

a. What is the probability that at most five cases of *E. coli* per 100,000 are reported in Colorado in a given year?

b. What is the probability that more than five cases of *E. coli* per 100,000 are reported in a given year?

c. Approximately 95% of occurrences of *E. coli* involve at most how many cases?

THE HYPERGEOMETRIC PROBABILITY DISTRIBUTION

5.4

Suppose you are selecting a sample of elements from a population and you record whether or not each element possesses a certain characteristic. You are recording the typical "success" or "failure" data found in the binomial experiment. The sample survey of Example 5.1 and the sampling for defectives of Example 5.2 are practical illustrations of these sampling situations.

If the number of elements in the population is large relative to the number in the sample (as in Example 5.1), the probability of selecting a success on a single trial is equal to the proportion p of successes in the population. Because the population is large in relation to the sample size, this probability will remain constant (for all practical purposes) from trial to trial, and the number x of successes in the sample will follow a binomial probability distribution. However, if the number of elements in the population is small in relation to the sample size ($n/N \geq .05$), the probability of a success for a given trial is dependent on the outcomes of preceding trials. Then the number x of successes follows what is known as a **hypergeometric probability distribution.**

It is easy to visualize the **hypergeometric random variable x** by thinking of a bowl containing M red balls and $N - M$ white balls, for a *total of N* balls in the bowl. You select n balls from the bowl and record x, the number of red balls that you see. If you now define a "success" to be a red ball, you have an example of the hypergeometric random variable x.

The formula for calculating the probability of exactly k successes in n trials is given next.

THE HYPERGEOMETRIC PROBABILITY DISTRIBUTION

A population contains M successes and $N - M$ failures. The probability of exactly k successes in a random sample of size n is

$$P(x = k) = \frac{C_k^M C_{n-k}^{N-M}}{C_n^N}$$

for values of k that depend on N, M, and n with

$$C_n^N = \frac{N!}{n!(N-n)!}$$

The mean and variance of a hypergeometric random variable are very similar to those of a binomial random variable with a correction for the finite population size:

$$\mu = n\left(\frac{M}{N}\right)$$

$$\sigma^2 = n\left(\frac{M}{N}\right)\left(\frac{N-M}{N}\right)\left(\frac{N-n}{N-1}\right)$$

EXAMPLE 5.11

A case of wine has 12 bottles, 3 of which contain spoiled wine. A sample of 4 bottles is randomly selected from the case.

1. Find the probability distribution for x, the number of bottles of spoiled wine in the sample.
2. What are the mean and variance of x?

Solution For this example, $N = 12$, $n = 4$, $M = 3$, and $(N - M) = 9$. Then

$$p(x) = \frac{C_x^3 C_{4-x}^9}{C_4^{12}}$$

1. The possible values for x are 0, 1, 2, and 3, with probabilities

$$p(0) = \frac{C_0^3 C_4^9}{C_4^{12}} = \frac{1(126)}{495} = .25$$

draw 4

3S
9G

$$p(1) = \frac{C_1^3 C_3^9}{C_4^{12}} = \frac{3(84)}{495} = .51$$

$$p(2) = \frac{C_2^3 C_2^9}{C_4^{12}} = \frac{3(36)}{495} = .22$$

$$p(3) = \frac{C_3^3 C_1^9}{C_4^{12}} = \frac{1(9)}{495} = .02$$

2. The mean is given by

$$\mu = 4\left(\frac{3}{12}\right) = 1$$

and the variance is

$$\sigma^2 = 4\left(\frac{3}{12}\right)\left(\frac{9}{12}\right)\left(\frac{12-4}{11}\right) = .5455$$

EXAMPLE 5.12

A particular industrial product is shipped in lots of 20. Testing to determine whether an item is defective is costly; hence, the manufacturer samples production rather than using a 100% inspection plan. A sampling plan constructed to minimize the number of defectives shipped to customers calls for sampling five items from each lot and rejecting the lot if more than one defective is observed. (If the lot is rejected, each item in the lot is then tested.) If a lot contains four defectives, what is the probability that it will be accepted?

Solution Let x be the number of defectives in the sample. Then $N = 20$, $M = 4$, $(N - M) = 16$, and $n = 5$. The lot will be rejected if $x = 2$, 3, or 4. Then

$$P(\text{accept the lot}) = P(x \le 1) = p(0) + p(1) = \frac{C_0^4 C_5^{16}}{C_5^{20}} + \frac{C_1^4 C_4^{16}}{C_5^{20}}$$

$$= \frac{\left(\dfrac{4!}{0!4!}\right)\left(\dfrac{16!}{5!11!}\right)}{\dfrac{20!}{5!15!}} + \frac{\left(\dfrac{4!}{1!3!}\right)\left(\dfrac{16!}{4!12!}\right)}{\dfrac{20!}{5!15!}}$$

$$= \frac{91}{323} + \frac{455}{969} = .2817 + .4696 = .7513$$

5.4 EXERCISES

BASIC TECHNIQUES

5.49 Evaluate these probabilities:

a. $\dfrac{C_1^3 C_1^2}{C_2^5}$ b. $\dfrac{C_2^4 C_1^3}{C_3^7}$ c. $\dfrac{C_4^5 C_0^3}{C_4^8}$

5.50 Let x be the number of successes observed in a sample of $n = 5$ items selected from $N = 10$. Suppose that, of the $N = 10$ items, 6 are considered "successes."

a. Find the probability of observing no successes.

b. Find the probability of observing at least two successes.

c. Find the probability of observing exactly two successes.

5.51 Let x be a hypergeometric random variable with $N = 15$, $n = 3$, and $M = 4$.

a. Calculate $p(0)$, $p(1)$, $p(2)$, and $p(3)$.

b. Construct the probability histogram for x.

c. Use the formulas given in Section 5.4 to calculate $\mu = E(x)$ and σ^2.

d. What proportion of the population of measurements fall into the interval $(\mu \pm 2\sigma)$? Into the interval $(\mu \pm 3\sigma)$? Do these results agree with those given by Tchebysheff's Theorem?

5.52 Candy Choices A candy dish contains five blue and three red candies. A child reaches up and selects three candies without looking.

a. What is the probability that there are two blue and one red candies in the selection?

b. What is the probability that the candies are all red?

c. What is the probability that the candies are all blue?

APPLICATIONS

5.53 Defective Computer Chips A piece of electronic equipment contains six computer chips, two of which are defective. Three computer chips are randomly chosen for inspection, and the number of defective chips is recorded. Find the probability distribution for x, the number of defective computer chips. Compare your results with the answers obtained in Exercise 4.90.

5.54 Gender Bias? A company has five applicants for two positions: two women and three men. Suppose that the five applicants are equally qualified and that no preference is given for choosing either gender. Let x equal the number of women chosen to fill the two positions.

a. Write the formula for $p(x)$, the probability distribution of x.

b. What are the mean and variance of this distribution?

c. Construct a probability histogram for x.

5.55 Teaching Credentials In southern California, a growing number of persons pursuing a teaching credential are choosing paid internships over traditional student teaching programs. A group of eight candidates for three local teaching positions consisted of five candidates who had enrolled in paid internships and three candidates who had enrolled in traditional student teaching programs. Let us assume that all eight candidates are equally qualified for the positions. Let x represent the number of internship-trained candidates who are hired for these three positions.

a. Does x have a binomial distribution or a hypergeometric distribution? Support your answer.

b. Find the probability that three internship-trained candidates are hired for these positions.

c. What is the probability that none of the three hired was internship-trained?

d. Find $P(x \leq 1)$.

5.56 Seed Treatments Seeds are often treated with a fungicide for protection in poor-draining, wet environments. In a small-scale trial prior to a large-scale experiment to determine what dilution of the fungicide to apply, five treated seeds and five untreated seeds were planted in clay soil and the number of plants emerging from the treated and untreated seeds were recorded. Suppose the dilution was not effective and only four plants emerged. Let x represent the number of plants that emerged from treated seeds.

a. Find the probability that $x = 4$.

b. Find $P(x \leq 3)$.

c. Find $P(2 \leq x \leq 3)$.

CHAPTER REVIEW

Key Concepts and Formulas

I. The Binomial Random Variable

1. **Five characteristics:** n identical independent trials, each resulting in either *success S* or *failure F*; probability of success is p and remains constant from trial to trial; and x is the number of successes in n trials

2. **Calculating binomial probabilities**

   a. Formula: $P(x = k) = C_k^n p^k q^{n-k}$

   b. Cumulative binomial tables

   c. Individual and cumulative probabilities using *MINITAB*

3. Mean of the binomial random variable: $\mu = np$

4. Variance and standard deviation: $\sigma^2 = npq$ and $\sigma = \sqrt{npq}$

II. The Poisson Random Variable

1. The number of events that occur in a period of time or space, during which an average of μ such events are expected to occur

2. **Calculating Poisson probabilities**

   a. Formula: $P(x = k) = \dfrac{\mu^k e^{-\mu}}{k!}$

   b. Cumulative Poisson tables

   c. Individual and cumulative probabilities using *MINITAB*

3. Mean of the Poisson random variable: $E(x) = \mu$

4. Variance and standard deviation: $\sigma^2 = \mu$ and $\sigma = \sqrt{\mu}$

5. Binomial probabilities can be approximated with Poisson probabilities when $np < 7$, using $\mu = np$.

III. The Hypergeometric Random Variable

1. The number of successes in a sample of size n from a finite population containing M successes and $N - M$ failures

2. Formula for the probability of k successes in n trials:
$$P(x = k) = \frac{C_k^M C_{n-k}^{N-M}}{C_n^N}$$

3. Mean of the hypergeometric random variable:
$$\mu = n\left(\frac{M}{N}\right)$$

4. Variance and standard deviation:
$$\sigma^2 = n\left(\frac{M}{N}\right)\left(\frac{N-M}{N}\right)\left(\frac{N-n}{N-1}\right) \quad \text{and} \quad \sigma = \sqrt{\sigma^2}$$

MINITAB

Binomial and Poisson Probabilities

For a random variable that has either a binomial or a Poisson probability distribution, *MINITAB* has been programmed to calculate either exact probabilities—$P(x = k)$—for a given value of k or the cumulative probabilities—$P(x \le k)$—for a given value of k. You must specify which distribution you are using and the necessary parameters: n and p for the binomial distribution and μ for the Poisson distribution. Also, you have the option of specifying only one single value of k or several values of k, which should be stored in a column (say, C1) of the *MINITAB* worksheet.

Consider a binomial distribution with $n = 16$ and $p = .25$. Neither n nor p appears in the tables in Appendix I. Since the possible values of x for this binomial random variable range from 0 to 16, we can generate the entire probability distribution as well as the cumulative probabilities by entering the numbers 0 to 16 in C1, named "x."

One way to quickly enter a set of consecutive integers in a column is to do the following:

- Enter the first two integers to create a pattern.
- Use your mouse to highlight the first two integers.
- Use your mouse to grab the square handle in the lower right corner of the highlighted area. Drag the handle down to continue the pattern.
- You will see an integer appear in a small yellow square. Release the mouse when you have the desired number of integers—in this case, $\boxed{16}$.

Once the necessary values of x have been entered, use **Calc** → **Probability Distributions** → **Binomial** to generate the Dialog box shown in Figure 5.7. Type the number of trials and the value of p in the appropriate boxes, and select "x" for the input column. (If you do not type a column number for storage, *MINITAB* will display the results in the Session window. If you type C2 or p(x) in the box marked "Optional

FIGURE 5.7

Binomial Distribution	☒

C1   x
C2   p(x)

- Probability
- Cumulative probability
- Inverse cumulative probability

Number of trials: 16
Probability of success: .25

- Input column: x
  Optional storage: |

- Input constant:
  Optional storage:

[Select]

[Help]   [OK]   [Cancel]

storage," the results will appear in column C2 and also in the Session window.) Make sure that the radio button marked "Probability" is selected. The probability density function appears in the Session window when you click **OK** (a portion is shown in Figure 5.8). What is the probability that x equals 4? That x is either 3 or 4?

To calculate cumulative probabilities, make sure that the dot marked "Cumulative probability" is selected, and enter the appropriate values of x in C1. If you have only one value of x, it is simpler to select the Input constant box and enter the appropriate value. For example, for a Poisson random variable with $\mu = 5$, use **Calc → Probability Distributions → Poisson** and enter a mean of 5. If the number 6 is entered into the Input constant box, the probability that x is less than or equal to 6 appears in the Session window (see Figure 5.9).

What value k is such that only 5% of the values of x exceed this value (and 95% are less than or equal to k)? If you enter the probability .95 into the Input constant box and select the option marked "Inverse cumulative probability" (see Figure 5.10), then the values of x on either side of the ".95 mark" are shown in the Session window as in Figure 5.11. Hence, if you observed a value of $x = 10$, this would be an unusual observation because $P(x > 9) = 1 - .968172 = .031828$.

FIGURE 5.8

FIGURE 5.9

FIGURE 5.10

FIGURE 5.11

Supplementary Exercises

5.57 List the five identifying characteristics of the binomial experiment.

5.58 Under what conditions can the Poisson random variable be used to approximate the probabilities associated with the binomial random variable? What application does the Poisson distribution have other than to estimate certain binomial probabilities?

5.59 Under what conditions would you use the hypergeometric probability distribution to evaluate the probability of x successes in n trials?

5.60 Tossing a Coin A balanced coin is tossed three times. Let x equal the number of heads observed.

a. Use the formula for the binomial probability distribution to calculate the probabilities associated with $x = 0, 1, 2,$ and 3.

b. Construct the probability distribution.

c. Find the mean and standard deviation of x, using these formulas:

$$\mu = np$$
$$\sigma = \sqrt{npq}$$

d. Using the probability distribution in part b, find the fraction of the population measurements lying within one standard deviation of the mean. Repeat for two standard deviations. How do your results agree with Tchebysheff's Theorem and the Empirical Rule?

5.61 Coins, continued Refer to Exercise 5.60. Suppose the coin is definitely unbalanced and the probability of a head is equal to $p = .1$. Follow the instructions in parts a, b, c, and d. Note that the probability distribution loses its symmetry and becomes skewed when p is not equal to 1/2.

5.62 Cancer Survivor Rates The 10-year survival rate for bladder cancer is approximately 50%. If 20 people who have bladder cancer are properly treated for the disease, what is the probability that:

a. At least 1 will survive for 10 years?

b. At least 10 will survive for 10 years?

c. At least 15 will survive for 10 years?

5.63 Garbage Collection A city commissioner claims that 80% of all people in the city favor garbage collection by contract to a private concern (in contrast to collection by city employees). To check the theory that the proportion of people in the city favoring private collection is .8, you randomly sample 25 people and find that x, the number of people who support the commissioner's claim, is 22.

a. What is the probability of observing at least 22 who support the commissioner's claim if, in fact, $p = .8$?

b. What is the probability that x is exactly equal to 22?

c. Based on the results of part a, what would you conclude about the claim that 80% of all people in the city favor private collection? Explain.

5.64 Integers If a person is given the choice of an integer from 0 to 9, is it more likely that he or she will choose an integer near the middle of the sequence than one at either end?

a. If the integers are equally likely to be chosen, find the probability distribution for x, the number chosen.

b. What is the probability that a person will choose a 4, 5, or 6?

c. What is the probability that a person will not choose a 4, 5, or 6?

5.65 Integers II Refer to Exercise 5.64. Twenty people are asked to select a number from 0 to 9. Eight of them choose a 4, 5, or 6.

a. If the choice of any one number is as likely as any other, what is the probability of observing eight or more choices of the numbers 4, 5, or 6?

b. What conclusions would you draw from the results of part a?

5.66 College Costs A *USA Today* snapshot reports that 52% of households with children under 18 have started to save for college. Of this number, approximately 50% have saved at least $5,000 toward college expense.[6] Suppose that we randomly select $n = 15$ families that have begun saving for college, and let x be the number that have saved $5,000 or more.

a. What is the probability distribution for x?

b. What is $P(x \le 8)$?

c. Find the probability that x exceeds 8?

d. What is the largest value of c for which $P(x \le c) \le .10$?

5.67 Vacation Homes As Americans start the 21st century, the number one status symbol is no longer being a top executive. Approximately 60% of Americans rank "owning a vacation home nestled on a beach or near a mountain resort" number one as a status symbol. A sample of $n = 400$ Americans is randomly selected.

a. What is the average number in the sample who would rank owning a vacation home number one?

b. What is the standard deviation of the number in the sample who would rank owning a vacation home number one?

c. Within what range would you expect to find the number in the sample who would rank having a vacation home as the number one status symbol?

d. If only 200 in a sample of 400 people ranked owning a vacation home as the top status symbol, would you consider this unusual? Explain. What conclusions might you draw from this sample information?

5.68 Reality TV Reality TV (*Survivor, Fear Factor*, etc.) is a new phenomenon in television programming, with contestants escaping to remote locations, taking dares, breaking world records, or racing across the country. Of those who watch reality TV, 50% say that their favorite reality show involves escaping to remote locations.[7] If 20 reality-TV fans are randomly selected, find the following probabilities:

a. Exactly 16 say that their favorite reality show involves escaping to remote locations.

b. From 15 to 18 say that their favorite reality show involves escaping to remote locations.

c. Five or fewer say that their favorite reality show involves escaping to remote locations. Would this be an unlikely occurrence?

5.69 Psychosomatic Problems A psychiatrist believes that 80% of all people who visit doctors have problems of a psychosomatic nature. She decides to select 25 patients at random to test her theory.

a. Assuming that the psychiatrist's theory is true, what is the expected value of x, the number of the 25 patients who have psychosomatic problems?

b. What is the variance of x, assuming that the theory is true?

c. Find $P(x \le 14)$. (Use tables and assume that the theory is true.)

d. Based on the probability in part c, if only 14 of the 25 sampled had psychosomatic problems, what conclusions would you make about the psychiatrist's theory? Explain.

5.70 Student Fees A student government states that 80% of all students favor an increase in student fees to subsidize a new recreational area. A random sample of $n = 25$ students produced 15 in favor of increased fees. What is the probability that 15 or fewer in the sample would favor the issue if student government is correct? Do the data support the student government's assertion, or does it appear that the percentage favoring an increase in fees is less than 80%?

5.71 Gray Hair on Campus College campuses are graying! According to a recent article, one in four college students is aged 30 or older. Many of these students are women updating their job skills. Assume that the 25% figure is accurate, that your college is representative of colleges at large, and that you sample $n = 200$ students, recording x, the number of students age 30 or older.

a. What are the mean and standard deviation of x?

b. If there are 35 students in your sample who are age 30 or older, would you be willing to assume that the 25% figure is representative of your campus? Explain.

5.72 Probability of Rain Most weather forecasters protect themselves very well by attaching probabilities to their forecasts, such as "The probability of rain today is 40%." Then, if a particular forecast is incorrect, you are expected to attribute the error to the random behavior of the weather rather than to the inaccuracy of the forecaster. To check the accuracy of a particular forecaster, records were checked only for those days when the forecaster predicted rain "with 30% probability." A check of 25 of those days indicated that it rained on 10 of the 25.

a. If the forecaster is accurate, what is the appropriate value of p, the probability of rain on one of the 25 days?

b. What are the mean and standard deviation of x, the number of days on which it rained, assuming that the forecaster is accurate?

c. Calculate the z-score for the observed value, $x = 10$. [HINT: Recall from Section 2.6 that z-score $= (x - \mu)/\sigma$.]

d. Do these data disagree with the forecast of a "30% probability of rain"? Explain.

5.73 What's for Breakfast? A packaging experiment is conducted by placing two different package designs for a breakfast food side by side on a supermarket shelf. The objective of the experiment is to see whether buyers indicate a preference for one of the two package designs. On a given day, 25 customers purchased a package from the supermarket. Let x equal the number of buyers who choose the second package design.

a. If there is no preference for either of the two designs, what is the value of p, the probability that a buyer chooses the second package design?

b. If there is no preference, use the results of part a to calculate the mean and standard deviation of x.

c. If 5 of the 25 customers choose the first package design and 20 choose the second design, what do you conclude about the customers' preference for the second package design?

5.74 Plant Density One model for plant competition assumes that there is a zone of resource depletion around each plant seedling. Depending on the size of the zones and the density of the plants, the zones of resource depletion may overlap with those of other seedlings in the vicinity. When the seeds are randomly dispersed over a wide area, the number of neighbors that a seedling may have usually follows a Poisson distribution with a mean equal to the density of seedlings per unit area. Suppose that the density of seedlings is four per square meter (m^2).

a. What is the probability that a given seedling has no neighbors within 1 m^2?

b. What is the probability that a seedling has at most three neighbors per m^2?

c. What is the probability that a seedling has five or more neighbors per m^2?

d. Use the fact that the mean and variance of a Poisson random variable are equal to find the proportion of neighbors that would fall into the interval $\mu \pm 2\sigma$. Comment on this result.

5.75 Plant Genetics A peony plant with red petals was crossed with another plant having streaky petals. The probability that an offspring from this cross has red flowers is .75. Let x be the number of plants with red petals resulting from ten seeds from this cross that were collected and germinated.

a. Does the random variable x have a binomial distribution? If not, why not? If so, what are the values of n and p?

b. Find $P(x \geq 9)$.

c. Find $P(x \leq 1)$.

d. Would it be unusual to observe one plant with red petals and the remaining nine plants with streaky petals? If these experimental results actually occurred, what conclusions could you draw?

5.76 Dominant Traits The alleles for black (B) and white (b) feather color in chickens show incomplete dominance; individuals with the gene pair Bb have "blue" feathers. When one individual that is homozygous dominant (BB) for this trait is mated with an individual that is homozygous recessive (bb) for this trait, 1/4 of the offspring will carry the gene pair BB, 1/2 will carry the gene pair Bb, and 1/4 will carry the gene pair bb. Let x be the number of chicks with "blue" feathers in a sample of $n = 20$ chicks resulting from crosses involving homozygous dominant chickens (BB) with homozygous recessive chickens (bb).

a. Does the random variable x have a binomial distribution? If not, why not? If so, what are the values of n and p?

b. What is the mean number of chicks with "blue" feathers in the sample?

c. What is the probability of observing fewer than five chicks with "blue" feathers?

d. What is the probability that the number of chicks with "blue" feathers is greater than or equal to 10 but less than or equal to 12?

5.77 Football Coin Tosses During the 1992 football season, the Los Angeles Rams (now the St. Louis Rams) had a bizarre streak of coin-toss losses. In fact, they lost the call 11 weeks in a row.[8]

a. The Rams' computer system manager said that the odds against losing 11 straight tosses are 2047 to 1. Is he correct?

b. After these results were published, the Rams lost the call for the next two games, for a total of 13 straight losses. What is the probability of this happening if, in fact, the coin was fair?

5.78 Diabetes in Children Insulin-dependent diabetes (IDD) is a common chronic disorder of children. This disease occurs most frequently in persons of northern European descent, but the incidence ranges from a low of 1–2 cases per 100,000 per year to a high of more than 40 per 100,000 in parts of Finland.[9] Let us assume that an area in Europe has an incidence of 5 cases per 100,000 per year.

a. Can the distribution of the number of cases of IDD in this area be approximated by a Poisson distribution? If so, what is the mean?

b. What is the probability that the number of cases of IDD in this area is less than or equal to 3 per 100,000?

c. What is the probability that the number of cases is greater than or equal to 3 but less than or equal to 7 per 100,000?

d. Would you expect to observe 10 or more cases of IDD per 100,000 in this area in a given year? Why or why not?

5.79 Defective Videotapes A manufacturer of videotapes ships them in lots of 1200 tapes per lot. Before shipment, 20 tapes are randomly selected from each lot and tested. If none is defective, the lot is shipped. If one or more are defective, every tape in the lot is tested.

a. What is the probability distribution for x, the number of defective tapes in the sample of 20?

b. What distribution can be used to approximate probabilities for the random variable x in part a?

c. What is the probability that a lot will be shipped if it contains 10 defectives? 20 defectives? 30 defectives?

5.80 Eating Out More American families with two working parents are eating out these days. What do Americans use as criteria in selecting the places where they will eat? According to *USA Today,* 56% choose a restaurant because of its great food, while 22% make their choice because of reasonable prices.[10] Suppose that in a random sample of $n = 5$ families who eat out, x is the number who choose a restaurant because of its great food.

a. What is the probability that all five families say they select a restaurant because of its great food?

b. What is the probability that exactly three of the five families select a restaurant because of its great food?

c. What is probability that at least one family selects a restaurant because of its great food?

5.81 Tay–Sachs Disease Tay–Sachs disease is a genetic disorder that is usually fatal in young children. If both parents are carriers of the disease, the probability that their offspring will develop the disease is approximately .25. Suppose a husband and wife are both carriers of the disease and the wife is pregnant on three different occasions. If the occurrence of Tay–Sachs in any one offspring is independent of the occurrence in any other, what are the probabilities of these events?

a. All three children will develop Tay–Sachs disease.

b. Only one child will develop Tay–Sachs disease.

c. The third child will develop Tay–Sachs disease, given that the first two did not.

5.82 After College What will you do once you graduate? Where will you live? A report in *American Demographics* indicates that 60% of all college students plan to move back home after graduation.[11] A group of 12 college students are randomly selected and asked whether or not they plan to move back home after graduation.

a. What is the probability that more than six of the students plan to move back home?

b. What is the probability that fewer than five of the students plan to move back home?

c. What is the probability that exactly ten of the students plan to move back home?

5.83 The Triangle Test A procedure often used to control the quality of name-brand food products utilizes a panel of five "tasters." Each member of the panel tastes three samples, two of which are from

batches of the product known to have the desired taste and the other from the latest batch. Each taster selects the sample that is different from the other two. Assume that the latest batch does have the desired taste, and that there is no communication between the tasters.

a. If the latest batch tastes the same as the other two batches, what is the probability that the taster picks it as the one that is different?

b. What is the probability that exactly one of the tasters picks the latest batch as different?

c. What is the probability that at least one of the tasters picks the latest batch as different?

5.84 Do You Return Your Questionnaires? The president of a company specializing in public opinion surveys claims that approximately 70% of all people to whom the agency sends questionnaires respond by filling out and returning the questionnaire. Twenty such questionnaires are sent out, and assume that the president's claim is correct.

a. What is the probability that exactly ten of the questionnaires are filled out and returned?

b. What is the probability that at least 12 of the questionnaires are filled out and returned?

c. What is the probability that at most ten of the questionnaires are filled out and returned?

5.85 Questionnaires, continued Refer to Exercise 5.84. If $n = 20$ questionnaires are sent out,

a. What is the average number of questionnaires that will be returned?

b. What is the standard deviation of the number of questionnaires that will be returned?

c. If $x = 10$ of the 20 questionnaires are returned to the company, would you consider this to be an unusual response? Explain.

5.86 Poultry Problems A preliminary investigation reported that approximately 30% of locally grown poultry were infected with an intestinal parasite that, though not harmful to those consuming the poultry, decrease the usual weight growth rates in the birds. A diet supplement believed to be effective against this parasite was added to the bird's food. Twenty-five birds were examined after having the supplement for at least two weeks, and three birds were still found to be infested with the parasite.

a. If the diet supplement is ineffective, what is the probability of observing three or fewer birds infected with the intestinal parasite?

b. If in fact the diet supplement was effective and reduced the infection rate to 10%, what is the probability observing three or fewer infected birds?

5.87 Machine Breakdowns In a food processing and packaging plant, there are, on the average, two packaging machine breakdowns per week. Assume the weekly machine breakdowns follow a Poisson distribution.

a. What is the probability that there are no machine breakdowns in a given week?

b. Calculate the probability that there are no more than two machine breakdowns in a given week.

5.88 Safe Drivers? Evidence shows that the probabilty that a driver will be involved in a serious auotmobile accident during a given year is .01. A particular corporation employs 100 full-time traveling sales reps. Based on this evidence, use the Poisson approximation to the binomial distribution to find the probability that exactly two of the sales reps will be involved in a serious automobile accident during the coming year.

5.89 Stressed Out A subject is taught to do a task in two different ways. Studies have shown that when subjected to mental strain and asked to perform the task, the subject most often reverts to the method first learned, regardless of whether it was easier or more difficult. If the probability that a subject returns to the first method learned is .8 and six subjects are tested, what is the probability that at least five of the subjects revert to their first learned method when asked to perform their task under stress?

5.90 Enrolling in College A West Coast university has found that about 90% of its accepted applicants for enrollment in the freshman class will actually enroll. In 2004, 1360 applicants were accepted to the university. Within what limits would you expect to find the size of the freshman class at this university in the fall of 2004?

5.91 Earthquakes! Suppose that one out of every ten homeowners in the state of California has invested in earthquake insurance. If 15 homeowners are randomly chosen to be interviewed,

a. What is the probability that at least one had earthquake insurance?

b. What is the probability that four or more have earthquake insurance?

c. Within what limits would you expect the number of homeowners insured against earthquakes to fall?

5.92 Bad Wiring Improperly wired control panels were mistakenly installed on two of eight large automated machine tools. It is uncertain which of the machine tools have the defective panels, and a sample of four tools is randomly chosen for inspection. What is the probability that the sample will include no defective panels? Both defective panels?

5.93 Eating on the Run How do you survive when there's no time to eat—fast food, no food, a protein bar, candy? A snapshot in *USA Today* indicates that 36% of women aged 25–55 say that, when they are too busy to eat, they get fast food from a drive-thru.[12] A random sample of 100 women aged 25–55 is selected.

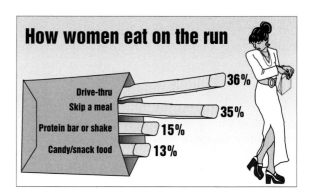

a. What is the average number of women who say they eat fast food when they're too busy to eat?

b. What is the standard deviation for the number of women who say they eat fast food when they're too busy to eat?

c. If 49 of the women in the sample said they eat fast food when they're too busy to eat, would this be an unusual occurrence? Explain.

MY APPLET Exercises

Use the **Calculating Binomial Probabilities** applet for the following set of exercises.

5.94 Refer to Exercise 5.8 and 5.9.

a. Use the applet to construct the probability histogram for a binomial random variable x with $n = 6$ and $p = .2$.

b. Use the applet to construct the probability histogram for a binomial random variable x with $n = 6$ and $p = .8$. How would you describe the shapes of the distributions in parts a and b?

c. Use the applet to construct the probability histogram for a binomial random variable x with $n = 6$ and $p = .5$. How would you describe the shape of this distribution?

5.95 Use the applet to find the following:

a. $P(x < 6)$ for $n = 22$, $p = .65$

b. $P(x = 8)$ for $n = 12$, $p = .4$

c. $P(x > 14)$ for $n = 20$, $p = .5$

d. $P(2 < x < 6)$ for $n = 15$, $p = .3$

e. $P(x \geq 6)$ for $n = 50$, $p = .7$

5.96 Successful Surgeries A new surgical procedure is said to be successful 80% of the time. Suppose the operation is performed five times and the results are assumed to be independent of one another. What are the probabilities of these events?

a. All five operations are successful.

b. Exactly four are successful.

c. Less than two are successful.

5.97 Surgery, continued Refer to Exercise 5.96. If less than two operations were successful, how would you feel about the performance of the surgical team?

5.98 Engine Failure Suppose the four engines of a commercial aircraft are arranged to operate independently and that the probability of in-flight failure of a single engine is .01. What is the probability of these events on a given flight?

a. No failures are observed.

b. No more than one failure is observed.

5.99 McDonald's or Burger King? Suppose that 50% of all young adults prefer McDonald's to Burger King when asked to state a preference. A group of 100 young adults were randomly selected and their preferences recorded.

a. What is the probability that more than 60 preferred McDonald's?

b. What is the probability that between 40 and 60 (inclusive) preferred McDonald's?

c. What is the probability that between 40 and 60 (inclusive) preferred Burger King?

5.100 After Graduation Most of today's college seniors want to start earning money as soon as they graduate from college. In fact, only 14% of college seniors say that they are likely to take some time off to travel and relax after graduation.[13] Suppose that 50 college seniors were randomly selected.

a. What is the average value of x, the number of college seniors in the group who say they will take some time off after graduation? What is the standard deviation of x?

b. Would it be unlikely to find 15 or more in the group who say they will take some time off after graduation? Use the applet to find the probability of this event.

c. How many standard deviations from the mean is the value $x = 15$? Does this confirm your answer in part b?

CASE STUDY **A Mystery: Cancers Near a Reactor**

How safe is it to live near a nuclear reactor? Men who lived in a coastal strip that extends 20 miles north from a nuclear reactor in Plymouth, Massachusetts, developed some forms of cancer at a rate 50% higher than the statewide rate, according to a study endorsed by the Massachusetts Department of Public Health and reported in the May 21, 1987, edition of the *New York Times.*[14]

The cause of the cancers is a mystery, but it was suggested that the cancer was linked to the Pilgrim I reactor, which had been shut down for 13 months because of management problems. Boston Edison, the owner of the reactor, acknowledged radiation releases in the mid-1970s that were just above permissible levels. If the reactor was in fact responsible for the excessive cancer rate, then the currently acknowledged level of radiation required to cause cancer would have to change. However, confounding the mystery was the fact that women in this same area were seemingly unaffected.

In his report, Dr. Sidney Cobb, an epidemiologist, noted the connection between the radiation releases at the Pilgrim I reactor and 52 cases of hematopoietic cancers. The report indicated that this unexpectedly large number might be attributable to airborne radioactive effluents from Pilgrim I, concentrated along the coast by wind patterns and not dissipated, as assumed by government regulators. How unusual was this number of cancer cases? That is, statistically speaking, is 52 a highly improbable number of cases? If the answer is yes, then either some external factor (possibly radiation) caused this unusually large number, or we have observed a very rare event!

The Poisson probability distribution provides a good approximation to the distributions of variables such as the number of deaths in a region due to a rare disease, the number of accidents in a manufacturing plant per month, or the number of airline crashes per month. Therefore, it is reasonable to assume that the Poisson distribution provides an appropriate model for the number of cancer cases in this instance.

1. If the 52 reported cases represented a rate 50% higher than the statewide rate, what is a reasonable estimate of μ, the average number of such cancer cases statewide?

2. Based on your estimate of μ, what is the estimated standard deviation of the number of cancer cases statewide?

3. What is the z-score for the $x = 52$ observed cases of cancer? How do you interpret this z-score in light of the concern about an elevated rate of hematopoietic cancers in this area?

6

The Normal Probability Distribution

© AFP/Getty Images

GENERAL OBJECTIVES

In Chapters 4 and 5, you learned about discrete random variables and their probability distributions. In this chapter, you will learn about continuous random variables and their probability distributions and about one very important continuous random variable—the normal. You will learn how to calculate normal probabilities and, under certain conditions, how to use the normal probability distribution to approximate the binomial probability distribution. Then, in Chapter 7 and in the chapters that follow, you will see how the normal probability distribution plays a central role in statistical inference.

CHAPTER INDEX

MY PERSONAL TRAINER

How Do I Use Table 3 to Calculate Probabilities under the Standard Normal Curve?
How Do I Calculate Binomial Probabilities Using the Normal Approximation?

The Long and Short of It

If you were the boss, would height play a role in your selection of a successor for your job? Would you purposely choose a successor who was shorter than you? The case study at the end of this chapter examines how the normal curve can be used to investigate the height distribution of Chinese men eligible for a very prestigious job.

PROBABILITY DISTRIBUTIONS FOR CONTINUOUS RANDOM VARIABLES

6.1

When a random variable x is discrete, you can assign a positive probability to each value that x can take and get the probability distribution for x. The sum of all the probabilities associated with the different values of x is 1. However, not all experiments result in random variables that are discrete. **Continuous random variables,** such as heights and weights, length of life of a particular product, or experimental laboratory error, can assume the infinitely many values corresponding to points on a line interval. If you try to assign a positive probability to each of these uncountable values, the probabilities will no longer sum to 1, as with discrete random variables. Therefore, you must use a different approach to generate the probability distribution for a continuous random variable.

Suppose you have a set of measurements on a continuous random variable, and you create a relative frequency histogram to describe their distribution. For a small number of measurements, you could use a small number of classes; then as more and more measurements are collected, you can use more classes and reduce the class width. The outline of the histogram will change slightly, for the most part becoming less and less irregular, as shown in Figure 6.1. As the number of measurements becomes very large and the class widths become very narrow, the relative frequency histogram appears more and more like the smooth curve shown in Figure 6.1(d). This smooth curve describes the **probability distribution of the continuous random variable.**

FIGURE 6.1

Relative frequency histograms for increasingly large sample sizes

(a)

(b)

(c)

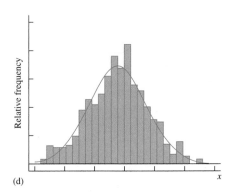

(d)

How can you create a model for this probability distribution? A continuous random variable can take on any of an infinite number of values on the real line, much like the infinite number of grains of sand on a beach. The probability distribution is created by distributing one unit of probability along the line, much as you might distribute a handful of sand. The probability—grains of sand or measurements—will pile up in certain places, and the result is the probability distribution shown in Figure 6.2. The depth or **density** of the probability, which varies with x, may be described by a mathematical formula $f(x)$, called the **probability distribution** or **probability density function** for the random variable x.

FIGURE 6.2

The probability distribution $f(x)$; $P(a < x < b)$ is equal to the shaded area under the curve

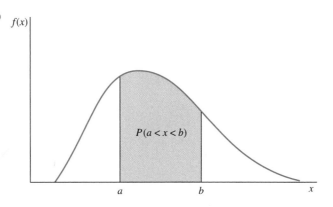

Several important properties of continuous probability distributions parallel their discrete counterparts. Just as the sum of discrete probabilities (or the sum of the relative frequencies) is equal to 1, and the probability that x falls into a certain interval can be found by summing the probabilities in that interval, continuous probability distributions have the characteristics listed next.

 TIP

For continuous random variables,

area = probability.

- The area under a continuous probability distribution is equal to 1.
- The probability that x will fall into a particular interval—say, from a to b—is equal to the area under the curve between the two points a and b. This is the shaded area in Figure 6.2.

 TIP

Area under the curve equals 1.

There is also one important difference between discrete and continuous random variables. Consider the probability that x equals some particular value—say, a. Since there is no area above a single point—say, $x = a$—in the probability distribution for a continuous random variable, our definition implies that the probability is 0.

- $P(x = a) = 0$ for continuous random variables.
- This implies that $P(x \geq a) = P(x > a)$ and $P(x \leq a) = P(x < a)$.
- This is *not* true in general for discrete random variables.

How do you choose the model—that is, the probability distribution $f(x)$—appropriate for a given experiment? Many types of continuous curves are available for modeling. Some are mound-shaped, like the one in Figure 6.1(d), but others are not. In general, try to pick a model that meets these criteria:

- It fits the accumulated body of data.
- It allows you to make the best possible inferences using the data.

EXAMPLE 6.1

The *uniform random variable* is used to model the behavior of a continuous random variable whose values are uniformly or evenly distributed over a given interval. For example, the error x introduced by rounding an observation to the nearest inch would probably have a uniform distribution over the interval from $-.5$ to $.5$. The probability density function $f(x)$ would be "flat" as shown in Figure 6.3. The height of the rectangle is set at 1, so that the total area under the probability distribution is 1.

FIGURE 6.3

A uniform probability distribution

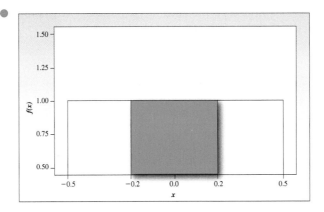

What is the probability that the rounding error is less than $.2$ in magnitude?

Solution This probability corresponds to the area under the distribution between $x = -.2$ and $x = .2$. Since the height of the rectangle is 1,

$$P(-.2 < x < .2) = [.2 - (-.2)] \times 1 = .4$$

EXAMPLE 6.2

The *exponential random variable* is used to model continuous random variables such as waiting times or lifetimes associated with electronic components. For example, the waiting time at a supermarket checkout counter has an exponential distribution with an average waiting time of 5 minutes. The probability density function $f(x) = .2e^{-.2x}$, is shown in Figure 6.4. To find areas under this curve, you can use the fact that $P(x > a) = e^{-.2a}$ for $a > 0$. What is the probability that you have to wait ten minutes or more at the checkout counter?

Solution The probability to be calculated is the area shaded in Figure 6.4. Use the general formula for $P(x > a)$ to find

$$P(x > 10) = e^{-.2(10)} = .135$$

FIGURE 6.4

An exponential probability distribution

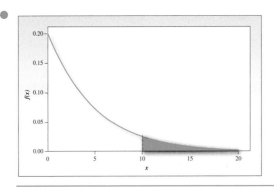

Your model may not always fit the experimental situation perfectly, but you should try to choose a model that *best fits* the population relative frequency histogram. The better the model approximates reality, the better your inferences will be. Fortunately, many continuous random variables have mound-shaped frequency distributions, such as the data in Figure 6.1(d). The **normal probability distribution** provides a good model for describing this type of data.

THE NORMAL PROBABILITY DISTRIBUTION

Continuous probability distributions can assume a variety of shapes. However, a large number of random variables observed in nature possess a frequency distribution that is approximately mound-shaped or, as the statistician would say, is approximately a normal probability distribution. The formula that generates this distribution is shown next.

NORMAL PROBABILITY DISTRIBUTION

$$f(x) = \frac{1}{\sigma\sqrt{2\pi}} e^{-(x-\mu)^2/(2\sigma^2)} \qquad -\infty < x < \infty$$

The symbols e and π are mathematical constants given approximately by 2.7183 and 3.1416, respectively; μ and σ ($\sigma > 0$) are parameters that represent the population mean and standard deviation, respectively.

The graph of a normal probability distribution with mean μ and standard deviation σ is shown in Figure 6.5. The mean μ locates the *center* of the distribution, and the distribution is *symmetric* about its mean μ. Since the total area under the normal probability distribution is equal to 1, the symmetry implies that the area to the right of μ is .5 and the area to the left of μ is also .5. The *shape* of the distribution is determined by σ, the population standard deviation. As you can see in Figure 6.6, large values of σ reduce the height of the curve and increase the spread; small values of σ increase the height of the curve and reduce the spread. Figure 6.6 shows three normal probability distributions with different means and standard deviations. Notice the differences in shape and location.

FIGURE 6.5

Normal probability distribution

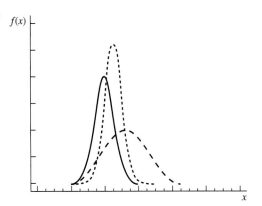

MY APPLET

The Java applet called **Visualizing Normal Curves** gives a visual display of the normal distribution for values of μ between -10 and $+8$ and for values of σ between 0.5 and 1.8. The dark teal curve is the standard normal z with mean 0 and standard deviation 1. You can use this applet to compare its shape to the shape of other normal curves (the red curve on your monitor, light teal in Figure 6.7) by moving the sliders to change the mean and standard deviation. What happens when you change the mean? When you change the standard deviation?

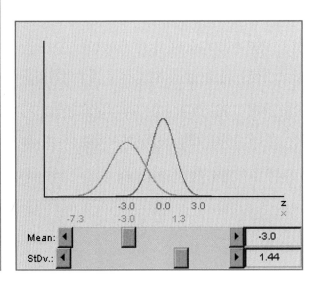

You rarely find a variable with values that are infinitely small $(-\infty)$ or infinitely large $(+\infty)$. Even so, many *positive* random variables (such as heights, weights, and times) have distributions that are well approximated by a normal distribution. According to the Empirical Rule, almost all values of a normal random variable lie in the interval $\mu \pm 3\sigma$. As long as the values within three standard deviations of the mean are *positive,* the normal distribution provides a good model to describe the data.

TABULATED AREAS OF THE NORMAL PROBABILITY DISTRIBUTION

To find the probability that a normal random variable x lies in the interval from a to b, we need to find the area under the normal curve between the points a and b (see Figure 6.2). However (see Figure 6.6), there are an infinitely large number of normal distributions—one for each different mean and standard deviation. A separate table of areas for each of these curves is obviously impractical. Instead, we use a standardization procedure that allows us to use the same table for all normal distributions.

The Standard Normal Random Variable

A normal random variable x is **standardized** by expressing its value as the number of standard deviations (σ) it lies to the left or right of its mean μ. This is really just a change in the units of measure that we use, as if we were measuring in inches rather than in feet! The standardized normal random variable, z, is defined as

$$z = \frac{x - \mu}{\sigma}$$

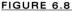

MY TIP

Area under the z-curve equals 1.

or equivalently,

$$x = \mu + z\sigma$$

From the formula for z, we can draw these conclusions:

- When x is less than the mean μ, the value of z is negative.
- When x is greater than the mean μ, the value of z is positive.
- When $x = \mu$, the value of $z = 0$.

The probability distribution for z, shown in Figure 6.8, is called the **standardized normal distribution** because its mean is 0 and its standard deviation is 1. Values of z on the left side of the curve are negative, while values on the right side are positive. The area under the standard normal curve to the left of a specified value of z—say, z_0—is the probability $P(z \leq z_0)$. This **cumulative area** is recorded in Table 3 of Appendix I and is shown as the shaded area in Figure 6.8. An abbreviated version of Table 3 is given in Table 6.1. Notice that the table contains both positive and negative values of z. The left-hand column of the table gives the value of z correct to the tenth place; the second decimal place for z, corresponding to hundredths, is given across the top row.

FIGURE 6.8

Standardized normal distribution

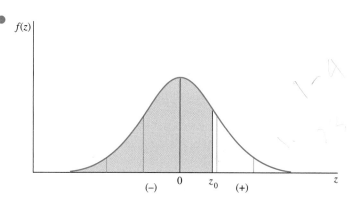

TABLE 6.1

Abbreviated Version of Table 3 in Appendix I
Table 3. Areas Under the Normal Curve

z	.00	.01	.02	.03	. . .	.09
−3.4	.0003	.0003	.0003	.0003		
−3.3	.0005	.0005	.0005	.0004		
−3.2	.0007	.0007	.0006	.0006		
−3.1	.0010	.0009	.0009	.0009		
−3.0	.0013	.0013	.0013	.0012	. . .	.0010
−2.9	.0019	.	.	.		
−2.8	.0026	.	.	.		
−2.7	.0035	.	.	.		
−2.6	.0047					
−2.5	.0062					
.	.					
.	.					
.	.					
−2.0	.0228					
.	.					
.	.					
.	.					

Table 3. Areas Under the Normal Curve (*continued*)

z	.00	.01	.02	.03	.04	. . .	.09
0.0	.5000	.5040	.5080	.5120	.5160		
0.1	.5398	.5438	.5478	.5517	.5557		
0.2	.5793	.5832	.5871	.5910	.5948		
0.3	.6179	.6217	.6255	.6293	.6331		
0.4	.6554	.6591	.6628	.6664	.6700	. . .	.6879
0.5	.6915	.	.	.			
0.6	.7257	.	.	.			
0.7	.7580	.	.	.			
0.8	.7881						
0.9	.8159						
.	.						
.	.						
.	.						
2.0	.9772						

EXAMPLE 6.3 — Find $P(z \leq 1.63)$. This probability corresponds to the area to the left of a point $z = 1.63$ standard deviations to the right of the mean (see Figure 6.9).

Solution   The area is shaded in Figure 6.9. Since Table 3 in Appendix I gives areas under the normal curve to the left of a specified value of z, you simply need to find the tabled value for $z = 1.63$. Proceed down the left-hand column of the table to $z = 1.6$ and across the top of the table to the column marked .03. The intersection of this row and column combination gives the area .9484, which is $P(z \leq 1.63)$.

MY TIP

$P(z \leq 1.63) = P(z < 1.63)$

FIGURE 6.9

Area under the standard
normal curve for
Example 6.3

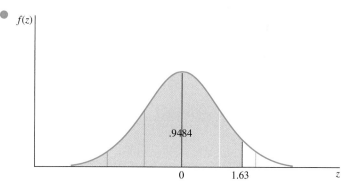

Areas to the left of $z = 0$ are found using negative values of z.

EXAMPLE 6.4 Find $P(z \geq -0.5)$. This probability corresponds to the area to the *right* of a point $z = -0.5$ standard deviation to the left of the mean (see Figure 6.10).

FIGURE 6.10

Area under the standard
normal curve for
Example 6.4

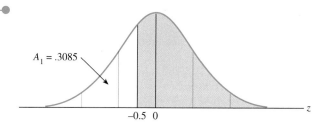

Solution The area given in Table 3 is the area to the left of a specified value of z. Indexing $z = -0.5$ in Table 3, we can find the area A_1 to the *left* of -0.5 to be .3085. Since the area under the curve is 1, we find $P(z \geq -0.5) = 1 - A_1 = 1 - .3085 = .6915$.

EXAMPLE 6.5 Find $P(-.5 \leq z \leq 1.0)$. This probability is the area between $z = -0.5$ and $z = 1.0$, as shown in Figure 6.11.

FIGURE 6.11

Area under the standard
normal curve for
Example 6.5

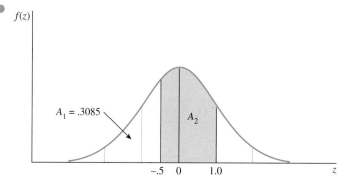

Solution The area required is the shaded area A_2 in Figure 6.11. From Table 3 in Appendix I, you can find the area to the left of $z = -0.5$ ($A_1 = .3085$) and the area to the left of $z = 1.0$ $(A_1 + A_2 = .8413)$. To find the area marked A_2, we subtract the two entries:

$$A_2 = (A_1 + A_2) - A_1 = .8413 - .3085 = .5328$$

That is, $P(-.5 \le z \le 1.0) = .5328$.

(MY) **PERSONAL TRAINER**

How Do I Use Table 3 to Calculate Probabilities under the Standard Normal Curve?

- To calculate the area to the left of a z-value, find the area directly from Table 3.

- To calculate the area to the right of a z-value, find the area in Table 3, and subtract from 1.

- To calculate the area between two values of z, find the two areas in Table 3, and subtract one area from the other.

Exercise Reps

Consider a standard random variable with mean $\mu = 0$ and standard deviation $\sigma = 1$. Use Table 3 and fill in the probabilities below.

The Interval	Write the Probability	Rewrite the Probability (if needed)	Find the Probability
Less than 1.5	$P(z <$ ____ $)$		
Greater than 2	$P(z >$ ____ $)$		
Greater than 2.33	$P(z >$ ____ $)$		
Between -1.96 and 1.96	$P($ ____ $< z <$ ____ $)$		
Between -1.24 and 2.37	$P($ ____ $< z <$ ____ $)$		
Less than or equal to -1	$P(z \le$ ____ $)$		

Progress Report

- Still having trouble? Try again using the Exercise Reps at the end of this section.

- Mastered the z-table? You can skip the Exercises Reps at the end of this section!

Answers are located on the perforated card at the back of this book.

EXAMPLE 6.6

Find the probability that a normally distributed random variable will fall within these ranges:

1. One standard deviation of its mean
2. Two standard deviations of its mean

Solution

1. Since the standard normal random variable z measures the distance from the mean in units of standard deviations, you need to find

$$P(-1 \le z \le 1) = .8413 - .1587 = .6826$$

   Remember that you calculate the area between two z-values by subtracting the tabled entries for the two values.

2. As in part 1, $P(-2 \le z \le 2) = .9772 - .0228 = .9544$.

These probabilities agree with the approximate values of 68% and 95% in the Empirical Rule from Chapter 2.

EXAMPLE 6.7

Find the value of z—say z_0—such that .95 of the area is within $\pm z_0$ standard deviations of the mean.

 TIP

We know the area. Work from the inside of the table out.

Solution The shaded area in Figure 6.12 is the area within $\pm z_0$ standard deviations of the mean, which needs to be equal to .95. The "tail areas" under the curve are not shaded, and have a combined area of $1 - .95 = .05$. Because of the symmetry of the normal curve, these two tail areas have the same area, so that $A_1 = .05/2 = .025$ in Figure 6.12. Thus, the entire *cumulative area* to the left of z_0 to equal $A_1 + A_2 = .95 + .025 = .9750$. This area is found in the interior of Table 3 in Appendix I in the row corresponding to $z = 1.9$ and the .06 column. Hence, $z_0 = 1.96$. Note that this result is very close to the approximate value, $z = 2$, used in the Empirical Rule.

FIGURE 6.12

Area under the standard normal curve for Example 6.7

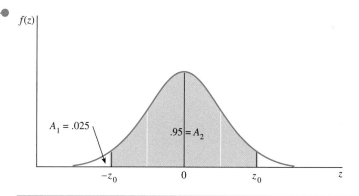

Calculating Probabilities for a General Normal Random Variable

Most of the time, the probabilities you are interested in will involve x, a normal random variable with mean μ and standard deviation σ. You must then *standardize* the interval of interest, writing it as the equivalent interval in terms of z, the standard normal random variable. Once this is done, the probability of interest is the area that you find using the *standard normal probability distribution*.

EXAMPLE 6.8

Let x be a normally distributed random variable with a mean of 10 and a standard deviation of 2. Find the probability that x lies between 11 and 13.6.

Solution The interval from $x = 11$ to $x = 13.6$ must be standardized using the formula for z. When $x = 11$,

$$z = \frac{x - \mu}{\sigma} = \frac{11 - 10}{2} = .5$$

 TIP

Always draw a picture—
it helps!

and when $x = 13.6$,

$$z = \frac{x - \mu}{\sigma} = \frac{13.6 - 10}{2} = 1.8$$

The desired probability is therefore $P(.5 \le z \le 1.8)$, the area lying between $z = .5$ and $z = 1.8$, as shown in Figure 6.13. From Table 3 in Appendix I, you find that the area to the left of $z = .5$ is .6915, and the area to the left of $z = 1.8$ is .9641. The desired probability is the difference between these two probabilities, or

$$P(.5 \le z \le 1.8) = .9641 - .6915 = .2726$$

FIGURE 6.13

Area under the standard
normal curve for
Example 6.8

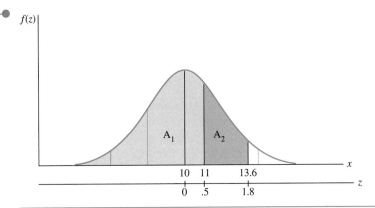

MY **APPLET**

The Java applet called **Normal Distribution Probabilities** allows you to calculate areas under a normal distribution for any values of μ and σ you select. Simply type the appropriate mean and standard deviation into the boxes at the top of the applet, type the interval of interest into the boxes at the bottom of the applet, and press "Enter" at each step to record your changes. (The "Tab" key will move your cursor from box to box.) The necessary area will be shaded in red on your monitor (teal in Figure 6.14) and the probability is given to the left of the curve.

- If you need an area under the standard normal distribution, use $\mu = 0$ and $\sigma = 1$.
- In Example 6.8, we need an area under a normal distribution with $\mu = 10$ and $\sigma = 2$. Notice the values of x and z located along the horizontal axis. Find the probability, $P(11 \le x \le 13.6) = P(0.5 \le z \le 1.8) = .2726$, in Figure 6.14.

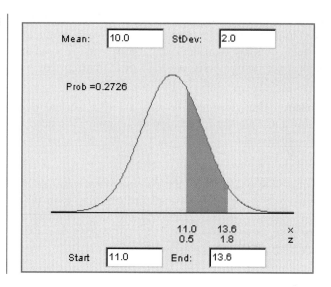

EXAMPLE 6.9

Studies show that gasoline use for compact cars sold in the United States is normally distributed, with a mean of 25.5 miles per gallon (mpg) and a standard deviation of 4.5 mpg. What percentage of compacts get 30 mpg or more?

Solution The proportion of compacts that get 30 mpg or more is given by the shaded area in Figure 6.15. To solve this problem, you must first find the z-value corresponding to $x = 30$. Substituting into the formula for z, you get

$$z = \frac{x - \mu}{\sigma} = \frac{30 - 25.5}{4.5} = 1.0$$

The area A_1 to the left of $z = 1.0$, is .8413 (from Table 3 in Appendix I). Then the proportion of compacts that get 30 mpg or more is equal to:

$$P(x \geq 30) = 1 - P(z < 1) = 1 - .8413 = .1587$$

The percentage exceeding 30 mpg is

$$100(.1587) = 15.87\%$$

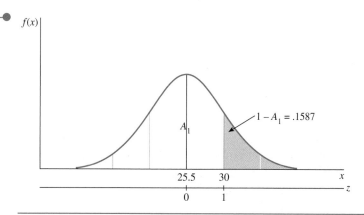

EXAMPLE 6.10

Refer to Example 6.9. In times of scarce energy resources, a competitive advantage is given to an automobile manufacturer who can produce a car that has substantially better fuel economy than the competitors' cars. If a manufacturer wishes to develop a compact car that outperforms 95% of the current compacts in fuel economy, what must the gasoline use rate for the new car be?

Solution The gasoline use rate x has a normal distribution with a mean of 25.5 mpg and a standard deviation of 4.5 mpg. You need to find a particular value—say, x_0—such that

$$P(x \le x_0) = .95$$

This is the 95th percentile of the distribution of gasoline use rate x. Since the only information you have about normal probabilities is in terms of the standard normal random variable z, start by standardizing the value of x_0:

$$z_0 = \frac{x_0 - 25.5}{4.5}$$

Since the value of z_0 corresponds to x_0, it must *also* have area .95 to its left, as shown in Figure 6.16. If you look in the interior of Table 3 in Appendix I, you will find that the area .9500 is exactly halfway between the areas for $z = 1.64$ and $z = 1.65$. Thus, z_0 must be exactly halfway between 1.64 and 1.65, or

$$z_0 = \frac{x_0 - 25.5}{4.5} = 1.645$$

Solving for x_0, you obtain

$$x_0 = \mu + z_0\sigma = 25.5 + (1.645)(4.5) = 32.9$$

FIGURE 6.16

Area under the standard normal curve for Example 6.10

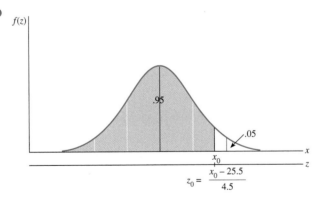

The manufacturer's new compact car must therefore get 32.9 mpg to outperform 95% of the compact cars currently available on the U.S. market.

MY APPLET

The Java applet called **Normal Probabilities and z-scores** allows you to calculate areas under a normal distribution for any values of μ and σ you select. Once you specify one value for x, the applet calculates the value of z, and one of four types of areas, which you can select from the dropdown list at the bottom of the applet:

- Cumulative ⇒ area to the left of z
- One-tailed ⇒ area to the right of z
- Two-tailed ⇒ area in two tails cut off by $-z$ and z
- Middle ⇒ area between $-z$ and z

You can also work backward as we did to solve the problem in Example 6.10. We entered the mean and standard deviation, and then selected "Cumulative" with a probability of **.95.** If the boxes for x and z are left blank, pressing "Enter" will solve for these values, as shown in Figure 6.17. What is the value of x, correct to 6 decimal places?

FIGURE 6.17

Normal Probabilities and z-scores applet

6.3 EXERCISES

EXERCISE REPS

6.1 Consider a standard random variable with $\mu = 0$ and standard deviation $\sigma = 1$. Use Table 3 and fill in the probabilities below.

The Interval	Write the Probability	Rewrite the Probability (if needed)	Find the Probability
Less than −2	$P(z < \underline{\quad})$		
Greater than 1.16	$P(z > \underline{\quad})$		
Greater than 1.645	$P(z > \underline{\quad})$		
Between −2.33 and 2.33	$P(\underline{\quad} < z < \underline{\quad})$		
Between 1.24 and 2.58	$P(\underline{\quad} < z < \underline{\quad})$		
Less than or equal to 1.88	$P(z \leq \underline{\quad})$		

6.2 Repeat Exercise 6.1. Use Table 3 and fill in the probabilities below.

The Interval	Write the Probability	Rewrite the Probability (if needed)	Find the Probability
Greater than 5	$P(z > \_\_\_\_)$		
Between -3 and 3	$P(\_\_\_\_ < z < \_\_\_\_)$		
Between -0.5 and 1.5	$P(\_\_\_\_ < z < \_\_\_\_)$		
Less than or equal to -6.7	$P(z \leq \_\_\_\_)$		
Less than 2.81	$P(z < \_\_\_\_)$		
Greater than 2.81	$P(z > \_\_\_\_)$		

BASIC TECHNIQUES

6.3 Calculate the area under the standard normal curve to the left of these values:

a. $z = 1.6$ **b.** $z = 1.83$

c. $z = .90$ **d.** $z = 4.18$

6.4 Calculate the area under the standard normal curve between these values:

a. $z = -1.4$ and $z = 1.4$ **b.** $z = -3.0$ and $z = 3.0$

6.5 Find the following probabilities for the standard normal random variable z:

a. $P(-1.43 < z < .68)$ **b.** $P(.58 < z < 1.74)$

c. $P(-1.55 < z < -.44)$ **d.** $P(z > 1.34)$

e. $P(z < -4.32)$

6.6 Find these probabilities for the standard normal random variable z:

a. $P(z < 2.33)$ **b.** $P(z < 1.645)$

c. $P(z > 1.96)$ **d.** $P(-2.58 < z < 2.58)$

6.7 a. Find a z_0 such that $P(z > z_0) = .025$.

b. Find a z_0 such that $P(z < z_0) = .9251$.

6.8 Find a z_0 such that $P(-z_0 < z < z_0) = .8262$.

6.9 a. Find a z_0 that has area .9505 to its left.

b. Find a z_0 that has area .05 to its left.

6.10 a. Find a z_0 such that $P(-z_0 < z < z_0) = .90$.

b. Find a z_0 such that $P(-z_0 < z < z_0) = .99$.

6.11 Find the following *percentiles* for the standard normal random variable z:

a. 90th percentile   **b.** 95th percentile

c. 98th percentile   **d.** 99th percentile

6.12 A normal random variable x has mean $\mu = 10$ and standard deviation $\sigma = 2$. Find the probabilities of these x-values:

a. $x > 13.5$ **b.** $x < 8.2$ **c.** $9.4 < x < 10.6$

6.13 A normal random variable x has mean $\mu = 1.20$ and standard deviation $\sigma = .15$. Find the probabilities of these x-values:

a. $1.00 < x < 1.10$ **b.** $x > 1.38$

c. $1.35 < x < 1.50$

6.14 A normal random variable x has an unknown mean μ and standard deviation $\sigma = 2$. If the probability that x exceeds 7.5 is .8023, find μ.

6.15 A normal random variable x has mean 35 and standard deviation 10. Find a value of x that has area .01 to its right. This is the *99th percentile* of this normal distribution.

6.16 A normal random variable x has mean 50 and standard deviation 15. Would it be unusual to observe the value $x = 0$? Explain your answer.

6.17 A normal random variable x has an unknown mean and standard deviation. The probability that x exceeds 4 is .9772, and the probability that x exceeds 5 is .9332. Find μ and σ.

APPLICATIONS

6.18 Hamburger Meat The meat department at a local supermarket specifically prepares its "1-pound" packages of ground beef so that there will be a variety of weights, some slightly more and some slightly less than 1 pound. Suppose that the weights of these "1-pound" packages are normally distributed with a mean of 1.00 pound and a standard deviation of .15 pound.

a. What proportion of the packages will weigh more than 1 pound?

b. What proportion of the packages will weigh between .95 and 1.05 pounds?

c. What is the probability that a randomly selected package of ground beef will weigh less than .80 pound?

d. Would it be unusual to find a package of ground beef that weighs 1.45 pounds? How would you explain such a large package?

6.19 Human Heights Human heights are one of many biological random variables that can be modeled by the normal distribution. Assume the heights of men have a mean of 69 inches with a standard deviation of 3.5 inches.

a. What proportion of all men will be taller than 6′0″? (HINT: Convert the measurements to inches.)

b. What is the probability that a randomly selected man will be between 5′8″ and 6′1″ tall?

c. President George W. Bush is 5′11″ tall. Is this an unusual height?

d. An article in *Chance* magazine indicates that, of the 36 elected presidents from 1856 to year 2000, 17 were 6′0″ or taller.[1] Would you consider this to be unusual, given the proportion found in part a?

6.20 Christmas Trees The diameters of Douglas firs grown at a Christmas tree farm are normally distributed with a mean of 4 inches and a standard deviation of 1.5 inches.

a. What proportion of the trees will have diameters between 3 and 5 inches?

b. What proportion of the trees will have diameters less than 3 inches?

c. Your Christmas tree stand will expand to a diameter of 6 inches. What proportion of the trees will not fit in your Christmas tree stand?

6.21 Cerebral Blood Flow Cerebral blood flow (CBF) in the brains of healthy people is normally distributed with a mean of 74 and a standard deviation of 16.

a. What proportion of healthy people will have CBF readings between 60 and 80?

b. What proportion of healthy people will have CBF readings above 100?

c. If a person has a CBF reading below 40, he is classified as at risk for a stroke. What proportion of healthy people will mistakenly be diagnosed as "at risk"?

6.22 Braking Distances For a car traveling 30 miles per hour (mph), the distance required to brake to a stop is normally distributed with a mean of 50 feet and a standard deviation of 8 feet. Suppose you are traveling 30 mph in a residential area and a car moves abruptly into your path at a distance of 60 feet.

a. If you apply your brakes, what is the probability that you will brake to a stop within 40 feet or less? Within 50 feet or less?

b. If the only way to avoid a collision is to brake to a stop, what is the probability that you will avoid the collision?

6.23 Elevator Capacities Suppose you must establish regulations concerning the maximum number of people who can occupy an elevator. A study of elevator occupancies indicates that, if eight people occupy the elevator, the probability distribution of the total weight of the eight people has a mean equal to 1200 pounds and a variance equal to 9800 lbs.[2] What is the probability that the total weight of eight people exceeds 1300 pounds? 1500 pounds? (Assume that the probability distribution is approximately normal.)

6.24 A Phosphate Mine The discharge of suspended solids from a phosphate mine is normally distributed, with a mean daily discharge of 27 milligrams per liter (mg/l) and a standard deviation of 14 mg/l. What proportion of days will the daily discharge exceed 50 mg/l?

6.25 Sunflowers An experimenter publishing in the *Annals of Botany* investigated whether the stem diameters of the dicot sunflower would change depending on whether the plant was left to sway freely in the wind or was artificially supported.[2] Suppose that the unsupported stem diameters at the base of a particular species of sunflower plant have a normal distribution with an average diameter of 35 millimeters (mm) and a standard deviation of 3 mm.

a. What is the probability that a sunflower plant will have a basal diameter of more than 40 mm?

b. If two sunflower plants are randomly selected, what is the probability that both plants will have a basal diameter of more than 40 mm?

c. Within what limits would you expect the basal diameters to lie, with probability .95?

d. What diameter represents the 90th percentile of the distribution of diameters?

6.26 Breathing Rates The number of times x an adult human breathes per minute when at rest depends on the age of the human and varies greatly from person to person. Suppose the probability distribution for x is approximately normal, with the mean equal to 16 and the standard deviation equal to 4. If a person is selected at random and the number x of breaths per minute while at rest is recorded, what is the probability that x will exceed 22?

6.27 Economic Forecasts One method of arriving at economic forecasts is to use a consensus approach. A forecast is obtained from each of a large number of analysts, and the average of these individual forecasts is the consensus forecast. Suppose the individual 2006 January prime interest rate forecasts of all economic analysts are approximately normally distributed, with the mean equal to 4.5% and the standard deviation equal to 0.1%. If a single analyst is randomly selected from among this group, what is the probability that the analyst's forecast of the prime interest rate will take on these values?

a. Exceed 4.75%

b. Be less than 4.375%

6.28 Tax Audit How does the IRS decide on the percentage of income tax returns to audit for each state? Suppose they do it by randomly selecting 50 values from a normal distribution with a mean equal to 1.55% and a standard deviation equal to .45%. (Computer programs are available for this type of sampling.)

a. What is the probability that a particular state will have more than 2.5% of its income tax returns audited?

b. What is the probability that a state will have less than 1% of its income tax returns audited?

6.29 Bacteria in Drinking Water Suppose the numbers of a particular type of bacteria in samples of 1 milliliter (ml) of drinking water tend to be approximately normally distributed, with a mean of 85 and a standard deviation of 9. What is the probability that a given 1-ml sample will contain more than 100 bacteria?

6.30 Loading Grain A grain loader can be set to discharge grain in amounts that are normally distributed, with mean μ bushels and standard deviation equal to 25.7 bushels. If a company wishes to use the loader to fill containers that hold 2000 bushels of grain and wants to overfill only one container in 100, at what value of μ should the company set the loader?

6.31 How Many Words? A publisher has discovered that the numbers of words contained in a new manuscript are normally distributed, with a mean equal to 20,000 words in excess of that specified in the author's contract and a standard deviation of 10,000 words. If the publisher wants to be almost certain (say, with a probability of .95) that the manuscript will have less than 100,000 words, what number of words should the publisher specify in the contract?

6.32 Tennis Anyone? A stringer of tennis rackets has found that the actual string tension achieved for any individual racket stringing will vary as much as 6 pounds per square inch from the desired tension set on the stringing machine. If the stringer wishes to string at a tension lower than that specified by a customer only 5% of the time, how much above or below the customer's specified tension should the stringer set the stringing machine? (NOTE: Assume that the distribution of string tensions produced by the stringing machine is normally distributed, with a mean equal to the tension set on the machine and a standard deviation equal to 2 pounds per square inch.)

6.33 Mall Rats An article in *American Demographics* claims that more than twice as many shoppers are out shopping on the weekends than during the week.[3] Not only that, but they spend more money on their purchases on Saturdays and Sundays! Suppose that the amount of money spent at shopping centers between 4 and 6 P.M. on Sundays has a normal distribution with a mean of $85 and with a standard deviation of $10. A shopper is randomly selected on a Sunday between 4 and 6 P.M. and asked about his spending patterns.

a. What is the probability that he has spent more than $90 at the mall?

b. What is the probability that he has spent between $90 and $100 at the mall?

c. If two shoppers are randomly selected, what is the probability that both shoppers have spent more than $100 at the mall?

6.34 Pulse Rates Your pulse rate is a measure of the number of heartbeats per minute. It can be measured in several places on your body, where an artery passes close to the skin. Once you find the pulse, count the number of beats per minute, or, count for 30 seconds and multiply by two. What's a *normal* pulse rate? That depends on a variety of factors. Pulse rates between 60 and 100 beats per minute are considered normal for children over 10 and adults.[4] Suppose that these pulse rates are approximately normally distributed with a mean of 78 and a standard deviation of 12.

a. What proportion of adults will have pulse rates between 60 and 100?

b. What is the 95th percentile for the pulse rates of adults?

c. Would a pulse rate of 110 be considered unusual? Explain.

THE NORMAL APPROXIMATION TO THE BINOMIAL PROBABILITY DISTRIBUTION (OPTIONAL)

6.4

In Chapter 5, you learned three ways to calculate probabilities for the binomial random variable x:

- Using the binomial formula, $P(x = k) = C_k^n p^k q^{n-k}$
- Using the cumulative binomial tables
- Using the Java applets

The binomial formula produces lengthy calculations, and the tables are available for only certain values of n and p. There is another option available when $np < 7$; the Poisson probabilities can be used to approximate $P(x = k)$. When this approximation *does not work* and n is large, the normal probability distribution provides another approximation for binomial probabilities.

THE NORMAL APPROXIMATION TO THE BINOMIAL PROBABILITY DISTRIBUTION

Let x be a binomial random variable with n trials and probability p of success. The probability distribution of x is approximated using a normal curve with

$$\mu = np \quad \text{and} \quad \sigma = \sqrt{npq}$$

This approximation is adequate as long as n is large and p is not too close to 0 or 1.

Since the normal distribution is continuous, the area under the curve at any single point is equal to 0. Keep in mind that this result applies only to continuous random variables. Because the binomial random variable x is a discrete random variable, the probability that x takes some specific value—say, $x = 11$—will not necessarily equal 0.

Figures 6.18 and 6.19 show the binomial probability histograms for $n = 25$ with $p = .5$ and $p = .1$, respectively. The distribution in Figure 6.18 is exactly symmetric.

FIGURE 6.18

The binomial probability distribution for $n = 25$ and $p = .5$ and the approximating normal distribution with $\mu = 12.5$ and $\sigma = 2.5$

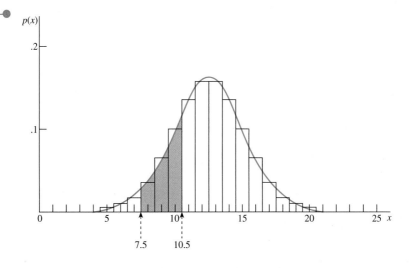

FIGURE 6.19 ●

The binomial probability distribution and the approximating normal distribution for $n = 25$ and $p = .1$

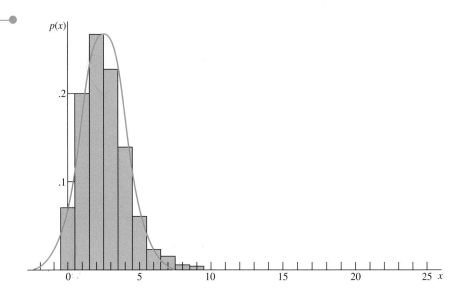

If you superimpose a normal curve with the same mean, $\mu = np$, and the same standard deviation, $\sigma = \sqrt{npq}$, over the top of the bars, it "fits" quite well; that is, the areas under the curve are almost the same as the areas under the bars. However, when the probability of success, p, gets small and the distribution is skewed, as in Figure 6.19, the symmetric normal curve no longer fits very well. If you try to use the normal curve areas to approximate the area under the bars, your approximation will not be very good.

EXAMPLE (6.11)

Use the normal curve to approximate the probability that $x = 8$, 9, or 10 for a binomial random variable with $n = 25$ and $p = .5$. Compare this approximation to the exact binomial probability.

Solution You can find the exact binomial probability for this example because there are cumulative binomial tables for $n = 25$. From Table 1 in Appendix I,

$$P(x = 8, 9, \text{ or } 10) = P(x \le 10) - P(x \le 7) = .212 - .022 = .190$$

To use the normal approximation, first find the appropriate mean and standard deviation for the normal curve:

$$\mu = np = 25(.5) = 12.5$$

$$\sigma = \sqrt{npq} = \sqrt{25(.5)(.5)} = 2.5$$

MY TIP

Only use the continuity correction if x has a **binomial** distribution!

The probability that you need corresponds to the area of the three rectangles lying over $x = 8$, 9, and 10. The equivalent area under the normal curve lies between $x = 7.5$ (the lower edge of the rectangle for $x = 8$) and $x = 10.5$ (the upper edge of the rectangle for $x = 10$). This area is shaded in Figure 6.18.

To find the normal probability, follow the procedures of Section 6.3. First you standardize each interval endpoint:

$$z = \frac{x - \mu}{\sigma} = \frac{7.5 - 12.5}{2.5} = -2.0$$

$$z = \frac{x - \mu}{\sigma} = \frac{10.5 - 12.5}{2.5} = -.8$$

Then the approximate probability (shaded in Figure 6.20) is found from Table 3 in Appendix I:

$$P(-2.0 < z < -.8) = .2119 - .0228 = .1891$$

You can compare the approximation, .1891, to the actual probability, .190. They are quite close!

FIGURE 6.20

Area under the normal curve for Example 6.11

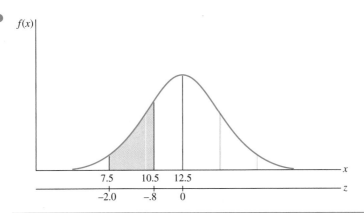

MY APPLET

You can use the Java applet called **Normal Approximation to Binomial Probabilities** shown in Figure 6.21 to compare the actual and approximate probabilities for the binomial distribution in Example 6.11. Enter the appropriate values of n and p in the boxes at the top left corner of the applet, and press "Enter" to record each entry. The exact binomial distribution on the left of the applet will change depending on the value of n you have entered. Now change the value of k in the box at the bottom left corner of the applet, and press "Enter." The applet will calculate the exact binomial probability $P(x \le k)$ in the box marked "Prob:" It will also calculate the approximate probability using the area under the normal curve. The z-value, with the continuity correction is shown at the top right, and the approximate probability is shown to the left of the normal curve. For Example 6.11, the applet calculates the normal approximation as $P(x \le 10) \approx .2119$. What is the exact value of $P(x \le 10)$? If you change k to 7 and press "Enter," what is the approximate value for $P(x \le 7)$? Now calculate $P(8 \le x \le 10)$. Does it match the answer we got in Example 6.11? You will use this applet again for the MyApplet Exercises section at the end of the chapter.

FIGURE 6.21 ●

Normal Approximation
to Binomial Probabilities
applet

You must be careful not to exclude half of the two extreme probability rectangles when you use the normal approximation to the binomial probability distribution. This adjustment, called the **continuity correction,** helps account for the fact that you are approximating a *discrete random variable* with a *continuous* one. If you forget the correction, your approximation will not be very good! Use this correction only for *binomial probabilities;* do not try to use it when the random variable is already continuous, such as a height or weight.

How can you tell when it is appropriate to use the normal approximation to binomial probabilities? The normal approximation works well when the binomial histogram is roughly symmetric. This happens when the binomial distribution is not "bunched up" near 0 or *n*—that is, when it can spread out at least two standard deviations from its mean without exceeding its limits, 0 and *n*. Using this criterion, you can derive this simple rule of thumb:

RULE OF THUMB

The normal approximation to the binomial probabilities will be adequate if both

$$np > 5 \quad \text{and} \quad nq > 5$$

MY) **PERSONAL TRAINER**

How Do I Calculate Binomial Probabilities Using the Normal Approximation?

- Find the necessary values of *n* and *p*. Calculate $\mu = np$ and $\sigma = \sqrt{npq}$.

- Write the probability you need in terms of *x* and locate the appropriate area on the curve.

- Correct the value of *x* by $\pm.5$ to include the entire block of probability for that value. This is the *continuity correction.*

- Convert the necessary x-values to z-values using

$$z = \frac{x \pm .5 - np}{\sqrt{npq}}$$

- Use Table 3 in Appendix I to calculate the approximate probability.

Exercise Reps

Consider a binomial random variable with $n = 30$ and $p = .4$. Fill in the blanks below to find some probabilities using the normal approximation.

A. Preliminary Steps:

1. Can we use the normal approximation? Calculate $np =$ _____ and $nq =$ _____

2. Are np and nq both greater than 5?   Yes _____   No _____

3. If the answer to Question 2 is yes, calculate $\mu = np =$ _____ and
   $\sigma = \sqrt{npq} =$ _____

B. Calculate the Probability:

1. To find the probability of 20 or more successes, what values of x should be included? $x =$ _____

2. To include the entire block of probability for the first value of $x =$ _____, start at _____.

3. Calculate $z = \dfrac{x \pm .5 - np}{\sqrt{npq}} =$ _____.

4. Calculate $P(x \geq 20) \approx P(z >$ _____$) = 1 -$ _____ $=$ _____.

Progress Report

- Still having trouble? Try again using the Exercise Reps at the end of this section.

- Mastered the normal approximation? You can skip the Exercise Reps at the end of this section!

Answers are located on the perforated card at the back of this book.

EXAMPLE 6.12

The reliability of an electrical fuse is the probability that a fuse, chosen at random from production, will function under its designed conditions. A random sample of 1000 fuses was tested and $x = 27$ defectives were observed. Calculate the approximate probability of observing 27 or more defectives, assuming that the fuse reliability is .98.

Solution   The probability of observing a defective when a single fuse is tested is $p = .02$, given that the fuse reliability is .98. Then

$$\mu = np = 1000(.02) = 20$$
$$\sigma = \sqrt{npq} = \sqrt{1000(.02)(.98)} = 4.43$$

The probability of 27 or more defective fuses, given $n = 1000$, is

$$P(x \geq 27) = p(27) + p(28) + p(29) + \cdots + p(999) + p(1000)$$

MY TIP

If np and nq are both greater than 5, you can use the normal approximation.

It is appropriate to use the normal approximation to the binomial probability because

$$np = 1000(.02) = 20 \quad \text{and} \quad nq = 1000(.98) = 980$$

are both greater than 5. The normal area used to approximate $P(x \geq 27)$ is the area under the normal curve to the right of 26.5, so that the entire rectangle for $x = 27$ is included. Then, the z-value corresponding to $x = 26.5$ is

$$z = \frac{x - \mu}{\sigma} = \frac{26.5 - 20}{4.43} = \frac{6.5}{4.43} = 1.47$$

and the area to the left of $z = 1.47$ is equal to .9292, as shown in Figure 6.22. Since the total area under the curve is 1, you have

$$P(x \geq 27) \approx P(z \geq 1.47) = 1 - .9292 = .0708$$

FIGURE 6.22

Normal approximation to the binomial for Example 6.12

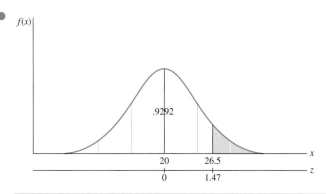

EXAMPLE 6.13

A producer of soft drinks was fairly certain that her brand had a 10% share of the soft drink market. In a market survey involving 2500 consumers of soft drinks, $x = 211$ expressed a preference for her brand. If the 10% figure is correct, find the probability of observing 211 or fewer consumers who prefer her brand of soft drink.

Solution If the producer is correct, then the probability that a consumer prefers her brand of soft drink is $p = .10$. Then

$$\mu = np = 2500(.10) = 250$$
$$\sigma = \sqrt{npq} = \sqrt{2500(.10)(.90)} = 15$$

The probability of observing 211 or fewer who prefer her brand is

$$P(x \leq 211) = p(0) + p(1) + \cdots + p(210) + p(211)$$

The normal approximation to this probability is the area to the left of 211.5 under a normal curve with a mean of 250 and a standard deviation of 15. First calculate

$$z = \frac{x - \mu}{\sigma} = \frac{211.5 - 250}{15} = -2.57$$

Then

$$P(x \leq 211) \approx P(z < -2.57) = .0051$$

The probability of observing a sample value of 211 or less when $p = .10$ is so small that you can conclude that one of two things has occurred: Either you have observed an unusual sample even though really $p = .10$, *or* the sample reflects that the actual value of p is less than .10 and perhaps closer to the observed sample proportion, $211/2500 = .08$.

6.4 EXERCISES

EXERCISE REPS

6.35 Consider a binomial random varible with $n = 25$ and $p = .6$. Fill in the blanks below to find some probabilities using the normal approximation.

a. Can we use the normal approximation? Calculate $np = $ _____ and $nq = $ _____

b. Are np and nq both greater than 5?   Yes _____   No _____

c. If the answer to part b is yes, calculate $\mu = np = $ _____ and $\sigma = \sqrt{npq} = $ _____

d. To find the probability of more than 9 successes, what values of x should be included? $x = $ _____

e. To include the entire block of probability for the first value of $x = $ _____, start at _____.

f. Calculate $z = \dfrac{x \pm .5 - np}{\sqrt{npq}} = $ _____.

g. Calculate $P(x > 9) \approx P(z > $ _____$) = 1 - $ _____ $= $ _____.

6.36 Consider a binomial random variable with $n = 45$ and $p = .05$. Fill in the blanks below to find some probabilities using the normal approximation.

a. Can we use the normal approximation? Calculate $np = $ _____ and $nq = $ _____

b. Are np and nq both greater than 5?   Yes _____   No _____

c. If the answer to part b is yes, calculate $\mu = np = $ _____ and $\sigma = \sqrt{npq} = $ _____

d. To find the probability of 10 or fewer successes, what values of x should be included? $x = $ _____

e. To include the entire block of probability for the first value of $x = $ _____, start at _____.

f. Calculate $z = \dfrac{x \pm .5 - np}{\sqrt{npq}} = $ _____.

g. Calculate $P(x \leq 10) \approx P(z < $ _____$) = $ _____.

BASIC TECHNIQUES

6.37 Let x be a binomial random variable with $n = 25$ and $p = .3$.

a. Is the normal approximation appropriate for this binomial random variable?

b. Find the mean and standard deviation for x.

c. Use the normal approximation to find $P(6 \leq x \leq 9)$.

d. Use Table 1 in Appendix I to find the exact probability $P(6 \leq x \leq 9)$. Compare the results of parts c and d. How close was your approximation?

6.38 Let x be a binomial random variable with $n = 15$ and $p = .5$.

a. Is the normal approximation appropriate?

b. Find $P(x \geq 6)$ using the normal approximation.

c. Find $P(x > 6)$ using the normal approximation.

d. Find the exact probabilities for parts b and c, and compare these with your approximations.

6.39 Let x be a binomial random variable with $n = 100$ and $p = .2$. Find approximations to these probabilities:

a. $P(x > 22)$ **b.** $P(x \geq 22)$

c. $P(20 < x < 25)$ **d.** $P(x \leq 25)$

6.40 Let x be a binomial random variable for $n = 25$, $p = .2$.

a. Use Table 1 in Appendix I to calculate $P(4 \leq x \leq 6)$.

b. Find μ and σ for the binomial probability distribution, and use the normal distribution to approximate the probability $P(4 \leq x \leq 6)$. Note that this value is a good approximation to the exact value of $P(4 \leq x \leq 6)$ even though $np = 5$.

6.41 Suppose the random variable x has a binomial distribution corresponding to $n = 20$ and $p = .30$. Use Table 1 of Appendix I to calculate these probabilities:

a. $P(x = 5)$ **b.** $P(x \geq 7)$

6.42 Refer to Exercise 6.41. Use the normal approximation to calculate $P(x = 5)$ and $P(x \geq 7)$. Compare with the exact values obtained from Table 1 in Appendix I.

6.43 Consider a binomial experiment with $n = 20$ and $p = .4$. Calculate $P(x \geq 10)$ using each of these methods:

a. Table 1 in Appendix I

b. The normal approximation to the binomial probability distribution

6.44 Find the normal approximation to $P(355 \leq x \leq 360)$ for a binomial probability distribution with $n = 400$ and $p = .9$.

APPLICATIONS

6.45 Life Savers What is your favorite LifeSaver flavor? The data from a *USA Today* Snapshot claims that 32% of Americans prefer cherry LifeSavers.[5]

Source: USA Today. Copyright 2004. Reprinted with permission.

Does this hold true for you and your statistics classmates? Assume that it does and that your class contains 50 students. What are the approximate probabilities for these events?

a. More than 20 students prefer cherry LifeSavers.

b. Fewer than 15 students prefer cherry LifeSavers.

c. Fewer than 28 students *do not* prefer cherry LifeSavers.

d. Are you willing to assume that you and your classmates are a representative sample of all Americans when it comes to this question? How does your answer affect the probabilities in parts a to c?

6.46 Genetic Defects Data collected over a long period of time show that a particular genetic defect occurs in 1 of every 1000 children. The records of a medical clinic show $x = 60$ children with the defect in a total of 50,000 examined. If the 50,000 children were a random sample from the population of children represented by past records, what is the probability of observing a value of x equal to 60 or more? Would you say that the observation of $x = 60$ children with genetic defects represents a rare event?

6.47 No Shows Airlines and hotels often grant reservations in excess of capacity to minimize losses due to no-shows. Suppose the records of a hotel show that, on the average, 10% of their prospective guests will not claim their reservation. If the hotel accepts 215 reservations and there are only 200 rooms in the hotel, what is the probability that all guests who arrive to claim a room will receive one?

6.48 Lung Cancer Compilation of large masses of data on lung cancer shows that approximately 1 of every 40 adults acquires the disease. Workers in a certain occupation are known to work in an air-polluted environment that may cause an increased rate of lung cancer. A random sample of $n = 400$ workers shows 19 with identifiable cases of lung cancer. Do the data provide sufficient evidence to indicate a higher rate of lung cancer for these workers than for the national average?

6.49 Tall or Short? Is a tall president better than a short one? Do Americans tend to vote for the taller of the two candidates in a presidential election? In 31 of our presidential elections between 1856 and 2000, 17 of the winners were taller than their opponents.[1] Assume that Americans are not biased by a candidate's height and that the winner is just as likely to be taller or shorter than his opponent. Is the observed number of taller winners in the U.S. presidential elections unusual?

a. Find the approximate probability of finding 17 or more of the 31 pairs in which the taller candidate wins.

b. Based on your answer to part a, can you conclude that Americans might consider a candidate's height when casting their ballot?

6.50 The Rh Factor   In a certain population, 15% of the people have Rh-negative blood. A blood bank serving this population receives 92 blood donors on a particular day.

a. What is the probability that 10 or fewer are Rh-negative?

b. What is the probability that 15 to 20 (inclusive) of the donors are Rh-negative?

c. What is the probability that more than 80 of the donors are Rh-positive?

6.51 Pepsi's Market Share   Two of the biggest soft drink rivals, Pepsi and Coke, are very concerned about their market share. The following pie chart, which appeared on the company website (http://www.pepsico.com) in March 2004, claims that PepsiCo's share of the U.S. beverage market is 28%.[6] Assume that this proportion will be *close* to the probability that a person selected at random indicates a preference for a Pepsi product when choosing a soft drink.

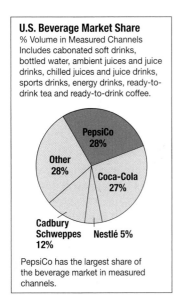

U.S. Beverage Market Share
% Volume in Measured Channels
Includes cabonated soft drinks, bottled water, ambient juices and juice drinks, chilled juices and juice drinks, sports drinks, energy drinks, ready-to-drink tea and ready-to-drink coffee.

PepsiCo 28%
Other 28%
Coca-Cola 27%
Cadbury Schweppes 12%
Nestlé 5%

PepsiCo has the largest share of the beverage market in measured channels.

Source: © 2004 PepsiCo, Inc. Used with permission.

A test group of 500 consumers is randomly selected. Use the normal curve to approximate the following binomial probabilities:

a. Exactly 160 consumers prefer a Pepsi product.

b. Between 120 and 150 consumers (inclusive) prefer a Pepsi product.

c. Fewer than 150 consumers prefer a Pepsi product.

d. Would it be unusual to find that 232 of the 500 consumers preferred a Pepsi product? If this were to occur, what conclusions would you draw?

6.52 Ready, Set, Relax!   The typical American family spends lots of time driving to and from various activities, and lots of time in the drive-thru lines at fast-food restaurants. There is a rising amount of evidence suggesting that we are beginning to burn out! In fact, in a study conducted for the Center for a New American Dream, *Time* magazine reports that 60% of Americans felt pressure to work too much, and 80% wished for more family time.[7] Assume that these percentages are correct for all Americans, and that a random sample of 25 Americans is selected.

a. Use Table 1 in Appendix I to find the probability that more than 20 felt pressure to work too much.

b. Use the normal approximation to the binomial distribution to aproximate the probability in part a. Compare your answer with the exact value from part a.

c. Use Table 1 in Appendix I to find the probability that between 15 and 20 (inclusive) wished for more family time.

d. Use the normal approximation to the binomial distribution to approximate the probability in part c. Compare your answer with the exact value from part c.

6.53 We said, "Relax!"   The article in *Time* magazine[7] (Exercise 6.52) also reported that 80% of men and 62% of women put in more than 40 hours a week on the job. Assume that these percentages are correct for all Americans, and that a random sample of 50 working women is selected.

a. What is the average number of women who put in more than 40 hours a week on the job?

b. What is the standard deviation for the number of women who put in more than 40 hours a week on the job?

c. Suppose that in our sample of 50 working women, there are 25 who work more than 40 hours a week. Would you consider this to be an unusual occurrence? Explain.

CHAPTER REVIEW

Key Concepts and Formulas

I. Continuous Probability Distributions

1. Continuous random variables

2. Probability distributions or probability density functions

   a. Curves are smooth.

   b. Area under the curve equals 1.

   c. The area under the curve between a and b represents the probability that x falls between a and b.

   d. $P(x = a) = 0$ for continuous random variables.

II. The Normal Probability Distribution

1. Symmetric about its mean μ

2. Shape determined by its standard deviation σ

III. The Standard Normal Distribution

1. The standard normal random variable z has mean 0 and standard deviation 1.

2. Any normal random variable x can be transformed to a standard normal random variable using

$$z = \frac{x - \mu}{\sigma}$$

3. Convert necessary values of x to z.

4. Use Table 3 in Appendix I to compute standard normal probabilities.

5. Several important z-values have right-tail areas as follows:

Right-Tail Area	.005	.01	.025	.05	.10
z-Value	2.58	2.33	1.96	1.645	1.28

 MINITAB

Normal Probabilities

When the random variable of interest has a normal probability distribution, you can generate either of these probabilities:

- Cumulative probabilities—$P(x \le k)$—for a given value of k
- Inverse cumulative probabilities—the value of k such that the area to its left under the normal probability distribution is equal to a

You must specify which normal distribution you are using and the necessary parameters: the mean μ and the standard deviation σ. As in Chapter 5, you have the option of specifying only one single value of k (or a) or several values of k (or a), which should be stored in a column (say, C1) of the *MINITAB* worksheet.

Suppose that the average birth weights of babies born at hospitals owned by a major health maintenance organization (HMO) are approximately normal with mean 6.75 pounds and standard deviation .54 pound. What proportion of babies born at these hospitals weigh between 6 and 7 pounds? To use *MINITAB* to find $P(6 < x < 7)$, enter the critical values $x = 6$ and $x = 7$ into a column (say, C1) of a *MINITAB* worksheet. Use **Calc → Probability Distributions → Normal** to generate the Dialog box, as shown in Figure 6.23.

Type the values for μ and σ in the appropriate boxes (the default values generate probabilities for the standard normal z distribution), and select C1 for the input column. (If you do not type a column number for storage, *MINITAB* will display the

FIGURE 6.23

FIGURE 6.24

results in the Session window.) Make sure that the radio button marked "Cumulative probability" is selected. The cumulative distribution function for $x = 6$ and $x = 7$ appears in the Session window when you click **OK** (see Figure 6.24). To find $P(6 < x < 7)$, remember that the cumulative probability is the area to the left of the given value of x. Hence,

$$P(6 < x < 7) = P(x < 7) - P(x < 6) = .678305 - .082433 = .595872$$

You can check this calculation using Table 3 in Appendix I if you wish!

To calculate inverse cumulative probabilities, make sure that the radio button marked "Inverse cumulative probability" is selected. Then enter the appropriate values of a in C1, or if you have only a single value, enter the value in the Input constant box. For example, to find the 95th percentile of the birth weights, you look for a value k such that only 5% of the values of x exceed this value (and 95% are less than or equal to k). If you enter the probability **.95** into the Input constant box and select the option marked "Inverse cumulative probability," the 95th percentile will appear in the Session window, as in Figure 6.25. That is, 95% of all babies born at these hospitals weigh 7.63822 pounds or less. Would you consider a baby who weighs 9 pounds to be unusually large?

FIGURE 6.25

Supplementary Exercises

6.54 Using Table 3 in Appendix I, calculate the area under the standard normal curve to the left of:

a. $z = 1.2$ **b.** $z = -.9$

c. $z = 1.46$ **d.** $z = -.42$

6.55 Find the following probabilities for the standard normal random variable:

a. $P(.3 < z < 1.56)$ **b.** $P(-.2 < z < .2)$

6.56 a. Find the probability that z is greater than $-.75$.

b. Find the probability that z is less than 1.35.

6.57 Find z_0 such that $P(z > z_0) = .5$.

6.58 Find the probability that z lies between $z = -1.48$ and $z = 1.48$.

6.59 Find z_0 such that $P(-z_0 < z < z_0) = .5$. What percentiles do $-z_0$ and z_0 represent?

6.60 Drill Bits The life span of oil-drilling bits depends on the types of rock and soil that the drill encounters, but it is estimated that the mean length of life is 75 hours. Suppose an oil exploration company purchases drill bits that have a life span that is approximately normally distributed, with a mean equal to 75 hours and a standard deviation equal to 12 hours.

a. What proportion of the company's drill bits will fail before 60 hours of use?

b. What proportion will last at least 60 hours?

c. What proportion will have to be replaced after more than 90 hours of use?

6.61 Faculty Ages The influx of new ideas into a college or university, introduced primarily by new young faculty, is becoming a matter of concern because of the increasing ages of faculty members; that is, the distribution of faculty ages is shifting upward due most likely to a shortage of vacant positions and an oversupply of PhDs. Thus, faculty members are reluctant to move and give up a secure position. If the retirement age at most universities is 65, would you expect the distribution of faculty ages to be normal? Explain.

6.62 Bearing Diameters A machine operation produces bearings whose diameters are normally distributed, with mean and standard deviation equal to .498 and .002, respectively. If specifications require that the bearing diameter equal .500 inch ± .004 inch, what fraction of the production will be unacceptable?

6.63 Used Cars A used-car dealership has found that the length of time before a major repair is required on the cars it sells is normally distributed, with a mean equal to 10 months and a standard deviation of 3 months. If the dealer wants only 5% of the cars to fail before the end of the guarantee period, for how many months should the cars be guaranteed?

6.64 Restaurant Sales The daily sales total (excepting Saturday) at a small restaurant has a probability distribution that is approximately normal, with a mean μ equal to $1230 per day and a standard deviation σ equal to $120.

a. What is the probability that the sales will exceed $1400 for a given day?

b. The restaurant must have at least $1000 in sales per day to break even. What is the probability that on a given day the restaurant will not break even?

6.65 Washers The life span of a type of automatic washer is approximately normally distributed, with mean and standard deviation equal to 3.1 and 1.2 years, respectively. If this type of washer is guaranteed for 1 year, what fraction of original sales will require replacement?

6.66 Garage Door Openers Most users of automatic garage door openers activate their openers at distances that are normally distributed, with a mean of 30 feet and a standard deviation of 11 feet. To minimize interference with other remote-controlled devices, the manufacturer is required to limit the operating distance to 50 feet. What percentage of the time will users attempt to operate the opener outside its operating limit?

6.67 How Long is the Test? The average length of time required to complete a college achievement test was found to equal 70 minutes, with a standard deviation of 12 minutes. When should the test be terminated if you wish to allow sufficient time for 90% of the students to complete the test? (Assume that the time required to complete the test is normally distributed.)

6.68 Servicing Automobiles The length of time required for the periodic maintenance of an automobile will usually have a probability distribution that is mound-shaped and, because some long service times will occur occasionally, is skewed to the right. The length of time required to run a 5000-mile check and to service an automobile has a mean equal to 1.4 hours and a standard deviation of .7 hour. Suppose that the service department plans to service 50 automobiles per 8-hour day and that, in order to do so, it must spend no more than an average of 1.6 hours per automobile. What proportion of all days will the service department have to work overtime?

6.69 TV Viewers An advertising agency has stated that 20% of all television viewers watch a particular program. In a random sample of 1000 viewers, $x = 184$ viewers were watching the program. Do these data present sufficient evidence to contradict the advertiser's claim?

6.70 Forecasting Earnings A researcher notes that senior corporation executives are not very accurate forecasters of their own annual earnings. He states that his studies of a large number of company executive forecasts "showed that the average estimate missed the mark by 15%."

a. Suppose the distribution of these forecast errors has a mean of 15% and a standard deviation of 10%. Is it likely that the distribution of forecast errors is approximately normal?

b. Suppose the probability is .5 that a corporate executive's forecast error exceeds 15%. If you were to sample the forecasts of 100 corporate executives, what is the probability that more than 60 would be in error by more than 15%?

6.71 Filling Soda Cups A soft drink machine can be regulated to discharge an average of μ ounces per cup. If the ounces of fill are normally distributed, with standard deviation equal to .3 ounce, give the setting for μ so that 8-ounce cups will overflow only 1% of the time.

6.72 Light Bulbs A manufacturing plant uses 3000 electric light bulbs whose life spans are normally distributed, with mean and standard deviation equal to 500 and 50 hours, respectively. In order to minimize the number of bulbs that burn out during operating hours, all the bulbs are replaced after a given period of operation. How often should the bulbs be replaced if we wish no more than 1% of the bulbs to burn out between replacement periods?

6.73 The Freshman Class The admissions office of a small college is asked to accept deposits from a number of qualified prospective freshmen so that, with probability about .95, the size of the freshman class will be less than or equal to 120. Suppose the applicants constitute a random sample from a population of applicants, 80% of whom would actually enter the freshman class if accepted.

a. How many deposits should the admissions counselor accept?

b. If applicants in the number determined in part a are accepted, what is the probability that the freshman class size will be less than 105?

6.74 No Shows An airline finds that 5% of the persons making reservations on a certain flight will not show up for the flight. If the airline sells 160 tickets for a flight that has only 155 seats, what is the probability that a seat will be available for every person holding a reservation and planning to fly?

6.75 Long Distance It is known that 30% of all calls coming into a telephone exchange are long-distance calls. If 200 calls come into the exchange, what is the probability that at least 50 will be long-distance calls?

6.76 Plant Genetics In Exercise 5.75, a cross between two peony plants—one with red petals and one with streaky petals—produced offspring plants with red petals 75% of the time. Suppose that 100 seeds from this cross were collected and germinated, and x, the number of plants with red petals, was recorded.

a. What is the exact probability distribution for x?

b. Is it appropriate to approximate the distribution in part a using the normal distribution? Explain.

c. Use an appropriate method to find the approximate probability that between 70 and 80 (inclusive) offspring plants have red flowers.

d. What is the probability that 53 or fewer offspring plants had red flowers? Is this an unusual occurrence?

e. If you actually observed 53 of 100 offspring plants with red flowers, and if you were certain that the genetic ratio 3:1 was correct, what other explanation could you give for this unusual occurrence?

6.77 Suppliers A or B? A purchaser of electric relays buys from two suppliers, A and B. Supplier A supplies two of every three relays used by the company. If 75 relays are selected at random from those in use by the company, find the probability that at most 48 of these relays come from supplier A. Assume that the company uses a large number of relays.

6.78 Snacking and TV Is television dangerous to your diet? Psychologists believe that excessive eating may be associated with emotional states (being upset or bored) and environmental cues (watching television, reading, and so on). To test this theory, suppose you randomly selected 60 overweight persons and matched them by weight and gender in pairs. For a period of 2 weeks, one of each pair is required to spend evenings reading novels of interest to him or her. The other member of each pair spends each evening watching television. The calorie count for all snack and drink intake for the evenings is recorded for each person, and you record $x = 19$, the number of pairs for which the television watchers' calorie intake exceeded the intake of the readers. If there is no difference in the effects of television and reading on calorie intake, the probability p that the calorie intake of one member of a pair exceeds that of the other member is .5. Do these data provide sufficient evidence to indicate a difference between the effects of television watching and reading on calorie intake? (HINT: Calculate the z-score for the observed value, $x = 19$.)

6.79 Gestation Times *The Biology Data Book* reports that the gestation time for human babies averages 278 days with a standard deviation of 12 days.[8] Suppose that these gestation times are normally distributed.

a. Find the upper and lower quartiles for the gestation times.

b. Would it be unusual to deliver a baby after only 6 months of gestation? Explain.

6.80 Tax Audits In Exercise 6.28 we suggested that the IRS assign auditing rates per state by randomly selecting 50 auditing percentages from a normal distribution with a mean equal to 1.55% and a standard deviation of .45%.

a. What is the probability that a particular state would have more than 2% of its tax returns audited?

b. What is the expected value of x, the number of states that will have more than 2% of their income tax returns audited?

c. Is it likely that as many as 15 of the 50 states will have more than 2% of their income tax returns audited?

6.81 Your Favorite Sport There is a difference in sports preferences between men and women, according to a recent survey. Among the 10 most popular sports, men include competition-type sports—pool and billiards, basketball, and softball—whereas women include aerobics, running, hiking, and calisthenics. However, the top recreational activity for men was still the relaxing sport of fishing, with 41% of those surveyed indicating that they had fished during the year. Suppose 180 randomly selected men are asked whether they had fished in the past year.

a. What is the probability that fewer than 50 had fished?

b. What is the probability that between 50 and 75 had fished?

c. If the 180 men selected for the interview were selected by the marketing department of a sporting-goods company based on information obtained from their mailing lists, what would you conclude about the reliability of their survey results?

6.82 Introvert or Extrovert? A psychological introvert–extrovert test produced scores that had a normal distribution with a mean and standard deviation of 75 and 12, respectively. If we wish to designate the *highest* 15% as extroverts, what would be the proper score to choose as the cutoff point?

6.83 Curving the Grades Students very often ask their professors whether they will be "curving the grades." The traditional interpretation of "curving grades" required that the grades have a normal distribution, and that the grades will be assigned in these proportions:

Letter Grade	A	B	C	D	F
Proportion of Students	10%	20%	40%	20%	10%

a. If the average "C" grade is centered at the average grade for all students, and if we assume that the grades are normally distributed, how many standard deviations on either side of the mean will constitute the "C" grades?

b. How many deviations on either side of the mean will be the cutoff points for the "B" and "D" grades?

6.84 Curving the Grades, continued Refer to Exercise 6.83. For ease of calculation, round the number of standard deviations for "C" grades to $\pm.5$ standard deviations, and for "B" and "D" grades to ±1.5 standard deviations. Suppose that the distribution of grades for a large class of students has an average of 78 with a standard deviation of 11. Find the appropriate cutoff points for the grades A, B, C, D, and F.

6.85 Normal Temperatures In Exercise 1.67, Allen Shoemaker derived a distribution of human body temperatures, which has a distinct mound-shape.[9] Suppose we assume that the temperatures of healthy humans is approximately normal with a mean of 98.6 degrees and a standard deviation of 0.8 degrees.

a. If a healthy person is selected at random, what is the probability that the person has a temperature above 99.0 degrees?

b. What is the 95th percentile for the body temperatures of healthy humans?

MY APPLET Exercises

Use one of the three applets (**Normal Distribution Probabilities, Normal Probabilities and z-scores,** or **Normal Approximation to Binomial Probabilities**) described in this chapter to solve the following exercises.

6.86 Calculate the area under the standard normal curve to the left of these values:

a. $z = -.90$ **b.** $z = 2.34$ **c.** $z = 5.4$

6.87 Calculate the area under the standard normal curve between these values:

a. $z = -2.0$ and $z = 2.0$ **b.** $z = -2.3$ and -1.5

6.88 Find the following probabilities for the standard normal random variable z:

a. $P(-1.96 \leq z \leq 1.96)$ **b.** $P(z > 1.96)$
c. $P(z < -1.96)$

6.89 a. Find a z_0 such that $P(z > z_0) = .9750$
b. Find a z_0 such that $P(z > z_0) = .3594$

6.90 a. Find a z_0 such that $P(-z_0 \leq z \leq z_0) = .95$.
b. Find a z_0 such that $P(-z_0 \leq z \leq z_0) = .98$.

6.91 A normal random variable x has mean $\mu = 5$ and $\sigma = 2$. Find the following probabilities of these x-values:

a. $1.2 < x < 10$ **b.** $x > 7.5$ **c.** $x \leq 0$

6.92 Let x be a binomial random variable with $n = 36$ and $p = .54$. Use the normal approximation to find:

a. $P(x \leq 25)$ **b.** $P(15 \leq x \leq 20)$
c. $P(x > 30)$

6.93 Stress A snapshot in *USA Today* indicates that forty-nine percent of adult Americans list money issues as the top cause of stress for them in their daily lives.[10] Suppose that 100 adult Americans are randomly selected.

a. Use the **Calculating Binomial Probabilities** applet from Chapter 5 to find the exact probability that 60 or more adults would give money reasons as the top cause of stress.

b. Use the **Normal Approximation to Binomial Probabilities** applet to approximate the probability in part a. Compare your answers.

6.94 Stamps Philatelists (stamp collectors) often buy stamps at or near retail prices, but, when they sell, the price is considerably lower. For example, it may be reasonable to assume that (depending on the mix of a collection, condition, demand, economic conditions, etc.) a collection will sell at x% of the retail price, where x is normally distributed with a mean equal to 45% and a standard deviation of 4.5%. If a philatelist has a collection to sell that has a retail value of $30,000, what is the probability that the philatelist receives these amounts for the collection?

a. More than $15,000   **b.** Less than $15,000

c. Less than $12,000

6.95 Test Scores The scores on a national achievement test were approximately normally distributed, with a mean of 540 and a standard deviation of 110.

a. If you achieved a score of 680, how far, in standard deviations, did your score depart from the mean?

b. What percentage of those who took the examination scored higher than you?

6.96 Faculty Salaries Although faculty salaries at colleges and universities in the United States continue to rise, they do not always keep pace with the cost of living. During the 2002–2003 academic year, *Academe* reports that female assistant professors with PhDs earned an average of $53,851 per year.[11] Suppose that these salaries are normally distributed, with a standard deviation of $4000.

a. What proportion of female assistant professors will have salaries less than $45,000?

b. What proportion of female assistant professors will have salaries between $45,000 and $50,000?

6.97 Transplanting Cells Briggs and King developed the technique of nuclear transplantation, in which the nucleus of a cell from one of the later stages of the development of an embryo is transplanted into a zygote (a single-cell fertilized egg) to see whether the nucleus can support normal development. If the probability that a single transplant from the early gastrula stage will be successful is .65, what is the probability that more than 70 transplants out of 100 will be successful?

CASE STUDY **The Long and Short of It**

If you were the boss, would height play a role in your selection of a successor for your job? In his *Fortune* column, Daniel Seligman discussed his ideas concerning height as a factor in Deng Xiaoping's choice of Hu Yaobang as his replacement as Chairman of the Chinese Communist Party.[12] As Seligman notes, the facts surrounding the case arouse suspicions when examined in the light of statistics.

Deng, it seemed, was only 5 feet tall, a height that is short even in China. Therefore, the choice of Hu Yaobang, who was also 5 feet tall, raised (or lowered) some eyebrows because, as Seligman notes, "the odds against a 'height-blind' decision producing a chairman as short as Deng are about 40 to 1." In other words, if we had the relative frequency distribution of the heights of all Chinese men, only 1 in 41 (i.e., 2.4%) would be 5 feet tall or shorter. To calculate these odds, Seligman notes that the Chinese equivalent of the U.S. Health Service does not exist and hence that health statistics on the current population of China are difficult to acquire. He says, however, that "it is generally held that a boy's length at birth represents 28.6% of his final height" and that, in prerevolutionary China, the average length of a Chinese boy at birth was 18.9 inches. From this, Seligman deduces that the mean height of mature Chinese men is

$$\frac{18.9}{.286} = 66.08 \text{ inches, or 5 feet 6.08 inches}$$

He then assumes that the distribution of the heights of men in China follows a normal distribution ("as it does in the U.S."), with a mean of 66 inches and a standard deviation equal to 2.7 inches, "a figure that looks about right for that mean."

1. Using Seligman's assumptions, calculate the probability that a single adult Chinese man, chosen at random, will be less than or equal to 5 feet tall, or equivalently, 60 inches tall.

2. Do the results in part 1 agree with Seligman's odds?

3. Comment on the validity of Seligman's assumptions. Are there any basic flaws in his reasoning?

4. Based on the results of parts 1 and 3, do you think that Deng Xiaoping took height into account in selecting his successor?

Sampling Distributions

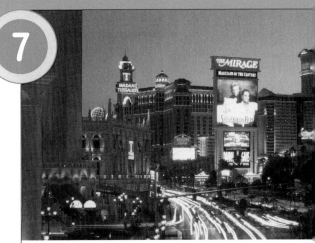

© PictureNet/Royalty-Free/CORBIS

MY PERSONAL TRAINER

How Do I Calculate Probabilities for the Sample Mean $\bar{x}$?

How Do I Calculate Probabilities for the Sample Proportion $\hat{p}$?

Sampling the Roulette at Monte Carlo

How would you like to try your hand at gambling without the risk of losing? You could do it by simulating the gambling process, making imaginary bets, and observing the results. This technique, called a Monte Carlo procedure, is the topic of the case study at the end of this chapter.

INTRODUCTION

In the previous three chapters, you have learned a lot about probability distributions, such as the binomial and normal distributions. The shape of the normal distribution is determined by its mean μ and its standard deviation σ, whereas the shape of the binomial distribution is determined by p. These numerical descriptive measures— called **parameters**—are needed to calculate the probability of observing sample results.

In practical situations, you may be able to decide which *type* of probability distribution to use as a model, but the values of the *parameters* that specify its *exact form* are unknown. Here are two examples:

- A pollster is sure that the responses to his "agree/disagree" questions will follow a binomial distribution, but p, the proportion of those who "agree" in the population, is unknown.
- An agronomist believes that the yield per acre of a variety of wheat is approximately normally distributed, but the mean μ and standard deviation σ of the yields are unknown.

In these cases, you must rely on the *sample* to learn about these parameters. The proportion of those who "agree" in the pollster's sample provides information about the actual value of p. The mean and standard deviation of the agronomist's sample approximate the actual values of μ and σ. If you want the sample to provide *reliable information* about the population, however, you must select your sample in a certain way!

SAMPLING PLANS AND EXPERIMENTAL DESIGNS

The way a sample is selected is called the **sampling plan** or **experimental design** and determines the quantity of information in the sample. Knowing the sampling plan used in a particular situation will often allow you to measure the reliability or goodness of your inference.

Simple random sampling is a commonly used sampling plan in which every sample of size n has the same chance of being selected. For example, suppose you want to select a sample of size $n = 2$ from a population containing $N = 4$ objects. If the four objects are identified by the symbols x_1, x_2, x_3, and x_4, there are six distinct pairs that could be selected, as listed in Table 7.1. If the sample of $n = 2$ observations is selected so that each of these six samples has the same chance of selection, given by $1/6$, then the resulting sample is called a **simple random sample,** or just a **random sample.**

TABLE 7.1 ● **Ways of Selecting a Sample of Size 2 from 4 Objects**

Sample	Observations in Sample
1	x_1, x_2
2	x_1, x_3
3	x_1, x_4
4	x_2, x_3
5	x_2, x_4
6	x_3, x_4

Definition If a sample of n elements is selected from a population of N elements using a sampling plan in which each of the possible samples has the same chance of selection, then the sampling is said to be **random** and the resulting sample is a **simple random sample.**

Perfect random sampling is difficult to achieve in practice. If the size of the population N is small, you might write each of N numbers on a poker chip, mix the chips, and select a sample of n chips. The numbers that you select correspond to the n measurements that appear in the sample. Since this method is not always very practical, a simpler and more reliable method uses **random numbers**—digits generated so that the values 0 to 9 occur randomly and with equal frequency. These numbers can be generated by computer or may even be available on your scientific calculator. Alternatively, Table 10 in Appendix I is a table of random numbers that you can use to select a *random sample.*

EXAMPLE 7.1

A computer database at a downtown law firm contains files for $N = 1000$ clients. The firm wants to select $n = 5$ files for review. Select a simple random sample of 5 files from this database.

Solution You must first label each file with a number from 1 to 1000. Perhaps the files are stored alphabetically, and the computer has already assigned a number to each. Then generate a sequence of ten three-digit random numbers. If you are using Table 10 of Appendix I, select a random starting point and use a portion of the table similar to the one shown in Table 7.2. The random starting point ensures that you will not use the same sequence over and over again. The first three digits of Table 7.2 indicate the number of the first file to be reviewed. The random number 001 corresponds to file #1, and the last file, #1000, corresponds to the random number 000. Using Table 7.2, you would choose the five files numbered 155, 450, 32, 882, and 350 for review. Alternately, you might choose to read across the lines, and choose files 155, 350, 989, 450 and 369 for review.

TABLE 7.2 ● **Portion of a Table of Random Numbers**

15574	35026	98924
45045	36933	28630
03225	78812	50856
88292	26053	21121

The situation described in Example 7.1 is called an **observational study** because the data already existed before you decided to *observe* or describe their characteristics. Most sample surveys, in which information is gathered with a questionnaire, fall into this category. Computer databases make it possible to assign identification numbers to each element even when the population is large and to select a simple random sample. You must be careful when conducting a *sample survey,* however, to watch for these frequently occurring problems:

- **Nonresponse:** You have carefully selected your random sample and sent out your questionnaires, but only 50% of those surveyed return their questionnaires. Are the responses you received still representative of the entire population, or are they **biased** because only those people who were particularly opinionated about the subject chose to respond?

- **Undercoverage:** You have selected your random sample using telephone records as a database. Does the database you used systematically exclude certain segments of the population—perhaps those who do not have telephones?

- **Wording bias:** Your questionnaire may have questions that are too complicated or tend to confuse the reader. Possibly the questions are sensitive in nature—for example, "Have you ever used drugs?" or "Have you ever cheated on your income tax?"—and the respondents will not answer truthfully.

Methods have been devised to solve some of these problems, but only if you know that they exist. If your survey is *biased* by any of these problems, then your conclusions will not be very reliable, even though you did select a random sample!

Some research involves **experimentation,** in which an experimental condition or *treatment* is imposed on the *experimental units.* Selecting a simple random sample is more difficult in this situation.

EXAMPLE 7.2

A research chemist is testing a new method for measuring the amount of titanium (Ti) in ore samples. She chooses 10 ore samples of the same weight for her experiment. Five of the samples will be measured using a standard method, and the other 5 using the new method. Use random numbers to assign the 10 ore samples to the new and standard groups. Do these data represent a simple random sample from the population?

Solution There are really two populations in this experiment. They consist of titanium measurements, using either the new or standard method, for *all possible* ore samples of this weight. These populations do not exist in fact; they are **hypothetical populations,** envisioned in the mind of the researcher. Thus, it is impossible to select a simple random sample using the methods of Example 7.1. Instead, the researcher selects what she believes are 10 *representative* ore samples and hopes that these samples will *behave as if* they had been randomly selected from the two populations.

The researcher can, however, randomly select the five samples to be measured with each method. Number the samples from 1 to 10. The five samples selected for the new method may correspond to five one-digit random numbers. Use this sequence of random digits generated on a scientific calculator:

948247817184610

Since you cannot select the same ore sample twice, you must skip any digit that has already been chosen. Ore samples 9, 4, 8, 2, and 7 will be measured using the new method. The other samples—1, 3, 5, 6, and 10—will be measured using the standard method.

In addition to *simple random sampling,* there are other sampling plans that involve randomization and therefore provide a probabilistic basis for inference making. Three such plans are based on *stratified, cluster,* and *systematic sampling.*

When the population consists of two or more subpopulations, called **strata,** a sampling plan that ensures that each subpopulation is represented in the sample is called a **stratified random sample.**

Definition **Stratified random sampling** involves selecting a simple random sample from each of a given number of subpopulations, or **strata.**

Citizens' opinions about the construction of a performing arts center could be collected using a stratified random sample with city voting wards as strata. National polls usually involve some form of stratified random sampling with states as strata.

Another form of random sampling is used when the available sampling units are groups of elements, called **clusters.** For example, a household is a *cluster* of individuals living together. A city block or a neighborhood might be a convenient sampling unit and might be considered a *cluster* for a given sampling plan.

Definition A **cluster sample** is a simple random sample of clusters from the available clusters in the population.

When a particular cluster is included in the sample, a census of every element in the cluster is taken.

Sometimes the population to be sampled is ordered, such as an alphabetized list of people with driver's licenses, a list of utility users arranged by service addresses, or a list of customers by account numbers. In these and other situations, one element is chosen at random from the first k elements, and then every kth element thereafter is included in the sample.

Definition A **1-in-k systematic random sample** involves the random selection of one of the first k elements in an ordered population, and then the systematic selection of every kth element thereafter.

Not all sampling plans, however, involve random selection. You have probably heard of the nonrandom telephone polls in which those people who wish to express support for a question call one "900 number" and those opposed call a second "900 number." Each person must pay for his or her call. It is obvious that those people who call do not represent the population at large. This type of sampling plan is one form of a **convenience sample**—a sample that can be easily and simply obtained without random selection. Advertising for subjects who will be paid a fee for participating in an experiment produces a convenience sample. **Judgment sampling** allows the sampler to decide who will or will not be included in the sample. **Quota sampling,** in which the makeup of the sample must reflect the makeup of the population on some preselected characteristic, often has a nonrandom component in the selection process. **Remember that nonrandom samples can be described but cannot be used for making inferences!**

 TIP

All sampling plans used for making inferences must involve *randomization*!

7.2 **EXERCISES**

BASIC TECHNIQUES

7.1 A population consists of $N = 500$ experimental units. Use a random number table to select a random sample of $n = 20$ experimental units. (HINT: Since you need to use three-digit numbers, you can assign two three-digit numbers to each of the sampling units in the manner shown in the table.) What is the probability that each experimental unit is selected for inclusion in the sample?

Experimental Units	Random Numbers
1	001, 501
2	002, 502
3	003, 503
4	004, 504
.	.
.	.
.	.
499	499, 999
500	500, 000

7.2 A political analyst wishes to select a sample of $n = 20$ people from a population of 2000. Use the random number table to identify the people to be included in the sample.

7.3 A population contains 50,000 voters. Use the random number table to identify the voters to be included in a random sample of $n = 15$.

7.4 A small city contains 20,000 voters. Use the random number table to identify the voters to be included in a random sample of $n = 15$.

7.5 Every 10th Person A random sample of public opinion in a small town was obtained by selecting every 10th person who passed by the busiest corner in the downtown area. Will this sample have the characteristics of a random sample selected from the town's citizens? Explain.

7.6 Parks and Recreation A questionnaire was mailed to 1000 registered municipal voters selected at random. Only 500 questionnaires were returned, and of the 500 returned, 360 respondents were strongly opposed to a surcharge proposed to support the city Parks and Recreation Department. Are you willing to accept the 72% figure as a valid estimate of the percentage in the city who are opposed to the surcharge? Why or why not?

7.7 DMV Lists In many states, lists of possible jurors are assembled from voter registration lists and Department of Motor Vehicles records of licensed drivers and car owners. In what ways might this list not cover certain sectors of the population adequately?

7.8 Sex and Violence One question on a survey questionnaire is phrased as follows: "Don't you agree that there is too much sex and violence during prime TV viewing hours?" Comment on possible problems with the responses to this question. Suggest a better way to pose the question.

APPLICATIONS

7.9 Cancer in Rats The *Press Enterprise* identified a byproduct of chlorination called MX that has been linked to cancer in rats.[1] A scientist wants to conduct a validation study using 25 rats in the experimental group, each to receive a fixed dose of MX, and 25 rats in a control group that will receive no MX. Determine a randomization scheme to assign the 50 individual rats to the two groups.

7.10 Racial Bias? Does the race of an interviewer matter? This question was investigated by Chris Gilberg and colleagues and reported in an issue of *Chance* magazine.[2] The interviewer asked, "Do you feel that affirmative action should be used as an occupation selection criteria?" with possible answers of yes or no.

a. What problems might you expect with responses to this question when asked by interviewers of different ethnic origins?

b. When people were interviewed by an African-American, the response was about 70% in favor of affirmative action, approximately 35% when interviewed by an Asian, and approximately 25% when interviewed by a Caucasian. Do these results support your answer in part a?

7.11 MRIs In a study described in the *American Journal of Sports Medicine,* Peter D. Franklin and colleagues reported on the accuracy of using magnetic resonance imaging (MRI) to evaluate ligament sprains and tears on 35 patients.[3] Consecutive patients with acute or chronic knee pain were selected from the clinical practice of one of the authors and agreed to participate in the study.

a. Describe the sampling plan used to select study participants.

b. What chance mechanism was used to select this sample of 35 individuals with knee pain?

c. Can valid inferences be made using the results of this study? Why or why not?

d. Devise an alternative sampling plan. What would you change?

7.12 Blood Thinner A study of an experimental blood thinner was conducted to determine whether it works better than the simple aspirin tablet in warding off heart attacks and strokes.[4] The study reported in the *Press Enterprise* involved 19,185 people who had suffered heart attacks, strokes, or pain from clogged arteries. Each person was randomly assigned to take either aspirin or the experimental drug for 1 to 3 years. Assume that each person was equally likely to be assigned one of the two medications.

a. Devise a randomization plan to assign the medications to the patients.

b. Will there be an equal number of patients in each treatment group? Explain.

7.13 Going to the Moon Two different Gallup Polls were conducted for *CNN/USA Today,* both of which involved people's feelings about the U.S. space program.[5] Here is a question from each poll, along with the responses of the sampled Americans:

Space Exploration

CNN/USA Today/Gallup Poll. Dec. 5–7, 2003. Nationwide:

"Would you favor or oppose a new U.S. space program that would send astronauts to the moon?" Form A (N = 510, MoE ± 5)

	Favor %	Oppose %	No Opinion %
12/03	53	45	2

"Would you favor or oppose the U.S. government spending billions of dollars to send astronauts to the moon?" Form B (N = 494, MoE ± 5)

	Favor %	Oppose %	No Opinion %
12/03	31	67	2

---

a. Read the two poll questions. Which of the two wordings is more unbiased? Explain.

b. Look at the responses for the two different polls. How would you explain the large differences in the percentages either favoring or opposing the new program?

7.14 Ask America A nationwide policy survey titled "Ask America" was sent by the National Republican Congressional Committee to voters in the Forty-fourth Congressional District, asking for opinions on a variety of political issues.[6] Here are some questions from the survey:

- In recent years has the federal government grown more or less intrusive in your personal and business affairs?

- Is President Bush right in trying to rein in the size and scope of the federal government against the wishes of the big government Democrats?

- Do you believe the death penalty is a deterrrent to crime?

- Do you agree that the obstructionist Democrats should not be allowed to gain control of the U.S. Congress in the 2004 elections?

Comment on the effect of wording bias on the responses gathered using this survey.

7.3 STATISTICS AND SAMPLING DISTRIBUTIONS

When you select a random sample from a population, the numerical descriptive measures you calculate from the sample are called **statistics.** These statistics vary or change for each different random sample you select; that is, they are *random variables.* The probability distributions for statistics are called **sampling distributions** because, in repeated sampling, they provide this information:

- What values of the statistic can occur

- How often each value occurs

Definition The **sampling distribution of a statistic** is the probability distribution for the possible values of the statistic that results when random samples of size *n* are repeatedly drawn from the population.

There are three ways to find the sampling distribution of a statistic:

1. Derive the distribution *mathematically* using the laws of probability.

2. Use a *simulation* to approximate the distribution. That is, draw a large number of samples of size *n,* calculating the value of the statistic for each sample, and tabulate the results in a relative frequency histogram. When the number of

samples is large, the histogram will be very close to the theoretical sampling distribution.

3. Use *statistical theorems* to derive exact or approximate sampling distributions.

The next example demonstrates how to derive the sampling distributions of two statistics for a very small population.

EXAMPLE 7.3 — A population consists of $N = 5$ numbers: 3, 6, 9, 12, 15. If a random sample of size $n = 3$ is selected without replacement, find the sampling distributions for the sample mean $\bar{x}$ and the sample median m.

Solution You are sampling from the population shown in Figure 7.1. It contains five distinct numbers and each is equally likely, with probability $p(x) = 1/5$. You can easily find the population mean and median as

$$\mu = \frac{3 + 6 + 9 + 12 + 15}{5} = 9 \quad \text{and} \quad M = 9$$

FIGURE 7.1

Probability histogram for the $N = 5$ population values in Example 7.3

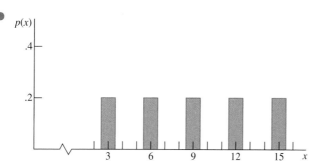

MY TIP

Sampling distributions can be either discrete or continuous.

There are 10 possible random samples of size $n = 3$ and each is equally likely, with probability 1/10. These samples, along with the calculated values of $\bar{x}$ and m for each, are listed in Table 7.3. You will notice that some values of $\bar{x}$ are more likely than others because they occur in more than one sample. For example,

$$P(\bar{x} = 8) = \frac{2}{10} = .2 \quad \text{and} \quad P(m = 6) = \frac{3}{10} = .3$$

The values in Table 7.3 are tabulated, and the sampling distributions for $\bar{x}$ and m are shown in Table 7.4 and Figure 7.2.

Since the population of $N = 5$ values is symmetric about the value $x = 9$, both the *population mean* and the *median* equal 9. It would seem reasonable, therefore, to consider using either $\bar{x}$ or m as a possible estimator of $M = \mu = 9$. Which estimator would you choose? From Table 7.3, you see that, in using m as an estimator, you would be in error by $9 - 6 = 3$ with probability .3 or by $9 - 12 = -3$ with probability .3. That is, the error in estimation using m would be 3 with probability .6. In using $\bar{x}$, however, an error of 3 would occur with probability only .2. On these grounds alone, you may wish to use $\bar{x}$ as an estimator in preference to m.

TABLE 7.3

Values of $\bar{x}$ and m for Simple Random Sampling when $n = 3$ and $N = 5$

Sample	Sample Values	$\bar{x}$	m
1	3, 6, 9	6	6
2	3, 6, 12	7	6
3	3, 6, 15	8	6
4	3, 9, 12	8	9
5	3, 9, 15	9	9
6	3, 12, 15	10	12
7	6, 9, 12	9	9
8	6, 9, 15	10	9
9	6, 12, 15	11	12
10	9, 12, 15	12	12

TABLE 7.4

Sampling Distributions for (a) the Sample Mean and (b) the Sample Median

(a)	$\bar{x}$	$p(\bar{x})$	(b)	m	$p(m)$
	6	.1		6	.3
	7	.1		9	.4
	8	.2		12	.3
	9	.2			
	10	.2			
	11	.1			
	12	.1			

FIGURE 7.2

Probability histograms for the sampling distributions of the sample mean, $\bar{x}$, and the sample median, m, in Example 7.3

It was not too difficult to derive these sampling distributions in Example 7.3 because the number of elements in the population was very small. When this is not the case, you may need to use one of these methods:

- Use a simulation to approximate the sampling distribution empirically.
- Rely on statistical theorems and theoretical results.

MY TIP

Almost every statistic has a mean and a standard deviation (or *standard error*) describing its center and spread.

One important statistical theorem that describes the sampling distribution of statistics that are sums or averages is presented in the next section.

THE CENTRAL LIMIT THEOREM

7.4

The **Central Limit Theorem** states that, under rather general conditions, sums and means of random samples of measurements drawn from a population tend to have an approximately normal distribution. Suppose you toss a balanced die $n = 1$ time. The random variable x is the number observed on the upper face. This familiar random variable can take six values, each with probability 1/6, and its probability distribution is shown in Figure 7.3. The shape of the distribution is *flat* or *uniform* and symmetric about the mean $\mu = 3.5$.

FIGURE 7.3

Probability distribution for x, the number appearing on a single toss of a die

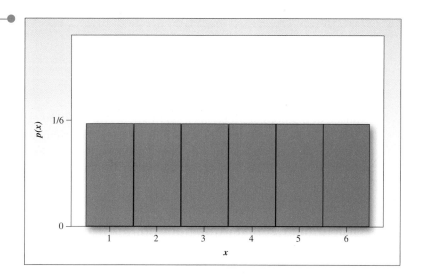

Now, take a sample of size $n = 2$ from this population; that is, toss two dice and record the sum of the numbers on the two upper faces, $\Sigma x_i = x_1 + x_2$. Table 7.5 shows the 36 possible outcomes, each with probability 1/36. The sums are tabulated, and each of the possible sums is divided by $n = 2$ to obtain an average. The result is the **sampling distribution** of $\bar{x} = \Sigma x_i/n$, shown in Figure 7.4. You should notice the dramatic difference in the shape of the sampling distribution. It is now roughly mound-shaped but still symmetric about the mean $\mu = 3.5$.

TABLE 7.5 ● **Sums of the Upper Faces of Two Dice**

Probability = 0/36 = 0.0

	First Die					
Second Die	1	2	3	4	5	6
1	2	3	4	5	6	7
2	3	4	5	6	7	8
3	4	5	6	7	8	9
4	5	6	7	8	9	10
5	6	7	8	9	10	11
6	7	8	9	10	11	12

Reset All

FIGURE 7.4
Sampling distribution of $\bar{x}$
for $n = 2$ dice

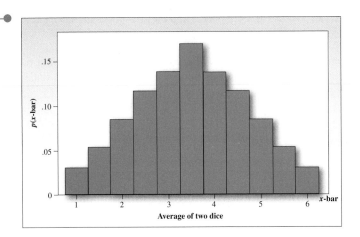

Using *MINITAB,* we generated the sampling distributions of $\bar{x}$ when $n = 3$ and $n = 4$. For $n = 3$, the sampling distribution in Figure 7.5 clearly shows the mound shape of the normal probability distribution, still centered at $\mu = 3.5$. Figure 7.6 dramatically shows that the distribution of $\bar{x}$ is approximately normally distributed based on a sample as small as $n = 4$. This phenomenon is the result of an important statistical theorem called the **Central Limit Theorem (CLT).**

FIGURE 7.5
MINITAB sampling distribution of $\bar{x}$ for $n = 3$ dice

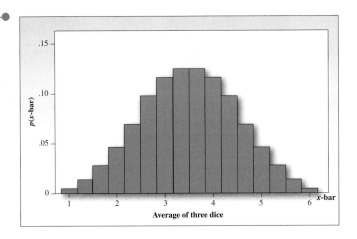

FIGURE 7.6
MINITAB sampling distribution of $\bar{x}$ for $n = 4$ dice

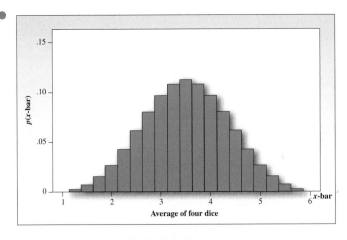

Central Limit Theorem

If random samples of n observations are drawn from a nonnormal population with finite mean μ and standard deviation σ, then, when n is large, the sampling distribution of the sample mean $\bar{x}$ is approximately normally distributed, with mean μ and standard deviation

$$\frac{\sigma}{\sqrt{n}}$$

The approximation becomes more accurate as n becomes large.

Regardless of its shape, the sampling distribution of $\bar{x}$ always has a mean identical to the mean of the sampled population and a standard deviation equal to the population standard deviation σ divided by $\sqrt{n}$. Consequently, *the spread of the distribution of sample means is considerably less than the spread of the sampled population.*

The Central Limit Theorem can be restated to apply to the **sum of the sample measurements** Σx_i, which, as n becomes large, also has an approximately normal distribution with mean $n\mu$ and standard deviation $\sigma\sqrt{n}$.

MY APPLET

The Java applet called **The Central Limit Theorem** can be used to perform a *simulation* for the sampling distributions of the average of one, two, three or four dice. Figure 7.7 shows the applet after the pair of dice ($n = 2$) has been tossed 2500 times. This is not as hard as it seems, since you need only press the [Roll 100 Sets] button 25 times. The simulation shows the possible values for $\bar{x} = \Sigma x_i/10$ and also shows the mean and standard deviation for these 2500 measurements. The mean, 3.5, is exactly equal to $\mu = 3.5$. What is the standard deviation for these 2500 measurements? Is it close to the theoretical value, $\sigma/\sqrt{n}$? You will use this applet again for the MyApplet Exercises at the end of the chapter.

FIGURE 7.7

Central Limit Theorem applet

TIP

The sampling distribution of $\bar{x}$ always has a mean μ and standard deviation $\sigma/\sqrt{n}$. The CLT helps describe its **shape**.

The important contribution of the Central Limit Theorem is in statistical inference. Many estimators that are used to make inferences about population parameters are sums or averages of the sample measurements. When the sample size is sufficiently large, you can expect these estimators to have sampling distributions that are approximately normal. You can then use the normal distribution to describe the behavior of these estimators in repeated sampling and evaluate the probability of observing certain sample results. As in Chapter 6, these probabilities are calculated using the standard normal random variable

$$z = \frac{\text{Estimator} - \text{Mean}}{\text{Standard deviation}}$$

As you reread the Central Limit Theorem, you may notice that the approximation is valid as long as the sample size n is "large"—but how large is "large"? Unfortunately, there is no clear answer to this question. The appropriate value of n depends on the shape of the population from which you sample as well as on how you want to use the approximation. However, these guidelines will help:

HOW DO I DECIDE WHEN THE SAMPLE SIZE IS LARGE ENOUGH?

- If the sampled population is **normal,** then the sampling distribution of $\bar{x}$ will also be normal, no matter what sample size you choose. This result can be proven theoretically, but it should not be too difficult for you to accept without proof.

- When the sampled population is approximately **symmetric,** the sampling distribution of $\bar{x}$ becomes approximately normal for relatively small values of n. Remember how rapidly ($n = 3$) the "flat" distribution in the dice example became mound-shaped.

- When the sampled population is **skewed,** the sample size n must be larger, with n at least 30 before the sampling distribution of $\bar{x}$ becomes approximately normal.

These guidelines suggest that, for many populations, the sampling distribution of $\bar{x}$ will be approximately normal for moderate sample sizes; an exception to this rule occurs in sampling a binomial population when either p or $q = (1 - p)$ is very small. As specific applications of the Central Limit Theorem arise, we will give you the appropriate sample size n.

THE SAMPLING DISTRIBUTION OF THE SAMPLE MEAN

 7.5

If the population mean μ is unknown, you might choose several *statistics* as an estimator; the sample mean $\bar{x}$ and the sample median m are two that readily come to mind. Which should you use? Consider these criteria in choosing the estimator for μ:

- Is it easy or hard to calculate?
- Does it produce estimates that are consistently too high or too low?
- Is it more or less variable than other possible estimators?

The sampling distributions for $\bar{x}$ and m with $n = 3$ for the small population in Example 7.3 showed that, in terms of these criteria, the sample mean performed better than the sample median as an estimator of μ. In many situations, the sample mean $\bar{x}$ has desirable properties as an estimator that are not shared by other competing estimators; therefore, it is more widely used.

THE SAMPLING DISTRIBUTION OF THE SAMPLE MEAN, $\bar{x}$

- If a random sample of n measurements is selected from a population with mean μ and standard deviation σ, the sampling distribution of the sample mean $\bar{x}$ will have mean μ and standard deviation[†]

$$\frac{\sigma}{\sqrt{n}}$$

- If the population has a *normal* distribution, the sampling distribution of $\bar{x}$ will be *exactly* normally distributed, *regardless of the sample size, n.*
- If the population distribution is *nonnormal,* the sampling distribution of $\bar{x}$ will be *approximately* normally distributed for large samples (by the Central Limit Theorem).

Standard Error

Definition   The standard deviation of a statistic used as an estimator of a population parameter is also called the **standard error of the estimator** (abbreviated **SE**) because it refers to the precision of the estimator. Therefore, the standard deviation of $\bar{x}$—given by $\sigma/\sqrt{n}$—is referred to as the **standard error of the mean** (abbreviated as $SE(\bar{x})$ or just SE).

[†]When repeated samples of size n are randomly selected from a *finite* population with N elements whose mean is μ and whose variance is σ^2, the standard deviation of $\bar{x}$ is

$$\frac{\sigma}{\sqrt{n}} \sqrt{\frac{N-n}{N-1}}$$

where σ^2 is the population variance. When N is large relative to the sample size n, $\sqrt{(N-n)(N-1)}$ is approximately equal to 1, and the standard deviation of $\bar{x}$ is

$$\frac{\sigma}{\sqrt{n}}$$

PERSONAL TRAINER

How Do I Calculate Probabilities for the Sample Mean $\bar{x}$?

If you know that the sampling distribution of $\bar{x}$ is *normal* or *approximately normal*, you can describe the behavior of the sample mean $\bar{x}$ by calculating the probability of observing certain values of $\bar{x}$ in repeated sampling.

1. Find μ and calculate SE $(\bar{x}) = \sigma/\sqrt{n}$.

2. Write down the event of interest in terms of $\bar{x}$, and locate the appropriate area on the normal curve.

3. Convert the necessary values of $\bar{x}$ to z-values using

$$z = \frac{\bar{x} - \mu}{\sigma/\sqrt{n}}$$

4. Use Table 3 in Appendix I to calculate the probability.

Exercise Reps (Fill in the Blanks)

A. You take a random sample of size $n = 36$ from a distribution with mean $\mu = 75$ and $\sigma = 12$. The sampling distribution of $\bar{x}$ will be approximately _____ with a mean of _____ and a standard deviation (or standard error) of _____.

B. To find the probability that the sample mean exceeds 80, write down the event of interest. _____
    When $\bar{x} = 80$,

$$z = \frac{\bar{x} - \mu}{\sigma/\sqrt{n}} = \text{_____}$$

Find the probability:

$$P(\bar{x} > \text{\_\_\_\_}) = P(z > \text{\_\_\_\_}) = 1 - \text{\_\_\_\_} = \text{\_\_\_\_}$$

C. To find the probability that the sample mean is between 70 and 72, write down the event of interest. _____
    When $\bar{x} = 70$ and $\bar{x} = 72$,

$$z = \frac{\bar{x} - \mu}{\sigma/\sqrt{n}} = \text{\_\_\_\_} \quad \text{and} \quad z = \frac{\bar{x} - \mu}{\sigma/\sqrt{n}} = \text{\_\_\_\_}$$

Find the probability:

$$P(\text{\_\_\_\_} < \bar{x} < \text{\_\_\_\_}) = P(\text{\_\_\_\_} < z < \text{\_\_\_\_})$$

$$= \text{\_\_\_\_} - \text{\_\_\_\_} = \text{\_\_\_\_}$$

Progress Report

*   Still having trouble? Try again using the Exercise Reps at the end of this section.

*   No problems? You can skip the Exercise Reps at the end of this section!

Answers are located on the perforated card at the back of this book.

EXAMPLE 7.4

The duration of Alzheimer's disease from the onset of symptoms until death ranges from 3 to 20 years; the average is 8 years with a standard deviation of 4 years. The administrator of a large medical center randomly selects the medical records of 30 deceased Alzheimer's patients from the medical center's database and records the average duration. Find the approximate probabilities for these events:

1. The average duration is less than 7 years.
2. The average duration exceeds 7 years.
3. The average duration lies within 1 year of the population mean $\mu = 8$.

Solution   Since the administrator has selected a random sample from the database at this medical center, he can draw conclusions about only past, present, or future patients with Alzheimer's disease at this medical center. If, on the other hand, this medical center can be considered representative of other medical centers in the country, it may be possible to draw more far-reaching conclusions.

What can you say about the shape of the sampled population? It is not symmetric because the mean $\mu = 8$ does not lie halfway between the maximum and minimum values. Since the mean is closer to the minimum value, the distribution is skewed to the right, with a few patients living a long time after the onset of the disease. Regardless of the shape of the population distribution, however, the sampling distribution of $\bar{x}$ has a mean $\mu = 8$ and standard deviation $\sigma/\sqrt{n} = 4/\sqrt{30} = .73$. In addition, because the sample size is $n = 30$, the Central Limit Theorem ensures the approximate normality of the sampling distribution of $\bar{x}$.

1. The probability that $\bar{x}$ is less than 7 is given by the shaded area in Figure 7.8. To find this area, you need to calculate the value of z corresponding to $\bar{x} = 7$:

$$z = \frac{\bar{x} - \mu}{\sigma/\sqrt{n}} = \frac{7 - 8}{.73} = -1.37$$

From Table 3 in Appendix I, you can find the cumulative area corresponding to $z = -1.37$ and

$$P(\bar{x} < 7) = P(z < -1.37) = .0853$$

FIGURE 7.8

The probability that $\bar{x}$ is less than 7 for Example 7.4

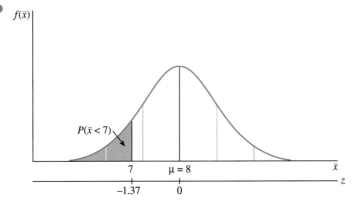

[NOTE: You must use $\sigma/\sqrt{n}$ (not σ) in the formula for z because you are finding an area under the sampling distribution for $\bar{x}$, not under the probability distribution for x.]

TIP

Remember that, for
continuous random
variables, there is no
probability
assigned to a single
point. Therefore
$P(\bar{x} \le 7) = P(\bar{x} < 7)$.

2. The event that $\bar{x}$ exceeds 7 is the complement of the event that $\bar{x}$ is less than 7. Thus, the probability that $\bar{x}$ exceeds 7 is

$$P(\bar{x} > 7) = 1 - P(\bar{x} \le 7)$$
$$= 1 - .0853 = .9147$$

3. The probability that $\bar{x}$ lies within 1 year of $\mu = 8$ is the shaded area in Figure 7.9. The z-value corresponding to $\bar{x} = 7$ is $z = -1.37$, from part 1, and the z-value for $\bar{x} = 9$ is

$$z = \frac{\bar{x} - \mu}{\sigma/\sqrt{n}} = \frac{9 - 8}{.73} = 1.37$$

The probability of interest is

$$P(7 < \bar{x} < 9) = P(-1.37 < z < 1.37)$$
$$= .9147 - .0853 = .8294$$

FIGURE 7.9

The probability that $\bar{x}$ lies
within 1 year of $\mu = 8$ for
Example 7.4

 APPLET

Example 7.4 can be solved using the **Normal Probabilities for Means** applet. If you enter the values for $\bar{x}$, σ, μ, and n (press "Enter" to record each change) and adjust the dropdown list at the bottom of the applet, you can calculate a tail area, a cumulative area, or the area between $-z_0$ and z_0. Conversely, if you need to find the value of $\bar{x}$ that cuts off a certain area under the curve, enter the area in the box marked "prob:" at the bottom of the applet, and the applet will provide the value of $\bar{x}$. The applet in Figure 7.10 is set to calculate $P(7 < \bar{x} < 9) = .829$, correct to three decimal places. You will use this applet for the MyApplet Exercises at the end of the chapter.

FIGURE 7.10

Normal Probabilities for Means applet

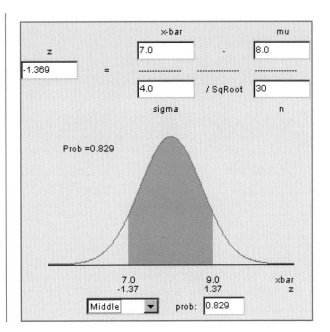

EXAMPLE   7.5

To avoid difficulties with the Federal Trade Commission or state and local consumer protection agencies, a beverage bottler must make reasonably certain that 12-ounce bottles actually contain 12 ounces of beverage. To determine whether a bottling machine is working satisfactorily, one bottler randomly samples 10 bottles per hour and measures the amount of beverage in each bottle. The mean $\bar{x}$ of the 10 fill measurements is used to decide whether to readjust the amount of beverage delivered per bottle by the filling machine. If records show that the amount of fill per bottle is normally distributed, with a standard deviation of .2 ounce, and if the bottling machine is set to produce a mean fill per bottle of 12.1 ounces, what is the approximate probability that the sample mean $\bar{x}$ of the 10 test bottles is less than 12 ounces?

Solution   The mean of the sampling distribution of the sample mean $\bar{x}$ is identical to the mean of the population of bottle fills—namely, $\mu = 12.1$ ounces—and the standard error of $\bar{x}$ is

$$\text{SE} = \frac{\sigma}{\sqrt{n}} = \frac{.2}{\sqrt{10}} = .063$$

(NOTE: σ is the standard deviation of the population of bottle fills, and n is the number of bottles in the sample.) Since the amount of fill is normally distributed, $\bar{x}$ is also normally distributed, as shown in Figure 7.11.

To find the probability that $\bar{x}$ is less than 12 ounces, express the value $\bar{x} = 12$ in units of standard deviations:

$$z = \frac{\bar{x} - \mu}{\sigma/\sqrt{n}} = \frac{12 - 12.1}{.063} = -1.59$$

Then

$$P(\bar{x} < 12) = P(z < -1.59) = .0559 \approx .056$$

FIGURE 7.11

Probability distribution of $\bar{x}$, the mean of the $n = 10$ bottle fills, for Example 7.5

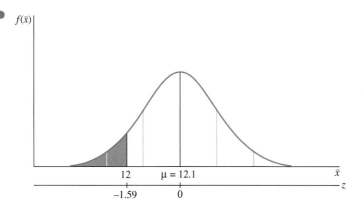

Thus, if the machine is set to deliver an average fill of 12.1 ounces, the mean fill $\bar{x}$ of a sample of 10 bottles will be less than 12 ounces with a probability equal to .056. When this danger signal occurs ($\bar{x}$ is less than 12), the bottler takes a larger sample to recheck the setting of the filling machine.

7.5 EXERCISES

EXERCISE REPS (FILL IN THE BLANKS)

7.15 You take a random sample of size $n = 49$ from a distribution with mean $\mu = 53$ and $\sigma = 21$. The sampling distribution of $\bar{x}$ will be approximately _____ with a mean of _____ and a standard deviation (or standard error) of _____.

7.16 Refer to Exercise 7.15. To find the probability that the sample mean is greater than 55, write down the event of interest. _____ When $\bar{x} = 55$,

$$z = \frac{\bar{x} - \mu}{\sigma/\sqrt{n}} = \underline{\quad}$$

Find the probability:

$$P(\bar{x} > \underline{\quad}) = P(z > \underline{\quad}) = 1 - \underline{\quad} = \underline{\quad}$$

7.17 You take a random sample of size $n = 40$ from a distribution with mean $\mu = 100$ and $\sigma = 20$. The sampling distribution of $\bar{x}$ will be approximately _____ with a mean of _____ and a standard deviation (or standard error) of _____.

7.18 Refer to Exercise 7.17. To find the probability that the sample mean is between 105 and 110, write down the event of interest. _____ When $\bar{x} = 105$ and $\bar{x} = 110$,

$$z = \frac{\bar{x} - \mu}{\sigma/\sqrt{n}} = \underline{\quad} \text{ and } z = \frac{\bar{x} - \mu}{\sigma/\sqrt{n}} = \underline{\quad}$$

Find the probability:

$$P(\underline{\quad} < \bar{x} < \underline{\quad}) = P(\underline{\quad} < z < \underline{\quad})$$

$$= \underline{\quad} - \underline{\quad} = \underline{\quad}$$

BASIC TECHNIQUES

7.19 Random samples of size n were selected from populations with the means and variances given here. Find the mean and standard deviation of the sampling distribution of the sample mean in each case:

a. $n = 36$, $\mu = 10$, $\sigma^2 = 9$

b. $n = 100$, $\mu = 5$, $\sigma^2 = 4$

c. $n = 8$, $\mu = 120$, $\sigma^2 = 1$

7.20 Refer to Exercise 7.19.

a. If the sampled populations are normal, what is the sampling distribution of $\bar{x}$ for parts a, b, and c?

b. According to the Central Limit Theorem, if the sampled populations are *not* normal, what can be said about the sampling distribution of $\bar{x}$ for parts a, b, and c?

7.21 Refer to Exercise 7.19, part b.

a. Sketch the sampling distribution for the sample mean and locate the mean and the interval $\mu \pm 2\sigma/\sqrt{n}$ along the $\bar{x}$-axis.

b. Shade the area under the curve that corresponds to the probability that $\bar{x}$ lies within .15 unit of the population mean μ.

c. Find the probability described in part b.

7.22 A random sample of n observations is selected from a population with standard deviation $\sigma = 1$. Calculate the standard error of the mean (SE) for these values of n:

a. $n = 1$ **b.** $n = 2$ **c.** $n = 4$

d. $n = 9$ **e.** $n = 16$ **f.** $n = 25$

g. $n = 100$

7.23 Refer to Exercise 7.22. Plot the standard error of the mean (SE) versus the sample size n and connect the points with a smooth curve. What is the effect of increasing the sample size on the standard error?

7.24 Suppose a random sample of $n = 25$ observations is selected from a population that is normally distributed, with mean equal to 106 and standard deviation equal to 12.

a. Give the mean and the standard deviation of the sampling distribution of the sample mean $\bar{x}$.

b. Find the probability that $\bar{x}$ exceeds 110.

c. Find the probability that the sample mean deviates from the population mean $\mu = 106$ by no more than 4.

APPLICATIONS

7.25 Measurement Error When research chemists perform experiments, they may obtain slightly different results on different replications, even when the experiment is performed identically each time. These differences are due to a phenomenon called "measurement error."

a. List some variables in a chemical experiment that might cause some small changes in the final response measurement.

b. If you want to make sure that your measurement error is small, you can replicate the experiment and take the sample average of all the measurements. To decrease the amount of variability in your average measurement, should you use a large or a small number of replications? Explain.

7.26 Tomatoes Explain why the weight of a package of one dozen tomatoes should be approximately normally distributed if the dozen tomatoes represent a random sample.

7.27 Bacteria in Water Use the Central Limit Theorem to explain why a Poisson random variable—say, the number of a particular type of bacteria in a cubic foot of water—has a distribution that can be approximated by a normal distribution when the mean μ is large. (HINT: One cubic foot of water contains 1728 cubic inches of water.)

7.28 Faculty Salaries Suppose that college faculty with the rank of professor at 2-year institutions earn an average of \$65,608 per year[7] with a standard deviation of \$4000. In an attempt to verify this salary level, a random sample of 60 professors was selected from a personnel database for all 2-year institutions in the United States.

a. Describe the sampling distribution of the sample mean $\bar{x}$.

b. Within what limits would you expect the sample average to lie, with probability .95?

c. Calculate the probability that the sample mean $\bar{x}$ is greater than \$67,000.

d. If your random sample actually produced a sample mean of \$67,000, would you consider this unusual? What conclusion might you draw?

7.29 Tax Savings An important expectation of a federal income tax reduction is that consumers will reap a substantial portion of the tax savings. Suppose estimates of the portion of total tax saved, based on a random sampling of 35 economists, have a mean of 26% and a standard deviation of 12%.

a. What is the approximate probability that a sample mean, based on a random sample of $n = 35$ economists, will lie within 1% of the mean of the population of the estimates of all economists?

b. Is it necessarily true that the mean of the population of estimates of all economists is equal to the percentage of tax savings that will actually be achieved? Why?

7.30 Paper Strength A manufacturer of paper used for packaging requires a minimum strength of 20 pounds per square inch. To check on the quality of the paper, a random sample of 10 pieces of paper is selected each hour from the previous hour's production and a strength measurement is recorded for each. The standard deviation σ of the strength measurements, computed by pooling the sum of squares of deviations of many samples, is known to equal 2 pounds per square inch, and the strength measurements are normally distributed.

a. What is the approximate sampling distribution of the sample mean of $n = 10$ test pieces of paper?

b. If the mean of the population of strength measurements is 21 pounds per square inch, what is the approximate probability that, for a random sample of $n = 10$ test pieces of paper, $\bar{x} < 20$?

c. What value would you select for the mean paper strength μ in order that $P(\bar{x} < 20)$ be equal to .001?

7.31 Potassium Levels The normal daily human potassium requirement is in the range of 2000 to 6000 milligrams (mg), with larger amounts required during hot summer weather. The amount of potassium in food varies, depending on the food. For example, there are approximately 7 mg in a cola drink, 46 mg in a beer, 630 mg in a banana, 300 mg in a carrot, and 440 mg in a glass of orange juice. Suppose the distribution of potassium in a banana is normally distributed, with mean equal to 630 mg and standard deviation equal to 40 mg per banana. You eat $n = 3$ bananas per day, and T is the total number of milligrams of potassium you receive from them.

a. Find the mean and standard deviation of T.

b. Find the probability that your total daily intake of potassium from the three bananas will exceed 2000 mg. (HINT: Note that T is the sum of three random variables, x_1, x_2, and x_3, where x_1 is the amount of potassium in banana number 1, etc.)

7.32 Deli Sales The total daily sales, x, in the deli section of a local market is the sum of the sales gener-

ated by a fixed number of customers who make purchases on a given day.

a. What kind of probability distribution do you expect the total daily sales to have? Explain.

b. For this particular market, the average sale per customer in the deli section is $8.50 with $\sigma = 2.50. If 30 customers make deli purchases on a given day, give the mean and standard deviation of the probability distribution of the total daily sales, x.

7.33 Normal Temperatures In Exercise 1.67, Allen Shoemaker derived a distribution of human body temperatures with a distinct mound shape.[8] Suppose we assume that the temperatures of healthy humans is approximately normal with a mean of 98.6 degrees and a standard deviation of 0.8 degrees.

a. If 130 healthy people are selected at random, what is the probability that the average temperature for these people is 98.25 degrees or lower?

b. Would you consider an average temperature of 98.25 degrees to be an unlikely occurrence, given that the true average temperature of healthy people is 98.6 degrees? Explain.

7.34 Sports and Achilles Tendon Injuries Some sports that involve a significant amount of running, jumping, or hopping put participants at risk for Achilles tendinopathy (AT), an inflammation and thickening of the Achilles tendon. A study in *The American Journal of Sports Medicine* looked at the diameter (in mm) of the affected and nonaffected tendons for patients who participated in these types of sports activities.[9] Suppose that the Achilles tendon diameters in the general population have a mean of 5.97 millimeters (mm) with a standard deviation of 1.95 mm.

a. What is the probability that a randomly selected sample of 31 patients would produce an average diameter of 6.5 mm or less for the nonaffected tendon?

b. When the diameters of the affected tendon was measured for a sample of 31 patients, the average diameter was 9.80. If the average tendon diameter in the population of patients with AT is no different than the average diameter of the nonaffected tendons (5.97 mm), what is the probability of observing an average diameter of 9.80 or higher?

c. What conclusions might you draw from the results of part b?

THE SAMPLING DISTRIBUTION OF THE SAMPLE PROPORTION

7.6

There are many practical examples of the binomial random variable x. One common application involves consumer preference or opinion polls, in which we use a random sample of n people to estimate the proportion p of people in the population who have a specified characteristic. If x of the sampled people have this characteristic, then the sample proportion

$$\hat{p} = \frac{x}{n}$$

 TIP

Q: How do you know if it's binomial or not?
A: Look to see if the measurement taken on a single experimental unit in the sample is a "success/failure" type. If so, it's probably binomial.

can be used to estimate the population proportion p (Figure 7.12).[†]

The binomial random variable x has a probability distribution $p(x)$, described in Chapter 5, with mean np and standard deviation $\sqrt{npq}$. Since $\hat{p}$ is simply the value of x, expressed as a proportion $\left(\hat{p} = \dfrac{x}{n}\right)$, the sampling distribution of $\hat{p}$ is identical to the probability distribution of x, except that it has a new scale along the horizontal axis.

FIGURE 7.12

Sampling distribution of the binomial random variable x and the sample proportion $\hat{p}$.

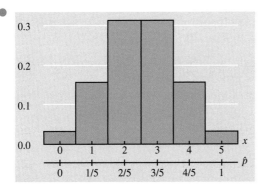

Because of this change of scale, the mean and standard deviation of $\hat{p}$ are also rescaled, so that the mean of the sampling distribution of $\hat{p}$ is p, and its standard error is

$$SE(\hat{p}) = \sqrt{\frac{pq}{n}} \qquad \text{where } q = 1 - p$$

Finally, just as we can approximate the probability distribution of x with a normal distribution when the sample size n is large, we can do the same with the sampling distribution of $\hat{p}$.

[†]A "hat" placed over the symbol of a population parameter denotes a statistic used to estimate the population parameter. For example, the symbol $\hat{p}$ denotes the sample proportion.

PROPERTIES OF THE SAMPLING DISTRIBUTION OF THE SAMPLE PROPORTION, $\hat{p}$

- If a random sample of n observations is selected from a binomial population with parameter p, then the sampling distribution of the sample proportion

$$\hat{p} = \frac{x}{n}$$

will have a mean

$$p$$

and a standard deviation

$$SE(\hat{p}) = \sqrt{\frac{pq}{n}} \qquad \text{where } q = 1 - p$$

- When the sample size n is large, the sampling distribution of $\hat{p}$ can be approximated by a normal distribution. The approximation will be adequate if $np > 5$ and $nq > 5$.

EXAMPLE 7.6 In a survey, 500 mothers and fathers were asked about the importance of sports for boys and girls. Of the parents interviewed, 60% agreed that the genders are equal and should have equal opportunities to participate in sports. Describe the sampling distribution of the sample proportion $\hat{p}$ of parents who agree that the genders are equal and should have equal opportunities.

Solution You can assume that the 500 parents represent a random sample of the parents of all boys and girls in the United States and that the true proportion in the population is equal to some unknown value that you can call p. The sampling distribution of $\hat{p}$ can be approximated by a normal distribution,[†] with mean equal to p (see Figure 7.13) and standard error

$$SE(\hat{p}) = \sqrt{\frac{pq}{n}}$$

FIGURE 7.13

The sampling distribution for $\hat{p}$ based on a sample of $n = 500$ parents for Example 7.6

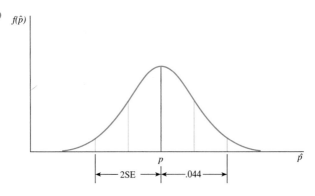

[†]Checking the conditions that allow the normal approximation to the distribution of $\hat{p}$, you can see that $n = 500$ is adequate for values of p near .60 because $n\hat{p} = 300$ and $n\hat{q} = 200$ are both greater than 5.

You can see from Figure 7.13 that the sampling distribution of $\hat{p}$ is centered over its mean p. Even though you do not know the exact value of p (the sample proportion $\hat{p} = .60$ may be larger or smaller than p), an approximate value for the standard deviation of the sampling distribution can be found using the sample proportion $\hat{p} = .60$ to approximate the unknown value of p. Thus,

$$SE = \sqrt{\frac{pq}{n}} \approx \sqrt{\frac{\hat{p}\hat{q}}{n}}$$

$$= \sqrt{\frac{(.60)(.40)}{500}} = .022$$

Therefore, approximately 95% of the time, $\hat{p}$ will fall within $2SE \approx .044$ of the (unknown) value of p.

MY PERSONAL TRAINER

How Do I Calculate Probabilities for the Sample Proportion $\hat{p}$?

1. Find the necessary values of n and p.
2. Check whether the normal approximation to the binomial distribution is appropriate ($np > 5$ and $nq > 5$).
3. Write down the event of interest in terms of $\hat{p}$, and locate the appropriate area on the normal curve.
4. Convert the necessary values of $\hat{p}$ to z-values using

$$z = \frac{\hat{p} - p}{\sqrt{\dfrac{pq}{n}}}$$

5. Use Table 3 in Appendix I to calculate the probability.

Exercise Reps (Fill in the Blanks)

A. You take a random sample of size $n = 36$ from a binomial distribution with mean $p = .4$. The sampling distribution of $\hat{p}$ will be approximately _____ with a mean of _____ and a standard deviation (or standard error) of _____.

B. To find the probability that the sample proportion exceeds .5, write down the event of interest. _____
When $\hat{p} = .5$,

$$z = \frac{\hat{p} - p}{\sqrt{\dfrac{pq}{n}}} = \underline{\hspace{1cm}}$$

Find the probability:

$$P(\hat{p} > \underline{\hspace{0.8cm}}) = P(z > \underline{\hspace{0.8cm}}) = 1 - \underline{\hspace{0.8cm}} = \underline{\hspace{0.8cm}}$$

C. To find the probability that the sample proportion is between .5 and .6, write down the event of interest. _____
When $\hat{p} = .5$ and $\hat{p} = .6$,

$$z = \frac{\hat{p} - p}{\sqrt{\dfrac{pq}{n}}} = \underline{\quad} \text{ and } z = \frac{\hat{p} - p}{\sqrt{\dfrac{pq}{n}}} = \underline{\quad}$$

Find the probability:

$$P(\underline{\quad} < \hat{p} < \underline{\quad}) = P(\underline{\quad} < z < \underline{\quad}) = \underline{\quad} - \underline{\quad} = \underline{\quad}$$

Progress Report

- Still having trouble? Try again using the Exercise Reps at the end of this section.

- No problems? You can skip the Exercise Reps at the end of this section!

Answers are located on the perforated card at the back of this book.

EXAMPLE 7.7

Refer to Example 7.6. Suppose the proportion p of parents in the population is actually equal to .55. What is the probability of observing a sample proportion as large as or larger than the observed value $\hat{p} = .60$?

Solution Figure 7.14 shows the sampling distribution of $\hat{p}$ when $p = .55$, with the observed value $\hat{p} = .60$ located on the horizontal axis. The probability of observing a sample proportion $\hat{p}$ equal to or larger than .60 is approximated by the shaded area in the upper tail of this normal distribution with

$$p = .55$$

and

$$SE = \sqrt{\frac{pq}{n}} = \sqrt{\frac{(.55)(.45)}{500}} = .0222$$

FIGURE 7.14

The sampling distribution of $\hat{p}$ for $n = 500$ and $p = .55$ for Example 7.7

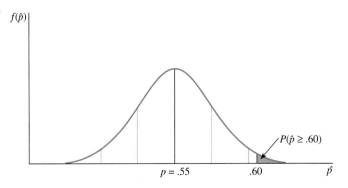

To find this shaded area, first calculate the z-value corresponding to $\hat{p} = .60$:

$$z = \frac{\hat{p} - p}{\sqrt{pq/n}} = \frac{.60 - .55}{.0222} = 2.25$$

Using Table 3 in Appendix I, you find

$$P(\hat{p} > .60) \approx P(z > 2.25) = 1 - .9878 = .0122$$

That is, if you were to select a random sample of $n = 500$ observations from a population with proportion p equal to .55, the probability that the sample proportion $\hat{p}$ would be as large as or larger than .60 is only .0122.

When the normal distribution was used in Chapter 6 to approximate the binomial probabilities associated with x, a correction of $\pm.5$ was applied to improve the approximation. The equivalent correction here is $\pm(.5/n)$. For example, for $\hat{p} = .60$ the value of z with the correction is

$$z_1 = \frac{(.60 - .001) - .55}{\sqrt{\dfrac{(.55)(.45)}{500}}} = 2.20$$

with $P(\hat{p} > .60) \approx .0139$. To two-decimal-place accuracy, this value agrees with the earlier result. When n is large, the effect of using the correction is generally negligible. You should solve problems in this and the remaining chapters *without* the correction factor unless you are specifically instructed to use it.

7.6 EXERCISES

EXERCISE REPS (FILL IN THE BLANKS)

7.35 You take a random sample of size $n = 50$ from a binomial distribution with a mean of $p = .7$. The sampling distribution of $\hat{p}$ will be approximately _____ with a mean of ____ and a standard deviation (or standard error) of ____.

7.36 To find the probability that the sample proportion is less than .8, write down the event of interest. ____
When $\hat{p} = .8$,

$$z = \frac{\hat{p} - p}{\sqrt{\dfrac{pq}{n}}} = \underline{\quad}$$

Find the probability:

$$P(\hat{p} < \underline{\quad}) = P(z < \underline{\quad}) = \underline{\quad}$$

BASIC TECHNIQUES

7.37 Random samples of size n were selected from binomial populations with population parameters p given here. Find the mean and the standard deviation of the sampling distribution of the sample proportion $\hat{p}$ in each case:

a. $n = 100$, $p = .3$

b. $n = 400$, $p = .1$

c. $n = 250$, $p = .6$

7.38 Sketch each of the sampling distributions in Exercise 7.37. For each, locate the mean p and the interval $p \pm 2$ SE along the $\hat{p}$-axis of the graph.

7.39 Refer to the sampling distribution in Exercise 7.37, part a.

a. Sketch the sampling distribution for the sample proportion and shade the area under the curve that corresponds to the probability that $\hat{p}$ lies within .08 of the population proportion p.

b. Find the probability described in part a.

7.40 Random samples of size $n = 500$ were selected from a binomial population with $p = .1$.

a. Is it appropriate to use the normal distribution to approximate the sampling distribution of $\hat{p}$? Check to make sure the necessary conditions are met.

Using the results of part a, find these probabilities:

b. $\hat{p} > .12$

c. $\hat{p} < .10$

d. $\hat{p}$ lies within .02 of p

7.41 Calculate $SE(\hat{p})$ for $n = 100$ and these values of p:

a. $p = .01$ **b.** $p = .10$ **c.** $p = .30$

d. $p = .50$ **e.** $p = .70$ **f.** $p = .90$

g. $p = .99$

h. Plot $SE(\hat{p})$ versus p on graph paper and sketch a smooth curve through the points. For what value of p is the standard deviation of the sampling distribution of $\hat{p}$ a maximum? What happens to the standard error when p is near 0 or near 1.0?

7.42 a. Is the normal approximation to the sampling distribution of $\hat{p}$ appropriate when $n = 400$ and $p = .8$?

b. Use the results of part a to find the probability that $\hat{p}$ is greater than .83.

c. Use the results of part a to find the probability that $\hat{p}$ lies between .76 and .84.

APPLICATIONS

7.43 Stress and Sweets One of the ways most Americans relieve stress is to reward themselves with sweets. According to one study in *Food Technology*, 46% admit to overeating sweet foods when stressed.[10] Suppose that the 46% figure is correct and that a random sample of $n = 100$ Americans is selected.

a. Does the distribution of $\hat{p}$, the sample proportion of Americans who relieve stress by overeating sweet foods, have an approximately normal distribution? If so, what are its mean and standard deviation?

b. What is the probability that the sample proportion, $\hat{p}$, exceeds .5?

c. What is the probability that $\hat{p}$ lies within the interval .35 to .55?

d. What might you conclude if the sample proportion were as small as 30%?

7.44 Surfing the Net Do you use the Internet to gather information for a project? The *Press Enterprise* reports that the percentage of students who used the Internet as their major resource for a school project in a recent year was 66%.[11] Suppose that you take a sample of $n = 1000$ students, and record the number of students who used the Internet as their major resource for their school project during the past year. Let $\hat{p}$ be the proportion of students surveyed who used the Internet as a major resource in the past year.

a. What is the exact distribution of $\hat{p}$? How can you approximate the distribution of $\hat{p}$?

b. What is the probability that the sample proportion $\hat{p}$ exceeds 68%?

c. What is the probability that the sample proportion lies between 64% and 68%?

d. Would a sample proportion of 70% contradict the reported value of 66%?

7.45 M&Ms According to the M&M® website, the average percentage of brown M&M® candies in a package of milk chocolate M&Ms is 13%.[12] (This percentage varies, however, among the different types of packaged M&Ms.) Suppose you randomly select a package of milk chocolate M&Ms that contains 55 candies and determine the proportion of brown candies in the package.

a. What is the approximate distribution of the sample proportion of brown candies in a package that contains 55 candies?

b. What is the probability that the sample proportion of brown candies is less than 20%?

c. What is the probability that the sample proportion exceeds 35%?

d. Within what range would you expect the sample proportion to lie about 95% of the time?

7.46 The "Cheeseburger Bill" In the spring of 2004, the U.S. Congress considered a bill that would prevent Americans from suing fast-food giants like McDonald's for making them overweight. Although the fast-food industry may not be to blame, a study by Children's Hospital in Boston reports that about two-thirds of adult Americans and about 15% of children and adolescents are overweight.[13] A random sample of 100 children is selected.

a. What is the probability that the sample proportion of overweight children exceeds 25%?

b. What is the probability that the sample proportion of overweight children is less than 12%?

c. Would it be unusual to find that 30% of the sampled children were overweight? Explain.

7.47 Oh, Nuts! Are you a chocolate "purist," or do you like other ingredients in your chocolate? *American Demographics* reports that almost 75% of consumers like traditional ingredients such as nuts or caramel in their chocolate. They are less enthusiastic about the taste of mint or coffee that provide more distinctive flavors.[14] A random sample of 200 consumers is selected and the number who like nuts or caramel in their chocolate is recorded.

a. What is the approximate sampling distribution for the sample proportion $\hat{p}$? What are the mean and standard deviation for this distribution?

b. What is the probability that the sample proportion is greater than 80%?

c. Within what limits would you expect the sample proportion to lie about 95% of the time?

A SAMPLING APPLICATION: STATISTICAL PROCESS CONTROL (OPTIONAL)

Statistical process control (SPC) methodology was developed to monitor, control, and improve products and services. Steel bearings must conform to size and hardness specifications, industrial chemicals must have a low prespecified level of impurities, and accounting firms must minimize and ultimately eliminate incorrect bookkeeping entries. It is often said that statistical process control consists of 10% statistics, 90% engineering and common sense. We can statistically monitor a process mean and tell when the mean falls outside preassigned limits, but we cannot tell *why* it is out of control. Answering this last question requires knowledge of the process and problem-solving ability—the other 90%!

Product quality is usually monitored using statistical control charts. Measurements on a process variable to be monitored change over time. The cause of a change in the variable is said to be *assignable* if it can be found and corrected. Other variation—small haphazard changes due to alteration in the production environment—that is not controllable is regarded as *random variation*. If the variation in a process variable is solely random, the process is said to be *in control*. The first objective in statistical process control is to eliminate assignable causes of variation in the process variable and then get the process in control. The next step is to reduce variation and get the measurements on the process variable within *specification limits,* the limits within which the measurements on usable items or services must fall.

Once a process is in control and is producing a satisfactory product, the process variables are monitored with **control charts.** Samples of *n* items are drawn from the process at specified intervals of time, and a sample statistic is computed. These statistics are plotted on the control chart, so that the process can be checked for shifts in the process variable that might indicate control problems.

A Control Chart for the Process Mean: The $\bar{x}$ Chart

Assume that *n* items are randomly selected from the production process at equal intervals and that measurements are recorded on the process variable. If the process is in control, the sample means should vary about the population mean μ in a random manner. Moreover, according to the Central Limit Theorem, the sampling distribution of $\bar{x}$ should be approximately normal, so that almost all of the values of $\bar{x}$ fall into the interval $(\mu \pm 3 \text{ SE}) = \mu \pm 3(\sigma/\sqrt{n})$. Although the exact values of μ and σ are unknown, you can obtain accurate estimates by using the sample measurements.

Every control chart has a *centerline* and *control limits*. The centerline for the **$\bar{x}$ chart** is the estimate of μ, the grand average of all the sample statistics calculated from the measurements on the process variable. The upper and lower *control limits* are placed three standard deviations above and below the centerline. If you monitor the process mean based on k samples of size n taken at regular intervals, the centerline is $\bar{\bar{x}}$, the average of the sample means, and the control limits are at $\bar{\bar{x}} \pm 3(\sigma/\sqrt{n})$, with σ estimated by s, the standard deviation of the nk measurements.

EXAMPLE 7.8

A statistical process control monitoring system samples the inside diameters of $n = 4$ bearings each hour. Table 7.6 provides the data for $k = 25$ hourly samples. Construct an $\bar{x}$ chart for monitoring the process mean.

Solution The sample mean was calculated for each of the $k = 25$ samples. For example, the mean for sample 1 is

$$\bar{x} = \frac{.992 + 1.007 + 1.016 + .991}{4} = 1.0015$$

TABLE 7.6

25 Hourly Samples of Bearing Diameters, $n = 4$ Bearings per Sample

Sample	Sample Measurements				Sample Mean, $\bar{x}$
1	.992	1.007	1.016	.991	1.00150
2	1.015	.984	.976	1.000	.99375
3	.988	.993	1.011	.981	.99325
4	.996	1.020	1.004	.999	1.00475
5	1.015	1.006	1.002	1.001	1.00600
6	1.000	.982	1.005	.989	.99400
7	.989	1.009	1.019	.994	1.00275
8	.994	1.010	1.009	.990	1.00075
9	1.018	1.016	.990	1.011	1.00875
10	.997	1.005	.989	1.001	.99800
11	1.020	.986	1.002	.989	.99925
12	1.007	.986	.981	.995	.99225
13	1.016	1.002	1.010	.999	1.00675
14	.982	.995	1.011	.987	.99375
15	1.001	1.000	.983	1.002	.99650
16	.992	1.008	1.001	.996	.99925
17	1.020	.988	1.015	.986	1.00225
18	.993	.987	1.006	1.001	.99675
19	.978	1.006	1.002	.982	.99200
20	.984	1.009	.983	.986	.99050
21	.990	1.012	1.010	1.007	1.00475
22	1.015	.983	1.003	.989	.99750
23	.983	.990	.997	1.002	.99300
24	1.011	1.012	.991	1.008	1.00550
25	.987	.987	1.007	.995	.99400

The sample means are shown in the last column of Table 7.6. The centerline is located at the average of the sample means, or

$$\bar{\bar{x}} = \frac{24.9675}{25} = .9987$$

The calculated value of s, the sample standard deviation of all $nk = 4(25) = 100$ observations, is $s = .011458$, and the estimated standard error of the mean of $n = 4$ observations is

$$\frac{s}{\sqrt{n}} = \frac{.011458}{\sqrt{4}} = .005729$$

The upper and lower control limits are found as

$$\text{UCL} = \bar{\bar{x}} + 3\frac{s}{\sqrt{n}} = .9987 + 3(.005729) = 1.015887$$

and

$$\text{LCL} = \bar{\bar{x}} - 3\frac{s}{\sqrt{n}} = .9987 - 3(.005729) = .981513$$

Figure 7.15 shows a *MINITAB* printout of the $\bar{x}$ chart constructed from the data. If you assume that the samples used to construct the $\bar{x}$ chart were collected when the process was in control, the chart can now be used to detect changes in the process mean. Sample means are plotted periodically, and if a sample mean falls outside the control limits, a warning should be conveyed. The process should be checked to locate the cause of the unusually large or small mean.

FIGURE 7.15

MINITAB $\bar{x}$ chart for Example 7.8

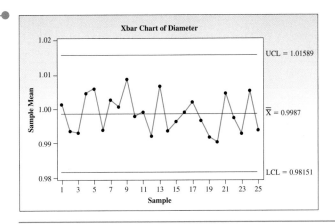

A Control Chart for the Proportion Defective: The *p* Chart

Sometimes the observation made on an item is simply whether or not it meets specifications; thus, it is judged to be defective or nondefective. If the fraction defective produced by the process is p, then x, the number of defectives in a sample of n items, has a binomial distribution.

To monitor a process for defective items, samples of size n are selected at periodic intervals and the sample proportion $\hat{p}$ is calculated. When the process is in control, $\hat{p}$ should fall into the interval $p \pm 3\text{SE}$, where p is the proportion of defectives in the population (or the process fraction defective) with standard error

$$\text{SE} = \sqrt{\frac{pq}{n}} = \sqrt{\frac{p(1-p)}{n}}$$

The process fraction defective is unknown but can be estimated by the average of the k sample proportions:

$$\bar{p} = \frac{\Sigma \hat{p}_i}{k}$$

and the standard error is estimated by

$$SE = \sqrt{\frac{\bar{p}(1 - \bar{p})}{n}}$$

The centerline for the **p chart** is located at $\bar{p}$, and the upper and lower control limits are

$$UCL = \bar{p} + 3\sqrt{\frac{\bar{p}(1 - \bar{p})}{n}}$$

and

$$LCL = \bar{p} - 3\sqrt{\frac{\bar{p}(1 - \bar{p})}{n}}$$

EXAMPLE 7.9

A manufacturer of ballpoint pens randomly samples 400 pens per day and tests each to see whether the ink flow is acceptable. The proportions of pens judged defective each day over a 40-day period are listed in Table 7.7. Construct a control chart for the proportion $\hat{p}$ defective in samples of $n = 400$ pens selected from the process.

TABLE 7.7

Proportions of Defectives in Samples of n = 400 Pens

Day	Proportion	Day	Proportion	Day	Proportion	Day	Proportion
1	.0200	11	.0100	21	.0300	31	.0225
2	.0125	12	.0175	22	.0200	32	.0175
3	.0225	13	.0250	23	.0125	33	.0225
4	.0100	14	.0175	24	.0175	34	.0100
5	.0150	15	.0275	25	.0225	35	.0125
6	.0200	16	.0200	26	.0150	36	.0300
7	.0275	17	.0225	27	.0200	37	.0200
8	.0175	18	.0100	28	.0250	38	.0150
9	.0200	19	.0175	29	.0150	39	.0150
10	.0250	20	.0200	30	.0175	40	.0225

Solution The estimate of the process proportion defective is the average of the $k = 40$ sample proportions in Table 7.7. Therefore, the centerline of the control chart is located at

$$\bar{p} = \frac{\Sigma \hat{p}_i}{k} = \frac{.0200 + .0125 + \cdots + .0225}{40} = \frac{.7600}{40} = .019$$

An estimate of SE, the standard error of the sample proportions, is

$$\sqrt{\frac{\bar{p}(1 - \bar{p})}{n}} = \sqrt{\frac{(.019)(.981)}{400}} = .00683$$

and $3SE = (3)(.00683) = .0205$. Therefore, the upper and lower control limits for the p chart are located at

$$UCL = \bar{p} + 3SE = .0190 + .0205 = .0395$$

and

$$LCL = \bar{p} - 3SE = .0190 - .0205 = -.0015$$

Or, since p cannot be negative, LCL $= 0$.

The p control chart is shown in Figure 7.16. Note that all 40 sample proportions fall within the control limits. If a sample proportion collected at some time in the future falls outside the control limits, the manufacturer should be concerned about an increase in the defective rate. He should take steps to look for the possible causes of this increase.

FIGURE 7.16

MINITAB p chart for Example 7.9

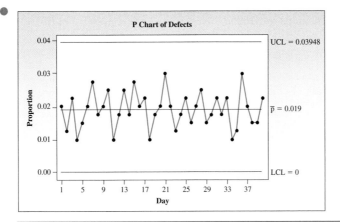

Other commonly used control charts are the *R chart,* which is used to monitor variation in the process variable by using the sample range, and the *c chart,* which is used to monitor the number of defects per item.

7.7 EXERCISES

BASIC TECHNIQUES

7.48 The sample means were calculated for 30 samples of size $n = 10$ for a process that was judged to be in control. The means of the 30 $\bar{x}$-values and the standard deviation of the combined 300 measurements were $\bar{\bar{x}} = 20.74$ and $s = .87$, respectively.

a. Use the data to determine the upper and lower control limits for an $\bar{x}$ chart.

b. What is the purpose of an $\bar{x}$ chart?

c. Construct an $\bar{x}$ chart for the process and explain how it can be used.

7.49 The sample means were calculated for 40 samples of size $n = 5$ for a process that was judged to be in control. The means of the 40 values and the standard deviation of the combined 200 measurements were $\bar{\bar{x}} = 155.9$ and $s = 4.3$, respectively.

a. Use the data to determine the upper and lower control limits for an $\bar{x}$ chart.

b. Construct an $\bar{x}$ chart for the process and explain how it can be used.

7.50 Explain the difference between an $\bar{x}$ chart and a p chart.

7.51 Samples of $n = 100$ items were selected hourly over a 100-hour period, and the sample proportion of defectives was calculated each hour. The mean of the 100 sample proportions was .035.

a. Use the data to find the upper and lower control limits for a p chart.

b. Construct a p chart for the process and explain how it can be used.

7.52 Samples of $n = 200$ items were selected hourly over a 100-hour period, and the sample proportion of defectives was calculated each hour. The mean of the 100 sample proportions was .041.

a. Use the data to find the upper and lower control limits for a p chart.

b. Construct a *p* chart for the process and explain how it can be used.

APPLICATIONS

7.53 Black Jack A gambling casino records and plots the mean daily gain or loss from five blackjack tables on an $\bar{x}$ chart. The overall mean of the sample means and the standard deviation of the combined data over 40 weeks were $\bar{\bar{x}} = \$10,752$ and $s = \$1605$, respectively.

a. Construct an $\bar{x}$ chart for the mean daily gain per blackjack table.

b. How can this $\bar{x}$ chart be of value to the manager of the casino?

7.54 Brass Rivets A producer of brass rivets randomly samples 400 rivets each hour and calculates the proportion of defectives in the sample. The mean sample proportion calculated from 200 samples was equal to .021. Construct a control chart for the proportion of defectives in samples of 400 rivets. Explain how the control chart can be of value to a manager.

7.55 Lumber Specs The manager of a building-supplies company randomly samples incoming lumber to see whether it meets quality specifications. From each shipment, 100 pieces of 2 × 4 lumber are inspected and judged according to whether they are first (acceptable) or second (defective) grade. The proportions of second-grade 2 × 4s recorded for 30 shipments were as follows:

.14	.21	.19	.18	.23	.20	.25	.19	.22	.17
.21	.15	.23	.12	.19	.22	.15	.26	.22	.21
.14	.20	.18	.22	.21	.13	.20	.23	.19	.26

Construct a control chart for the proportion of second-grade 2 × 4s in samples of 100 pieces of lumber. Explain how the control chart can be of use to the manager of the building-supplies company.

7.56 Coal Burning Power Plant A coal-burning power plant tests and measures three specimens of coal each day to monitor the percentage of ash in the coal. The overall mean of 30 daily sample means and the combined standard deviation of all the data were $\bar{\bar{x}} = 7.24$ and $s = .07$, respectively. Construct an $\bar{x}$ chart for the process and explain how it can be of value to the manager of the power plant.

7.57 Nuclear Power Plant The data in the table are measures of the radiation in air particulates at a nuclear power plant. Four measurements were recorded at weekly intervals over a 26-week

period. Use the data to construct an $\bar{x}$ chart and plot the 26 values of $\bar{x}$. Explain how the chart can be used.

Week	Radiation			
1	.031	.032	.030	.031
2	.025	.026	.025	.025
3	.029	.029	.031	.030
4	.035	.037	.034	.035
5	.022	.024	.022	.023
6	.030	.029	.030	.030
7	.019	.019	.018	.019
8	.027	.028	.028	.028
9	.034	.032	.033	.033
10	.017	.016	.018	.018
11	.022	.020	.020	.021
12	.016	.018	.017	.017
13	.015	.017	.018	.017
14	.029	.028	.029	.029
15	.031	.029	.030	.031
16	.014	.016	.016	.017
17	.019	.019	.021	.020
18	.024	.024	.024	.025
19	.029	.027	.028	.028
20	.032	.030	.031	.030
21	.041	.042	.038	.039
22	.034	.036	.036	.035
23	.021	.022	.024	.022
24	.029	.029	.030	.029
25	.016	.017	.017	.016
26	.020	.021	.020	.022

7.58 Baseball Bats A hardwoods manufacturing plant has several different production lines to make baseball bats of different weights. One such production line is designed to produce bats weighing 32 ounces. During a period of time when the production process was known to be in statistical control, the average bat weight was found to be 31.7 ounces. The observed data were gathered from 50 samples, each consisting of 5 measurements. The standard deviation of all samples was found to be $s = .2064$ ounces. Construct an $\bar{x}$-chart to monitor the 32-ounce bat production process.

7.59 More Baseball Bats Refer to Exercise 7.58 and suppose that during a day when the state of the 32-ounce bat production process was unknown, the following measurements were obtained at hourly intervals.

Hour	$\bar{x}$	Hour	$\bar{x}$
1	31.6	4	33.1
2	32.5	5	31.6
3	33.4	6	31.8

Each measurement represents a statistic computed from a sample of five bat weights selected from the production process during a certain hour. Use the control chart constructed in Exercise 7.58 to monitor the process.

CHAPTER REVIEW

Key Concepts and Formulas

I. Sampling Plans and Experimental Designs

1. Simple random sampling

   a. Each possible sample of size n is equally likely to occur.

   b. Use a computer or a table of random numbers.

   c. Problems are nonresponse, undercoverage, and wording bias.

2. Other sampling plans involving randomization

   a. Stratified random sampling

   b. Cluster sampling

   c. Systematic 1-in-k sampling

3. Nonrandom sampling

   a. Convenience sampling

   b. Judgment sampling

   c. Quota sampling

II. Statistics and Sampling Distributions

1. Sampling distributions describe the possible values of a statistic and how often they occur in repeated sampling.

2. Sampling distributions can be derived mathematically, approximated empirically, or found using statistical theorems.

3. The Central Limit Theorem states that sums and averages of measurements from a nonnormal population with finite mean μ and standard deviation σ have approximately normal distributions for large samples of size n.

III. Sampling Distribution of the Sample Mean

1. When samples of size n are randomly drawn from a normal population with mean μ and variance σ^2, the sample mean $\bar{x}$ has a normal distribution with mean μ and standard deviation $\sigma/\sqrt{n}$.

2. When samples of size n are randomly drawn from a nonnormal population with mean μ and variance σ^2, the Central Limit Theorem ensures that the sample mean $\bar{x}$ will have an approximately normal distribution with mean μ and standard deviation $\sigma/\sqrt{n}$ when n is large ($n \geq 30$).

3. Probabilities involving the sample mean can be calculated by standardizing the value of $\bar{x}$ using z:

$$z = \frac{\bar{x} - \mu}{\sigma/\sqrt{n}}$$

IV. Sampling Distribution of the Sample Proportion

1. When samples of size n are drawn from a binomial population with parameter p, the sample proportion $\hat{p}$ will have an approximately normal distribution with mean p and standard deviation $\sqrt{pq/n}$ as long as $np > 5$ and $nq > 5$.

2. Probabilities involving the sample proportion can be calculated by standardizing the value $\hat{p}$ using z:

$$z = \frac{\hat{p} - p}{\sqrt{\dfrac{pq}{n}}}$$

V. Statistical Process Control

1. To monitor a quantitative process, use an $\bar{x}$ chart. Select k samples of size n and calculate the overall mean $\bar{\bar{x}}$ and the standard deviation s of all nk measurements. Create upper and lower control limits as

$$\bar{\bar{x}} \pm 3\frac{s}{\sqrt{n}}$$

   If a sample mean exceeds these limits, the process is out of control.

2. To monitor a *binomial* process, use a p chart. Select k samples of size n and calculate the average of the sample proportions as

$$\bar{p} = \frac{\Sigma \hat{p}_i}{k}$$

   Create upper and lower control limits as

$$\bar{p} \pm 3\sqrt{\frac{\bar{p}(1 - \bar{p})}{n}}$$

   If a sample proportion exceeds these limits, the process is out of control.

MINITAB

The Central Limit Theorem at Work

MINITAB provides a perfect tool for exploring the way the Central Limit Theorem works in practice. Remember that, according to the Central Limit Theorem, if random samples of size n are drawn from a nonnormal population with mean μ and standard deviation σ, then when n is large, the sampling distribution of the sample mean $\bar{x}$ will be approximately normal with the same mean μ and with standard error $\sigma/\sqrt{n}$. Let's try sampling from a nonnormal population with the help of *MINITAB*.

In a new *MINITAB* worksheet, generate 100 samples of size $n = 30$ from a nonnormal distribution called the exponential distribution. Use **Calc → Random Data → Exponential.** Type **100** for the number of rows of data, and store the results in C1–C30 (see Figure 7.17). Leave the mean at the default of 1.0, the threshold at 0.0, and click **OK.** The data are generated and stored in the worksheet. Use **Graph → Histogram → Simple** to look at the distribution of some of the data—say, C1 (as in Figure 7.18). Notice that the distribution is not mound-shaped; it is highly skewed to the right.

FIGURE 7.17

Exponential Distribution	✖

Generate `100` rows of data

Store in column(s):

`C1-C30`

Scale: `1.0` (=Mean when Threshold=0)

Threshold: `0.0`

Select

Help   OK   Cancel

For the exponential distribution that we have used, the mean and standard deviation are $\mu = 1$ and $\sigma = 1$, respectively. Check the descriptive statistics for one of the columns (use **Stat → Basic Statistics → Display Descriptive Statistics**), and you will find that the 100 observations have a sample mean and standard deviation that are both *close to* but not exactly equal to 1. Now, generate 100 values of $\bar{x}$ based on samples of size $n = 30$ by creating a column of means for the 100 rows. Use **Calc → Row Statistics,** and select **Mean.** To average the entries in all 30 columns, select or type **C1–C30** in the Input variables box, and store the results in **C31** (see Figure 7.19). You can now look at the distribution of the sample means using **Graph → Histogram → Simple** and selecting **C31.** The distribution of the 100 sample means generated for our example is shown in Figure 7.20.

FIGURE 7.18

FIGURE 7.19

FIGURE 7.20

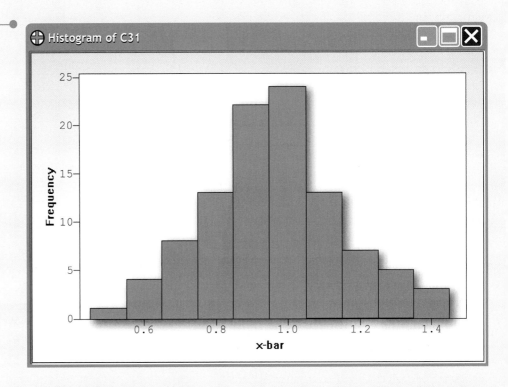

Notice the distinct mound shape of the distribution in Figure 7.20 compared to the original distribution in Figure 7.18. Also, if you check the descriptive statistics for C31, you will find that the mean and standard deviation of our 100 sample means are not too different from the theoretical values, $\mu = 1$ and $\sigma/\sqrt{n} = 1/\sqrt{30} = .18$. (For our data, the sample mean is .9645 and the standard deviation is .1830.) Since we had only 100 samples, our results are not *exactly* equal to the theoretical values. If we had generated an *infinite* number of samples, we would have gotten an exact match. This is the Central Limit Theorem at work!

Supplementary Exercises

7.60 A finite population consists of four elements: 6, 1, 3, 2.

a. How many different samples of size $n = 2$ can be selected from this population if you sample without replacement? (Sampling is said to be *without replacement* if an element cannot be selected twice for the same sample.)

b. List the possible samples of size $n = 2$.

c. Compute the sample mean for each of the samples given in part b.

d. Find the sampling distribution of $\bar{x}$. Use a probability histogram to graph the sampling distribution of $\bar{x}$.

e. If all four population values are equally likely, calculate the value of the population mean μ. Do any of the samples listed in part b produce a value of $\bar{x}$ exactly equal to μ?

7.61 Refer to Exercise 7.60. Find the sampling distribution for $\bar{x}$ if random samples of size $n = 3$ are selected *without replacement*. Graph the sampling distribution of $\bar{x}$.

7.62 Lead Pipes Studies indicate that drinking water supplied by some old lead-lined city piping systems may contain harmful levels of lead. An important study of the Boston water supply system showed that the distribution of lead content readings for individual water specimens had a mean and standard deviation of approximately .033 milligrams per liter (mg/l) and .10 mg/l, respectively.[15]

a. Explain why you believe this distribution is or is not normally distributed.

b. Because the researchers were concerned about the shape of the distribution in part a, they calculated the average daily lead levels at 40 different locations on each of 23 randomly selected days. What can you say about the shape of the distribution of the average daily lead levels from which the sample of 23 days was taken?

c. What are the mean and standard deviation of the distribution of average lead levels in part b?

7.63 Biomass The total amount of vegetation held by the earth's forests is important to both ecologists and politicians because green plants absorb carbon dioxide. An underestimate of the earth's vegetative mass, or biomass, means that much of the carbon dioxide emitted by human activities (primarily fossil-burning fuels) will not be absorbed, and a climate-altering buildup of carbon dioxide will occur. Studies[16] indicate that the biomass for tropical woodlands, thought to be about 35 kilograms per square meter (kg/m^2), may in fact be too high and that tropical biomass values vary regionally—from about 5 to 55 kg/m^2. Suppose you measure the tropical biomass in 400 randomly selected square-meter plots.

a. Approximate σ, the standard deviation of the biomass measurements.

b. What is the probability that your sample average is within two units of the true average tropical biomass?

c. If your sample average is $\bar{x} = 31.75$, what would you conclude about the overestimation that concerns the scientists?

7.64 Hard Hats The safety requirements for hard hats worn by construction workers and others, established by the American National Standards Institute (ANSI), specify that each of three hats pass the following test. A hat is mounted on an aluminum head form. An 8-pound steel ball is dropped on the hat from a height of 5 feet, and the resulting force is measured at the bottom of the head form. The force exerted on the head form by each of the three hats must be less than 1000 pounds, and the average of the three must be less than 850 pounds. (The relationship between this test and actual human head damage is unknown.) Suppose

the exerted force is normally distributed, and hence a sample mean of three force measurements is normally distributed. If a random sample of three hats is selected from a shipment with a mean equal to 900 and $\sigma = 100$, what is the probability that the sample mean will satisfy the ANSI standard?

7.65 Imagery and Memory A research psychologist is planning an experiment to determine whether the use of imagery—picturing a word in your mind—affects people's ability to memorize. He wants to use two groups of subjects: a group that memorizes a set of 20 words using the imagery technique, and a control group that does not use imagery.

a. Use a randomization technique to divide a group of 20 subjects into two groups of equal size.

b. How can the researcher randomly select the group of 20 subjects?

c. Suppose the researcher offers to pay subjects $50 each to participate in the experiment and uses the first 20 students who apply. Would this group behave as if it were a simple random sample of size $n = 20$?

7.66 Child Abuse A study of nearly 2000 women included questions dealing with child abuse and its effect on the women's adult life.[17] The study reported on the likelihood that a woman who was abused as a child would suffer either physical abuse or physical problems arising from depression, anxiety, low self-esteem, and drug abuse as an adult.

a. Is this an observational study or a designed experiment?

b. What problems might arise because of the sensitive nature of this study? What kinds of biases might occur?

7.67 Sprouting Radishes A biology experiment was designed to determine whether sprouting radish seeds inhibit the germination of lettuce seeds.[18] Three 10-centimeter petri dishes were used. The first contained 26 lettuce seeds, the second contained 26 radish seeds, and the third contained 13 lettuce seeds and 13 radish seeds.

a. Assume that the experimenter had a package of 50 radish seeds and another of 50 lettuce seeds. Devise a plan for randomly assigning the radish and lettuce seeds to the three treatment groups.

b. What assumptions must the experimenter make about the packages of 50 seeds in order to assure randomness in the experiment?

7.68 9/11 A study of about $n = 1000$ individuals in the United States during September 21–22, 2001, revealed that 43% of the respondents indicated that they were less willing to fly following the events of September 11, 2001.[19]

a. Is this an observational study or a designed experiment?

b. What problems might or could have occurred because of the sensitive nature of the subject? What kinds of biases might have occurred?

7.69 Telephone Service Suppose a telephone company executive wishes to select a random sample of $n = 20$ (a small number is used to simplify the exercise) out of 7000 customers for a survey of customer attitudes concerning service. If the customers are numbered for identification purposes, indicate the customers whom you will include in your sample. Use the random number table and explain how you selected your sample.

7.70 Rh Positive The proportion of individuals with an Rh-positive blood type is 85%. You have a random sample of $n = 500$ individuals.

a. What are the mean and standard deviation of $\hat{p}$, the sample proportion with Rh-positive blood type?

b. Is the distribution of $\hat{p}$ approximately normal? Justify your answer.

c. What is the probability that the sample proportion $\hat{p}$ exceeds 82%?

d. What is the probability that the sample proportion lies between 83% and 88%?

e. 99% of the time, the sample proportion would lie between what two limits?

7.71 What survey design is used in each of these situations?

a. A random sample of $n = 50$ city blocks is selected, and a census is done for each single-family dwelling on each block.

b. The highway patrol stops every tenth vehicle on a given city artery between 9:00 A.M. and 3:00 P.M. to perform a routine traffic safety check.

c. One hundred households in each of four city wards are surveyed concerning a pending city tax relief referendum.

d. Every 10th tree in a managed slash pine plantation is checked for pine needle borer infestation.

e. A random sample of $n = 1000$ taxpayers from the city of San Bernardino is selected by the Internal Revenue Service and their tax returns are audited.

7.72 Elevator Loads The maximum load (with a generous safety factor) for the elevator in an office building is 2000 pounds. The relative frequency distribution of the weights of all men and women using the elevator is mound-shaped (slightly skewed to the heavy weights), with mean μ equal to 150 pounds and standard deviation σ equal to 35 pounds. What is the largest number of people you can allow on the elevator if you want their total weight to exceed the maximum weight with a small probability (say, near .01)? (HINT: If $x_1, x_2, \ldots, x_n$ are independent observations made on a random variable x, and if x has mean μ and variance σ^2, then the mean and variance of Σx_i are $n\mu$ and $n\sigma^2$, respectively. This result was given in Section 7.4.)

7.73 Wiring Packages The number of wiring packages that can be assembled by a company's employees has a normal distribution, with a mean equal to 16.4 per hour and a standard deviation of 1.3 per hour.

a. What are the mean and standard deviation of the number x of packages produced per worker in an 8-hour day?

b. Do you expect the probability distribution for x to be mound-shaped and approximately normal? Explain.

c. What is the probability that a worker will produce at least 135 packages per 8-hour day?

7.74 Wiring Packages, continued Refer to Exercise 7.73. Suppose the company employs 10 assemblers of wiring packages.

a. Find the mean and standard deviation of the company's daily (8-hour day) production of wiring packages.

b. What is the probability that the company's daily production is less than 1280 wiring packages per day?

MY CD **7.75 Defective Lightbulbs** The table lists the
EX0775 number of defective 60-watt lightbulbs found in samples of 100 bulbs selected over 25 days from a manufacturing process. Assume that during these 25 days the manufacturing process was not producing an excessively large fraction of defectives.

Day	1	2	3	4	5	6	7	8	9	10
Defectives	4	2	5	8	3	4	4	5	6	1

Day	11	12	13	14	15	16	17	18	19	20
Defectives	2	4	3	4	0	2	3	1	4	0

Day	21	22	23	24	25
Defectives	2	2	3	5	3

a. Construct a p chart to monitor the manufacturing process, and plot the data.

b. How large must the fraction of defective items be in a sample selected from the manufacturing process before the process is assumed to be out of control?

c. During a given day, suppose a sample of 100 items is selected from the manufacturing process and 15 defective bulbs are found. If a decision is made to shut down the manufacturing process in an attempt to locate the source of the implied controllable variation, explain how this decision might lead to erroneous conclusions.

7.76 Lightbulbs, continued A hardware store chain purchases large shipments of lightbulbs from the manufacturer described in Exercise 7.75 and specifies that each shipment must contain no more than 4% defectives. When the manufacturing process is in control, what is the probability that the hardware store's specifications are met?

7.77 Lightbulbs, again Refer to Exercise 7.75. During a given week the number of defective bulbs in each of five samples of 100 were found to be 2, 4, 9, 7, and 11. Is there reason to believe that the production process has been producing an excessive proportion of defectives at any time during the week?

(MY) CD **7.78 Canned Tomatoes** During long produc-
EX0778 tion runs of canned tomatoes, the average weights (in ounces) of samples of five cans of standard-grade tomatoes in puree form were taken at 30 control points during an 11-day period. These results are shown in the table.[20] When the machine is performing normally, the average weight per can is 21 ounces with a standard deviation of 1.20 ounces.

a. Compute the upper and lower control limits and the centerline for the $\bar{x}$ chart.

b. Plot the sample data on the $\bar{x}$ chart and determine whether the performance of the machine is in control.

Sample Number	Average Weight	Sample Number	Average Weight
1	23.1	16	21.4
2	21.3	17	20.4
3	22.0	18	22.8
4	21.4	19	21.1
5	21.8	20	20.7
6	20.6	21	21.6
7	20.1	22	22.4
8	21.4	23	21.3
9	21.5	24	21.1
10	20.2	25	20.1
11	20.3	26	21.2
12	20.1	27	19.9
13	21.7	28	21.1
14	21.0	29	21.6
15	21.6	30	21.3

Source: Adapted from J. Hackl, *Journal of Quality Technology,* April 1991. Used with permission.

7.79 Pepsi or Coke? The battle for consumer preference continues between Pepsi and Coke. How can you make your preferences known? There is a webpage where you can vote for one of these colas if you click on the link that says PAY CASH for your opinion. Explain why the respondents do not represent a random sample of the opinions of purchasers or drinkers of these drinks. Explain the types of distortions that could creep into an Internet opinion poll.

7.80 Strawberries An experimenter wants to find an appropriate temperature at which to store fresh strawberries to minimize the loss of ascorbic acid. There are 20 storage containers, each with controllable temperature, in which strawberries can be stored. If two storage temperatures are to be used, how would the experimenter assign the 20 containers to one of the two storage temperatures?

7.81 Filling Soda Cans A bottler of soft drinks packages cans in six-packs. Suppose that the fill per can has an approximate normal distribution with a mean of 12 fluid ounces and a standard deviation of 0.2 fluid ounces.

a. What is the distribution of the total fill for a case of 24 cans?

b. What is the probability that the total fill for a case is less than 286 fluid ounces?

c. If a six-pack of soda can be considered a random sample of size $n = 6$ from the population, what is the probability that the average fill per can for a six-pack of soda is less than 11.8 fluid ounces?

7.82 Total Packing Weight Packages of food whose average weight is 16 ounces with a standard deviation of 0.6 ounces are shipped in boxes of 24 packages. If the package weights are approximately normally distributed, what is the probability that a box of 24 packages will weigh more than 392 ounces (24.5 pounds)?

7.83 Electronic Components A manufacturing process is designed to produce an electronic component for use in small portable television sets. The components are all of standard size and need not conform to any measurable characteristic, but are sometimes inoperable when emerging from the manufacturing process. Fifteen samples were selected from the process at times when the process was known to be in statistical control. Fifty components were observed within each sample, and the number of inoperable components was recorded.

$$6, 7, 3, 5, 6, 8, 4, 5, 7, 3, 1, 6, 5, 4, 5$$

Construct a p chart to monitor the manufacturing process.

MY APPLET Exercises

7.84 Dice Refer to the die-tossing experiment with $n = 1$ in Section 7.4 in which x is the number on the upper face of a single balanced die.

a. Use the formulas in Section 4.8 to verify that $\mu = 3.5$ and $\sigma = 1.71$ for this population.

b. Use the **Central Limit Theorem** applet to toss a single die at least 2000 times. (Your simulation can be done quickly by using the Roll 100 Sets button.) What are the mean and standard deviation of these 2000 observations? What is the shape of the histogram?

c. Compare the results of part b to the actual probability distribution shown in Figure 7.3 and the actual mean and standard deviation in part a. They should be similar!

7.85 Dice Two balanced dice are thrown, and the average number on the two upper faces is recorded.

a. Use the values $\mu = 3.5$ and $\sigma = 1.71$ from Exercise 7.84. What are the theoretical mean and standard deviation of the sampling distribution for $\bar{x}$?

b. Use the **Central Limit Theorem** applet to toss a single die at least 2000 times. (Your simulation can be done quickly by using the Roll 100 Sets button.) What are the mean and standard deviation of these 2000 observations? What is the shape of the histogram?

c. Compare the results of part b to the actual probability distribution shown in Figure 7.4 and the actual mean and standard deviation in part a.

7.86 Repeat the instructions in Exercise 7.85 when three dice are tossed.

7.87 Repeat the instructions in Exercise 7.85 when four dice are tossed.

7.88 Suppose a random sample of $n = 5$ observations is selected from a population that is normally distributed, with mean equal to 1 and standard deviation equal to .36.

a. Give the mean and the standard deviation of the sampling distribution of $\bar{x}$.

b. Find the probability that $\bar{x}$ exceeds 1.3, using the **Normal Probabilities for Means** applet.

c. Find the probability that the sample mean $\bar{x}$ is less than .5.

d. Find the probability that the sample mean deviates from the population mean $\mu = 1$ by more than .4.

7.89 Batteries A certain type of automobile battery is known to last an average of 1110 days with a standard deviation of 80 days. If 400 of these batteries are selected, use the **Normal Probabilities for Means** applet to find the following probabilities for the average length of life of the selected batteries:

a. The average is between 1100 and 1110.

b. The average is greater than 1120.

c. The average is less than 900.

CASE STUDY Sampling the Roulette at Monte Carlo

The technique of simulating a process that contains random elements and repeating the process over and over to see how it behaves is called a **Monte Carlo procedure.** It is widely used in business and other fields to investigate the properties of an operation that is subject to random effects, such as weather, human behavior, and so on. For example, you could model the behavior of a manufacturing company's inventory by creating, on paper, daily arrivals and departures of manufactured products from the company's warehouse. Each day a random number of items produced by the company would be received into inventory. Similarly, each day a random number of orders of varying random sizes would be shipped. Based on the input and output of items, you could calculate the inventory—that is, the number of items on hand at the end of each day. The values of the random variables, the number of items produced, the number of orders, and the number of items per order needed for each day's simulation would be obtained from theoretical distributions of observations that closely model the corresponding distributions of the variables that have been observed over time in the manufacturing operation. By repeating the simulation of the supply, the shipping, and the calculation of daily inventory for a large number of days (a sampling of what might really happen), you can observe the behavior of the plant's daily inventory. The Monte Carlo procedure is particularly valuable because it enables the manufacturer to see how the daily inventory would behave when certain changes are made in the supply pattern or in some other aspect of the operation that could be controlled.

In an article entitled "The Road to Monte Carlo," Daniel Seligman comments on the Monte Carlo method, noting that, although the technique is widely used in business schools to study capital budgeting, inventory planning, and cash flow management, no one seems to have used the procedure to study how well we might do if we were to gamble at Monte Carlo.[21]

To follow up on this thought, Seligman programmed his personal computer to simulate the game of roulette. Roulette involves a wheel with its rim divided into 38 pockets. Thirty-six of the pockets are numbered 1 to 36 and are alternately colored red and black. The two remaining pockets are colored green and are marked 0 and 00. To play the game, you bet a certain amount of money on one or more pockets. The wheel is spun and turns until it stops. A ball falls into a slot on the wheel to indicate the winning number. If you have money on that number, you win a specified amount. For example, if you were to play the number 20, the payoff is 35 to 1. If the wheel does not stop at that number, you lose your bet. Seligman decided to see how his nightly gains (or losses) would fare if he were to bet $5 on each turn of the wheel and repeat the process 200 times each night. He did this 365 times, thereby simulating the outcomes of 365 nights at the casino. Not surprisingly, the mean "gain" per $1000 evening for the 365 nights was a *loss* of $55, the average of the winnings retained by the gambling house. The surprise, according to Seligman, was the extreme variability of the nightly "winnings." Seven times out of the 365 evenings, the fictitious gambler lost the $1000 stake, and only once did he win a maximum of $1160. On 141 nights, the loss exceeded $250.

1. To evaluate the results of Seligman's Monte Carlo experiment, first find the probability distribution of the gain x on a single $5 bet.
2. Find the expected value and variance of the gain x from part 1.

3. Find the expected value and variance for the evening's gain, the sum of the gains or losses for the 200 bets of $5 each.

4. Use the results of part 2 to evaluate the probability of 7 out of 365 evenings resulting in a loss of the total $1000 stake.

5. Use the results of part 3 to evaluate the probability that the largest evening's winnings were as great as $1160.

Large-Sample Estimation

© AP/Wide World Photos

CHAPTER INDEX

 PERSONAL TRAINER

How Do I Estimate a Population Mean or Proportion?
How Do I Choose the Sample Size?

How Reliable Is That Poll?

Do the national polls conducted by the Gallup and Harris organizations, the news media, and others provide accurate estimates of the percentages of people in the United States who favor various propositions? The case study at the end of this chapter examines the reliability of a poll conducted by the Gallup organization using the theory of large-sample estimation.

WHERE WE'VE BEEN

The first seven chapters of this book have given you the building blocks you will need to understand statistical inference and how it can be applied in practical situations. The first three chapters were concerned with using descriptive statistics, both graphical and numerical, to describe and interpret sets of measurements. In the next three chapters, you learned about probability and probability distributions—the basic tools used to describe *populations* of measurements. The binomial and the normal distributions were emphasized as important for practical applications. The seventh chapter provided the link between probability and statistical inference. Many statistics are either sums or averages calculated from sample measurements. The Central Limit Theorem states that, even if the sampled populations are not normal, the sampling distributions of these *statistics* will be approximately normal when the sample size n is large. These statistics are the tools you use for *inferential statistics*—making inferences about a population using information contained in a sample.

WHERE WE'RE GOING— STATISTICAL INFERENCE

Inference—specifically, decision making and prediction—is centuries old and plays a very important role in most peoples' lives. Here are some applications:

- The government needs to predict short- and long-term interest rates.
- A broker wants to forecast the behavior of the stock market.
- A metallurgist wants to decide whether a new type of steel is more resistant to high temperatures than the old type.
- A consumer wants to estimate the selling price of her house before putting it on the market.

Parameter ⇔ Population
Statistic ⇔ Sample

There are many ways to make these decisions or predictions, some subjective and some more objective in nature. How good will your predictions or decisions be? Although you may feel that your own built-in decision-making ability is quite good, experience suggests that this may not be the case. It is the job of the mathematical statistician to provide methods of statistical inference making that are better and more reliable than just subjective guesses.

Statistical inference is concerned with making decisions or predictions about **parameters**—the numerical descriptive measures that characterize a population. Three parameters you encountered in earlier chapters are the population mean μ, the population standard deviation σ, and the binomial proportion p. In statistical inference, a practical problem is restated in the framework of a population with a specific parameter of interest. For example, the metallurgist could measure the *average* coefficients of expansion for both types of steel and then compare their values.

Methods for making inferences about population parameters fall into one of two categories:

- **Estimation:** Estimating or predicting the value of the parameter
- **Hypothesis testing:** Making a decision about the value of a parameter based on some preconceived idea about what its value might be

EXAMPLE 8.1   The circuits in computers and other electronics equipment consist of one or more printed circuit boards (PCB), and computers are often repaired by simply replacing one or more defective PCBs. In an attempt to find the proper setting of a plating process applied to one side of a PCB, a production supervisor might *estimate* the average thickness of copper plating on PCBs using samples from several days of operation. Since he has no knowledge of the average thickness μ before observing the production process, his is an *estimation* problem.

EXAMPLE 8.2   The supervisor in Example 8.1 is told by the plant owner that the thickness of the copper plating must not be less than .001 inch in order for the process to be in control. To decide whether or not the process is in control, the supervisor might formulate a test. He could *hypothesize* that the process is in control—that is, assume that the average thickness of the copper plating is .001 or greater—and use samples from several days of operation to decide whether or not his hypothesis is correct. The supervisor's decision-making approach is called a *test of hypothesis.*

Which method of inference should be used? That is, should the parameter be estimated, or should you test a hypothesis concerning its value? The answer is dictated by the practical question posed and is often determined by personal preference. Since both estimation and tests of hypotheses are used frequently in scientific literature, we include both methods in this and the next chapter.

A statistical problem, which involves planning, analysis, and inference making, is incomplete without a measure of the **goodness of the inference.** That is, how accurate or reliable is the method you have used? If a stockbroker predicts that the price of a stock will be $80 next Monday, will you be willing to take action to buy or sell your stock without knowing how reliable her prediction is? Will the prediction be within $1, $2, or $10 of the actual price next Monday? Statistical procedures are important because they provide two types of information:

- Methods for making the inference
- A numerical measure of the goodness or reliability of the inference

8.3 TYPES OF ESTIMATORS

To estimate the value of a population parameter, you can use information from the sample in the form of an **estimator.** Estimators are calculated using information from the sample observations, and hence, by definition they are also *statistics.*

Definition   An **estimator** is a rule, usually expressed as a formula, that tells us how to calculate an estimate based on information in the sample.

Estimators are used in two different ways.

- **Point estimation:** Based on sample data, a single number is calculated to estimate the population parameter. The rule or formula that describes this calculation is called the **point estimator,** and the resulting number is called a **point estimate.**

- **Interval estimation:** Based on sample data, two numbers are calculated to form an interval within which the parameter is expected to lie. The rule or formula that describes this calculation is called the **interval estimator,** and the resulting pair of numbers is called an **interval estimate** or **confidence interval.**

EXAMPLE 8.3

A veterinarian wants to estimate the average weight gain per month of 4-month-old golden retriever pups that have been placed on a lamb and rice diet. The *population* consists of the weight gains per month of all 4-month-old golden retriever pups that are given this particular diet. The veterinarian wants to estimate the unknown parameter μ, the average monthly weight gain for this *hypothetical* population. One possible *estimator* based on sample data is the sample mean, $\bar{x} = \Sigma x_i/n$. It could be used in the form of a single number or *point estimate*—for instance, 3.8 pounds—or you could use an *interval estimate* and estimate that the average weight gain will be between 2.7 and 4.9 pounds.

Both point and interval estimation procedures use information provided by the sampling distribution of the specific estimator you have chosen to use. We will begin by discussing *point estimation* and its use in estimating population means and proportions.

POINT ESTIMATION

8.4

In a practical situation, there may be several statistics that could be used as point estimators for a population parameter. To decide which of several choices is best, you need to know how the estimator behaves in repeated sampling, described by its *sampling distribution.*

By way of analogy, think of firing a revolver at a target. The parameter of interest is the bull's-eye, at which you are firing bullets. Each bullet represents a single sample estimate, fired by the revolver, which represents the estimator. Suppose your friend fires a single shot and hits the bull's-eye. Can you conclude that he is an excellent shot? Would you stand next to the target while he fires a second shot? Probably not, because you have no measure of how well he performs in repeated trials. Does he always hit the bull's-eye, or is he consistently too high or too low? Do his shots cluster closely around the target, or do they consistently miss the target by a wide margin? Figure 8.1 shows several target configurations. Which target would you pick as belonging to the best shot?

FIGURE 8.1

Which marksman is best?

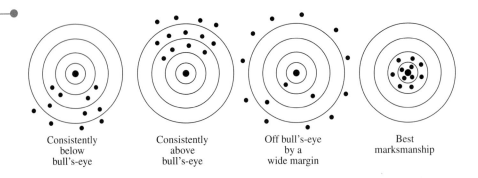

Consistently below bull's-eye | Consistently above bull's-eye | Off bull's-eye by a wide margin | Best marksmanship

Sampling distributions provide information that can be used to select the **best estimator.** What characteristics would be valuable? First, the **sampling distribution of the point estimator should be centered over the true value of the parameter to be estimated.** That is, the estimator should not consistently underestimate or overestimate the parameter of interest. Such an estimator is said to be **unbiased.**

Definition　An estimator of a parameter is said to be **unbiased** if the mean of its distribution is equal to the true value of the parameter. Otherwise, the estimator is said to be **biased.**

The sampling distributions for an unbiased estimator and a biased estimator are shown in Figure 8.2. The sampling distribution for the biased estimator is shifted to the right of the true value of the parameter. This biased estimator is more likely than an unbiased one to overestimate the value of the parameter.

FIGURE 8.2

Distributions for biased and unbiased estimators

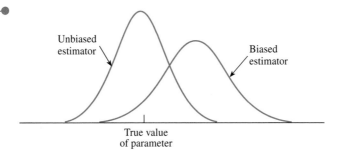

The second desirable characteristic of an estimator is that **the spread (as measured by the variance) of the sampling distribution should be as small as possible.** This ensures that, with a high probability, an individual estimate will fall close to the true value of the parameter. The sampling distributions for two unbiased estimators, one with a small variance[†] and the other with a larger variance, are shown in Figure 8.3.

FIGURE 8.3

Comparison of estimator variability

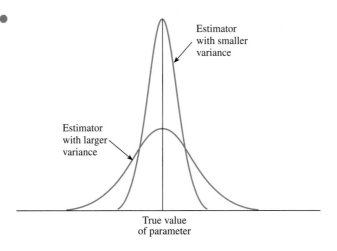

[†]Statisticians usually use the term *variance of an estimator* when in fact they mean the variance of the sampling distribution of the estimator. This contractive expression is used almost universally.

Naturally, you would prefer the estimator with the smaller variance because the estimates tend to lie closer to the true value of the parameter than in the distribution with the larger variance.

In real-life sampling situations, you may know that the sampling distribution of an estimator centers about the parameter that you are attempting to estimate, but all you have is the estimate computed from the n measurements contained in the sample. How far from the true value of the parameter will your estimate lie? How close is the marksman's bullet to the bull's-eye? The distance between the estimate and the true value of the parameter is called the **error of estimation.**

Definition   The distance between an estimate and the estimated parameter is called the **error of estimation.**

In this chapter, you may assume that the sample sizes are always large and, therefore, that the *unbiased* estimators you will study have sampling distributions that can be approximated by a normal distribution (because of the Central Limit Theorem). Remember that, for any point estimator with a normal distribution, the Empirical Rule states that approximately 95% of all the point estimates will lie within two (or more exactly, 1.96) standard deviations of the mean of that distribution. For *unbiased* estimators, this implies that the difference between the point estimator and the true value of the parameter will be less than 1.96 standard deviations or 1.96 standard errors (SE). This quantity, called the 95% **margin of error** (or simply the **"margin of error"**), provides a practical upper bound for the error of estimation (see Figure 8.4). It is possible that the error of estimation will exceed this margin of error, but that is very unlikely.

FIGURE 8.4

Sampling distribution of an unbiased estimator

 TIP

95% Margin of error = 1.96 × standard error

POINT ESTIMATION OF A POPULATION PARAMETER

Point estimator: a statistic calculated using sample measurements
95% Margin of error: 1.96 × Standard error of the estimator

The sampling distributions for two unbiased point estimators were discussed in Chapter 7. It can be shown that both of these point estimators have the *minimum variability* of all unbiased estimators and are thus the *best estimators* you can find in each situation.

The variability of the estimator is measured using its standard error. However, you might have noticed that the standard error usually depends on unknown parameters such as σ or p. These parameters must be estimated using sample statistics such as s and $\hat{p}$. Although not exactly correct, experimenters generally refer to the estimated standard error as *the standard error*.

(MY) **PERSONAL TRAINER**

How Do I Estimate a Population Mean or Proportion?

- To estimate the population mean μ for a quantitative population, the point estimator $\bar{x}$ is *unbiased* with standard error estimated as

$$SE = \frac{s}{\sqrt{n}}^\dagger$$

The 95% margin of error when $n \geq 30$ is estimated as

$$\pm 1.96\left(\frac{s}{\sqrt{n}}\right)$$

- To estimate the population proportion p for a binomial population, the point estimator $\hat{p} = x/n$ is *unbiased,* with standard error estimated as

$$SE = \sqrt{\frac{\hat{p}\hat{q}}{n}}$$

The 95% margin of error is estimated as

$$\pm 1.96\sqrt{\frac{\hat{p}\hat{q}}{n}}$$

Assumptions: $n\hat{p} > 5$ and $n\hat{q} > 5$.

EXAMPLE 8.4

An investigator is interested in the possibility of merging the capabilities of television and the Internet. A random sample of $n = 50$ Internet users who were polled about the time they spend watching television produced an average of 11.5 hours per week, with a standard deviation of 3.5 hours. Use this information to estimate the population mean time Internet users spend watching television.

Solution The random variable measured is the time spent watching television per week. This is a quantitative random variable best described by its mean μ. The point estimate of μ, the average time Internet users spend watching television, is $\bar{x} = 11.5$ hours. The margin of error is estimated as

$$1.96\ SE = 1.96\left(\frac{s}{\sqrt{n}}\right) = 1.96\left(\frac{3.5}{\sqrt{50}}\right) = .97 \approx 1$$

$\dagger$When you sample from a normal distribution, the statistic $(\bar{x} - \mu)/(s/\sqrt{n})$ has a t distribution, which will be discussed in Chapter 10. When the sample is *large,* this statistic is approximately normally distributed whether the sampled population is normal or nonnormal.

You can feel fairly confident that the sample estimate of 11.5 hours of television watching for Internet users is within ±1 hour of the population mean.

In reporting research results, investigators often attach either the sample standard deviation s (sometimes called SD) or the standard error $s/\sqrt{n}$ (usually called SE or SEM) to the estimates of population means. You should always look for an explanation somewhere in the text of the report that tells you whether the investigator is reporting $\bar{x} \pm$ SD or $\bar{x} \pm$ SE. In addition, the sample means and standard deviations or standard errors are often presented as "error bars" using the graphical format shown in Figure 8.5.

FIGURE 8.5

Plot of treatment means and their standard errors

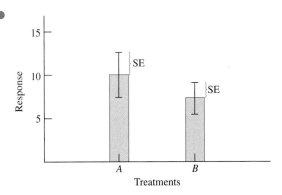

EXAMPLE 8.5

In addition to the average time Internet users spend watching television, the researcher from Example 8.4 is interested in estimating the proportion of individuals in the population at large who want to purchase a television that also acts as a computer. In a random sample of $n = 100$ adults, 45% in the sample indicated that they might buy one. Estimate the true population proportion of adults who are interested in buying a television that also acts as a computer, and find the margin of error for the estimate.

Solution The parameter of interest is now p, the proportion of individuals in the population who want to purchase a television that also acts as a computer. The best estimator of p is the sample proportion, $\hat{p}$, which for this sample is $\hat{p} = .45$. In order to find the margin of error, you can approximate the value of p with its estimate $\hat{p} = .45$:

$$1.96 \text{ SE} = 1.96 \sqrt{\frac{\hat{p}\hat{q}}{n}} = 1.96 \sqrt{\frac{.45(.55)}{100}} = .10$$

With this margin of error, you can be fairly confident that the estimate of .45 is within ± .10 of the true value of p. Hence, you can conclude that the true value of p could be as small as .35 or as large as .55. This margin of error is quite large when compared to the estimate itself and reflects the fact that large samples are required to achieve a small margin of error when estimating p.

TABLE 8.1 ● **Some Calculated Values of $\sqrt{pq}$**

p	pq	$\sqrt{pq}$	p	pq	$\sqrt{pq}$
.1	.09	.30	.6	.24	.49
.2	.16	.40	.7	.21	.46
.3	.21	.46	.8	.16	.40
.4	.24	.49	.9	.09	.30
.5	.25	.50			

Table 8.1 shows how the numerator of the standard error of $\hat{p}$ changes for various values of p. Notice that, for most values of p—especially when p is between .3 and .7—there is very little change in $\sqrt{pq}$, the numerator of SE, reaching its maximum value when $p = .5$. This means that the margin of error using the estimator $\hat{p}$ will also be a maximum when $p = .5$. Some pollsters routinely use the maximum margin of error—often called the **sampling error**—when estimating p, in which case they calculate

$$1.96 \text{ SE} = 1.96\sqrt{\frac{.5(.5)}{n}} \quad \text{or sometimes} \quad 2 \text{ SE} = 2\sqrt{\frac{.5(.5)}{n}}$$

Gallup, Harris, and Roper polls generally use sample sizes of approximately 1000, so their margin of error is

$$1.96\sqrt{\frac{.5(.5)}{1000}} = .031 \quad \text{or approximately 3\%}$$

In this case, the estimate is said to be within ± 3 percentage points of the true population proportion.

8.4 EXERCISES

BASIC TECHNIQUES

8.1 Explain what is meant by "margin of error" in point estimation.

8.2 What are two characteristics of the best point estimator for a population parameter?

8.3 Calculate the margin of error in estimating a population mean μ for these values:

a. $n = 30, \sigma^2 = .2$

b. $n = 30, \sigma^2 = .9$

c. $n = 30, \sigma^2 = 1.5$

8.4 Refer to Exercise 8.3. What effect does a larger population variance have on the margin of error?

8.5 Calculate the margin of error in estimating a population mean μ for these values:

a. $n = 50, s^2 = 4$

b. $n = 500, s^2 = 4$

c. $n = 5000, s^2 = 4$

8.6 Refer to Exercise 8.5. What effect does an increased sample size have on the margin of error?

8.7 Calculate the margin of error in estimating a binomial proportion for each of the following values of n. Use $p = .5$ to calculate the standard error of the estimator.

a. $n = 30$ **b.** $n = 100$

c. $n = 400$ **d.** $n = 1000$

8.8 Refer to Exercise 8.7. What effect does increasing the sample size have on the margin of error?

8.9 Calculate the margin of error in estimating a binomial proportion p using samples of size $n = 100$ and the following values for p:

a. $p = .1$ **b.** $p = .3$ **c.** $p = .5$

d. $p = .7$ **e.** $p = .9$

f. Which of the values of p produces the largest margin of error?

8.10 Suppose you are writing a questionnaire for a sample survey involving $n = 100$ individuals. The questionnaire will generate estimates for several different binomial proportions. If you want to report a single margin of error for the survey, which margin of error from Exercise 8.9 is the correct one to use?

8.11 A random sample of $n = 900$ observations from a binomial population produced $x = 655$ successes. Estimate the binomial proportion p and calculate the margin of error.

8.12 A random sample of $n = 50$ observations from a quantitative population produced $\bar{x} = 56.4$ and $s^2 = 2.6$. Give the best point estimate for the population mean μ, and calculate the margin of error.

APPLICATIONS

8.13 The San Andreas Fault Geologists are interested in shifts and movements of the earth's surface indicated by fractures (cracks) in the earth's crust. One of the most famous large fractures is the San Andreas fault in California. A geologist attempting to study the movement of the relative shifts in the earth's crust at a particular location found many fractures in the local rock structure. In an attempt to determine the mean angle of the breaks, she sampled $n = 50$ fractures and found the sample mean and standard deviation to be $39.8°$ and $17.2°$, respectively. Estimate the mean angular direction of the fractures and find the margin of error for your estimate.

8.14 Biomass Estimates of the earth's biomass, the total amount of vegetation held by the earth's forests, are important in determining the amount of unabsorbed carbon dioxide that is expected to remain in the earth's atmosphere.[1] Suppose a sample of 75 1-square-meter plots, randomly chosen in North America's boreal (northern) forests, produced a mean biomass of 4.2 kilograms per square meter (kg/m^2), with a standard deviation of 1.5 kg/m^2. Estimate the average biomass for the boreal forests of North America and find the margin of error for your estimate.

Source: Reprinted with permission from *Science News,* the weekly magazine of *Science,* copyright 1989 by Science Services Inc.

8.15 Consumer Confidence An increase in the rate of consumer savings is frequently tied to a lack of confidence in the economy and is said to be an indicator of a recessional tendency in the economy. A random sampling of $n = 200$ savings accounts in a local community showed a mean increase in savings account values of 7.2% over the past 12 months, with a standard deviation of 5.6%. Estimate the mean percent increase in savings account values over the past 12 months for depositors in the community. Find the margin of error for your estimate.

8.16 Male Teachers Although most school districts do not specifically recruit men to be elementary school teachers, those men who do choose a career in elementary education are highly valued and find the career very rewarding.[2] If there were 40 men in a random sample of 250 elementary school teachers, estimate the proportion of male elementary school teachers in the entire population. Give the margin of error for your estimate.

8.17 Sports Crazy Are you "sports crazy"? Most Americans love participating in or at least watching a multitude of sporting events, but many feel that sports have more than just an entertainment value. In a survey of 1000 adults conducted by KRC Research & Consulting, 78% feel that spectator sports have a positive effect on society.[3]

a. Find a point estimate for the proportion of American adults who feel that spectator sports have a positive effect on society. Calculate the margin of error.

b. The poll reports a margin of error of "plus or minus 3.1%." Does this agree with your results in part a? If not, what value of p produces the margin of error given in the poll?

8.18 Summer Vacations One of the major costs involved in planning a summer vacation is the cost of lodging. Even within a particular chain of hotels, costs can vary substantially depending on the type of room and the amenities offered.[4] Suppose that we randomly select 50 billing statements from each of the computer databases of the Marriott, Radisson, and Wyndham hotel chains, and record the nightly room rates.

	Marriott	Radisson	Wyndham
Sample average	$170	$145	$150
Sample standard deviation	17.5	10	16.5

a. Describe the sampled population(s).

b. Find point estimate for the average room rate for the Marriott hotel chain. Calculate the margin of error.

c. Find a point estimate for the average room rate for the Radisson hotel chain. Calculate the margin of error.

d. Find a point estimate for the average room rate for the Wyndham hotel chain. Calculate the margin of error.

e. Display the results of parts b, c, and d graphically, using the form shown in Figure 8.5. Use this display to compare the average room rates for the three hotel chains.

8.19 *"900" Numbers*   Radio and television stations often air controversial issues during broadcast time and ask viewers to indicate their agreement or disagreement with a given stand on the issue. A poll is conducted by asking those viewers who *agree* to call a certain 900 telephone number and those who *disagree* to call a second 900 telephone number. All respondents pay a fee for their calls.

a. Does this polling technique result in a random sample?

b. What can be said about the validity of the results of such a survey? Do you need to worry about a margin of error in this case?

8.20 *Men On Mars?*   The Mars twin rovers, Spirit and Opportunity, which roamed the surface of Mars in the winter of 2004, found evidence that there was once water on Mars, raising the possibility that there was once life on the planet. Do you think that the United States should pursue a program to send humans to

Mars? An opinion poll conducted by the Associated Press indicated that 49% of the 1034 adults surveyed think that we should pursue such a program.[5]

a. Estimate the true proportion of Americans who think that the United States should pursue a program to send humans to Mars. Calculate the margin of error.

b. The question posed in part a was only one of many questions concerning our space program that were asked in the opinion poll. If the Associated Press wanted to report one sampling error that would be valid for the entire poll, what value should they report?

8.21 *Hungry Rats*   In an experiment to assess the strength of the hunger drive in rats, 30 previously trained animals were deprived of food for 24 hours. At the end of the 24-hour period each animal was put into a cage where food was dispensed if the animal pressed a lever. The length of time the animal continued pressing the bar (although receiving no food) was recorded for each animal. If the data yielded a sample mean of 19.3 minutes with a standard deviation of 5.2 minutes, estimate the true mean time and calculate the margin of error.

INTERVAL ESTIMATION

An *interval estimator* is a rule for calculating two numbers—say, *a* and *b*—to create an interval that you are fairly certain contains the parameter of interest. The concept of "fairly certain" means "with high probability." We measure this probability using the **confidence coefficient,** designated by **$1 - \alpha$.**

Definition   The probability that a confidence interval will contain the estimated parameter is called the **confidence coefficient.**

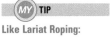

MY TIP

Like Lariat Roping:
Parameter = Fence post
Interval estimate = Lariat

For example, experimenters often construct 95% confidence intervals. This means that the confidence coefficient, or the probability that the interval will contain the estimated parameter, is .95. You can increase or decrease your amount of certainty by changing the confidence coefficient. Some values typically used by experimenters are .90, .95, .98, and .99.

Consider an analogy—this time, throwing a lariat at a fence post. The fence post represents the parameter that you wish to estimate, and the loop formed by the lariat represents the confidence interval. Each time you throw your lariat, you hope to rope the fence post; however, sometimes your lariat misses. In the same way, each time you draw a sample and construct a confidence interval for a parameter, you hope to include the parameter in your interval, but, just like the lariat, sometimes you miss. Your "success rate"—the proportion of intervals that "rope the post" in repeated sampling—is the confidence coefficient.

Constructing a Confidence Interval

When the sampling distribution of a point estimator is approximately normal, an interval estimator or **confidence interval** can be constructed using the following reasoning. For simplicity, assume that the confidence coefficient is .95 and refer to Figure 8.6.

FIGURE 8.6

Parameter ±1.96 SE

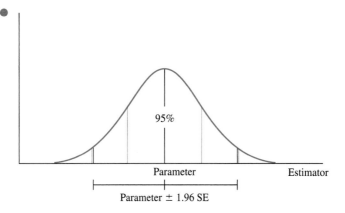

- We know that, of all possible values of the estimator that we might select, 95% of them will be in the interval

    Parameter ± 1.96SE

  shown in Figure 8.6.
- Since the value of the parameter is unknown, consider constructing the interval

    estimator ± 1.96 SE

  which has the same width as the first interval, but has a variable center.
- How often will this interval work properly and enclose the parameter of interest? Refer to Figure 8.7.

FIGURE 8.7

Some 95% confidence
intervals

TIP

Like a game of Ring Toss:
Parameter = Peg
Interval estimate = Ring

The first two intervals work properly—the parameter (marked with a dotted line) is contained within both intervals. The third interval does not work, since it fails to enclose the parameter. This happened because the value of the estimator at the center of the interval was too far away from the parameter. Fortunately, values of the estimator only fall this far away 5% of the time—our procedure will work properly 95% of the time!

You may want to change the *confidence coefficient* from $(1 - \alpha) = .95$ to another confidence level $(1 - \alpha)$. To accomplish this, you need to change the value $z = 1.96$, which locates an area .95 in the center of the standard normal curve, to a value of z that locates the area $(1 - \alpha)$ in the center of the curve, as shown in Figure 8.8. Since the total area under the curve is 1, the remaining area in the two tails is α, and each tail contains area $\alpha/2$. The value of z that has "tail area" $\alpha/2$ to its right is called $z_{\alpha/2}$, and the area between $-z_{\alpha/2}$ and $z_{\alpha/2}$ is the confidence coefficient $(1 - \alpha)$. Values of $z_{\alpha/2}$ that are typically used by experimenters will become familiar to you as you begin to construct confidence intervals for different practical situations. Some of these values are given in Table 8.2.

FIGURE 8.8

Location of $z_{\alpha/2}$

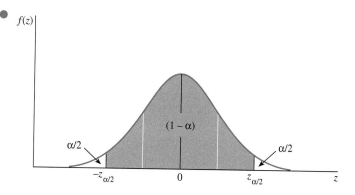

A $(1 - \alpha)100\%$ LARGE-SAMPLE CONFIDENCE INTERVAL

(Point estimator) $\pm z_{\alpha/2} \times$ (Standard error of the estimator)

where $z_{\alpha/2}$ is the z-value with an area $\alpha/2$ in the right tail of a standard normal distribution. This formula generates two values; the **lower confidence limit (LCL)** and the **upper confidence limit (UCL).**

TABLE 8.2

Values of z Commonly Used for Confidence Intervals

Confidence coefficient, $(1 - \alpha)$	α	$\alpha/2$	$z_{\alpha/2}$
.90	.10	.05	1.645
.95	.05	.025	1.96
.98	.02	.01	2.33
.99	.01	.005	2.58

Large-Sample Confidence Interval for a Population Mean μ

Practical problems very often lead to the estimation of μ, the mean of a population of quantitative measurements. Here are some examples:

- The average achievement of college students at a particular university
- The average strength of a new type of steel
- The average number of deaths per age category
- The average demand for a new cosmetics product

When the sample size n is large, the sample mean $\bar{x}$ is the best point estimator for the population mean μ. Since its sampling distribution is approximately normal, it can be used to construct a confidence interval according to the general approach given earlier.

A $(1 - \alpha)100\%$ LARGE-SAMPLE CONFIDENCE INTERVAL FOR A POPULATION MEAN μ

$$\bar{x} \pm z_{\alpha/2} \frac{\sigma}{\sqrt{n}}$$

where $z_{\alpha/2}$ is the z-value corresponding to an area $\alpha/2$ in the upper tail of a standard normal z distribution, and

$n = $ Sample size
$\sigma = $ Standard deviation of the sampled population

If σ is unknown, it can be approximated by the sample standard deviation s when the sample size is large ($n \geq 30$) and the approximate confidence interval is

$$\bar{x} \pm z_{\alpha/2} \frac{s}{\sqrt{n}}$$

Another way to find the large-sample confidence interval for a population mean μ is to begin with the statistic

$$z = \frac{\bar{x} - \mu}{\sigma/\sqrt{n}}$$

which has a standard normal distribution. If you write $z_{\alpha/2}$ as the value of z with area $\alpha/2$ to its right, then you can write

$$P\left(-z_{\alpha/2} < \frac{\bar{x} - \mu}{\sigma/\sqrt{n}} < z_{\alpha/2}\right) = 1 - \alpha$$

You can rewrite this inequality as

$$-z_{\alpha/2}\frac{\sigma}{\sqrt{n}} < \bar{x} - \mu < z_{\alpha/2}\frac{\sigma}{\sqrt{n}}$$

$$-\bar{x} - z_{\alpha/2}\frac{\sigma}{\sqrt{n}} < -\mu < -\bar{x} + z_{\alpha/2}\frac{\sigma}{\sqrt{n}}$$

so that

$$P\left(\bar{x} - z_{\alpha/2}\frac{\sigma}{\sqrt{n}} < \mu < \bar{x} + z_{\alpha/2}\frac{\sigma}{\sqrt{n}}\right) = 1 - \alpha$$

Both $\bar{x} - z_{\alpha/2}(\sigma/\sqrt{n})$ and $\bar{x} + z_{\alpha/2}(\sigma/\sqrt{n})$, the lower and upper confidence limits, are actually random quantities that depend on the sample mean $\bar{x}$. Therefore, in repeated sampling, the random interval, $\bar{x} \pm z_{\alpha/2}(\sigma/\sqrt{n})$, will contain the population mean μ with probability $(1 - \alpha)$.

EXAMPLE 8.6 A scientist interested in monitoring chemical contaminants in food, and thereby the accumulation of contaminants in human diets, selected a random sample of $n = 50$ male adults. It was found that the average daily intake of dairy products was $\bar{x} = 756$ grams per day with a standard deviation of $s = 35$ grams per day. Use this sample information to construct a 95% confidence interval for the mean daily intake of dairy products for men.

Solution Since the sample size of $n = 50$ is large, the distribution of the sample mean $\bar{x}$ is approximately normally distributed with mean μ and standard error estimated by $s/\sqrt{n}$. The approximate 95% confidence interval is

$$\bar{x} \pm 1.96\left(\frac{s}{\sqrt{n}}\right)$$

$$756 \pm 1.96\left(\frac{35}{\sqrt{50}}\right)$$

$$756 \pm 9.70$$

Hence, the 95% confidence interval for μ is from 746.30 to 765.70 grams per day.

MY TIP

A 95% confidence interval tells you that, if you were to construct many of these intervals (all of which would have slightly different endpoints), 95% of them would enclose the population mean.

Interpreting the Confidence Interval

What does it mean to say you are "95% confident" that the true value of the population mean μ is within a given interval? If you were to construct 20 such intervals, each using different sample information, your intervals might look like those shown in Figure 8.9. Of the 20 intervals, you might expect that 95% of them, or 19 out of 20, will perform as planned and contain μ within their upper and lower bounds.

FIGURE 8.9

Twenty confidence intervals for the mean for Example 8.6

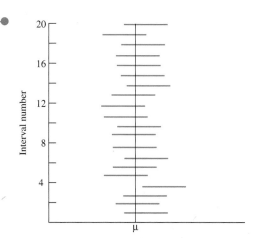

Remember that you cannot be absolutely sure that any one particular interval contains the mean μ. You will never know whether your particular interval is one of the 19 that "worked," or whether it is the one interval that "missed." Your confidence in the estimated interval follows from the fact that when repeated intervals are calculated, 95% of these intervals will contain μ.

 APPLET

You can try this experiment on your own using the Java applet called **Interpreting Confidence Intervals.** The applet shown in Figure 8.10(a) shows the calculation of a 95% confidence interval for μ when $n = 50$ and $\sigma = 35$. For this particular confidence interval, we used the *One Sample* button. You can see the value of μ shown as a vertical green line on your monitor (gray in Figure 8.10). Notice that this confidence interval worked properly and enclosed the vertical line between its upper and lower limits. Figure 8.10(b) shows the calculation of 100 such intervals, using the *100 Samples* button. The intervals that fail to work properly are shown in red on your monitor (black in Figure 8.10). How many intervals fail to work? Is it close to the 95% confidence that we claim to have? You will use this applet again for the MyApplet Exercises section at the end of the chapter.

FIGURE 8.10

Interpeting Confidence Intervals applet

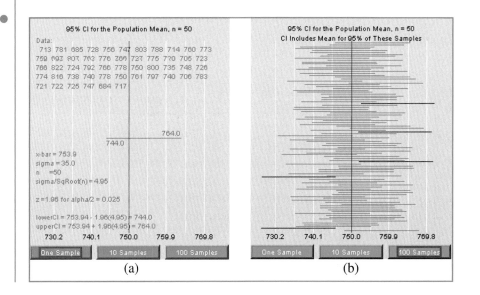

A good confidence interval has two desirable characteristics:

- It is as narrow as possible. The narrower the interval, the more exactly you have located the estimated parameter.
- It has a large confidence coefficient, near 1. The larger the confidence coefficient, the more likely it is that the interval will contain the estimated parameter.

EXAMPLE 8.7

Construct a 99% confidence interval for the mean daily intake of dairy products for adult men in Example 8.6.

Solution   To change the confidence level to .99, you must find the appropriate value of the standard normal z that puts area $(1 - \alpha) = .99$ in the center of the curve. This value, with tail area $\alpha/2 = .005$ to its right, is found from Table 8.2 to be $z = 2.58$ (see Figure 8.11). The 99% confidence interval is then

$$\bar{x} \pm 2.58\left(\frac{s}{\sqrt{n}}\right)$$

$$756 \pm 2.58(4.95)$$
$$756 \pm 12.77$$

FIGURE 8.11

Standard normal values for a 99% confidence interval

   TIP

Right Tail Area	z-Value
.05	1.645
.025	1.96
.01	2.33
.005	2.58

or 743.23 to 768.77 grams per day. This confidence interval is *wider* than the 95% confidence interval in Example 8.6. The increased width is necessary to increase the confidence, just as you might want a wider loop on your lariat to ensure roping the fence post! The only way to *increase the confidence* without increasing the width of the interval is to *increase the sample size, n.*

The standard error of $\bar{x}$,

$$SE = \frac{\sigma}{\sqrt{n}}$$

measures the variability or spread of the values of $\bar{x}$. The more variable the population data, measured by σ, the more variable will be $\bar{x}$, and the standard error will be larger. On the other hand, if you increase the sample size n, more information is available for estimating μ. The estimates should fall closer to μ and the standard error will be smaller. You can use the **Exploring Confidence Intervals** applet, shown in Figure 8.12, to see the effect of changing the sample size n, the standard deviation σ, and the confidence coefficient $1 - \alpha$ on the width of the confidence interval.

The confidence intervals of Examples 8.6 and 8.7 are approximate because you substituted s as an approximation for σ. That is, instead of the confidence coefficient being .95, the value specified in the example, the true value of the coefficient may be .92, .94, or .97. But this discrepancy is of little concern from a practical point of view; as far as your "confidence" is concerned, there is little difference among these confidence coefficients. Most interval estimators used in statistics yield approximate confidence intervals because the assumptions upon which they are based are not satisfied exactly. Having made this point, we will not continue to refer to confidence intervals as "approximate." It is of little practical concern as long as the actual confidence coefficient is near the value specified.

FIGURE 8.12

Exploring Confidence
Intervals applet

Large-Sample Confidence Interval for a Population Proportion *p*

Many research experiments or sample surveys have as their objective the estimation of the proportion of people or objects in a large group that possess a certain characteristic. Here are some examples:

- The proportion of sales that can be expected in a large number of customer contacts
- The proportion of seeds that germinate
- The proportion of "likely" voters who plan to vote for a particular political candidate

Each is a practical example of the binomial experiment, and the parameter to be estimated is the binomial proportion *p*.

When the sample size is large, the sample proportion,

$$\hat{p} = \frac{x}{n} = \frac{\text{Total number of successes}}{\text{Total number of trials}}$$

is the best point estimator for the population proportion *p*. Since its sampling distribution is approximately normal, with mean *p* and standard error $SE = \sqrt{pq/n}$, $\hat{p}$ can be used to construct a confidence interval according to the general approach given in this section.

A $(1 - \alpha)100\%$ LARGE-SAMPLE CONFIDENCE INTERVAL FOR A POPULATION PROPORTION *p*

$$\hat{p} \pm z_{\alpha/2}\sqrt{\frac{pq}{n}}$$

where $z_{\alpha/2}$ is the *z*-value corresponding to an area $\alpha/2$ in the right tail of a standard normal *z* distribution. Since *p* and *q* are unknown, they are estimated using the best point estimators: $\hat{p}$ and $\hat{q}$. The sample size is considered large when the normal approximation to the binomial distribution is adequate—namely, when $n\hat{p} > 5$ and $n\hat{q} > 5$.

EXAMPLE (8.8) A random sample of 985 "likely" voters—those who are likely to vote in the upcoming election—were polled during a phone-athon conducted by the Republican party. Of those surveyed, 592 indicated that they intended to vote for the Republican candidate in the upcoming election. Construct a 90% confidence interval for p, the proportion of likely voters in the population who intend to vote for the Republican candidate. Based on this information, can you conclude that the candidate will win the election?

Solution The point estimate for p is

$$\hat{p} = \frac{x}{n} = \frac{592}{985} = .601$$

and the standard error is

$$\sqrt{\frac{\hat{p}\hat{q}}{n}} = \sqrt{\frac{(.601)(.399)}{985}} = .016$$

The z-value for a 90% confidence interval is the value that has area $\alpha/2 = .05$ in the upper tail of the z distribution, or $z_{.05} = 1.645$ from Table 8.2. The 90% confidence interval for p is thus

$$\hat{p} \pm 1.645 \sqrt{\frac{\hat{p}\hat{q}}{n}}$$

$$.601 \pm .026$$

or $.575 < p < .627$. You estimate that the percentage of likely voters who intend to vote for the Republican candidate is between 57.5% and 62.7%. Will the candidate win the election? Assuming that she needs more than 50% of the vote to win, and since both the upper and lower confidence limits exceed this minimum value, you can say with 90% confidence that the candidate will win.

There are some problems, however, with this type of sample survey. What if the voters who consider themselves "likely to vote" do not actually go to the polls? What if a voter changes his or her mind between now and election day? What if a surveyed voter does not respond truthfully when questioned by the campaign worker? The 90% confidence interval you have constructed gives you 90% confidence only if you have selected a *random sample from the population of interest.* You can no longer be assured of "90% confidence" if your sample is biased, or if the population of voter responses changes before the day of the election!

You may have noticed that the point estimator with its 95% margin of error looks very similar to a 95% confidence interval for the same parameter. This close relationship exists for most of the parameters estimated in this book, but it is not true in general. Sometimes the best point estimator for a parameter *does not* fall in the middle of the best confidence interval; the best confidence interval may not even be a function of the best point estimator. Although this is a theoretical distinction, you should remember that there is a difference between point and interval estimation, and that the choice between the two depends on the preference of the experimenter.

8.5 **EXERCISES**

BASIC TECHNIQUES

8.22 Find and interpret a 95% confidence interval for a population mean μ for these values:

a. $n = 36, \bar{x} = 13.1, s^2 = 3.42$

b. $n = 64, \bar{x} = 2.73, s^2 = .1047$

8.23 Find a 90% confidence interval for a population mean μ for these values:

a. $n = 125, \bar{x} = .84, s^2 = .086$

b. $n = 50, \bar{x} = 21.9, s^2 = 3.44$

c. Interpret the intervals found in parts a and b.

8.24 Find a $(1 - \alpha)100\%$ confidence interval for a population mean μ for these values:

a. $\alpha = .01, n = 38, \bar{x} = 34, s^2 = 12$

b. $\alpha = .10, n = 65, \bar{x} = 1049, s^2 = 51$

c. $\alpha = .05, n = 89, \bar{x} = 66.3, s^2 = 2.48$

8.25 A random sample of $n = 300$ observations from a binomial population produced $x = 263$ successes. Find a 90% confidence interval for p and interpret the interval.

8.26 Suppose the number of successes observed in $n = 500$ trials of a binomial experiment is 27. Find a 95% confidence interval for p. Why is the confidence interval narrower than the confidence interval in Exercise 8.25?

8.27 A random sample of n measurements is selected from a population with unknown mean μ and known standard deviation $\sigma = 10$. Calculate the width of a 95% confidence interval for μ for these values of n:

a. $n = 100$ **b.** $n = 200$ **c.** $n = 400$

8.28 Compare the confidence intervals in Exercise 8.27. What effect does each of these actions have on the width of a confidence interval?

a. Double the sample size

b. Quadruple the sample size

8.29 Refer to Exercise 8.28.

a. Calculate the width of a 90% confidence interval for μ when $n = 100$.

b. Calculate the width of a 99% confidence interval for μ when $n = 100$.

c. Compare the widths of 90%, 95%, and 99% confidence intervals for μ. What effect does increasing the confidence coefficient have on the width of the confidence interval?

APPLICATIONS

8.30 A Chemistry Experiment Due to a variation in laboratory techniques, impurities in materials, and other unknown factors, the results of an experiment in a chemistry laboratory will not always yield the same numerical answer. In an electrolysis experiment, a class measured the amount of copper precipitated from a saturated solution of copper sulfate over a 30-minute period. The $n = 30$ students calculated a sample mean and standard deviation equal to .145 and .0051 mole, respectively. Find a 90% confidence interval for the mean amount of copper precipitated from the solution over a 30-minute period.

8.31 Acid Rain Acid rain, caused by the reaction of certain air pollutants with rainwater, appears to be a growing problem in the northeastern United States. (Acid rain affects the soil and causes corrosion on exposed metal surfaces.) Pure rain falling through clean air registers a pH value of 5.7 (pH is a measure of acidity: 0 is acid; 14 is alkaline). Suppose water samples from 40 rainfalls are analyzed for pH, and $\bar{x}$ and s are equal to 3.7 and .5, respectively. Find a 99% confidence interval for the mean pH in rainfall and interpret the interval. What assumption must be made for the confidence interval to be valid?

8.32 'Tweens When it comes to advertising, "'tweens" (kids aged 10 to 13) are not ready for the hardline messages that advertisers often use to reach teenagers. The Geppetto Group study found that 78% of 'tweens understand and enjoy ads that are silly in nature. Unlike teenagers, 'tweens would much rather see dancing (69%) and "boyfriends and girlfriends" (63%) than "sexy looking people" or "kissing."[6] Suppose that these results are based on a sample of size $n = 1030$ 'tweens.

a. Construct a 95% confidence interval estimate of the proportion of 'tweens who understand and enjoy ads that are silly in nature.

b. Construct a 95% confidence interval for the proportion of 'tweens who would rather see dancing.

8.33 Hamburger Meat The meat department of a local supermarket chain packages ground beef using meat trays of two sizes: one designed to hold approximately 1 pound of meat, and one that holds approximately 3 pounds. A random sample of 35 packages in the smaller meat trays produced weight measurements with an average of 1.01 pounds and a standard deviation of .18 pound.

a. Construct a 99% confidence interval for the average weight of all packages sold in the smaller meat trays by this supermarket chain.

b. What does the phrase "99% confident" mean?

c. Suppose that the quality control department of this supermarket chain intends that the amount of ground beef in the smaller trays should be 1 pound on average. Should the confidence interval in part a concern the quality control department? Explain.

8.34 Movie Ratings Is America's romance with movies on the wane? In a poll of $n = 800$ randomly chosen adults, 45% indicated that movies were getting better, while 43% indicated that movies were getting worse.[7] However, these results vary by age: 69% of the 18-to-29-year-olds say that movies are getting better, 39% of 50-to-64-year-olds and just 15% of those over 60 agree. Construct a 90% confidence interval for the overall proportion of adults who say that movies are getting better.

8.35 SUVs A sample survey is designed to estimate the proportion of sports utility vehicles being driven in the state of California. A random sample of 500 registrations are selected from a Department of Motor Vehicles database, and 68 are classified as sports utility vehicles.

a. Use a 95% confidence interval to estimate the proportion of sports utility vehicles in California.

b. How can you estimate the proportion of sports utility vehicles in California with a higher degree of accuracy? (HINT: There are two answers.)

8.36 e-Shopping In a report of why e-shoppers abandon their online sales transactions, Alison Stein Wellner[8] found that "pages took too long to load" and "site was so confusing that I couldn't find the product"

were the two complaints heard most often. Based on customers' responses, the average time to complete an online order form will take 4.5 minutes. Suppose that $n = 50$ customers responded and that the standard deviation of the time to complete an online order is 2.7 minutes.

a. Do you think that x, the time to complete the online order form, has a mound-shaped distribution? If not, what shape would you expect?

b. If the distribution of the completion times is not normal, you can still use the standard normal distribution to construct a confidence interval for μ, the mean completion time for online shoppers. Why?

c. Construct a 95% confidence interval for μ, the mean completion time for online orders.

8.37 What's Normal? What *is* normal, when it comes to people's body temperatures? A random sample of 130 human body temperatures, provided by Allen Shoemaker[9] in the *Journal of Statistical Education,* had a mean of 98.25 degrees and a standard deviation of 0.73 degrees.

a. Construct a 99% confidence interval for the average body temperature of healthy people.

b. Does the confidence interval constructed in part a contain the value 98.6 degrees, the usual average temperature cited by physicians and others? If not, what conclusions can you draw?

8.38 Rocking the Vote How likely are you to vote in the next presidential election? A random sample of 300 adults was taken, and 192 of them said that they always vote in presidential elections.

a. Construct a 95% confidence interval for the proportion of adult Americans who say they always vote in presidential elections.

b. An article in *American Demographics* reports this percentage of 67%.[10] Based on the interval constructed in part a, would you disagree with their reported percentage? Explain.

c. Can we use the interval estimate from part a to estimate the actual proportion of adult Americans who vote in the 2004 presidential election? Why or why not?

ESTIMATING THE DIFFERENCE BETWEEN TWO POPULATION MEANS

8.6

A problem equally as important as the estimation of a single population mean μ for a quantitative population is the comparison of two population means. You may want to make comparisons like these:

- The average scores on the Medical College Admission Test (MCAT) for students whose major was biochemistry and those whose major was biology
- The average yields in a chemical plant using raw materials furnished by two different suppliers
- The average stem diameters of plants grown on two different types of nutrients

For each of these examples, there are two populations: the first with mean and variance μ_1 and σ_1^2 and the second with mean and variance μ_2 and σ_2^2. A random sample of n_1 measurements is drawn from population 1 and a second random sample of size n_2 is independently drawn from population 2. Finally, the estimates of the population parameters are calculated from the sample data using the estimators $\bar{x}_1$, s_1^2, $\bar{x}_2$, and s_2^2 as shown in Table 8.3.

TABLE 8.3 ● **Samples From Two Quantitative Populations**

	Population 1	Population 2
Mean	μ_1	μ_2
Variance	σ_1^2	σ_2^2

	Sample 1	Sample 2
Mean	$\bar{x}_1$	$\bar{x}_2$
Variance	s_1^2	s_2^2
Sample size	n_1	n_2

Intuitively, the difference between two sample means would provide the maximum information about the actual difference between two population means, and this is in fact the case. The best point estimator of the difference $(\mu_1 - \mu_2)$ between the population means is $(\bar{x}_1 - \bar{x}_2)$. The sampling distribution of this estimator is not difficult to derive, but we state it here without proof.

PROPERTIES OF THE SAMPLING DISTRIBUTION OF $(\bar{x}_1 - \bar{x}_2)$, THE DIFFERENCE BETWEEN TWO SAMPLE MEANS

When independent random samples of n_1 and n_2 observations have been selected from populations with means μ_1 and μ_2 and variances σ_1^2 and σ_2^2, respectively, the sampling distribution of the difference $(\bar{x}_1 - \bar{x}_2)$ has the following properties:

1. The mean of $(\bar{x}_1 - \bar{x}_2)$ is

$$\mu_1 - \mu_2$$

and the standard error is

$$SE = \sqrt{\frac{\sigma_1^2}{n_1} + \frac{\sigma_2^2}{n_2}}$$

which can be estimated as

$$SE = \sqrt{\frac{s_1^2}{n_1} + \frac{s_2^2}{n_2}} \quad \text{when the sample sizes are large.}$$

2. **If the sampled populations are normally distributed,** then the sampling distribution of $(\bar{x}_1 - \bar{x}_2)$ is **exactly** normally distributed, regardless of the sample size.

3. **If the sampled populations are not normally distributed,** then the sampling distribution of $(\bar{x}_1 - \bar{x}_2)$ is **approximately** normally distributed when n_1 and n_2 are both 30 or more, due to the Central Limit Theorem.

Since $(\mu_1 - \mu_2)$ is the mean of the sampling distribution, it follows that $(\bar{x}_1 - \bar{x}_2)$ is an unbiased estimator of $(\mu_1 - \mu_2)$ with an approximately normal distribution when n_1 and n_2 are large. That is, the statistic

$$z = \frac{(\bar{x}_1 - \bar{x}_2) - (\mu_1 - \mu_2)}{\sqrt{\dfrac{s_1^2}{n_1} + \dfrac{s_2^2}{n_2}}}$$

has an approximately standard normal z distribution, and the general procedures of Section 8.5 can be used to construct point and interval estimates. Although the choice between point and interval estimation depends on your personal preference, most experimenters choose to construct confidence intervals for two-sample problems. The appropriate formulas for both methods are given next.

LARGE-SAMPLE POINT ESTIMATION OF $(\mu_1 - \mu_2)$

Point estimator: $(\bar{x}_1 - \bar{x}_2)$

95% Margin of error: $\pm 1.96\ SE = \pm\ 1.96\sqrt{\dfrac{s_1^2}{n_1} + \dfrac{s_2^2}{n_2}}$

A $(1 - \alpha)100\%$ LARGE-SAMPLE CONFIDENCE INTERVAL FOR $(\mu_1 - \mu_2)$

$$(\bar{x}_1 - \bar{x}_2) \pm z_{\alpha/2}\sqrt{\frac{s_1^2}{n_1} + \frac{s_2^2}{n_2}}$$

Right Tail Area	z-Value
.05	1.645
.025	1.96
.01	2.33
.005	2.58

EXAMPLE 8.9

The wearing qualities of two types of automobile tires were compared by road-testing samples of $n_1 = n_2 = 100$ tires for each type. The number of miles until wearout was defined as a specific amount of tire wear. The test results are given in Table 8.4. Estimate $(\mu_1 - \mu_2)$, the difference in mean miles to wearout, using a 99% confidence interval. Is there a difference in the average wearing quality for the two types of tires?

TABLE 8.4 ● **Sample Data Summary for Two Types of Tires**

Tire 1	Tire 2
$\bar{x}_1 = 26{,}400$ miles	$\bar{x}_2 = 25{,}100$ miles
$s_1^2 = 1{,}440{,}000$	$s_2^2 = 1{,}960{,}000$

Solution The point estimate of $(\mu_1 - \mu_2)$ is

$$(\bar{x}_1 - \bar{x}_2) = 26{,}400 - 25{,}100 = 1300 \text{ miles}$$

and the standard error of $(\bar{x}_1 - \bar{x}_2)$ is estimated as

$$\text{SE} = \sqrt{\frac{s_1^2}{n_1} + \frac{s_2^2}{n_2}} = \sqrt{\frac{1{,}440{,}000}{100} + \frac{1{,}960{,}000}{100}} = 184.4 \text{ miles}$$

The 99% confidence interval is calculated as

$$(\bar{x}_1 - \bar{x}_2) \pm 2.58 \sqrt{\frac{s_1^2}{n_1} + \frac{s_2^2}{n_2}}$$

$$1300 \pm 2.58(184.4)$$
$$1300 \pm 475.8$$

or $824.2 < (\mu_1 - \mu_2) < 1775.8$. The difference in the average miles to wearout for the two types of tires is estimated to lie between LCL = 824.2 and UCL = 1775.8 miles of wear.

Based on this confidence interval, can you conclude that there is a difference in the average miles to wearout for the two types of tires? If there were no difference in the two population means, then μ_1 and μ_2 would be equal and $(\mu_1 - \mu_2) = 0$. If you look at the confidence interval you constructed, you will see that 0 is not one of the possible values for $(\mu_1 - \mu_2)$. Therefore, it is not likely that the means are the same; you can conclude that there is a difference in the average miles to wearout for the two types of tires. The confidence interval has allowed you to *make a decision* about the equality of the two population means.

 MY TIP

If 0 is not in the interval, you *can* conclude that there is a difference in the population means.

EXAMPLE 8.10 The scientist in Example 8.6 wondered whether there was a difference in the average daily intakes of dairy products between men and women. He took a sample of $n = 50$ adult women and recorded their daily intakes of dairy products in grams per day. He did the same for adult men. A summary of his sample results is listed in Table 8.5. Construct a 95% confidence interval for the difference in the average daily intakes of dairy products for men and women. Can you conclude that there is a difference in the average daily intakes for men and women?

TABLE 8.5 ● **Sample Values for Daily Intakes of Dairy Products**

	Men	Women
Sample size	50	50
Sample mean	756	762
Sample standard deviation	35	30

Solution   The confidence interval is constructed using a value of z with tail area $\alpha/2 = .025$ to its right; that is, $z_{.025} = 1.96$. Using the sample standard deviations to approximate the unknown population standard deviations, the 95% confidence interval is

$$(\bar{x}_1 - \bar{x}_2) \pm 1.96\sqrt{\frac{s_1^2}{n_1} + \frac{s_2^2}{n_2}}$$

$$(756 - 762) \pm 1.96\sqrt{\frac{35^2}{50} + \frac{30^2}{50}}$$

$$-6 \pm 12.78$$

or $-18.78 < (\mu_1 - \mu_2) < 6.78$. Look at the possible values for $(\mu_1 - \mu_2)$ in the confidence interval. It is possible that the difference $(\mu_1 - \mu_2)$ could be negative (indicating that the average for women exceeds the average for men), it could be positive (indicating that men have the higher average), or it could be 0 (indicating no difference between the averages). Based on this information, you *should not be willing to conclude* that there is a difference in the average daily intakes of dairy products for men and women.

Examples 8.9 and 8.10 deserve further comment with regard to using sample estimates in place of unknown parameters. The sampling distribution of

$$\frac{(\bar{x}_1 - \bar{x}_2) - (\mu_1 - \mu_2)}{\sqrt{\frac{\sigma_1^2}{n_1} + \frac{\sigma_2^2}{n_2}}}$$

has a standard normal distribution for all sample sizes when both sampled populations are normal and an *approximate* standard normal distribution when the sampled populations are not normal but the sample sizes are large (≥ 30). When σ_1^2 and σ_2^2 are not known and are estimated by the sample estimates s_1^2 and s_2^2, the resulting statistic will still have an approximate standard normal distribution when the sample sizes are large. The behavior of this statistic when the population variances are unknown and the sample sizes are small will be discussed in Chapter 10.

8.6   EXERCISES

BASIC TECHNIQUES

8.39 Independent random samples were selected from populations 1 and 2. The sample sizes, means, and variances are as follows:

	Population	
	1	2
Sample size	35	49
Sample mean	12.7	7.4
Sample variance	1.38	4.14

a. Find a 95% confidence interval for estimating the difference in the population means $(\mu_1 - \mu_2)$.

b. Based on the confidence interval in part a, can you conclude that there is a difference in the means for the two populations? Explain.

8.40 Independent random samples were selected from populations 1 and 2. The sample sizes, means, and variances are as follows:

	Population	
	1	2
Sample size	64	64
Sample mean	2.9	5.1
Sample variance	0.83	1.67

a. Find a 90% confidence interval for the difference in the population means. What does the phrase "90% confident" mean?

b. Find a 99% confidence interval for the difference in the population means. Can you conclude that there is a difference in the two population means? Explain.

APPLICATIONS

8.41 Selenium A small amount of the trace element selenium, 50–200 micrograms (μg) per day, is considered essential to good health. Suppose that random samples of $n_1 = n_2 = 30$ adults were selected from two regions of the United States and that a day's intake of selenium, from both liquids and solids, was recorded for each person. The mean and standard deviation of the selenium daily intakes for the 30 adults from region 1 were $\bar{x}_1 = 167.1$ and $s_1 = 24.3$ μg, respectively. The corresponding statistics for the 30 adults from region 2 were $\bar{x}_2 = 140.9$ and $s_2 = 17.6$. Find a 95% confidence interval for the difference in the mean selenium intakes for the two regions. Interpret this interval.

8.42 9-1-1 A study was conducted to compare the mean numbers of police emergency calls per 8-hour shift in two districts of a large city. Samples of 100 8-hour shifts were randomly selected from the police records for each of the two regions, and the number of emergency calls was recorded for each shift. The sample statistics are listed here:

	Region	
	1	2
Sample size	100	100
Sample mean	2.4	3.1
Sample variance	1.44	2.64

Find a 90% confidence interval for the difference in the mean numbers of police emergency calls per shift between the two districts of the city. Interpret the interval.

8.43 Teaching Biology In developing a standard for assessing the teaching of precollege sciences in the United States, an experiment was conducted to evaluate a teacher-developed curriculum, "Biology: A Community Context" (BACC) that was standards-based, activity-oriented, and inquiry-centered. This approach was compared to the historical presentation through lecture, vocabulary, and memorized facts. Students were tested on biology concepts that featured biological knowledge and process skills in the traditional

sense. The perhaps not-so-startling results from a test on biology concepts, published in *The American Biology Teacher*, are shown in the following table.[11]

	Mean	Sample Size	Standard Deviation
Pretest: All BACC Classes	13.38	372	5.59
Pretest: All Traditional	14.06	368	5.45
Posttest: All BACC Classes	18.5	365	8.03
Posttest: All Traditional	16.5	298	6.96

a. Find a 95% confidence interval for the mean score for the posttest for all BACC classes.

b. Find a 95% confidence interval for the mean score for the posttest for all traditional classes.

c. Find a 95% confidence interval for the difference in mean scores for the posttest BACC classes and the posttest traditional classes.

d. Does the confidence interval in c provide evidence that there is a real difference in the posttest BACC and traditional class scores? Explain.

Source: From "Performance Assessment of a Standards-Based High School Biology Curriculum," by W. Leonard, B. Speziale, and J. Pernick, *The American Biology Teacher*, 2001, 63(5), 310–316. Reprinted by permission of National Association of Biology Teachers.

8.44 Are you Dieting? An experiment was conducted to compare two diets A and B designed for weight reduction. Two groups of 30 overweight dieters each were randomly selected. One group was placed on diet A and the other on diet B, and their weight losses were recorded over a 30-day period. The means and standard deviations of the weight-loss measurements for the two groups are shown in the table. Find a 95% confidence interval for the difference in mean weight loss for the two diets. Interpret your confidence interval.

Diet A	Diet B
$\bar{x}_A = 21.3$	$\bar{x}_B = 13.4$
$s_A = 2.6$	$s_B = 1.9$

8.45 Starting Salaries In an attempt to compare the starting salaries of college graduates majoring in education and social sciences, random samples of 50 recent college graduates in each major were selected and the following information was obtained:

Major	Mean	SD
Education	40,554	2225
Social science	38,348	2375

a. Find a point estimate for the difference in the average starting salaries of college students majoring in education and the social sciences. What is the margin of error for your estimate?

b. Based on the results of part a, do you think that there is a significant difference in the means for the two groups in the general population? Explain.

8.46 Biology Skills Refer to Exercise 8.43. In addition to tests involving biology concepts, students were also tested on process skills. The results of pretest and posttest scores, published in *The American Biology Teacher*, are given below.[11]

	Mean	Sample Size	Standard Deviation
Pretest: All BACC Classes	10.52	395	4.79
Pretest: All Traditional	11.97	379	5.39
Posttest: All BACC Classes	14.06	376	5.65
Posttest: All Traditional	12.96	308	5.93

a. Find a 95% confidence interval for the mean score on process skills for the posttest for all BACC classes.

b. Find a 95% confidence interval for the mean score on process skills for the posttest for all traditional classes.

c. Find a 95% confidence interval for the difference in mean scores on process skills for the posttest BACC classes and the posttest traditional classes.

d. Does the confidence interval in c provide evidence that there is a real difference in the mean process skills scores between posttest BACC and traditional class scores? Explain.

Source: From "Peformance Assessment of a Standards-Based High School Biology Curriculum," by W. Leonard, B. Speziale, and J. Pernick, *The American Biology Teacher*, 2001, 63(5), 310–316. Reprinted by permission of National Association of Biology Teachers.

8.47 Hotel Costs Refer to Exercise 8.18. The means and standard deviations for 50 billing statements from each of the computer databases of each of the three hotel chains are given in the table:[4]

	Marriott	Radisson	Wyndham
Sample average	$170	$145	$160
Sample standard deviation	17.5	10	16.5

a. Find a 95% confidence interval for the difference in the average room rates for the Marriott and the Wyndham hotel chains.

b. Find a 99% confidence interval for the difference in the average room rates for the Radisson and the Wyndham hotel chains.

c. Do the intervals in parts a and b contain the value $(\mu_1 - \mu_2) = 0$? Why is this of interest to the researcher?

d. Do the data indicate a difference in the average room rates between the Marriott and the Wyndham chains? Between the Radisson and the Wyndham chains?

8.48 Noise and Stress To compare the effect of stress in the form of noise on the ability to perform a simple task, 70 subjects were divided into two groups. The first group of 30 subjects acted as a control, while the second group of 40 were the experimental group. Although each subject performed the task in the same control room, each of the experimental group subjects had to perform the task while loud rock music was played. The time to finish the task was recorded for each subject and the following summary was obtained:

	Control	Experimental
n	30	40
$\bar{x}$	15 minutes	23 minutes
s	4 minutes	10 minutes

a. Find a 99% confidence interval for the difference in mean completion times for these two groups.

b. Based on the confidence interval in part a, is there sufficient evidence to indicate a difference in the average time to completion for the two groups? Explain.

8.49 What's Normal II Of the 130 people in Exercise 8.37, 65 were female and 65 were male.[9] The means and standard deviation of their temperatures are shown below.

	Men	Women
Sample Mean	98.11	98.39
Standard Deviation	0.70	0.74

Find a 95% confidence interval for the difference in the average body temperatures for males versus females. Based on this interval, can you conclude that there is a difference in the average temperatures for males versus females? Explain.

ESTIMATING THE DIFFERENCE BETWEEN TWO BINOMIAL PROPORTIONS

8.7

A simple extension of the estimation of a binomial proportion p is the estimation of the difference between two binomial proportions. You may wish to make comparisons like these:

- The proportion of defective items manufactured in two production lines
- The proportion of female voters and the proportion of male voters who favor an equal rights amendment
- The germination rates of untreated seeds and seeds treated with a fungicide

These comparisons can be made using the difference $(p_1 - p_2)$ between two binomial proportions, p_1 and p_2. Independent random samples consisting of n_1 and n_2 trials are drawn from populations 1 and 2, respectively, and the sample estimates $\hat{p}_1$ and $\hat{p}_2$ are calculated. The unbiased estimator of the difference $(p_1 - p_2)$ is the sample difference $(\hat{p}_1 - \hat{p}_2)$.

PROPERTIES OF THE SAMPLING DISTRIBUTION OF THE DIFFERENCE $(\hat{p}_1 - \hat{p}_2)$ BETWEEN TWO SAMPLE PROPORTIONS

Assume that independent random samples of n_1 and n_2 observations have been selected from binomial populations with parameters p_1 and p_2, respectively. The sampling distribution of the difference between sample proportions

$$(\hat{p}_1 - \hat{p}_2) = \left(\frac{x_1}{n_1} - \frac{x_2}{n_2} \right)$$

has these properties:

1. The mean of $(\hat{p}_1 - \hat{p}_2)$ is

   $$p_1 - p_2$$

   and the standard error is

   $$SE = \sqrt{\frac{p_1 q_1}{n_1} + \frac{p_2 q_2}{n_2}}$$

   which is estimated as

   $$SE = \sqrt{\frac{\hat{p}_1 \hat{q}_1}{n_1} + \frac{\hat{p}_2 \hat{q}_2}{n_2}}$$

2. The sampling distribution of $(\hat{p}_1 - \hat{p}_2)$ can be approximated by a normal distribution when n_1 and n_2 are large, due to the Central Limit Theorem.

Although the range of a single proportion is from 0 to 1, the difference between two proportions ranges from -1 to 1. To use a normal distribution to approximate the distribution of $(\hat{p}_1 - \hat{p}_2)$, both $\hat{p}_1$ and $\hat{p}_2$ should be approximately normal; that is, $n_1 \hat{p}_1 > 5$, $n_1 \hat{q}_1 > 5$, and $n_2 \hat{p}_2 > 5$, $n_2 \hat{q}_2 > 5$.

The appropriate formulas for point and interval estimation are given next.

LARGE-SAMPLE POINT ESTIMATION OF $(p_1 - p_2)$

Point estimator: $(\hat{p}_1 - \hat{p}_2)$

95% Margin of error: $\pm\, 1.96\ \text{SE} = \pm 1.96 \sqrt{\dfrac{\hat{p}_1 \hat{q}_1}{n_1} + \dfrac{\hat{p}_2 \hat{q}_2}{n_2}}$

A $(1 - \alpha)100\%$ LARGE-SAMPLE CONFIDENCE INTERVAL FOR $(p_1 - p_2)$

$(\hat{p}_1 - \hat{p}_2) \pm z_{\alpha/2} \sqrt{\dfrac{\hat{p}_1 \hat{q}_1}{n_1} + \dfrac{\hat{p}_2 \hat{q}_2}{n_2}}$

Assumption: n_1 and n_2 must be sufficiently large so that the sampling distribution of $(\hat{p}_1 - \hat{p}_2)$ can be approximated by a normal distribution—namely, if $n_1 \hat{p}_1$, $n_1 \hat{q}_1$, $n_2 \hat{p}_2$, and $n_2 \hat{q}_2$ are all greater than 5.

EXAMPLE 8.11

A bond proposal for school construction will be submitted to the voters at the next municipal election. A major portion of the money derived from this bond issue will be used to build schools in a rapidly developing section of the city, and the remainder will be used to renovate and update school buildings in the rest of the city. To assess the viability of the bond proposal, a random sample of $n_1 = 50$ residents in the developing section and $n_2 = 100$ residents from the other parts of the city were asked whether they plan to vote for the proposal. The results are tabulated in Table 8.6.

TABLE 8.6 • **Sample Values for Opinion on Bond Proposal**

	Developing Section	Rest of the City
Sample size	50	100
Number favoring proposal	38	65
Proportion favoring proposal	.76	.65

1. Estimate the difference in the true proportions favoring the bond proposal with a 99% confidence interval.
2. If both samples were pooled into one sample of size $n = 150$, with 103 in favor of the proposal, provide a point estimate of the proportion of city residents who will vote for the bond proposal. What is the margin of error?

Solution

1. The best point estimate of the difference $(p_1 - p_2)$ is given by

$(\hat{p}_1 - \hat{p}_2) = .76 - .65 = .11$

and the standard error of $(\hat{p}_1 - \hat{p}_2)$ is estimated as

$\sqrt{\dfrac{\hat{p}_1 \hat{q}_1}{n_1} + \dfrac{\hat{p}_2 \hat{q}_2}{n_2}} = \sqrt{\dfrac{(.76)(.24)}{50} + \dfrac{(.65)(.35)}{100}} = .0770$

For a 99% confidence interval, $z_{.005} = 2.58$, and the approximate 99% confidence interval is found as

$$(\hat{p}_1 - \hat{p}_2) \pm z_{.005} \sqrt{\frac{\hat{p}_1 \hat{q}_1}{n_1} + \frac{\hat{p}_2 \hat{q}_2}{n_2}}$$

$$.11 \pm (2.58)(.0770)$$
$$.11 \pm .199$$

or $(-.089, .309)$. Since this interval contains the value $(p_1 - p_2) = 0$, it is possible that $p_1 = p_2$, which implies that there may be no difference in the proportions favoring the bond issue in the two sections of the city.

2. If there is no difference in the two proportions, then the two samples are not really different and might well be combined to obtain an overall estimate of the proportion of the city residents who will vote for the bond issue. If both samples are pooled, then $n = 150$ and

$$\hat{p} = \frac{103}{150} = .69$$

Therefore, the point estimate of the overall value of p is .69, with a margin of error given by

$$\pm 1.96 \sqrt{\frac{(.69)(.31)}{150}} = \pm 1.96(.0378) = \pm .074$$

Notice that $.69 \pm .074$ produces the interval .62 to .76, which includes only proportions greater than .5. Therefore, if voter attitudes do not change adversely prior to the election, the bond proposal should pass by a reasonable majority.

8.7 EXERCISES

BASIC TECHNIQUES

8.50 Independent random samples of $n_1 = 500$ and $n_2 = 500$ observations were selected from binomial populations 1 and 2, and $x_1 = 120$ and $x_2 = 147$ successes were observed.

a. What is the best point estimator for the difference $(p_1 - p_2)$ in the two binomial proportions?

b. Calculate the approximate standard error for the statistic used in part a.

c. What is the margin of error for this point estimate?

8.51 Independent random samples of $n_1 = 800$ and $n_2 = 640$ observations were selected from binomial populations 1 and 2, and $x_1 = 337$ and $x_2 = 374$ successes were observed.

a. Find a 90% confidence interval for the difference $(p_1 - p_2)$ in the two population proportions. Interpret the interval.

b. What assumptions must you make for the confidence interval to be valid? Are these assumptions met?

8.52 Independent random samples of $n_1 = 1265$ and $n_2 = 1688$ observations were selected from binomial populations 1 and 2, and $x_1 = 849$ and $x_2 = 910$ successes were observed.

a. Find a 99% confidence interval for the difference $(p_1 - p_2)$ in the two population proportions. What does "99% confidence" mean?

b. Based on the confidence interval in part a, can you conclude that there is a difference in the two binomial proportions? Explain.

APPLICATIONS

8.53 M&Ms Does the M&M®/Mars corporation use the same proportion of red candies in its plain and peanut varieties? A random sample of 56 plain M&Ms contained 12 red candies, and another random sample of 32 peanut M&Ms contained 8 red candies.

a. Construct a 95% confidence interval for the difference in the proportions of red candies for the plain and peanut varieties.

b. Based on the confidence interval in part a, can you conclude that there is a difference in the proportions of red candies for the plain and peanut varieties? Explain.

8.54 Public Service Jobs Instead of paying to support welfare recipients, many Californians want them to find jobs; if necessary, they want the state to create public service jobs for those who cannot find jobs in private industry.[12] In a survey of 500 registered voters reported in *The Press-Enterprise*—250 Republicans and 250 Democrats—70% of the Republicans and 86% of the Democrats favored the creation of public service jobs. Use a large-sample estimation procedure to compare the proportions of Republicans and Democrats in the population of registered voters in California who favor creating public service jobs. Explain your conclusions.

8.55 Baseball Fans The first day of baseball comes in late March, and ends in October with the World Series. Does fan support grow as the season goes on? Two CNN/*USA Today*/Gallup polls, one conducted in March and one in November, both involved random samples of 1001 adults aged 18 and older. In the March sample 45% of the adults claimed to be fans of professional baseball, while 51% of the adults in the November sample claimed to be fans.[13]

a. Construct a 99% confidence interval for the difference in the proportion of adults who claim to be fans in March versus November.

b. Does the data indicate that the proportion of adults who claim to be fans increases in November, around the time of the World Series? Explain.

8.56 Baseball and Steroids Refer to Exercise 8.55. In the March opinion poll, suppose that 451 adults identified themselves as baseball fans, while the other 550 were not fans. The following question was posed:

Should major league baseball players be tested for steroids or other performance enhancing drugs, or not?

Suppose that 410 of the baseball fans and 505 of those who are not fans answered yes to this question.

a. Construct a 95% confidence interval for the difference in the proportion of adults (fans versus nonfans) who favor mandatory drug testing for professional baseball players.

b. Does the data indicate that there is a difference in the proportion of fans versus nonfans who favor mandatory drug testing for professional baseball players? Explain.

8.57 Catching a Cold Do well-rounded people get fewer colds? A study on the *Chronicle of Higher Education* was conducted by scientists at Carnegie Mellon University, the University of Pittsburgh, and the University of Virginia. They found that people who have only a few social outlets get more colds than those who are involved in a variety of social activities.[14] Suppose that of the 276 healthy men and women tested, $n_1 = 96$ had only a few social outlets and $n_2 = 105$ were busy with six or more activities. When these people were exposed to a cold virus, the following results were observed:

	Few Social Outlets	Many Social Outlets
Sample size	96	105
Percent with colds	62%	35%

a. Construct a 99% confidence interval for the difference in the two population proportions.

b. Does there appear to be a difference in the population proportions for the two groups?

c. You might think that coming into contact with more people would lead to more colds, but the data show the opposite effect. How can you explain this unexpected finding?

8.58 Union, Yes! A sampling of political candidates—200 randomly chosen from the West and 200 from the East—was classified according to whether the candidate received backing by a national labor union and whether the candidate won. In the West, 120 winners had union backing, and in the East, 142 winners were backed by a national union. Find a 95% confidence interval for the difference between the proportions of union-backed winners in the West versus the East. Interpret this interval.

8.59 Birth Order and College Success In a study of the relationship between birth order and college success, an investigator found that 126 in a sample of 180 college graduates were firstborn or only children. In a sample of 100 nongraduates of comparable age and socioeconomic background, the number of firstborn or only children was 54. Estimate the difference between the proportions of firstborn or only children in the two populations from which these samples were drawn. Use a 90% confidence interval and interpret your results.

8.60 Ads in Outer Space? Do you think that we should let Radio Shack film a commercial in outer space? The commercialism of our space program is a topic of great interest since Dennis Tito paid $20 million dollars to ride along with the Russians on the space shuttle.[15] In a survey of 500 men and 500 women, 20% of the men and 26% of the women responded that space should remain commercial-free.

a. Construct a 98% confidence interval for the difference in the proportions of men and women who think that space should remain commercial-free.

b. What does it mean to say that you are "98% confident"?

c. Based on the confidence interval in part a, can you conclude that there is a difference in the proportions of men and women who think space should remain commercial-free?

8.61 Excedrin or Tylenol? In a study to compare the effects of two pain relievers it was found that of $n_1 = 200$ randomly selectd individuals instructed to use the first pain reliever, 93% indicated that it relieved their pain. Of $n_2 = 450$ randomly selected individuals instructed to use the second pain reliever, 96% indicated that it relieved their pain.

a. Find a 99% confidence interval for the difference in the proportions experiencing relief from pain for these two pain relievers.

b. Based on the confidence interval in part a, is there sufficient evidence to indicate a difference in the proportions experiencing relief for the two pain relievers? Explain.

8.62 Auto Accidents Last year's records of auto accidents occurring on a given section of highway were classified according to whether the resulting damage was $1,000 or more and to whether a physical injury resulted from the accident. The data follows:

	Under $1000	$1000 or more
Number of accidents	32	41
Number involving injuries	10	23

a. Estimate the true proportion of accidents involving injuries when the damage was $1,000 or more for similar sections of highway and find the margin of error.

b. Estimate the true difference in the proportion of accidents involving injuries for accidents with damage under $1,000 and those with damage of $1,000 or more. Use a 95% confidence interval.

ONE-SIDED CONFIDENCE BOUNDS

The confidence intervals discussed in Sections 8.5–8.7 are sometimes called **two-sided confidence intervals** because they produce both an upper (UCL) and a lower (LCL) bound for the parameter of interest. Sometimes, however, an experimenter is interested in only one of these limits; that is, he needs only an upper bound (or possibly a lower bound) for the parameter of interest. In this case, you can construct a **one-sided confidence bound** for the parameter of interest, such as μ, p, $\mu_1 - \mu_2$ or $p_1 - p_2$.

When the sampling distribution of a point estimator is approximately normal, an argument similar to the one in Section 8.5 can be used to show that one-sided confidence bounds, constructed using the following equations *when the sample size is large,* will contain the true value of the parameter of interest $(1 - \alpha)100\%$ of the time in repeated sampling.

A $(1 - \alpha)100\%$ LOWER CONFIDENCE BOUND (LCB)

(Point estimator) $- z_\alpha \times$ (Standard error of the estimator)

A $(1 - \alpha)100\%$ UPPER CONFIDENCE BOUND (UCB)

(Point estimator) $+ z_\alpha \times$ (Standard error of the estimator)

The z-value used for a $(1 - \alpha)100\%$ one-sided confidence bound, z_α, locates an area α in a single tail of the normal distribution as shown in Figure 8.13.

FIGURE 8.13

z-value for a one-sided confidence bound

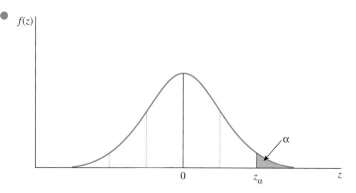

EXAMPLE 8.12

A corporation plans to issue some short-term notes and is hoping that the interest it will have to pay will not exceed 11.5%. To obtain some information about this problem, the corporation marketed 40 notes, one through each of 40 brokerage firms. The mean and standard deviation for the 40 interest rates were 10.3% and .31%, respectively. Since the corporation is interested in only an upper limit on the interest rates, find a 95% upper confidence bound for the mean interest rate that the corporation will have to pay for the notes.

Solution   Since the parameter of interest is μ, the point estimator is $\bar{x}$ with standard error $\text{SE} \approx \dfrac{s}{\sqrt{n}}$. The confidence coefficient is .95, so that $\alpha = .05$ and $z_{.05} = 1.645$. Therefore, the 95% upper confidence bound is

$$\text{UCB} = \bar{x} + 1.645\left(\frac{s}{\sqrt{n}}\right) = 10.3 + 1.645\left(\frac{.31}{\sqrt{40}}\right) = 10.3 + .0806 = 10.3806$$

Thus, you can estimate that the mean interest rate that the corporation will have to pay on its notes will be less than 10.3806%. The corporation should not be concerned about its interest rates exceeding 11.5%. How confident are you of this conclusion? Fairly confident, because intervals constructed in this manner contain μ 95% of the time.

8.9   CHOOSING THE SAMPLE SIZE

Designing an experiment is essentially a plan for buying a certain amount of information. Just as the price you pay for a video game varies depending on where and when you buy it, the price of statistical information varies depending on how and where the information is collected. As when you buy any product, you should buy as much statistical information as you can for the minimum possible cost.

The total amount of relevant information in a sample is controlled by two factors:

- The **sampling plan** or **experimental design:** the procedure for collecting the information
- The **sample size** n**:** the amount of information you collect

You can increase the amount of information you collect by *increasing* the sample size, or perhaps by *changing* the type of sampling plan or experimental design you are using. We will discuss the simplest sampling plan—random sampling from a relatively large population—and focus on ways to choose the sample size n needed to purchase a given amount of information.

A researcher makes little progress in planning an experiment before encountering the problem of sample size. **How many measurements should be included in the sample?** How much information does the researcher want to buy? The total amount of information in the sample will affect the reliability or goodness of the inferences made by the researcher, and it is this reliability that the researcher must specify. In a statistical estimation problem, the accuracy of the estimate is measured by the *margin of error* or the *width of the confidence interval.* Since both of these measures are a function of the sample size, specifying the accuracy determines the necessary sample size.

For instance, suppose you want to estimate the average daily yield μ of a chemical process and you need the margin of error to be less than 4 tons. This means that, approximately 95% of the time in repeated sampling, the distance between the sample mean $\bar{x}$ and the population mean μ will be less than 1.96 SE. You want this quantity to be less than 4. That is,

$$1.96 \text{ SE} < 4 \quad \text{or} \quad 1.96\left(\frac{\sigma}{\sqrt{n}}\right) < 4$$

Solving for n, you obtain

$$n > \left(\frac{1.96}{4}\right)^2 \sigma^2 \quad \text{or} \quad n > .24\sigma^2$$

If you know σ, the population standard deviation, you can substitute its value into the formula and solve for n. If σ is unknown—which is usually the case—you can use the best approximation available:

- An estimate s obtained from a previous sample
- A range estimate based on knowledge of the largest and smallest possible measurements: $\sigma \approx \text{Range}/4$

For this example, suppose that a prior study of the chemical process produced a sample standard deviation of $s = 21$ tons. Then

$$n > .24\sigma^2 = .24(21)^2 = 105.8$$

Using a sample of size $n = 106$ or larger, you could be reasonably certain (with probability approximately equal to .95) that your estimate of the average yield will be within ± 4 tons of the actual average yield.

The solution $n = 106$ is only approximate because you had to use an approximate value for σ to calculate the standard error of the mean. Although this may bother you, it is the best method available for selecting the sample size, and it is certainly better than guessing!

Sometimes researchers request a different confidence level than the 95% confidence specified by the margin of error. In this case, the half-width of the confidence interval provides the accuracy measure for your estimate; that is, the bound B on the error of your estimate is

$$z_{\alpha/2}\left(\frac{\sigma}{\sqrt{n}}\right) < B$$

This method for choosing the sample size can be used for all four estimation procedures presented in this chapter. The general procedure is described next.

 PERSONAL TRAINER

How Do I Choose the Sample Size?

Determine the parameter to be estimated and the standard error of its point estimator. Then proceed as follows:

1. Choose B, the bound on the error of your estimate, and a confidence coefficient $(1 - \alpha)$.

2. For a one-sample problem, solve this equation for the sample size n:

$$z_{\alpha/2} \times (\text{Standard error of the estimator}) \leq B$$

where $z_{\alpha/2}$ is the value of z having area $\alpha/2$ to its right.

3. For a two-sample problem, set $n_1 = n_2 = n$ and solve the equation in step 2.

[NOTE: For most estimators (all presented in this textbook), the standard error is a function of the sample size n.]

Exercise Reps

Fill in the blanks in the table below and find the necessary sample sizes.

Type of Data	One or Two Samples	Margin of Error	p or σ	Bound, B	Solve this Inequality	Sample Size
Binomial	One	_____	$p \approx .4$	.1	_____ $\leq .1$	$n \geq$ _____
_____	_____	$1.96\dfrac{\sigma}{\sqrt{n}}$	$\sigma \approx 6$	1	_____ ≤ 1	$n \geq$ _____
Quantitative	Two	_____	$\sigma_1 \approx \sigma_2 \approx 6$	2	_____ ≤ 2	$n_1 = n_2 \geq$ _____
_____	_____	$1.96\sqrt{\dfrac{p_1 q_1}{n_1} + \dfrac{p_2 q_2}{n_2}}$	$p_1 \approx p_2 \approx .4$	.05	_____ $\leq .05$	$n_1 = n_2 \geq$ _____

Progress Report

- Still having trouble with sample sizes? Try again using the Exercise Reps at the end of this section.

- Mastered the sample size problem? You can skip the Exercise Reps at the end of this section!

Answers are located on the perforated card at the back of this book.

EXAMPLE

Producers of polyvinyl plastic pipe want to have a supply of pipes sufficient to meet marketing needs. They wish to survey wholesalers who buy polyvinyl pipe in order to estimate the proportion who plan to increase their purchases next year. What sample size is required if they want their estimate to be within .04 of the actual proportion with probability equal to .90?

Solution For this particular example, the bound B on the error of the estimate is .04. Since the confidence coefficient is $(1 - \alpha) = .90$, α must equal .10 and $\alpha/2$ is .05. The z-value corresponding to an area equal to .05 in the upper tail of the z distribution is $z_{.05} = 1.645$. You then require

$$1.645 \text{ SE} = 1.645 \sqrt{\frac{pq}{n}} = .04$$

In order to solve this equation for n, you must substitute an approximate value of p into the equation. If you want to be certain that the sample is large enough, you should use $p = .5$ (substituting $p = .5$ will yield the largest possible solution for n because the maximum value of pq occurs when $p = q = .5$). Then

$$1.645 \sqrt{\frac{(.5)(.5)}{n}} \le .04$$

or

$$\sqrt{n} \ge \frac{(1.645)(.5)}{.04} = 20.56$$

$$n \ge (20.56)^2 = 422.7$$

Therefore, the producers must include at least 423 wholesalers in its survey if it wants to estimate the proportion p correct to within .04.

EXAMPLE 8.14

A personnel director wishes to compare the effectiveness of two methods of training industrial employees to perform a certain assembly operation. A number of employees are to be divided into two equal groups: the first receiving training method 1 and the second training method 2. Each will perform the assembly operation, and the length of assembly time will be recorded. It is expected that the measurements for both groups will have a range of approximately 8 minutes. For the estimate of the difference in mean times to assemble to be correct to within 1 minute with a probability equal to .95, how many workers must be included in each training group?

Solution Letting $B = 1$ minute, you get

$$1.96 \sqrt{\frac{\sigma_1^2}{n_1} + \frac{\sigma_2^2}{n_2}} \le 1$$

Since you wish n_1 to equal n_2, you can let $n_1 = n_2 = n$ and obtain the equation

$$1.96 \sqrt{\frac{\sigma_1^2}{n} + \frac{\sigma_2^2}{n}} \le 1$$

As noted above, the variability (range) of each method of assembly is approximately the same, and hence $\sigma_1^2 = \sigma_2^2 = \sigma^2$. Since the range, equal to 8 minutes, is approximately equal to 4σ, you have

$$4\sigma \approx 8 \quad \text{or} \quad \sigma \approx 2$$

Substituting this value for σ_1 and σ_2 in the earlier equation, you get

$$1.96\sqrt{\frac{(2)^2}{n} + \frac{(2)^2}{n}} \le 1$$

$$1.96\sqrt{\frac{8}{n}} \le 1$$

$$\sqrt{n} \ge 1.96\sqrt{8}$$

Solving, you have $n \ge 31$. Thus, each group should contain at least $n = 31$ workers.

Table 8.7 provides a summary of the formulas used to determine the sample sizes required for estimation with a given bound on the error of the estimate or confidence interval width W ($W = 2B$). Notice that to estimate p, the sample size formula uses $\sigma^2 = pq$, whereas to estimate $(p_1 - p_2)$, the sample size formula uses $\sigma_1^2 = p_1 q_1$ and $\sigma_2^2 = p_2 q_2$.

TABLE 8.7

Sample Size Formulas

Parameter	Estimator	Sample Size	Assumptions
μ	$\bar{x}$	$n \ge \dfrac{z_{\alpha/2}^2 \sigma^2}{B^2}$	
$\mu_1 - \mu_2$	$\bar{x}_1 - \bar{x}_2$	$n \ge \dfrac{z_{\alpha/2}^2(\sigma_1^2 + \sigma_2^2)}{B^2}$	$n_1 = n_2 = n$
p	$\hat{p}$	$\begin{cases} n \ge \dfrac{z_{\alpha/2}^2 pq}{B^2} \\ \text{or} \\ n \ge \dfrac{(.25)z_{\alpha/2}^2}{B^2} \end{cases}$	$p = .5$
$p_1 - p_2$	$\hat{p}_1 - \hat{p}_2$	$\begin{cases} n \ge \dfrac{z_{\alpha/2}^2(p_1 q_1 + p_2 q_2)}{B^2} \\ \text{or} \\ n \ge \dfrac{2(.25)z_{\alpha/2}^2}{B^2} \end{cases}$	$n_1 = n_2 = n$ $n_1 = n_2 = n$ and $p_1 = p_2 = .5$

8.9 EXERCISES

EXERCISE REPS

8.63 Fill in the blanks in the table below and find the necessary sample sizes.

Type of Data	One or Two Samples	Margin of Error	p or σ	Bound, B	Solve this Inequality	Sample Size
Binomial	One	_____	$p \approx .5$	.05	_____ $\le .05$	$n \ge$ _____
_____	_____	$1.96\dfrac{\sigma}{\sqrt{n}}$	$\sigma \approx 10$	2	_____ ≤ 2	$n \ge$ _____

8.64 Fill in the blanks in the table below and find the necessary sample sizes.

Type of Data	One or Two Samples	Margin of Error	p or σ	Bound, B	Solve this Inequality	Sample Size
Quantitative	Two	_____	$\sigma_1 \approx \sigma_2 \approx 10$	4	_____ ≤ 4	$n_1 = n_2 \geq$ _____
_____	_____	$1.96\sqrt{\dfrac{p_1 q_1}{n_1} + \dfrac{p_2 q_2}{n_2}}$	$p_1 \approx p_2 \approx .5$	.10	_____ $\leq .10$	$n_1 = n_2 \geq$ _____

BASIC TECHNIQUES

8.65 Find a 90% one-sided upper confidence bound for the population mean μ for these values:

a. $n = 40, s^2 = 65, \bar{x} = 75$

b. $n = 100, s = 2.3, \bar{x} = 1.6$

8.66 Find a 99% lower confidence bound for the binomial proportion p when a random sample of $n = 400$ trials produced $x = 196$ successes.

8.67 Independent random samples of size 50 are drawn from two quantitative populations, producing the sample information in the table. Find a 95% upper confidence bound for the difference in the two population means.

	Sample 1	Sample 2
Sample size	50	50
Sample mean	12	10
Sample standard deviation	5	7

8.68 Suppose you wish to estimate a population mean based on a random sample of n observations, and prior experience suggests that $\sigma = 12.7$. If you wish to estimate μ correct to within 1.6, with probability equal to .95, how many observations should be included in your sample?

8.69 Suppose you wish to estimate a binomial parameter p correct to within .04, with probability equal to .95. If you suspect that p is equal to some value between .1 and .3 and you want to be certain that your sample is large enough, how large should n be? (HINT: When calculating the standard error, use the value of p in the interval $.1 < p < .3$ that will give the largest sample size.)

8.70 Independent random samples of $n_1 = n_2 = n$ observations are to be selected from each of two populations 1 and 2. If you wish to estimate the difference between the two population means correct to within

.17, with probability equal to .90, how large should n_1 and n_2 be? Assume that you know $\sigma_1^2 \approx \sigma_2^2 \approx 27.8$.

8.71 Independent random samples of $n_1 = n_2 = n$ observations are to be selected from each of two binomial populations 1 and 2. If you wish to estimate the difference in the two population proportions correct to within .05, with probability equal to .98, how large should n be? Assume that you have no prior information on the values of p_1 and p_2, but you want to make certain that you have an adequate number of observations in the samples.

APPLICATIONS

8.72 Operating Expenses A random sampling of a company's monthly operating expenses for $n = 36$ months produced a sample mean of $5474 and a standard deviation of $764. Find a 90% upper confidence bound for the company's mean monthly expenses.

8.73 Spectator Sports Exercise 8.17 discussed a research poll conducted for *U.S. News & World Report* to determine the public's attitudes concerning the effect of spectator sports on society.[3] Suppose you were designing a poll of this type.

a. Explain how you would select your sample. What problems might you encounter in this process?

b. If you wanted to estimate the percentage of the population who agree with a particular statement in your survey questionnaire correct to within 1%, with probability .95, approximately how many people would have to be polled?

8.74 Political Corruption A questionnaire is designed to investigate attitudes about political corruption in government. The experimenter would like to survey two different groups—Republicans and Democrats—and compare the responses to various "yes–no" questions for the two groups. The experimenter requires that the sampling error for the difference in the proportion of yes responses for the two groups is no more than ±3 percentage points. If the two samples are the same size, how large should the samples be?

8.75 Less Red Meat! Americans are becoming more conscious of the importance of good nutrition, and some researchers believe that we may be altering our diets to include less red meat and more fruits and vegetables. To test this theory, a researcher decides to select hospital nutritional records for subjects surveyed 10 years ago and to compare the average amount of beef consumed per year to the amounts consumed by an equal number of subjects she will interview this year. She knows that the amount of beef consumed annually by Americans ranges from 0 to approximately 104 pounds. How many subjects should the researcher select for each group if she wishes to estimate the difference in the average annual per-capita beef consumption correct to within 5 pounds with 99% confidence?

8.76 Red Meat, continued Refer to Exercise 8.75. The researcher selects two groups of 400 subjects each and collects the following sample information on the annual beef consumption now and 10 years ago:

	Ten Years Ago	This Year
Sample mean	73	63
Sample standard deviation	25	28

a. The researcher would like to show that per-capita beef consumption has decreased in the last 10 years, so she needs to show that the difference in the averages is greater than 0. Find a 99% lower confidence bound for the difference in the average per-capita beef consumptions for the two groups.

b. What conclusions can the researcher draw using the confidence bound from part a?

8.77 Hunting Season If a wildlife service wishes to estimate the mean number of days of hunting per hunter for all hunters licensed in the state during a given season, with a bound on the error of estimation equal to 2 hunting days, how many hunters must be included in the survey? Assume that data collected in earlier surveys have shown σ to be approximately equal to 10.

8.78 Polluted Rain Suppose you wish to estimate the mean pH of rainfalls in an area that suffers heavy pollution due to the discharge of smoke from a power plant. You know that σ is in the neighborhood of .5 pH, and you wish your estimate to lie within .1 of μ, with a probability near .95. Approximately how many rainfalls must be included in your sample (one pH reading per rainfall)? Would it be valid to select all of your water specimens from a single rainfall? Explain.

8.79 pH in Rainfall Refer to Exercise 8.78. Suppose you wish to estimate the difference between the mean acidity for rainfalls at two different locations, one in a relatively unpolluted area along the ocean and the other in an area subject to heavy air pollution. If you wish your estimate to be correct to the nearest .1 pH, with probability near .90, approximately how many rainfalls (pH values) would have to be included in each sample? (Assume that the variance of the pH measurements is approximately .25 at both locations and that the samples will be of equal size.)

8.80 GPAs You want to estimate the difference in grade point averages between two groups of college students accurate to within .2 grade point, with probability approximately equal to .95. If the standard deviation of the grade point measurements is approximately equal to .6, how many students must be included in each group? (Assume that the groups will be of equal size.)

8.81 Selenium, again Refer to the comparison of the daily adult intake of selenium in two different regions of the United States in Exercise 8.41. Suppose you wish to estimate the difference in the mean daily intakes between the two regions correct to within 5 micrograms, with probability equal to .90. If you plan to select an equal number of adults from the two regions (i.e., $n_1 = n_2$), how large should n_1 and n_2 be?

CHAPTER REVIEW

Key Concepts and Formulas

I. Types of Estimators

1. Point estimator: a single number is calculated to estimate the population parameter.

2. Interval estimator: two numbers are calculated to form an interval that, with a certain amount of confidence, contains the parameter.

II. Properties of Good Estimators

1. Unbiased: the average value of the estimator equals the parameter to be estimated.

2. Minimum variance: of all the unbiased estimators, the best estimator has a sampling distribution with the smallest standard error.

3. The margin of error measures the maximum distance between the estimator and the true value of the parameter.

III. Large-Sample Point Estimators

To estimate one of four population parameters when the sample sizes are large, use the following point estimators with the appropriate margins of error.

Parameter	Point Estimator	95% Margin of Error
μ	$\bar{x}$	$\pm 1.96\left(\dfrac{s}{\sqrt{n}}\right)$
p	$\hat{p} = \dfrac{x}{n}$	$\pm 1.96\sqrt{\dfrac{\hat{p}\hat{q}}{n}}$
$\mu_1 - \mu_2$	$\bar{x}_1 - \bar{x}_2$	$\pm 1.96\sqrt{\dfrac{s_1^2}{n_1} + \dfrac{s_2^2}{n_2}}$
$p_1 - p_2$	$(\hat{p}_1 - \hat{p}_2) = \left(\dfrac{x_1}{n_1} - \dfrac{x_2}{n_2}\right)$	$\pm 1.96\sqrt{\dfrac{\hat{p}_1\hat{q}_1}{n_1} + \dfrac{\hat{p}_2\hat{q}_2}{n_2}}$

IV. Large-Sample Interval Estimators

To estimate one of four population parameters when the sample sizes are large, use the following interval estimators.

Parameter	$(1 - \alpha)100\%$ Confidence Interval
μ	$\bar{x} \pm z_{\alpha/2}\left(\dfrac{s}{\sqrt{n}}\right)$
p	$\hat{p} \pm z_{\alpha/2}\sqrt{\dfrac{\hat{p}\hat{q}}{n}}$
$\mu_1 - \mu_2$	$(\bar{x}_1 - \bar{x}_2) \pm z_{\alpha/2}\sqrt{\dfrac{s_1^2}{n_1} + \dfrac{s_2^2}{n_2}}$
$p_1 - p_2$	$(\hat{p}_1 - \hat{p}_2) \pm z_{\alpha/2}\sqrt{\dfrac{\hat{p}_1\hat{q}_1}{n_1} + \dfrac{\hat{p}_2\hat{q}_2}{n_2}}$

1. All values in the interval are possible values for the unknown population parameter.

2. Any values outside the interval are unlikely to be the value of the unknown parameter.

3. To compare two population means or proportions, look for the value 0 in the confidence interval. If 0 is in the interval, it is possible that the two population means or proportions are equal, and you should not declare a difference. If 0 is not in the interval, it is unlikely that the two means or proportions are equal, and you can confidently declare a difference.

V. One-Sided Confidence Bounds

Use either the upper $(+)$ or lower $(-)$ two-sided bound, with the critical value of z changed from $z_{\alpha/2}$ to z_α.

VI. Choosing the Sample Size

1. Determine the size of the margin of error, B, that you are willing to tolerate.

2. Choose the sample size by solving for n or $n = n_1 = n_2$ in the inequality: $z_{\alpha/2} \le B$, where SE is a function of the sample size n.

3. For quantitative populations, estimate the population standard deviation using a previously calculated value of s or the range approximation $\sigma \approx \text{Range}/4$.

4. For binomial populations, use the conservative approach and approximate p using the value $p = .5$.

Supplementary Exercises

8.82 State the Central Limit Theorem. Of what value is the Central Limit Theorem in large-sample statistical estimation?

8.83 A random sample of $n = 64$ observations has a mean $\bar{x} = 29.1$ and a standard deviation $s = 3.9$.

a. Give the point estimate of the population mean μ and find the margin of error for your estimate.

b. Find a 90% confidence interval for μ. What does "90% confident" mean?

c. Find a 90% lower confidence bound for the population mean μ. Why is this bound different from the lower confidence limit in part b?

d. How many observations do you need to estimate μ to within .5, with probability equal to .95?

8.84 Independent random samples of $n_1 = 50$ and $n_2 = 60$ observations were selected from populations 1 and 2, respectively. The sample sizes and computed sample statistics are given in the table:

	Population	
	1	2
Sample size	5	60
Sample mean	100.4	96.2
Sample standard deviation	0.8	1.3

Find a 90% confidence interval for the difference in population means and interpret the interval.

8.85 Refer to Exercise 8.84. Suppose you wish to estimate $(\mu_1 - \mu_2)$ correct to within .2, with probability equal to .95. If you plan to use equal sample sizes, how large should n_1 and n_2 be?

8.86 A random sample of $n = 500$ observations from a binomial population produced $x = 240$ successes.

a. Find a point estimate for p, and find the margin of error for your estimator.

b. Find a 90% confidence interval for p. Interpret this interval.

8.87 Refer to Exercise 8.86. How large a sample is required if you wish to estimate p correct to within .025, with probability equal to .90?

8.88 Independent random samples of $n_1 = 40$ and $n_2 = 80$ observations were selected from binomial populations 1 and 2, respectively. The number of successes in the two samples were $x_1 = 17$ and $x_2 = 23$. Find a 99% confidence interval for the difference between the two binomial population proportions. Interpret this interval.

8.89 Refer to Exercise 8.88. Suppose you wish to estimate $(p_1 - p_2)$ correct to within .06, with probability equal to .99, and you plan to use equal sample sizes—that is, $n_1 = n_2$. How large should n_1 and n_2 be?

8.90 Ethnic Cuisine Ethnic groups in America buy differing amounts of various food products because of their ethnic cuisine. Asians buy fewer canned vegetables than do other groups, and Hispanics purchase more cooking oil. A researcher interested in market segmentation for these two groups would like to estimate the proportion of households that select certain brands for various products. If the researcher wishes these estimates to be within .03 with probability .95,

how many households should she include in the samples?

8.91 Women on Wall Street Women on Wall Street can earn large salaries, but may need to make sacrifices in their personal lives. In fact, many women in the securities industry have to make significant personal sacrifices. A survey of 482 women and 356 men found that only half of the women have children, compared to three-quarters of the men surveyed.[16]

a. What are the values of $\hat{p}_1$ and $\hat{p}_2$ for the women and men in this survey?

b. Find a 95% confidence interval for the difference in the proportion of women and men on Wall Street who have children.

c. What conclusions can you draw regarding the groups compared in part b?

8.92 Smoking and Blood Pressure An experiment was conducted to estimate the effect of smoking on the blood pressure of a group of 35 cigarette smokers. The difference for each participant was obtained by taking the difference in the blood pressure readings at the beginning of the experiment and again five years later. The sample mean increase, measured in millimeters of mercury, was $\bar{x} = 9.7$. The sample standard deviation was $s = 5.8$. Estimate the mean increase in blood pressure that one would expect for cigarette smokers over the time span indicated by the experiment. Find the margin of error. Describe the population associated with the mean that you have estimated.

8.93 Blood Pressure, continued Using a confidence coefficient equal to .90, place a confidence interval on the mean increase in blood pressure for Exercise 8.92.

8.94 Iodine Concentration Based on repeated measurements of the iodine concentration in a solution, a chemist reports the concentration as 4.614, with an "error margin of .006."

a. How would you interpret the chemist's "error margin"?

b. If the reported concentration is based on a random sample of $n = 30$ measurements, with a sample standard deviation $s = .017$, would you agree that the chemist's "error margin" is .006?

8.95 Heights If it is assumed that the heights of men are normally distributed, with a standard deviation of 2.5 inches, how large a sample should be taken to be fairly sure (probability .95) that the sample mean does not differ from the true mean (population mean) by more than .50 in absolute value?

8.96 Chicken Feed An experimenter fed different rations, A and B, to two groups of 100 chicks each. Assume that all factors other than rations are the same for both groups. Of the chicks fed ration A, 13 died, and of the chicks fed ration B, 6 died.

a. Construct a 98% confidence interval for the true difference in mortality rates for the two rations.

b. Can you conclude that there is a difference in the mortality rates for the two rations?

8.97 Antibiotics You want to estimate the mean hourly yield for a process that manufactures an antibiotic. You observe the process for 100 hourly periods chosen at random, with the results $\bar{x} = 34$ ounces per hour and $s = 3$. Estimate the mean hourly yield for the process using a 95% confidence interval.

8.98 Cheese and Soda The average American has become accustomed to eating away from home, especially at fast-food restaurants. Partly as a result of this fast-food habit, the per-capita consumption of cheese (the main ingredient in pizza) and nondiet soft drinks has risen dramatically from a decade ago. A study in *American Demographics* reports that the average American consumes 25.7 pounds of cheese and drinks 40 gallons (or approximately 645 8-ounce servings) of nondiet soft drinks per year.[17] To test the accuracy of these reported averages, a random sample of 40 consumers is selected, and these summary statistics are recorded:

	Cheese (lb/yr)	Soft Drinks (gal/yr)
Sample mean	28.1	39.2
Sample standard deviation	3.8	4.5

Use your knowledge of statistical estimation to estimate the average per-capita annual consumption for these two products. Does this sample cause you to support or to question the accuracy of the reported averages? Explain.

8.99 Healthy Eating Don't Americans know that eating pizza and french fries leads to being overweight? In the same *American Demographics* article referenced in Exercise 8.98, a survey of women who are the main meal preparers in their households reported these results:

- 90% know that obesity causes health problems.
- 80% know that high fat intake may lead to health problems.

- 86% know that cholesterol is a health problem.
- 88% know that sodium may have negative effects on health.

a. Suppose that this survey was based on a random sample of 750 women. How accurate do you expect the percentages given above to be in estimating the actual population percentages? (HINT: If these are the only four percentages for which you need a margin of error, a conservative estimate for p is $p \approx .80$.)

b. If you want to decrease your sampling error to $\pm 1\%$, how large a sample should you take?

8.100 Sunflowers In an article in the *Annals of Botany,* a researcher reported the basal stem diameters of two groups of dicot sunflowers: those that were left to sway freely in the wind and those that were artificially supported.[18] A similar experiment was conducted for monocot maize plants. Although the authors measured other variables in a more complicated experimental design, assume that each group consisted of 64 plants (a total of 128 sunflower and 128 maize plants). The values shown in the table are the sample means plus or minus the standard error.

	Sunflower	Maize
Free-standing	35.3 ± .72	16.2 ± .41
Supported	32.1 ± .72	14.6 ± .40

Use your knowledge of statistical estimation to compare the free-standing and supported basal diameters for the two plants. Write a paragraph describing your conclusions, making sure to include a measure of the accuracy of your inference.

8.101 Are You Patriotic? In a CNN/*USA Today*/Gallup poll involving $n = 1000$ American citizens who were asked how the term "patriotic" described them, the following summary was obtained[19]:

	All	18–34	60+
Very Well	53%	35%	77%
Somewhat well	31%	41%	17%
Not very well	10%	16%	4%
Not well at all	6%	8%	2%

a. Construct a 95% confidence interval for the proportion of all American citizens who agreed that the term described them "very well."

b. If the 60+ age group consisted of $n = 150$ individuals, find a 95% confidence interval for the proportion of American citizens 60+ years of age who agreed that the term described them "somewhat well."

c. If the 18–34 age group consisted of $n = 340$ individuals, find a 98% confidence interval for the true difference in proportions of those aged 60+ and those 18–34 years of age who agreed that the term described them "very well."

8.102 College Costs A dean of men wishes to estimate the average cost of the freshman year at a particular college correct to within $500, with a probability of .95. If a random sample of freshmen is to be selected and each asked to keep financial data, how many must be included in the sample? Assume that the dean knows only that the range of expenditures will vary from approximately $4800 to $13,000.

8.103 Quality Control A quality-control engineer wants to estimate the fraction of defectives in a large lot of film cartridges. From previous experience, he feels that the actual fraction of defectives should be somewhere around .05. How large a sample should he take if he wants to estimate the true fraction to within .01, using a 95% confidence interval?

8.104 Circuit Boards Samples of 400 printed circuit boards were selected from each of two production lines A and B. Line A produced 40 defectives, and line B produced 80 defectives. Estimate the difference in the actual fractions of defectives for the two lines with a confidence coefficient of .90.

8.105 Circuit Boards II Refer to Exercise 8.104. Suppose 10 samples of $n = 400$ printed circuit boards were tested and a confidence interval was constructed for p for each of the ten samples. What is the probability that exactly one of the intervals will not contain the true value of p? That at least one interval will not contain the true value of p?

8.106 Ice Hockey The ability to accelerate rapidly is an important attribute for an ice hockey player. G. Wayne Marino investigated some of the variables related to the acceleration and speed of a hockey player from a stopped position.[20] Sixty-nine hockey players, varsity and intramural, from the University of Illinois were included in the experiment. Each player was required to move as rapidly as possible from a stopped position to cover a distance of 6 meters. The means and standard deviations of some of the variables recorded for each of the 69 skaters are shown in the table:

	Mean	SD
Weight (kilograms)	75.270	9.470
Stide length (meters)	1.110	.205
Stride rate (strides/second)	3.310	.390
Average acceleration (meters/second²)	2.962	.529
Instantaneous velocity (meters/second)	5.753	.892
Time to skate (seconds)	1.953	.131

a. Give the formula that you would use to construct a 95% confidence interval for one of the population means (e.g., mean time to skate the 6-meter distance).

b. Construct a 95% confidence interval for the mean time to skate. Interpret this interval.

8.107 Ice Hockey II Exercise 8.106 presented statistics from a study of fast starts by ice hockey skaters. The mean and standard deviation of the 69 individual average acceleration measurements over the 6-meter distance were 2.962 and .529 meters per second, respectively.

a. Find a 95% confidence interval for this population mean. Interpret the interval.

b. Suppose you were dissatisfied with the width of this confidence interval and wanted to cut the interval in half by increasing the sample size. How many skaters (total) would have to be included in the study?

8.108 Ice Hockey III The mean and standard deviation of the speeds of the sample of 69 skaters at the end of the 6-meter distance in Exercise 8.106 were 5.753 and .892 meters per second, respectively.

a. Find a 95% confidence interval for the mean velocity at the 6-meter mark. Interpret the interval.

b. Suppose you wanted to repeat the experiment and you wanted to estimate this mean velocity correct to within .1 second, with probability .99. How many skaters would have to be included in your sample?

8.109 School's Out! Summer school break adds more than 20 million teenagers to the American workforce, and in fact, July is the peak time for summer jobs for teenagers.[21] *American Demographics* reports that amusement and theme parks average 100 full-time employees, but at peak season average 665 workers per site. A group of $n = 40$ amusement and theme parks are randomly selected at peak season, and the average number of full-time workers is found to be 652 with a standard deviation of 32. Find a 99%

confidence interval estimate for the true mean number of full-time workers at amusement and theme parks at peak season. Does this contradict the reported average of 665 workers per site?

8.110 Recidivism An experimental rehabilitation technique was used on released convicts. It was shown that 79 of 121 men subjected to the technique pursued useful and crime-free lives for a three-year period following prison release. Find a 95% confidence interval for p, the probability that a convict subjected to the rehabilitation technique will follow a crime-free existence for at least three years after prison release.

8.111 Specific Gravity If 36 measurements of the specific gravity of aluminum had a mean of 2.705 and a standard deviation of .028, construct a 98% confidence interval for the actual specific gravity of aluminum.

8.112 Audiology Research In a study to establish the absolute threshold of hearing, 70 male college freshmen were asked to participate. Each subject was seated in a soundproof room and a 150 H tone was presented at a large number of stimulus levels in a randomized order. The subject was instructed to press a button if he detected the tone; the experimenter recorded the lowest stimulus level at which the tone was detected. The mean for the group was 21.6 db with $s = 2.1$. Estimate the mean absolute threshold for all college freshmen and calculate the margin of error.

8.113 Right- or Left-Handed A researcher classified his subjects as innately right-handed or left-handed by comparing thumbnail widths. He took a sample of 400 men and found that 80 men could be classified as left-handed according to his criterion. Estimate the proportion of all males in the population who would test to be left-handed using a 95% confidence interval.

8.114 The Citrus Red Mite An entomologist wishes to estimate the average development time of the citrus red mite correct to within .5 day. From previous experiments it is known that σ is in the neighborhood of 4 days. How large a sample should the entomologist take to be 95% confident of her estimate?

8.115 The Citrus Red Mite, continued A grower believes that one in five of his citrus trees are infected with the citrus red mite, mentioned in Exercise 8.114. How large a sample should be taken if the grower wishes to estimate the proportion of his trees that are infected with citrus red mite to within .08?

 APPLET Exercises

8.116 Refer to the **Interpreting Confidence Intervals** applet.

a. Suppose that you have a random sample of size $n = 50$ from a population with unknown mean μ and known standard deviation $\sigma = 35$. Calculate the half width of a 95% confidence interval for μ. What would the width of this interval be?

b. Use the [One Sample] button to create a single confidence interval for μ. What is the width of this interval? Compare your results to the calculation you did in part a.

8.117 Refer to the **Interpreting Confidence Intervals** applet.

a. Use the [10 Samples] button to create ten confidence intervals for μ.

b. What do you notice about the widths of these intervals?

c. How many of the intervals work properly and enclose the true value of μ?

d. Try this simulation again by clicking the [10 Samples] button a few more times and counting the number of intervals that work correctly. Is it close to our 95% confidence level?

8.118 Refer to the **Interpreting Confidence Intervals** applet.

a. Use the [100 Samples] button to create one hundred confidence intervals for μ.

b. What do you notice about the widths of these intervals?

c. How many of the intervals work properly and enclose the true value of μ?

d. Try this simulation again by clicking the [100 Samples] button a few more times and counting the number of intervals that work correctly. Is it close to our 95% confidence level?

8.119 Suppose that a random sample of size n is selected from a population with mean $\mu = 750$ and standard deviation σ. The **Exploring Confidence Intervals** applet shows the sampling distribution of $\bar{x}$ and a representative confidence interval, calculated as

$$\bar{x} \pm z_{\alpha/2}\frac{\sigma}{\sqrt{n}}$$

a. The applet loads with $n = 50$, $\sigma = 35$ and $\bar{x} = 756$. Calculate the half-width of a 95% confidence interval for μ.

b. Calculate the upper and lower confidence limits and compare these limits to the endpoints of the interval shown in the applet.

c. Does the confidence interval work properly? That is, does it enclose the true value of $\mu = 750$?

8.120 Refer to the **Exploring Confidence Intervals** applet.

a. Use the applet to find the values of $z_{\alpha/2}$ for a 99% confidence interval. For a 95% confidence interval? For a 90% confidence interval?

b. What effect does reducing the confidence level have on the width of the confidence interval?

c. A narrower interval indicates a more precise estimate of μ, consisting of a smaller range of values. To obtain a more precise estimate by using a smaller z-value, what has been sacrificed?

8.121 Refer to the **Exploring Confidence Intervals** applet.

a. Move the slider marked "n" from bottom to top.

b. What is the effect of increasing the sample size on the standard error of $\bar{x}$? On the width of the confidence interval?

c. Can you think of a practical explanation for the phenomena you observe in part b?

8.122 Refer to the **Exploring Confidence Intervals** applet.

a. Move the slider marked "sigma" from bottom to top.

b. What is the effect of increasing the variability on the standard error of $\bar{x}$? On the width of the confidence interval?

c. Can you think of a practical explanation for the phenomena you observe in part b?

CASE STUDY ## How Reliable Is That Poll?

In the days and weeks following important or controversial events, opinion polls saturate the media. In the years 2002–2004, public attention has been on the war in Iraq and its consequences to the American public. However, in March 2004 the attention of the nation focused on the fact that nationwide gas prices were at record highs. A recent CNN/*USA Today*/Gallup Poll[22] conducted March 26–28, 2004 involving $n = 1001$ adults 18 years or older reported that nearly seven in ten Americans say the cost of gasoline represents a major crisis/problem for the United States. Do national polls conducted by the Gallup and Harris organizations, the news media, and so on, really provide accurate estimates of the percentages of people in the United States who have various opinions? Let's look at some of the results of the poll survey methods that were used.

Although the level of gas prices is a major concern for the U.S. public, it ranks somewhere below a list of selected recent events rated as crisis or major problems in Gallup Polls, 2003–2004 as shown in the next table.

CNN/*USA Today*/Gallup Poll

Event	Year	Crisis %	Major Problem %	Crisis/Major Problem %
Iraq's military capabilities	2003	25	56	81
North Korea's capabilities	2003	23	58	81
Civilian aid for Iraqi civilians	2003	30	49	79
Federal budget deficit	2003	20	57	77
Cost of gasoline	2004	13	56	69

For several questions in the present poll, the sample of 1001 Americans 18 years and older was split into two groups, one with 510 and the second with 491 adults. The first group answered the following question (1): "Thinking about the cost of gasoline, would you say that the country—[rotated: is in a state of crisis, has major problems, has minor problems, (or) no problems at all]?"

Crisis Problem	Major Problem	Minor Problem	No Problem
13%	56%	26%	5%

The second split sample consisting of 491 adults were asked (2) "Do you think the current rise in gas prices represents—[rotated: a temporary fluctuation in prices, or a more permanent change in prices]?" The latest survey finds that 55% of Americans believe the recent price increases represent a "more permanent change in gas prices," while 42% say they are only a "temporary fluctuation in prices."

Temporary	More Permanent	No Opinion
42%	55%	3%

The following description of the "Survey Methods" that appears on the Gallup Poll website explains how the results of the poll should be viewed.

> The results are based upon telephone interviews with a randomly selected national sample of 1001 adults, aged 18 years or older, conducted March 26–28, 2004. For results based on this sample (the full sample of 1001 adults 18 years or older), one can say with 95% confidence that the maximum error attributable to sampling and other random effects if ±3 percentage points. In addition to sampling error, question wording and practical difficulties in conducting surveys can introduce error or bias into the findings of public opinion polls.

1. Verify the margin of error of ±3 percentage points given by the Gallup Poll for the full sample of 1001 adults. Find the margin of error for the two split samples of 510 adults answering the first question and of 491 adults answering the second question.

2. Do the numbers reported in the tables represent the number of people who fell into those categories? If not, what do those numbers represent?

3. When question (1) was asked, the pollster rotated the order of options given to the respondent. Why do you suppose this technique was used?

4. Construct 95% confidence intervals for the proportion of Americans who:

   a. report that the cost of gasoline in March of 2004 is a major problem.

   b. report that it is a crisis problem.

5. Compare the percentage of people in the first split sample who say that the cost of gasoline is a major problem with those in the second split sample who say that the prices may be more permanent. Is this a surprising difference?

6. If these questions were asked today, would you expect the responses to be similar to those reported here or would you expect them to differ substantially?

Large-Sample Tests of Hypotheses

9

© Getty Images

MY **PERSONAL TRAINER**

Rejection Regions, *p*-Values, and Conclusions
How Do I Calculate β?

An Aspirin a Day . . . ?

Will an aspirin a day reduce the risk of heart attack? A very large study of U.S. physicians showed that a single aspirin taken every other day reduced the risk of heart attack in men by one-half. However, 3 days later, a British study reported a completely opposite conclusion. How could this be? The case study at the end of this chapter looks at how the studies were conducted, and you will analyze the data using large-sample techniques.

TESTING HYPOTHESES ABOUT POPULATION PARAMETERS

In practical situations, statistical inference can involve either estimating a population parameter or making decisions about the value of the parameter. For example, if a pharmaceutical company is fermenting a vat of antibiotic, samples from the vat can be used to *estimate* the mean potency μ for all of the antibiotic in the vat. In contrast, suppose that the company is not concerned about the exact mean potency of the antibiotic, but is concerned only that it meet the minimum government potency standards. Then the company can use samples from the vat to decide between these two possibilities:

- The mean potency μ does not exceed the minimum allowable potency.
- The mean potency μ exceeds the minimum allowable potency.

The pharmaceutical company's problem illustrates a **statistical test of hypothesis.**

The reasoning used in a statistical test of hypothesis is similar to the process in a court trial. In trying a person for theft, the court must decide between innocence and guilt. As the trial begins, the accused person is assumed to be *innocent.* The prosecution collects and presents all available evidence in an attempt to contradict the innocent hypothesis and hence obtain a conviction. If there is enough evidence against innocence, the court will reject the innocence hypothesis and declare the defendant *guilty.* If the prosecution does not present enough evidence to prove the defendant guilty, the court will find him *not guilty.* Notice that this does not prove that the defendant is innocent, but merely that there was not enough evidence to conclude that the defendant was guilty.

We use this same type of reasoning to explain the basic concepts of hypothesis testing. These concepts are used to test the four population parameters discussed in Chapter 8: a single population mean or proportion (μ or p) and the difference between two population means or proportions ($\mu_1 - \mu_2$ or $p_1 - p_2$). When the sample sizes are large, the point estimators for each of these four parameters have normal sampling distributions, so that all four large-sample statistical tests follow the same general pattern.

A STATISTICAL TEST OF HYPOTHESIS

A statistical test of hypothesis consists of five parts:

1. The null hypothesis, denoted by H_0
2. The alternative hypothesis, denoted by H_a
3. The test statistic and its p-value
4. The rejection region
5. The conclusion

When you specify these five elements, you define a particular test; changing one or more of the parts creates a new test. Let's look at each part of the statistical test of hypothesis in more detail.

Definition   The two competing hypotheses are the **alternative hypothesis H_a,** generally the hypothesis that the researcher wishes to support, and the **null hypothesis H_0,** a contradiction of the alternative hypothesis.

As you will soon see, it is easier to show support for the alternative hypothesis by proving that the null hypothesis is false. Hence, the statistical researcher always begins by assuming that the null hypothesis H_0 is true. The researcher then uses the sample data to decide whether the evidence favors H_a rather than H_0 and draws one of these two **conclusions:**

- Reject H_0 and conclude that H_a is true.
- Accept (do not reject) H_0 as true.

EXAMPLE 9.1

You wish to show that the average hourly wage of construction workers in the state of California is different from $14, which is the national average. This is the alternative hypothesis, written as

2

$$H_a : \mu \neq 14$$

The null hypothesis is

1

$$H_0 : \mu = 14$$

You would like to reject the null hypothesis, thus concluding that the California mean is not equal to $14.

EXAMPLE 9.2

A milling process currently produces an average of 3% defectives. You are interested in showing that a simple adjustment on a machine will decrease p, the proportion of defectives produced in the milling process. Thus, the alternative hypothesis is

2

$$H_a : p < .03$$

and the null hypothesis is

1

$$H_0 : p = .03$$

If you can reject H_0, you can conclude that the adjusted process produces fewer than 3% defectives.

There is a difference in the forms of the alternative hypotheses given in Examples 9.1 and 9.2. In Example 9.1, no directional difference is suggested for the value of μ; that is, μ might be either larger or smaller than $14 if H_a is true. This type of test is called a **two-tailed test of hypothesis.** In Example 9.2, however, you are specifically interested in detecting a directional difference in the value of p; that is, if H_a is true, the value of p is less than .03. This type of test is called a **one-tailed test of hypothesis.**

The decision to reject or accept the null hypothesis is based on information contained in a sample drawn from the population of interest. This information takes these forms:

MY TIP

Two-tailed $\Leftrightarrow$ Look for a $\neq$ sign in H_a.
One-tailed $\Leftrightarrow$ Look for a $>$ or $<$ sign in H_a.

3

- **Test statistic:** a single number calculated from the sample data
- **p-value:** a probability calculated using the test statistic

Either or both of these measures act as decision makers for the researcher in deciding whether to reject or accept H_0.

EXAMPLE **9.3**

For the test of hypothesis in Example 9.1, the average hourly wage $\bar{x}$ for a random sample of 100 California construction workers might provide a good *test statistic* for testing

$$H_0 : \mu = 14 \quad \text{versus} \quad H_a : \mu \neq 14$$

If the null hypothesis H_0 is true, then the sample mean should not be too far from the population mean $\mu = 14$. Suppose that this sample produces a sample mean $\bar{x} = 15$ with standard deviation $s = 2$. Is this sample evidence likely or unlikely to occur, if in fact H_0 is true? You can use two measures to find out. Since the sample size is large, the sampling distribution of $\bar{x}$ is approximately normal with mean $\mu = 14$ and standard error $\sigma/\sqrt{n}$, estimated as

$$SE = \frac{s}{\sqrt{n}} = \frac{2}{\sqrt{100}} = .2$$

- The **test statistic** $\bar{x} = 15$ lies

3

$$z = \frac{\bar{x} - \mu}{\sigma/\sqrt{n}} \approx \frac{15 - 14}{.2} = 5$$

standard deviations from the population mean μ.

- The **p-value** is the probability of observing a test statistic as extreme as or more extreme than the observed value, if in fact H_0 is true. For this example, we define "extreme" as far below or far above what we would have expected. That is,

$$p\text{-value} = P(z > 5) + P(z < -5) \approx 0$$

The *large value of the test statistic* and the *small p-value* mean that you have observed a very unlikely event, if indeed H_0 is true and $\mu = 14$.

How do you decide whether to reject or accept H_0? The entire set of values that the test statistic may assume is divided into two sets, or regions. One set, consisting of values that support the alternative hypothesis and lead to rejecting H_0, is called the **rejection region.** The other, consisting of values that support the null hypothesis, is called the **acceptance region.**

4

For example, in Example 9.1, you would be inclined to believe that California's average hourly wage was different from $14 if the sample mean is either much less than $14 or much greater than $14. The two-tailed rejection region consists of very small and very large values of $\bar{x}$, as shown in Figure 9.1. In Example 9.2, since you want to prove that the percentage of defectives has *decreased,* you would be inclined to reject H_0 for values of $\hat{p}$ that are much smaller than .03. Only *small* values of $\hat{p}$ belong in the left-tailed rejection region shown in Figure 9.2. When the rejection region is in the left tail of the distribution, the test is called a **left-tailed test.** A test with its rejection region in the right tail is called a **right-tailed test.**

FIGURE 9.1

Rejection and acceptance regions for Example 9.1

FIGURE 9.2

Rejection and acceptance
regions for Example 9.2

If the test statistic falls into the rejection region, then the null hypothesis is rejected. If the test statistic falls into the acceptance region, then either the null hypothesis is accepted or the test is judged to be inconclusive. We will clarify the different types of conclusions that are appropriate as we consider several practical examples of hypothesis tests.

Finally, how do you decide on the **critical values** that separate the acceptance and rejection regions? That is, how do you decide how much statistical evidence you need before you can reject H_0? This depends on the amount of confidence that you, the researcher, want to attach to the test conclusions and the **significance level α,** the risk you are willing to take of making an incorrect decision.

Definition   A **Type I error** for a statistical test is the error of rejecting the null hypothesis when it is true. The **level of significance (significance level)** for a statistical test of hypothesis is

$$\alpha = P(\text{Type I error}) = P(\text{falsely rejecting } H_0) = P(\text{rejecting } H_0 \text{ when it is true})$$

This value α represents the *maximum tolerable risk* of incorrectly rejecting H_0. Once this significance level is fixed, the rejection region can be set to allow the researcher to reject H_0 with a fixed degree of confidence in the decision.

In the next section, we will show you how to use a test of hypothesis to test the value of a population mean μ. As we continue, we will clarify some of the computational details and add some additional concepts to complete your understanding of hypothesis testing.

A LARGE-SAMPLE TEST ABOUT A POPULATION MEAN

9.3

Consider a random sample of n measurements drawn from a population that has mean μ and standard deviation σ. You want to test a hypothesis of the form[†]

1 $H_0 : \mu = \mu_0$

 TIP

The null hypothesis will *always* have an "equals" sign attached.

where μ_0 is some hypothesized value for μ, versus a one-tailed alternative hypothesis:

2 $H_a : \mu > \mu_0$

The subscript zero indicates the value of the parameter specified by H_0. Notice that H_0 provides an exact value for the parameter to be tested, whereas H_a gives a range of possible values for μ.

[†]Note that if the test rejects the null hypothesis $\mu = \mu_0$ in favor of the alternative hypothesis $\mu > \mu_0$, then it will certainly reject a null hypothesis that includes $\mu < \mu_0$, since this is even more contradictory to the alternative hypothesis. For this reason, in this text we state the null hypothesis for a one-tailed test as $\mu = \mu_0$ rather than $\mu \leq \mu_0$.

The Essentials of the Test

The sample mean $\bar{x}$ is the best estimate of the actual value of μ, which is presently in question. What values of $\bar{x}$ would lead you to believe that H_0 is false and μ is, in fact, greater than the hypothesized value? The values of $\bar{x}$ that are extremely *large* would imply that μ is larger than hypothesized. Hence, you should reject H_0 if $\bar{x}$ is too large.

The next problem is to define what is meant by "too large." Values of $\bar{x}$ that lie too many standard deviations to the right of the mean are not very likely to occur. Those values have very little area to their right. Hence, you can define "too large" as being too many standard deviations away from μ_0. But what is "too many"? This question can be answered using the *significance level* α, the probability of rejecting H_0 when H_0 is *true*.

Remember that the standard error of $\bar{x}$ is estimated as

$$SE = \frac{s}{\sqrt{n}}$$

Since the sampling distribution of the sample mean $\bar{x}$ is approximately normal when **n is large,** the number of standard deviations that $\bar{x}$ lies from μ_0 can be measured using the **standardized test statistic:**

3

$$z = \frac{\bar{x} - \mu_0}{s/\sqrt{n}}$$

which has an approximate standard normal distribution when H_0 is true and $\mu = \mu_0$. The significance level α is equal to the area under the normal curve lying above the

4

rejection region. Thus, if you want $\alpha = .01$, you will reject H_0 when $\bar{x}$ is more than 2.33 standard deviations to the right of μ_0. Equivalently, you will reject H_0 if the standardized test statistic z is greater than 2.33 (see Figure 9.3).

FIGURE 9.3 ●

The rejection region for a right-tailed test with $\alpha = .01$

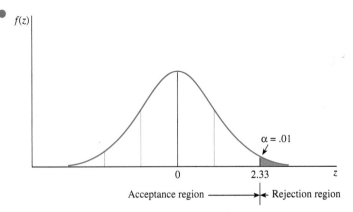

EXAMPLE 9.4 — The average weekly earnings for women in managerial and professional positions is $670. Do men in the same positions have average weekly earnings that are higher than those for women? A random sample of $n = 40$ men in managerial and professional positions showed $\bar{x} = \$725$ and $s = \$102$. Test the appropriate hypothesis using $\alpha = .01$.

MY TIP

For one-tailed tests, look for directional words like "greater," "less than," "higher," "lower," etc.

Solution You would like to show that the average weekly earnings for men are higher than \$670, the women's average. Hence, if μ is the average weekly earnings in managerial and professional positions for men, you can set out the formal test of hypothesis in steps:

1–2 **Null and alternative hypotheses:**

$$H_0: \mu = 670 \quad \text{versus} \quad H_a: \mu > 670$$

3 **Test statistic:** Using the sample information, with s as an estimate of the population standard deviation, calculate

$$z \approx \frac{\bar{x} - 670}{s/\sqrt{n}} = \frac{725 - 670}{102/\sqrt{40}} = 3.41$$

4 **Rejection region:** For this one-tailed test, values of $\bar{x}$ much larger than 670 would lead you to reject H_0; or, equivalently, values of the *standardized test statistic z* in the right tail of the standard normal distribution. To control the risk of making an incorrect decision as $\alpha = .01$, you must set the **critical value** separating the rejection and acceptance regions so that the area in the right tail is exactly $\alpha = .01$. This value is found in Table 3 of Appendix I to be $z = 2.33$, as shown in Figure 9.3. The null hypothesis will be rejected if the observed value of the test statistic, z, is greater than 2.33.

5 **Conclusion:** Compare the observed value of the test statistic, $z = 3.41$, with the critical value necessary for rejection, $z = 2.33$. Since the observed value of the test statistic falls in the rejection region, you can reject H_0 and conclude that the average weekly earnings for men in managerial and professional positions are higher than the average for women. The probability that you have made an incorrect decision is $\alpha = .01$.

If you wish to detect departures either greater or less than μ_0, then the alternative hypothesis is *two-tailed*, written as

$$H_a : \mu \neq \mu_0$$

which implies either $\mu > \mu_0$ or $\mu < \mu_0$. Values of $\bar{x}$ that are either "too large" or "too small" in terms of their distance from μ_0 are placed in the rejection region. If you choose $\alpha = .01$, the area in the rejection region is equally divided between the two tails of the normal distribution, as shown in Figure 9.4. Using the standardized test statistic z, you can reject H_0 if $z > 2.58$ or $z < -2.58$. For different values of α, the critical values of z that separate the rejection and acceptance regions will change accordingly.

 MY TIP

If the test is two-tailed, you will not see any directional words. The experimenter is only looking for a "difference" from the hypothesized value.

FIGURE 9.4

The rejection region for a two-tailed test with $\alpha = .01$

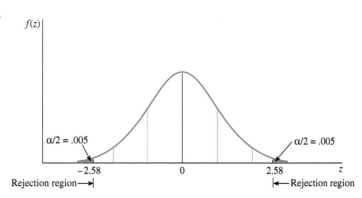

EXAMPLE 9.5

The daily yield for a local chemical plant has averaged 880 tons for the last several years. The quality control manager would like to know whether this average has changed in recent months. She randomly selects 50 days from the computer database and computes the average and standard deviation of the $n = 50$ yields as $\bar{x} = 871$ tons and $s = 21$ tons, respectively. Test the appropriate hypothesis using $\alpha = .05$.

Solution

1–2 **Null and alternative hypotheses:**

$$H_0: \mu = 880 \quad \text{versus} \quad H_a: \mu \neq 880$$

3 **Test statistic:** The point estimate for μ is $\bar{x}$. Therefore, the test statistic is

$$z \approx \frac{\bar{x} - \mu_0}{s/\sqrt{n}} = \frac{871 - 880}{21/\sqrt{50}} = -3.03$$

4 **Rejection region:** For this two-tailed test, you use values of z in both the right and left tails of the standard normal distribution. Using $\alpha = .05$, the **critical values** separating the rejection and acceptance regions cut off areas of $\alpha/2 = .025$ in the right and left tails. These values are $z = \pm 1.96$ and the null hypothesis will be rejected if $z > 1.96$ or $z < -1.96$.

5 **Conclusion:** Since $z = -3.03$ and the calculated value of z falls in the rejection region, the manager can reject the null hypothesis that $\mu = 880$ tons and conclude that it has changed. The probability of rejecting H_0 when H_0 is true and $\alpha = .05$, a fairly small probability. Hence, she is reasonably confident that the decision is correct.

LARGE-SAMPLE STATISTICAL TEST FOR μ

1. Null hypothesis: $H_0 : \mu = \mu_0$
2. Alternative hypothesis:

One-Tailed Test	**Two-Tailed Test**
$H_a : \mu > \mu_0$ (or, $H_a : \mu < \mu_0$)	$H_a : \mu \neq \mu_0$

3. Test statistic: $z = \dfrac{\bar{x} - \mu_0}{\sigma/\sqrt{n}}$ estimated as $z = \dfrac{\bar{x} - \mu_0}{s/\sqrt{n}}$

4. Rejection region: Reject H_0 when

One-Tailed Test	**Two-Tailed Test**
$z > z_\alpha$ (or $z < -z_\alpha$ when the alternative hypothesis is $H_a : \mu < \mu_0$)	$z > z_{\alpha/2}$ or $z < -z_{\alpha/2}$

 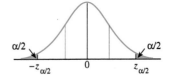

Assumptions: The n observations in the sample are randomly selected from the population and n is large—say, $n \geq 30$.

Calculating the p-Value

In the previous examples, the decision to reject or accept H_0 was made by comparing the calculated value of the test statistic with a critical value of z based on the significance level α of the test. However, different significance levels may lead to different conclusions. For example, if in a right-tailed test, the test statistic is $z = 2.03$, you can reject H_0 at the 5% level of significance because the test statistic exceeds $z = 1.645$. However, you cannot reject H_0 at the 1% level of significance because the test statistic is less than $z = 2.33$ (see Figure 9.5). To avoid any ambiguity in their conclusions, some experimenters prefer to use a variable level of significance called the **p-value** for the test.

Definition   The **p-value** or observed significance level of a statistical test is the smallest value of α for which H_0 can be rejected. It is the *actual risk* of committing a Type I error, if H_0 is rejected based on the observed value of the test statistic. The p-value measures the strength of the evidence against H_0.

In the right-tailed test with observed test statistic $z = 2.03$, the smallest critical value you can use and still reject H_0 is $z = 2.03$. For this critical value, the risk of an incorrect decision is

$$P(z \geq 2.03) = 1 - .9788 = .0212$$

This probability is the *p-value* for the test. *Notice that it is actually the area to the right of the calculated value of the test statistic.*

FIGURE 9.5

Variable rejection regions

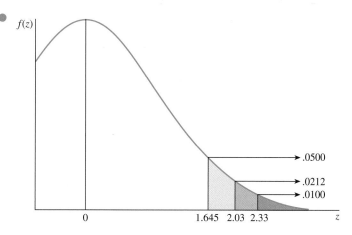

A *small p-value* indicates that the observed value of the test statistic lies far away from the hypothesized value of μ. This presents strong evidence that H_0 is false and should be rejected. *Large p-values* indicate that the observed test statistic is not far from the hypothesized mean and does not support rejection of H_0. How small does the p-value need to be before H_0 can be rejected?

MY TIP

p-value = tail area (one or two tails) "beyond" the observed value of the test statistic

Definition If the p-value is less than or equal to a preassigned significance level α, then the null hypothesis can be rejected, and you can report that the results are **statistically significant** at level α.

In the previous instance, if you choose $\alpha = .05$ as your significance level, H_0 can be rejected because the p-value is less than .05. However, if you choose $\alpha = .01$ as your significance level, the p-value (.0212) is not small enough to allow rejection of H_0. The results are significant at the 5% level, but not at the 1% level. You might see these results reported in professional journals as *significant* $(p < .05)$.[†]

EXAMPLE 9.6 Refer to Example 9.5. The quality control manager wants to know whether the daily yield at a local chemical plant—which has averaged 880 tons for the last several years—has changed in recent months. A random sample of 50 days gives an average yield of 871 tons with a standard deviation of 21 tons. Calculate the p-value for this two-tailed test of hypothesis. Use the p-value to draw conclusions regarding the statistical test.

Solution The rejection region for this two-tailed test of hypothesis is found in both tails of the normal probability distribution. Since the observed value of the test statistic is $z = -3.03$, the smallest rejection region that you can use and still reject H_0 is $|z| > 3.03$. For this rejection region, the value of α is the p-value:

$$p\text{-value} = P(z > 3.03) + P(z < -3.03) = (1 - .9988) + .0012 = .0024$$

Notice that the two-tailed p-value is actually twice the tail area corresponding to the calculated value of the test statistic. If this p-value = .0024 is less or equal to than the preassigned level of significance α, H_0 can be rejected. For this test, you can reject H_0 at either the 1% or the 5% level of significance.

If you are reading a research report, how small should the p-value be before you decide to reject H_0? Many researchers use a "sliding scale" to classify their results.

- If the p-value is less than .01, H_0 is rejected. The results are **highly significant.**
- If the p-value is between .01 and .05, H_0 is rejected. The results are **statistically significant.**
- If the p-value is between .05 and .10, H_0 is usually not rejected. The results are only **tending toward statistical significance.**
- If the p-value is greater than .10, H_0 is not rejected. The results are **not statistically significant.**

EXAMPLE 9.7 Standards set by government agencies indicate that Americans should not exceed an average daily sodium intake of 3300 milligrams (mg). To find out whether Americans are exceeding this limit, a sample of 100 Americans is selected, and the mean and standard deviation of daily sodium intake are found to be 3400 mg and 1100 mg, respectively. Use $\alpha = .05$ to conduct a test of hypothesis.

[†]In reporting statistical significance, many researchers write $(p < .05)$ or $(p < .05)$ to mean that the p-value of the test was smaller than .05, making the results significant at the 5% level. The symbol p or P in the expression has no connection with our notation for probability or with the binomial parameter p.

Solution   The hypotheses to be tested are

$$H_0 : \mu = 3300 \quad \text{versus} \quad H_a : \mu > 3300$$

and the test statistic is

$$z \approx \frac{\bar{x} - \mu_0}{s/\sqrt{n}} = \frac{3400 - 3300}{1100/\sqrt{100}} = .91$$

The two approaches developed in this section yield the same conclusions.

MY TIP

small p-value ⇔ large z-value

small p-value ⇒ reject H_0

How small? p-value $\leq \alpha$

- **The critical value approach:** Since the significance level is $\alpha = .05$ and the test is one-tailed, the rejection region is determined by a critical value with tail area equal to $\alpha = .05$; that is, H_0 can be rejected if $z > 1.645$. Since $z = .91$ is not greater than the critical value, H_0 is not rejected (see Figure 9.6).

- **The p-value approach:** Calculate the p-value, the probability that z is greater than or equal to $z = .91$:

$$p\text{-value} = P(z > .91) = 1 - .8186 = .1814$$

The null hypothesis can be rejected only if the p-value is less than or equal to the specified 5% significance level. Therefore, H_0 is not rejected and the results are *not statistically significant* (see Figure 9.6). There is not enough evidence to indicate that the average daily sodium intake exceeds 3300 mg.

FIGURE 9.6

Rejection region and p-value for Example 9.7

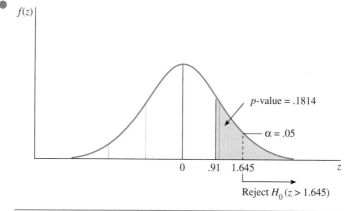

MY APPLET

You can use the **Large Sample Test of a Population Mean** applet to visualize the p-values for either one- or two-tailed tests of the population mean μ (Figure 9.7). Remember, however, that these large sample z-tests are restricted to samples of size $n \geq 30$. The applet does not prohibit you from entering a value of $n < 30$; you'll have to be careful to check the sample size before you start! The procedure follows the same pattern as with previous applets. You enter the values of $\bar{x}$, n, and s—remember to press "Enter" after each entry to record the changes. The applet will calculate z (using full accuracy) and give you the option of choosing one- or two-tailed p-values (as well as a *cumulative* and a *middle* area that you will not need.)

FIGURE 9.7 ●

Large Sample Test of a
Population Mean applet

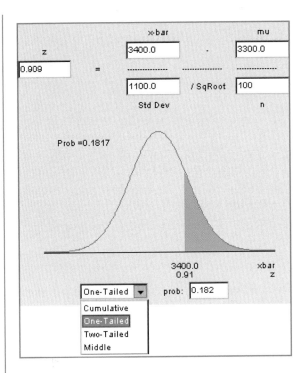

For the data of Example 9.7, the p-value is the one-tailed area to the right of $z = .909$. Do the results shown in the applet confirm our conclusions in Example 9.7? Remember that the applet uses full accuracy for the calculation of z and its corresponding probability. This means that the probability we calculate using Table 3 may be slightly different from the probability shown in the applet.

Notice that these two approaches are actually the same, as shown in Figure 9.6. As soon as the calculated value of the test statistic z becomes *larger than* the critical value, z_α, the p-value becomes *smaller than* the significance level α. You can use the most convenient of the two methods; the conclusions you reach will always be the same! The p-value approach does have two advantages, however:

- Statistical output from packages such as *MINITAB* usually reports the p-value of the test.
- Based on the p-value, your test results can be evaluated using any significance level you wish to use. Many researchers report the smallest possible significance level for which their results are *statistically significant.*

Sometimes it is easy to confuse the significance level α with the p-value (or observed significance level). They are both probabilities calculated as areas in the tails of the sampling distribution of the test statistic. However, the significance level α is preset by the experimenter before collecting the data. The p-value is linked directly to the data and actually describes how likely or unlikely the sample results are, assuming that H_0 is true. *The smaller the p-value, the more unlikely it is that H_0 is true!*

PERSONAL TRAINER

Rejection Regions, p-Values, and Conclusions

The significance level, α, lets you set the risk that you are willing to take of making an incorrect decision in a test of hypothesis.

- To set a rejection region, choose a **critical value** of z so that the area in the tail(s) of the z-distribution is (are) either α for a one-tailed test or $\alpha/2$ for a two-tailed test. Use the right tail for an upper-tailed test and the left tail for a lower-tailed test. Reject H_0 when the test statistic exceeds the critical value and falls in the rejection region.

- To find a **p-value,** find the area in the tail "beyond" the test statistic. If the test is one-tailed, this is the p-value. If the test is two-tailed, this is only half the p-value and must be doubled. Reject H_0 when the p-value is less than α.

Exercise Reps

A. **Critical value approach:** Fill in the blanks in the table below.

Test Statistic	Significance Level	One or Two-Tailed Test?	Critical Value	Rejection Region	Conclusion
$z = 1.4$	$\alpha = .05$	One-tailed (upper)			
$z = 2.46$	$\alpha = .01$	One-tailed (upper)			
$z = -0.74$	$\alpha = .05$	Two-tailed			
$z = -6.12$	$\alpha = .01$	Two-tailed			

B. **p-value approach:** Fill in the blanks in the table below.

Test Statistic	Significance Level	One or Two-Tailed Test?	p-Value	p-Value < α?	Conclusion
$z = 1.4$	$\alpha = .05$	One-tailed (upper)			
$z = 2.46$	$\alpha = .01$	One-tailed (upper)			
$z = -0.74$	$\alpha = .05$	Two-tailed			
$z = -6.12$	$\alpha = .01$	Two-tailed			

Progress Report

- Still having trouble with p-values and rejection regions? Try again using the Excercise Reps at the end of this section.

- Mastered p-values and rejection regions? You can skip the Exercise Reps at the end of this section!

Answers are located on the perforated card at the back of this book.

Two Types of Errors

You might wonder why, when H_0 was not rejected in the previous example, we did not say that H_0 was definitely true and $\mu = 3300$. This is because, if we choose to *accept* H_0, we must have a measure of the probability of error associated with this decision.

Since there are two choices in a statistical test, there are also two types of errors that can be made. In the courtroom trial, a defendant could be judged not guilty when he's really guilty, or vice versa—the same is true in a statistical test. In fact, the null hypothesis may be either true or false, regardless of the decision the experimenter makes. These two possibilities, along with the two decisions that can be made by the researcher, are shown in Table 9.1.

TABLE 9.1

Decision Table

Decision	Null Hypothesis	
	True	False
Reject H_0	Type I error	Correct decision
Accept H_0	Correct decision	Type II error

In addition to the Type I error with probability α defined earlier in this section, it is possible to commit a second error, called a **Type II error,** which has probability β.

Definition   A **Type I error** for a statistical test is the error of rejecting the null hypothesis when it is true. The probability of making a Type I error is denoted by the symbol α.

A **Type II error** for a statistical test is the error of accepting the null hypothesis when it is false and some alternative hypothesis is true. The probability of making a Type II error is denoted by the symbol β.

Notice that the probability of a Type I error is exactly the same as the **level of significance** α and is therefore controlled by the researcher. When H_0 is rejected, you have an accurate measure of the reliability of your inference; the probability of an incorrect decision is α. However, the probability β of a Type II error is not always controlled by the experimenter. In fact, when H_0 is false and H_a is true, you may not be able to specify an exact value for μ but only a range of values. This makes it difficult, if not impossible, to calculate β. Without a measure of reliability, it is not wise to conclude that H_0 is true. Rather than risk an incorrect decision, you should withhold judgment, concluding that you *do not have enough evidence to reject H_0.* Instead of *accepting H_0,* you should *not reject or fail to reject H_0.*

Keep in mind that *"accepting" a particular hypothesis means deciding in its favor. Regardless of the outcome of a test, you are never certain that the hypothesis you "accept" is true. There is always a risk of being wrong (measured by α or β).* Consequently, you never "accept" H_0 if β is unknown or its value is unacceptable to you. When this situation occurs, you should withhold judgment and collect more data.

 TIP

$\alpha = P(\text{reject } H_0 \text{ when } H_0 \text{ true})$

$\beta = P(\text{accept } H_0 \text{ when } H_0 \text{ false})$

The Power of a Statistical Test

The goodness of a statistical test is measured by the size of the two error rates: α, the probability of rejecting H_0 when it is true, and β, the probability of accepting H_0 when H_0 is false and H_a is true. A "good" test is one for which both of these error rates are small. The experimenter begins by selecting α, the probability of a Type I error. If he or she also decides to control the value of β, the probability of accepting H_0 when H_a is true, then an appropriate sample size is chosen.

Another way of evaluating a test is to look at the complement of a Type II error—that is, rejecting H_0 when H_a is true—which has probability

$$1 - \beta = P(\text{reject } H_0 \text{ when } H_a \text{ is true})$$

The quantity $(1 - \beta)$ is called the **power** of the test because it measures the probability of taking the action that we wish to have occur—that is, rejecting the null hypothesis when it is false and H_a is true.

Definition   The **power of a statistical test,** given as

$$1 - \beta = P(\text{reject } H_0 \text{ when } H_a \text{ is true})$$

measures the ability of the test to perform as required.

A graph of $(1 - \beta)$, the probability of rejecting H_0 when in fact H_0 is false, as a function of the true value of the parameter of interest is called the **power curve** for the statistical test. Ideally, you would like α to be small and the *power* $(1 - \beta)$ to be large.

EXAMPLE 9.8 Refer to Example 9.5. Calculate β and the power of the test $(1 - \beta)$ when μ is actually equal to 870 tons.

Solution   The acceptance region for the test of Example 9.5 is located in the interval $[\mu_0 \pm 1.96(s/\sqrt{n})]$. Substituting numerical values, you get

$$880 \pm 1.96\left(\frac{21}{\sqrt{50}}\right) \quad \text{or} \quad 874.18 \text{ to } 885.82$$

The probability of accepting H_0, given $\mu = 870$, is equal to the area under the sampling distribution for the test statistic $\bar{x}$ in the interval from 874.18 to 885.82. Since $\bar{x}$ is normally distributed with a mean of 870 and SE $= 21/\sqrt{50} = 2.97$, β is equal to the area under the normal curve with $\mu = 870$ located between 874.18 and 885.82 (see Figure 9.8). Calculating the z-values corresponding to 874.18 and 885.82, you get

FIGURE 9.8

Calculating β in Example 9.8

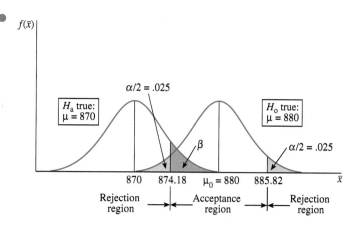

$$z_1 \approx \frac{\bar{x} - \mu}{s/\sqrt{n}} = \frac{874.18 - 870}{21/\sqrt{50}} = 1.41$$

$$z_2 \approx \frac{\bar{x} - \mu}{s/\sqrt{n}} = \frac{885.82 - 870}{21/\sqrt{50}} = 5.33$$

Then

$$\beta = P(\text{accept } H_0 \text{ when } \mu = 870) = P(874.18 < \bar{x} < 885.82 \text{ when } \mu = 870)$$
$$= P(1.41 < z < 5.33)$$

You can see from Figure 9.8 that the area under the normal curve with $\mu = 870$ above $\bar{x} = 885.82$ (or $z = 5.33$) is negligible. Therefore,

$$\beta = P(z > 1.41)$$

From Table 3 in Appendix I you can find

$$\beta = 1 - .9207 = .0793$$

Hence, the power of the test is

$$1 - \beta = 1 - .0793 = .9207$$

The probability of correctly rejecting H_0, given that μ is really equal to 870, is .9207, or approximately 92 chances in 100.

(MY) **APPLET**

You can use the **Power of a z-Test** applet to calculate the power for the hypothesis test in Example 9.8, and also for the same test when the sample size is changed. Refer to Figure 9.9. The applet in Figure 9.9 shows a sample size of $n = 50$. The slider at the bottom of the applet allows you to change the true value of μ; the power is recalculated as the mean changes. What is the true value of μ and the power of the test shown in the applet? Compare this to the value found in Table 9.2. The slider on the left side of the applet allows you to change α, and the slider on the right allows you to change the sample size n. Remember that n *must be* ≥ 30 for the z-test to be appropriate. You will use these applets to explore power using the MyApplet Exercises at the end of the chapter.

FIGURE 9.9

Power of a z-Test applet

Values of $(1 - \beta)$ can be calculated for various values of μ_a different from $\mu_0 = 880$ to measure the power of the test. For example, if $\mu_a = 885$,

$$\begin{aligned} \beta &= P(874.18 < \bar{x} < 885.82 \text{ when } \mu = 885) \\ &= P(-3.64 < z < .28) \\ &= .6103 - 0 = .6103 \end{aligned}$$

and the power is $(1 - \beta) = .3897$. Table 9.2 shows the power of the test for various values of μ_a, and a power curve is graphed in Figure 9.10. Note that the power of the test increases as the distance between μ_a and μ_0 increases. The result is a U-shaped curve for this two-tailed test.

TABLE 9.2

Value of $(1 - \beta)$ for Various Values of μ_a for Example 9.8

μ_a	$(1 - \beta)$	μ_a	$(1 - \beta)$
865	.9990	883	.1726
870	.9207	885	.3897
872	.7673	888	.7673
875	.3897	890	.9207
877	.1726	895	.9990
880	.0500		

FIGURE 9.10

Power curve for Example 9.8

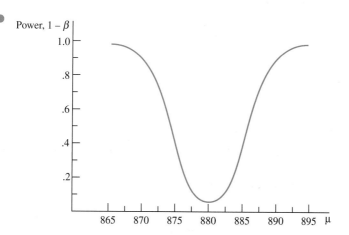

There are many important links among the two error rates, α and β, the power, $(1 - \beta)$, and the sample size, n. Look at the two curves shown in Figure 9.8.

- If α (the sum of the two tail areas in the curve on the right) is increased, the shaded area corresponding to β decreases, and vice versa.
- The only way to decrease β for a fixed α is to "buy" more information—that is, increase the sample size n.

What would happen to the area β as the curve on the left is moved closer to the curve on the right ($\mu = 880$)? With the rejection region in the right curve fixed, the value of β will *increase*. What effect does this have on the power of the test? Look at Figure 9.10.

You may also want to use the **Power of a z-Test** applet to help you visualize the following statements:

- As the distance between the true (μ_a) and hypothesized (μ_0) values of the mean increases, the power ($1 - \beta$) increases. The test is better at detecting *differences* when the distance is *large*.
- The closer the true value (μ_a) gets to the hypothesized value (μ_0), the less power ($1 - \beta$) the test has to detect the difference.
- The only way to increase the power ($1 - \beta$) for a fixed α is to "buy" more information—that is, increase the sample size, n.

The experimenter must decide on the values of α and β—measuring the risks of the possible errors he or she can tolerate. He or she also must decide how much power is needed to detect differences that are practically important in the experiment. Once these decisions are made, the sample size can be chosen by consulting the power curves corresponding to various sample sizes for the chosen test.

 PERSONAL TRAINER

How Do I Calculate β?

1. Find the critical value or values of $\bar{x}$ used to separate the acceptance and rejection regions.

2. Using one or more values for μ consistent with the alternative hypothesis H_a, calculate the probability that the sample mean $\bar{x}$ falls in the *acceptance region*. This produces the value $\beta = P(\text{accept } H_a \text{ when } \mu = \mu_a)$.

3. Remember that the **power** of the test is ($1 - \beta$).

9.3 EXERCISES

EXERCISE REPS

9.1 Critical Value Approach Fill in the blanks in the table below.

Test Statistic	Significance Level	One or Two-Tailed Test?	Critical Value	Rejection Region	Conclusion
$z = 0.88$	$\alpha = .05$	Two-tailed			
$z = -2.67$	$\alpha = .05$	One-tailed (lower)			
$z = 5.05$	$\alpha = .01$	Two-tailed			
$z = -1.22$	$\alpha = .01$	One-tailed (lower)			

9.2 *p*-value Approach Fill in the blanks in the table below.

Test Statistic	Significance Level	One or Two-Tailed Test?	*p*-Value	*p*-Value $< \alpha$?	Conclusion
$z = 3.01$	$\alpha = .05$	Two-tailed			
$z = 2.47$	$\alpha = .05$	One-tailed (upper)			
$z = -1.30$	$\alpha = .01$	Two-tailed			
$z = -2.88$	$\alpha = .01$	One-tailed (lower)			

BASIC TECHNIQUES

9.3 Find the appropriate rejection regions for the large-sample test statistic z in these cases:

a. A right-tailed test with $\alpha = .01$

b. A two-tailed test at the 5% significance level

c. A left-tailed test at the 1% significance level

d. A two-tailed test with $\alpha = .01$

9.4 Find the p-value for the following large-sample z tests:

a. A right-tailed test with observed $z = 1.15$

b. A two-tailed test with observed $z = -2.78$

c. A left-tailed test with observed $z = -1.81$

9.5 For the three tests given in Exercise 9.4, use the p-value to determine the significance of the results. Explain what "statistically significant" means in terms of rejecting or accepting H_0 and H_a.

9.6 A random sample of $n = 35$ observations from a quantitative population produced a mean $\bar{x} = 2.4$ and a standard deviation $s = .29$. Suppose your research objective is to show that the population mean μ exceeds 2.3.

a. Give the null and alternative hypotheses for the test.

b. Locate the rejection region for the test using a 5% significance level.

c. Find the standard error of the mean.

d. Before you conduct the test, use your intuition to decide whether the sample mean $\bar{x} = 2.4$ is likely or unlikely, assuming that $\mu = 2.3$. Now conduct the test. Do the data provide sufficient evidence to indicate that $\mu > 2.3$?

9.7 Refer to Exercise 9.6.

a. Calculate the p-value for the test statistic in part d.

b. Use the p-value to draw a conclusion at the 5% significance level.

c. Compare the conclusion in part b with the conclusion reached in part d of Exercise 9.6. Are they the same?

9.8 Refer to Exercise 9.6. You want to test $H_0 : \mu = 2.3$ against $H_a : \mu > 2.3$.

a. Find the critical value of $\bar{x}$ used for rejecting H_0.

b. Calculate $\beta = P(\text{accept } H_0 \text{ when } \mu = 2.4)$.

c. Repeat the calculation of β for $\mu = 2.3$, 2.5, and 2.6.

d. Use the values of β from parts b and c to graph the power curve for the test.

9.9 A random sample of 100 observations from a quantitative population produced a sample mean of 26.8 and a sample standard deviation of 6.5. Use the p-value approach to determine whether the population mean is different from 28. Explain your conclusions.

APPLICATIONS

9.10 Airline Occupancy Rates High airline occupancy rates on scheduled flights are essential to corporate profitability. Suppose a scheduled flight must average at least 60% occupancy in order to be profitable, and an examination of the occupancy rate for 120 10:00 A.M. flights from Atlanta to Dallas showed a mean occupancy per flight of 58% and a standard deviation of 11%.

a. If μ is the mean occupancy per flight and if the company wishes to determine whether or not this scheduled flight is unprofitable, give the alternative and the null hypotheses for the test.

b. Does the alternative hypothesis in part a imply a one- or two-tailed test? Explain.

c. Do the occupancy data for the 120 flights suggest that this scheduled flight is unprofitable? Test using $\alpha = .05$.

9.11 Hamburger Meat Exercise 8.33 involved the meat department of a local supermarket chain that packages ground beef in trays of two sizes. The smaller tray is intended to hold 1 pound of meat. A random sample of 35 packages in the smaller meat tray produced weight measurements with an average of 1.01 pounds and a standard deviation of .18 pound.

a. If you were the quality control manager and wanted to make sure that the average amount of ground beef was indeed 1 pound, what hypotheses would you test?

b. Find the p-value for the test and use it to perform the test in part a.

c. How would you, as the quality control manager, report the results of your study to a consumer interest group?

9.12 Advertising at the Movies "Welcome to the new movie pre-show!" Before you can see the newly released movie you have just paid to see, you must sit through a variety of trivia slides, snack bar ads, paid product advertising, and movie trailers. Although the total barrage of advertising may last up to 20 minutes or more, a particular theater chain claims that the average length of any one advertisement is no more than 3 minutes.[1] To test this claim, 50 theater advertisements were randomly selected and found to have an average duration of 3 minutes 15 seconds with a standard deviation of 30 seconds. Do the data provide sufficient evidence to indicate that the average duration of theater ads is more than that claimed by the theater? Test at the 1% level of significance. (HINT: Change "seconds" to fractions of a "minute.")

9.13 Potency of an Antibiotic A drug manufacturer claimed that the mean potency of one of its antibiotics was 80%. A random sample of $n = 100$ capsules were tested and produced a sample mean of $\bar{x} = 79.7\%$, with a standard deviation of $s = .8\%$. Do the data present sufficient evidence to refute the manufacturer's claim? Let $\alpha = .05$.

a. State the null hypothesis to be tested.

b. State the alternative hypothesis.

c. Conduct a statistical test of the null hypothesis and state your conclusion.

9.14 Flextime Many companies are becoming involved in *flextime*, in which a worker schedules his or her own work hours or compresses work weeks. A company that was contemplating the installation of a flextime schedule estimated that it needed a minimum mean of 7 hours per day per assembly worker in order to operate effectively. Each of a random sample of 80 of the company's assemblers was asked to submit a tentative flextime schedule. If the mean number of hours per day for Monday was 6.7 hours and the standard deviation was 2.7 hours, do the data provide sufficient evidence to indicate that the mean number of hours worked per day on Mondays, for all of the company's assemblers, will be less than 7 hours? Test using $\alpha = .05$.

9.15 Princeton Review In Exercise 1.44, we examined an advertising flyer for the *Princeton Review*, a review course designed for high school students taking the SAT tests.[2] The flyer claimed that the average score improvements for students who have taken the *Princeton Review* course is between 110 and 160 points. Are the claims made by the *Princeton Review* advertisers exaggerated? That is, is the average score improvement less than 110, the minimum claimed in the advertising flyer? A random sample of 100 students who took the *Princeton Review* course achieved an average score improvement of 107 points with a standard deviation of 13 points.

a. Use the *p*-value approach to test the *Princeton Review* claim. At which significance levels can you reject H_0?

b. If you were a competitor of the *Princeton Review*, how would you state your conclusions to put your company in the best possible light?

c. If you worked for the *Princeton Review*, how would you state your conclusions to protect your company's reputation?

9.16 What's Normal? What *is* normal, when it comes to people's body temperatures? A random sample of 130 human body temperatures, provided by Allen Shoemaker[3] in the *Journal of Statistical Education,* had a mean of 98.25 degrees and a standard deviation of 0.73 degrees. Does the data indicate that the average body temperature for healthy humans is different from 98.6 degrees, the usual average temperature cited by physicians and others? Test using both methods given in this section.

a. Use the *p*-value approach with $\alpha = .05$.

b. Use the critical value approach with $\alpha = .05$.

c. Compare the conclusions from parts a and b. Are they the same?

d. The 98.6 standard was derived by a German doctor in 1868, who claimed to have recorded 1 million temperatures in the course of his research.[4] What conclusions can you draw about his research in light of your conclusions in parts a and b?

9.17 Sports and Achilles Tendon Injuries Some sports that involve a significant amount of running, jumping, or hopping put participants at risk for Achilles tendinopathy (AT), an inflammation and thickening of the Achilles tendon. A study in *The American Journal of Sports Medicine* looked at the diameter (in mm) of the affected tendons for patients who participated in these types of sports activities.[5] Suppose that the Achilles tendon diameters in the general population have a mean of 5.97 millimeters (mm). When the diameters of the affected tendon were measured for a random sample of 31 patients, the average diameter was 9.80 with a standard deviation of 1.95 mm. Is there sufficient evidence to indicate that the average diameter of the tendon for patients with AT is greater than 5.97 mm? Test at the 5% level of significance.

A LARGE-SAMPLE TEST OF HYPOTHESIS FOR THE DIFFERENCE BETWEEN TWO POPULATION MEANS

9.4

In many situations, the statistical question to be answered involves a comparison of two population means. For example, the U.S. Postal Service is interested in reducing its massive 350 million gallons/year gasoline bill by replacing gasoline-powered trucks with electric-powered trucks. To determine whether significant savings in operating costs are achieved by changing to electric-powered trucks, a pilot study should be undertaken using, say, 100 conventional gasoline-powered mail trucks and 100 electric-powered mail trucks operated under similar conditions.

The statistic that summarizes the sample information regarding the difference in population means ($\mu_1 - \mu_2$) is the difference in sample means ($\bar{x}_1 - \bar{x}_2$). Therefore, in testing whether the difference in sample means indicates that the true difference in population means differs from a specified value, ($\mu_1 - \mu_2$) = D_0, you can use the standard error of ($\bar{x}_1 - \bar{x}_2$):

$$\sqrt{\frac{\sigma_1^2}{n_1} + \frac{\sigma_2^2}{n_2}} \quad \text{estimated by} \quad SE = \sqrt{\frac{s_1^2}{n_1} + \frac{s_2^2}{n_2}}$$

in the form of a z-statistic to measure how many standard deviations the difference ($\bar{x}_1 - \bar{x}_2$) lies from the hypothesized difference D_0. The formal testing procedure is described next.

LARGE-SAMPLE STATISTICAL TEST FOR ($\mu_1 - \mu_2$)

1. Null hypothesis: $H_0 : (\mu_1 - \mu_2) = D_0$, where D_0 is some specified difference that you wish to test. For many tests, you will hypothesize that there is no difference between μ_1 and μ_2; that is, $D_0 = 0$.

2. Alternative hypothesis:

   One-Tailed Test   **Two-Tailed Test**

   $H_a : (\mu_1 - \mu_2) > D_0$ $H_a : (\mu_1 - \mu_2) \neq D_0$
   [or $H_a : (\mu_1 - \mu_2) < D_0$]

3. Test statistic: $z \approx \dfrac{(\bar{x}_1 - \bar{x}_2) - D_0}{SE} = \dfrac{(\bar{x}_1 - \bar{x}_2) - D_0}{\sqrt{\dfrac{s_1^2}{n_1} + \dfrac{s_2^2}{n_2}}}$

4. Rejection region: Reject H_0 when

   One-Tailed Test   **Two-Tailed Test**

   $z > z_\alpha$ $z > z_{\alpha/2}$ or $z < -z_{\alpha/2}$
   [or $z < -z_\alpha$ when the
   alternative hypothesis is
   $H_a : (\mu_1 - \mu_2) < D_0$]

   or when p-value $< \alpha$

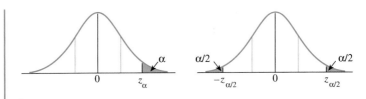

Assumptions: The samples are randomly and independently selected from the two populations and $n_1 \geq 30$ and $n_2 \geq 30$.

EXAMPLE 9.9

To determine whether car ownership affects a student's academic achievement, two random samples of 100 male students were each drawn from the student body. The grade point average for the $n_1 = 100$ non-owners of cars had an average and variance equal to $\bar{x}_1 = 2.70$ and $s_1^2 = .36$, while $\bar{x}_2 = 2.54$ and $s_2^2 = .40$ for the $n_2 = 100$ car owners. Do the data present sufficient evidence to indicate a difference in the mean achievements between car owners and nonowners of cars? Test using $\alpha = .05$.

Solution To detect a difference, if it exists, between the mean academic achievements for non-owners of cars μ_1 and car owners μ_2, you will test the null hypothesis that there is no difference between the means against the alternative hypothesis that $(\mu_1 - \mu_2) \neq 0$; that is,

$$H_0 : (\mu_1 - \mu_2) = D_0 = 0 \quad \text{versus} \quad H_a : (\mu_1 - \mu_2) \neq 0$$

Substituting into the formula for the test statistic, you get

$$z \approx \frac{(\bar{x}_1 - \bar{x}_2) - D_0}{\sqrt{\dfrac{s_1^2}{n_1} + \dfrac{s_2^2}{n_2}}} = \frac{2.70 - 2.54}{\sqrt{\dfrac{.36}{100} + \dfrac{.40}{100}}} = 1.84$$

MY TIP

|test statistic| > |critical value| ⟺ reject H_0

- **The critical value approach:** Using a two-tailed test with significance level $\alpha = .05$, you place $\alpha/2 = .025$ in each tail of the z distribution and reject H_0 if $z > 1.96$ or $z < -1.96$. Since $z = 1.84$ does not exceed 1.96 and is not less than -1.96, H_0 cannot be rejected (see Figure 9.11). That is, there is insufficient evidence to declare a difference in the average academic achievements for the two groups. Remember that you should not be willing to *accept* H_0—declare the two means to be the same—until β is evaluated for some meaningful values of $(\mu_1 - \mu_2)$.

FIGURE 9.11

Rejection region and p-value for Example 9.9

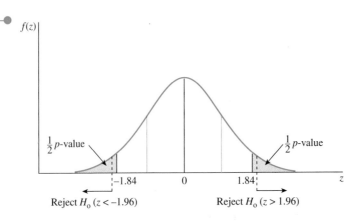

- **The p-value approach:** Calculate the p-value, the probability that z is greater than $z = 1.84$ plus the probability that z is less than $z = -1.84$, as shown in Figure 9.11:

$$p\text{-value} = P(z > 1.84) + P(z < -1.84) = (1 - .9671) + .0329 = .0658$$

The p-value lies between .10 and .05, so you can reject H_0 at the .10 level but not at the .05 level of significance. Since the p-value of .0658 exceeds the specified significance level $\alpha = .05$, H_0 cannot be rejected. Again, you should not be willing to *accept* H_0 until β is evaluated for some meaningful values of $(\mu_1 - \mu_2)$.

Hypothesis Testing and Confidence Intervals

Whether you use the critical value or the p-value approach for testing hypotheses about $(\mu_1 - \mu_2)$, you will always reach the same conclusion because the calculated value of the test statistic and the critical value are related *exactly* in the same way that the p-value and the significance level α are related. You might remember that the confidence intervals constructed in Chapter 8 could also be used to answer questions about the difference between two population means. In fact, for a two-tailed test, the $(1 - \alpha)100\%$ confidence interval for the parameter of interest can be used to test its value, just as you did informally in Chapter 8. The value of α indicated by the confidence coefficient in the confidence interval is equivalent to the significance level α in the statistical test. For a one-tailed test, the equivalent confidence interval approach would use the one-sided confidence bounds in Section 8.8 with confidence coefficient α. In addition, by using the confidence interval approach, you gain a range of possible values for the parameter of interest, regardless of the outcome of the test of hypothesis.

- If the confidence interval you construct *contains* the value of the parameter specified by H_0, then that value is one of the likely or possible values of the parameter and H_0 should not be rejected.
- If the hypothesized value *lies outside* of the confidence limits, the null hypothesis is rejected at the α level of significance.

EXAMPLE 9.10

Construct a 95% confidence interval for the difference in average academic achievements between car owners and non-owners. Using the confidence interval, can you conclude that there is a difference in the population means for the two groups of students?

Solution For the large-sample statistics discussed in Chapter 8, the 95% confidence interval is given as

Point estimator $\pm$ 1.96 $\times$ (Standard error of the estimator)

For the difference in two population means, the confidence interval is approximated as

$$(\bar{x}_1 - \bar{x}_2) \pm 1.96 \sqrt{\frac{s_1^2}{n_1} + \frac{s_2^2}{n_2}}$$

$$(2.70 - 2.54) \pm 1.96 \sqrt{\frac{.36}{100} + \frac{.40}{100}}$$

$$.16 \pm .17$$

or $-.01 < (\mu_1 - \mu_2) < .33$. This interval gives you a range of possible values for the difference in the population means. Since the hypothesized difference, $(\mu_1 - \mu_2) = 0$, is contained in the confidence interval, you should not reject H_0. Look at the signs of the possible values in the confidence interval. You cannot tell from the interval whether the difference in the means is negative $(-)$, positive $(+)$, or zero (0)—the latter of the three would indicate that the two means are the same. Hence, you can really reach no conclusion in terms of the question posed. There is not enough evidence to indicate that there is a difference in the average achievements for car owners versus non-owners. The conclusion is the same one reached in Example 9.9.

9.4 EXERCISES

BASIC TECHNIQUES

9.18 Independent random samples of 80 measurements were drawn from two quantitative populations, 1 and 2. Here is a summary of the sample data:

	Sample 1	Sample 2
Sample size	80	80
Sample mean	11.6	9.7
Sample variance	27.9	38.4

a. If your research objective is to show that μ_1 is larger than μ_2, state the alternative and the null hypotheses that you would choose for a statistical test.

b. Is the test in part a one- or two-tailed?

c. Calculate the test statistic that you would use for the test in part a. Based on your knowledge of the standard normal distribution, is this a likely or unlikely observation, assuming that H_0 is true and the two population means are the same?

d. *p-value approach:* Find the *p*-value for the test. Test for a significant difference in the population means at the 1% significance level.

e. *Critical value approach:* Find the rejection region when $\alpha = .01$. Do the data provide sufficient evidence to indicate a difference in the population means?

9.19 Independent random samples of 36 and 45 observations are drawn from two quantitative populations, 1 and 2, respectively. The sample data summary is shown here:

	Sample 1	Sample 2
Sample size	36	45
Sample mean	1.24	1.31
Sample variance	.0560	.0540

Do the data present sufficient evidence to indicate that the mean for population 1 is smaller than the mean for population 2? Use one of the two methods of testing presented in this section, and explain your conclusions.

9.20 Suppose you wish to detect a difference between μ_1 and μ_2 (either $\mu_1 > \mu_2$ or $\mu_1 < \mu_2$) and, instead of running a two-tailed test using $\alpha = .05$, you use the following test procedure. You wait until you have collected the sample data and have calculated $\bar{x}_1$ and $\bar{x}_2$. If $\bar{x}_1$ is larger than $\bar{x}_2$, you choose the alternative hypothesis $H_a : \mu_1 > \mu_2$ and run a one-tailed test placing $\alpha_1 = .05$ in the upper tail of the z distribution. If, on the other hand, $\bar{x}_2$ is larger than $\bar{x}_1$, you reverse the procedure and run a one-tailed test, placing $\alpha_2 = .05$ in the lower tail of the z distribution. If you use this procedure and if μ_1 actually equals μ_2, what is the probability α that you will conclude that μ_1 is not equal to μ_2 (i.e., what is the probability α that you will incorrectly reject H_0 when H_0 is true)? This exercise demonstrates why statistical tests should be formulated *prior* to observing the data.

APPLICATIONS

9.21 Cure for the Common Cold? An experiment was planned to compare the mean time (in days) required to recover from a common cold for persons given a daily dose of 4 milligrams (mg) of vitamin C versus those who were not given a vitamin supplement. Suppose that 35 adults were randomly selected for each treatment category and that the mean recovery times and standard deviations for the two groups were as follows:

	No Vitamin Supplement	4 mg Vitamin C
Sample size	35	35
Sample mean	6.9	5.8
Sample standard deviation	2.9	1.2

a. Suppose your research objective is to show that the use of vitamin C reduces the mean time required to recover from a common cold and its complications. Give the null and alternative hypotheses for the test. Is this a one- or a two-tailed test?

b. Conduct the statistical test of the null hypothesis in part a and state your conclusion. Test using $\alpha = .05$.

9.22 Healthy Eating Americans are becoming more conscious about the importance of good nutrition, and some researchers believe we may be altering our diets to include less red meat and more fruits and vegetables. To test the theory that the consumption of red meat has decreased over the last 10 years, a researcher decides to select hospital nutrition records for 400 subjects surveyed 10 years ago and to compare their average amount of beef consumed per year to amounts consumed by an equal number of subjects interviewed this year. The data are given in the table.

	Ten Years Ago	This Year
Sample mean	73	63
Sample standard deviation	25	28

a. Do the data present sufficient evidence to indicate that per-capita beef consumption has decreased in the last 10 years? Test at the 1% level of significance.

b. Find a 99% lower confidence bound for the difference in the average per-capita beef consumptions for the two groups. (This calculation was done as part of Exercise 8.76.) Does your confidence bound confirm your conclusions in part a? Explain. What additional information does the confidence bound give you?

9.23 Lead Levels in Drinking Water Analyses of drinking water samples for 100 homes in each of two different sections of a city gave the following means and standard deviations of lead levels (in parts per million):

	Section 1	Section 2
Sample size	100	100
Mean	34.1	36.0
Standard deviation	5.9	6.0

a. Calculate the test statistic and its p-value (observed significance level) to test for a difference in the two

population means. Use the p-value to evaluate the statistical significance of the results at the 5% level.

b. Use a 95% confidence interval to estimate the difference in the mean lead levels for the two sections of the city.

c. Suppose that the city environmental engineers will be concerned only if they detect a difference of more than 5 parts per million in the two sections of the city. Based on your confidence interval in part b, is the statistical significance in part a of *practical significance* to the city engineers? Explain.

9.24 Starting Salaries, again In an attempt to compare the starting salaries for college graduates who majored in education and the social sciences (see Exercise 8.45), random samples of 50 recent college graduates in each major were selected and the following information was obtained:

Major	Mean	SD
Education	40,554	2225
Social science	38,348	2375

a. Do the data provide sufficient evidence to indicate a difference in average starting salaries for college graduates who majored in education and the social sciences? Test using $\alpha = .05$.

b. Compare your conclusions in part a with the results of part b in Exercise 8.45. Are they the same? Explain.

9.25 Hotel Costs In Exercise 8.18, we explored the average cost of lodging at three different hotel chains.[6] We randomly select 50 billing statements from the computer databases of the Marriott, Radisson, and Wyndham hotel chains, and record the nightly room rates. A portion of the sample data is shown in the table.

	Marriott	Radisson
Sample average	$170	$145
Sample standard deviation	17.5	10

a. Before looking at the data, would you have any preconceived idea about the direction of the difference between the average room rates for these two hotels? If not, what null and alternative hypotheses should you test?

b. Use the *critical value* approach to determine if there is a significant difference in the average room rates for the Marriott and the Radisson hotel chains. Use $\alpha = .01$.

c. Find the p-value for this test. Does this p-value confirm the results of part b?

9.26 Hotel Costs II Refer to Exercise 9.25. The table below shows the sample data collected to compare the average room rates at the Wyndham and Radisson hotel chains.[6]

	Wyndham	Radisson
Sample average	$150	$145
Sample standard deviation	16.5	10

a. Do the data provide sufficient evidence to indicate a difference in the average room rates for the Wyndham and the Radisson hotel chains? Use $\alpha = .05$.

b. Construct a 95% confidence interval for the difference in the average room rates for the two chains. Does this interval confirm your conclusions in part a?

9.27 MMT in Gasoline The addition of MMT, a compound containing manganese (Mn), to gasoline as an octane enhancer has caused concern about human exposure to Mn because high intakes have been linked to serious health effects. In a study of ambient air concentrations of fine Mn, Wallace and Slonecker (*Journal of the Air and Waste Management Association*) presented the accompanying summary information about the amounts of fine Mn (in nanograms per cubic meter) in mostly rural national park sites and in mostly urban California sites.[7]

	National Parks	California
Mean	.94	2.8
Standard deviation	1.2	2.8
Number of sites	36	26

a. Is there sufficient evidence to indicate that the mean concentrations differ for these two types of sites at the $\alpha = .05$ level of significance? Use the large-sample z-test. What is the p-value of this test?

b. Construct a 95% confidence interval for $(\mu_1 - \mu_2)$. Does this interval confirm your conclusions in part a?

9.28 Noise and Stress In Exercise 8.48, you compared the effect of stress in the form of noise on the ability to perform a simple task. Seventy subjects were divided into two groups; the first group of 30 subjects acted as a control, while the second group of 40 was the experimental group. Although each subject performed the task in the same control room, each of the experimental group subjects had to perform the task while loud rock music was played. The time to finish the task was recorded for each subject and the following summary was obtained:

	Control	Experimental
n	30	40
$\bar{x}$	15 minutes	23 minutes
s	4 minutes	10 minutes

a. Is there sufficient evidence to indicate that the average time to complete the task was longer for the experimental "rock music" group? Test at the 1% level of significance.

b. Construct a 99% one-sided upper bound for the difference (control − experimental) in average times for the two groups. Does this interval confirm your conclusions in part a?

9.29 What's Normal II Of the 130 people in Exercise 9.16, 65 were female and 65 were male.[3] The means and standard deviations of their temperatures are shown below.

	Men	Women
Sample Mean	98.11	98.39
Standard deviation	0.70	0.74

a. Use the p-value approach to test for a significant difference in the average temperatures for males versus females.

b. Are the results significant at the 5% level? At the 1% level?

A LARGE-SAMPLE TEST OF HYPOTHESIS FOR A BINOMIAL PROPORTION

9.5

When a random sample of n identical trials is drawn from a binomial population, the sample proportion $\hat{p}$ has an approximately normal distribution when n is large, with mean p and standard error

$$SE = \sqrt{\frac{pq}{n}}$$

When you test a hypothesis about p, the proportion in the population possessing a certain attribute, the test follows the same general form as the large-sample tests in Sections 9.3 and 9.4. To test a hypothesis of the form

$$H_0 : p = p_0$$

versus a one- or two-tailed alternative

$$H_a : p > p_0 \quad \text{or} \quad H_a : p < p_0 \quad \text{or} \quad H_a : p \neq p_0$$

the test statistic is constructed using $\hat{p}$, the best estimator of the true population proportion p. The sample proportion $\hat{p}$ is standardized, using the hypothesized mean and standard error, to form a test statistic z, which has a standard normal distribution if H_0 is true. This large-sample test is summarized next.

LARGE-SAMPLE STATISTICAL TEST FOR p

1. Null hypothesis: $H_0 : p = p_0$
2. Alternative hypothesis:

One-Tailed Test	Two-Tailed Test
$H_a : p > p_0$	$H_a : p \neq p_0$
(or, $H_a : p < p_0$)	

3. Test statistic: $z = \dfrac{\hat{p} - p_0}{\text{SE}} = \dfrac{\hat{p} - p_0}{\sqrt{\dfrac{p_0 q_0}{n}}}$ with $\hat{p} = \dfrac{x}{n}$

   where x is the number of successes in n binomial trials.[†]

4. Rejection region: Reject H_0 when

   One-Tailed Test

   $z > z_\alpha$
   (or $z < -z_\alpha$ when the alternative hypothesis is $H_a : p < p_0$)

   or when p-value $< \alpha$

   Two-Tailed Test

   $z > z_{\alpha/2}$ or $z < -z_{\alpha/2}$

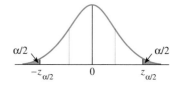

Assumption: The sampling satisfies the assumptions of a binomial experiment (see Section 5.2), and n is large enough so that the sampling distribution of $\hat{p}$ can be approximated by a normal distribution ($np_0 > 5$ and $nq_0 > 5$).

[†]An equivalent test statistic can be found by multiplying the numerator and denominator by z by n to obtain

$$z = \frac{x - np_0}{\sqrt{np_0 q_0}}$$

EXAMPLE 9.11

Regardless of age, about 20% of American adults participate in fitness activities at least twice a week. However, these fitness activities change as the people get older, and occasionally participants become nonparticipants as they age. In a local survey of $n = 100$ adults over 40 years old, a total of 15 people indicated that they participated in a fitness activity at least twice a week. Do these data indicate that the participation rate for adults over 40 years of age is significantly less than the 20% figure? Calculate the p-value and use it to draw the appropriate conclusions.

Solution   Assuming that the sampling procedure satisfies the requirements of a binomial experiment, you can answer the question posed using a one-tailed test of hypothesis:

$$H_0 : p = .2 \quad \text{versus} \quad H_a : p < .2$$

Begin by assuming that H_0 is true—that is, the true value of p is $p_0 = .2$. Then $\hat{p} = x/n$ will have an approximate normal distribution with mean p_0 and standard error $\sqrt{p_0 q_0/n}$. [This is different from the estimation procedure in which the unknown standard error is estimated by $\sqrt{\hat{p}\hat{q}/n}$.] The observed value of $\hat{p}$ is $15/100 = .15$ and the test statistic is

MY TIP

p-value $\le \alpha \Leftrightarrow$ reject H_0

p-value $> \alpha \Leftrightarrow$ do not reject H_0

$$z = \frac{\hat{p} - p_0}{\sqrt{\dfrac{p_0 q_0}{n}}} = \frac{.15 - .20}{\sqrt{\dfrac{(.20)(.80)}{100}}} = -1.25$$

The p-value associated with this test is found as the area under the standard normal curve to the left of $z = -1.25$ as shown in Figure 9.12. Therefore,

$$p\text{-value} = P(z < -1.25) = .1056$$

FIGURE 9.12

p-value for Example 9.11

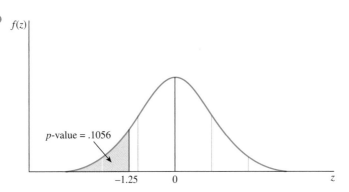

If you use the guidelines for evaluating p-values, then .1056 is greater than .10, and you would not reject H_0. There is insufficient evidence to conclude that the percentage of adults over age 40 who participate in fitness activities twice a week is less than 20%.

Statistical Significance and Practical Importance

It is important to understand the difference between results that are "significant" and results that are practically "important." In statistical language, the word *significant* does not necessarily mean "important," but only that the results could not have occurred by chance. For example, suppose that in Example 9.11, the researcher had used

$n = 400$ adults in her experiment and had observed the same sample proportion. The test statistic is now

$$z = \frac{\hat{p} - p_0}{\sqrt{\dfrac{p_0 q_0}{n}}} = \frac{.15 - .20}{\sqrt{\dfrac{(.20)(.80)}{400}}} = -2.50$$

with

$$p\text{-value} = P(z < -2.50) = .0062$$

Now the results are *highly significant:* H_0 is rejected, and there is sufficient evidence to indicate that the percentage of adults over age 40 who participate in physical fitness activities is less than 20%. However, is this drop in activity really *important?* Suppose that physicians would be concerned only about a drop in physical activity of more than 10%. If there had been a drop of more than 10% in physical activity, this would imply that the true value of p was less than .10. What is the largest possible value of p? Using a 95% upper one-sided confidence bound, you have

$$\hat{p} + 1.645 \sqrt{\frac{\hat{p}\hat{q}}{n}}$$

$$.15 + 1.645 \sqrt{\frac{(.15)(.85)}{400}}$$

$$.15 + .029$$

or $p < .179$. The physical activity for adults aged 40 and older has dropped from 20%, but you cannot say that it has dropped below 10%. So, the results, although *statistically significant,* are not *practically important.*

In this book, you will learn how to determine whether results are statistically significant. When you use these procedures in a practical situation, however, you must also make sure the results are practically important.

 9.5 **EXERCISES**

BASIC TECHNIQUES

9.30 A random sample of $n = 1000$ observations from a binomial population produced $x = 279$.

a. If your research hypothesis is that p is less than .3, what should you choose for your alternative hypothesis? Your null hypothesis?

b. What is the critical value that determines the rejection region for your test with $\alpha = .05$?

c. Do the data provide sufficient evidence to indicate that p is less than .3? Use a 5% significance level.

9.31 A random sample of $n = 1400$ observations from a binomial population produced $x = 529$.

a. If your research hypothesis is that p differs from .4, what hypotheses should you test?

b. Calculate the test statistic and its p-value. Use the p-value to evaluate the statistical significance of the results at the 1% level.

c. Do the data provide sufficient evidence to indicate that p is different from .4?

9.32 A random sample of 120 observations was selected from a binomial population, and 72 successes were observed. Do the data provide sufficient evidence to indicate that p is greater than .5? Use one of the two methods of testing presented in this section, and explain your conclusions.

APPLICATIONS

9.33 TV Habits An article in *American Demographics* reports that approximately 60% of U.S. households have two or more television sets and that at least half of Americans sometimes watch television alone.[8] Suppose that $n = 75$ U.S. households are sampled, and of those sampled, 49 had two or more television sets and 35 respondents sometimes watch television alone.

a. Two claims can be tested using the sample information. What are the two sets of hypotheses to be tested?

b. Do the data present sufficient evidence to contradict the claim that at least half of Americans sometimes watch television alone?

c. Do the data present sufficient evidence to show that the 60% figure claimed in the magazine article is incorrect?

9.34 Plant Genetics A peony plant with red petals was crossed with another plant having streaky petals. A geneticist states that 75% of the offspring resulting from this cross will have red flowers. To test this claim, 100 seeds from this cross were collected and germinated and 58 plants had red petals.

a. What hypothesis should you use to test the geneticist's claim?

b. Calculate the test statistic and its p-value. Use the p-value to evaluate the statistical significance of the results at the 1% level.

9.35 Early Detection of Breast Cancer Of those women who are diagnosed to have early-stage breast cancer, one-third eventually die of the disease. Suppose a community public health department instituted a screening program to provide for the early detection of breast cancer and to increase the survival rate p of those diagnosed to have the disease. A random sample of 200 women was selected from among those who were periodically screened by the program and who were diagnosed to have the disease. Let x represent the number of those in the sample who survive the disease.

a. If you wish to detect whether the community screening program has been effective, state the null hypothesis that should be tested.

b. State the alternative hypothesis.

c. If 164 women in the sample of 200 survive the disease, can you conclude that the community screening program was effective? Test using $\alpha = .05$ and explain the practical conclusions from your test.

d. Find the p-value for the test and interpret it.

9.36 Sweet Potato Whitefly Suppose that 10% of the fields in a given agricultural area are infested with the sweet potato whitefly. One hundred fields in this area are randomly selected, and 25 are found to be infested with whitefly.

a. Assuming that the experiment satisfies the conditions of the binomial experiment, do the data indicate that the proportion of infested fields is greater than expected? Use the p-value approach, and test using a 5% significance level.

b. If the proportion of infested fields is found to be significantly greater than .10, why is this of practical significance to the agronomist? What practical conclusions might she draw from the results?

9.37 Brown or Blue? An article in the *Washington Post* stated that nearly 45% of the U.S. population is born with brown eyes, although they don't necessarily stay that way.[9] To test the newspaper's claim, a random sample of 80 people was selected, and 32 had brown eyes. Is there sufficient evidence to dispute the newspaper's claim regarding the proportion of brown-eyed people in the United States? Use $\alpha = .01$.

9.38 Colored Contacts Refer to Exercise 9.37. Contact lenses, worn by about 26 million Americans, come in many styles and colors. Most Americans wear soft lenses, with the most popular colors being the blue varieties (25%), followed by greens (24%), and then hazel or brown. A random sample of 80 tinted contact lens wearers was checked for the color of their lenses. Of these people, 22 wore blue lenses and only 15 wore green lenses.[9]

a. Do the sample data provide sufficient evidence to indicate that the proportion of tinted contact lens wearers who wear blue lenses is different from 25%? Use $\alpha = .05$.

b. Do the sample data provide sufficient evidence to indicate that the proportion of tinted contact lens wearers who wear green lenses is different from 24%? Use $\alpha = .05$.

c. Is there any reason to conduct a one-tailed test for either part a or b? Explain.

9.39 A Cure for Insomnia An experimenter has prepared a drug-dose level that he claims will induce sleep for at least 80% of people suffering from insomnia. After examining the dosage we feel that his claims regarding the effectiveness of his dosage are inflated. In an attempt to disprove his claim, we administer his prescribed dosage to 50 insomniacs and observe that 37 of them have had sleep induced by the drug dose. Is there enough evidence to refute his claim at the 5% level of significance?

9.40 Rocking the Vote What proportion of adults in the United States always vote in the presidential elections? An article in *American Demographics* reports this percentage as 67%.[10] To test this claim, a random sample of 300 adults was taken, and 192 of them said that they always vote in presidential elections. Does this sample provide sufficient evidence to indicate that the percentage of adults who say that they always vote is different from their reported percentage? Test using $\alpha = .01$.

9.41 Man's Best Friend The Humane Society reports that there are approximately 68 million dogs owned in the United States and that 40% of all U.S. households own at least one dog.[11] In a random sample of 300 households, 114 households said that they owned at least one dog. Does this data provide sufficient evidence to indicate that the proportion of households with at least one dog is different from that reported by the Humane Society? Test using $\alpha = .05$.

A LARGE-SAMPLE TEST OF HYPOTHESIS FOR THE DIFFERENCE BETWEEN TWO BINOMIAL PROPORTIONS

9.6

When random and independent samples are selected from two *binomial* populations, the focus of the experiment may be the difference $(p_1 - p_2)$ in the proportions of individuals or items possessing a specified characteristic in the two populations. In this situation, you can use the difference in the sample proportions $(\hat{p}_1 - \hat{p}_2)$ along with its standard error,

$$\text{SE} = \sqrt{\frac{p_1 q_1}{n_1} + \frac{p_2 q_2}{n_2}}$$

in the form of a z-statistic to test for a significant difference in the two population proportions. The null hypothesis to be tested is usually of the form

$$H_0 : p_1 = p_2 \quad \text{or} \quad H_0 : (p_1 - p_2) = 0$$

MY TIP

Remember: Each trial results in one of two outcomes (S or F).

versus either a one- or two-tailed alternative hypothesis. The formal test of hypothesis is summarized in the next display. In estimating the standard error for the z-statistic, you should use the fact that when H_0 is true, the two population proportions are equal to some common value—say, p. To obtain the best estimate of this common value, the sample data are "pooled" and the estimate of p is

$$\hat{p} = \frac{\text{Total number of successes}}{\text{Total number of trials}} = \frac{x_1 + x_2}{n_1 + n_2}$$

Remember that, in order for the difference in the sample proportions to have an approximately normal distribution, the sample sizes must be large and the proportions should not be too close to 0 or 1.

LARGE-SAMPLE STATISTICAL TEST FOR $(p_1 - p_2)$

1. Null hypothesis: $H_0 : (p_1 - p_2) = 0$ or equivalently $H_0 : p_1 = p_2$
2. Alternative hypothesis:

One-Tailed Test   **Two-Tailed Test**

$$H_a : (p_1 - p_2) > 0 \qquad H_a : (p_1 - p_2) \neq 0$$
$$[\text{or } H_a : (p_1 - p_2) < 0]$$

3. Test statistic: $z = \dfrac{(\hat{p}_1 - \hat{p}_2) - 0}{\text{SE}} = \dfrac{\hat{p}_1 - \hat{p}_2}{\sqrt{\dfrac{p_1 q_1}{n_1} + \dfrac{p_2 q_2}{n_2}}} = \dfrac{\hat{p}_1 - \hat{p}_2}{\sqrt{\dfrac{pq}{n_1} + \dfrac{pq}{n_2}}}$

where $\hat{p}_1 = x_1/n_1$ and $\hat{p}_2 = x_2/n_2$. Since the common value of $p_1 = p_2 = p$ (used in the standard error) is unknown, it is estimated by

$$\hat{p} = \frac{x_1 + x_2}{n_1 + n_2}$$

and the test statistic is

$$z = \frac{(\hat{p}_1 - \hat{p}_2) - 0}{\sqrt{\dfrac{\hat{p}\hat{q}}{n_1} + \dfrac{\hat{p}\hat{q}}{n_2}}} \quad \text{or} \quad z = \frac{\hat{p}_1 - \hat{p}_2}{\sqrt{\hat{p}\hat{q}\left(\dfrac{1}{n_1} + \dfrac{1}{n_2}\right)}}$$

4. Rejection region: Reject H_0 when

One-Tailed Test   **Two-Tailed Test**

$z > z_\alpha$ $z > z_{\alpha/2}$ or $z < -z_{\alpha/2}$
[or $z < -z_\alpha$ when the
alternative hypothesis
is $H_a : (p_1 - p_2) < 0$]

or when p-value $< \alpha$

  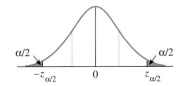

Assumptions: Samples are selected in a random and independent manner from two binomial populations, and n_1 and n_2 are large enough so that the sampling distribution of $(\hat{p}_1 - \hat{p}_2)$ can be approximated by a normal distribution. That is, $n_1\hat{p}_1$, $n_1\hat{q}_1$, $n_2\hat{p}_2$, and $n_2\hat{q}_2$ should all be greater than 5.

EXAMPLE 9.12 — The records of a hospital show that 52 men in a sample of 1000 men versus 23 women in a sample of 1000 women were admitted because of heart disease. Do these data present sufficient evidence to indicate a higher rate of heart disease among men admitted to the hospital? Use $\alpha = .05$.

Solution   Assume that the number of patients admitted for heart disease has an approximate binomial probability distribution for both men and women with parameters p_1 and p_2, respectively. Then, since you wish to determine whether $p_1 > p_2$, you will test the null hypothesis $p_1 = p_2$—that is, $H_0 : (p_1 - p_2) = 0$—against the alternative hypothesis $H_a : p_1 > p_2$ or, equivalently, $H_a : (p_1 - p_2) > 0$. To conduct this test, use the z-test statistic and approximate the standard error using the pooled estimate of p.

Since H_a implies a one-tailed test, you can reject H_0 only for large values of z. Thus, for $\alpha = .05$, you can reject H_0 if $z > 1.645$ (see Figure 9.13).

The pooled estimate of p required for the standard error is

$$\hat{p} = \frac{x_1 + x_2}{n_1 + n_2} = \frac{52 + 23}{1000 + 1000} = .0375$$

FIGURE 9.13

Location of the rejection region in Example 9.12

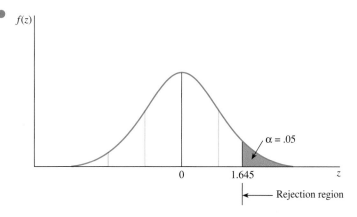

and the test statistic is

$$z = \frac{\hat{p}_1 - \hat{p}_2}{\sqrt{\hat{p}\hat{q}\left(\dfrac{1}{n_1} + \dfrac{1}{n_2}\right)}} = \frac{.052 - .023}{\sqrt{(.0375)(.9625)\left(\dfrac{1}{1000} + \dfrac{1}{1000}\right)}} = 3.41$$

Since the computed value of z falls in the rejection region, you can reject the hypothesis that $p_1 = p_2$. The data present sufficient evidence to indicate that the percentage of men entering the hospital because of heart disease is higher than that of women. (NOTE: This does not imply that the *incidence* of heart disease is higher in men. Perhaps fewer women enter the hospital when afflicted with the disease!)

How *much higher* is the proportion of men than women entering the hospital with heart disease? A 95% lower one-sided confidence bound will help you find the lowest likely value for the difference.

$$(\hat{p}_1 - \hat{p}_2) - 1.645 \sqrt{\frac{\hat{p}_1\hat{q}_1}{n_1} + \frac{\hat{p}_2\hat{q}_2}{n_2}}$$

$$(.052 - .023) - 1.645 \sqrt{\frac{.052(.948)}{1000} + \frac{.023(.977)}{1000}}$$

$$.029 - .014$$

or $(p_1 - p_2) > .015$. The proportion of men is roughly 1.5% higher than women. Is this of *practical importance*? This is a question for the researcher to answer.

In some situations, you may need to test for a difference D_0 (other than 0) between two binomial proportions. If this is the case, the test statistic is modified for testing

$H_0 : (p_1 - p_2) = D_0$, and a pooled estimate for a common p is no longer used in the standard error. The modified test statistic is

$$z = \frac{(\hat{p}_1 - \hat{p}_2) - D_0}{\sqrt{\dfrac{\hat{p}_1\hat{q}_1}{n_1} + \dfrac{\hat{p}_2\hat{q}_2}{n_2}}}$$

Although this test statistic is not used often, the procedure is no different from other large-sample tests you have already mastered!

9.6 EXERCISES

BASIC TECHNIQUES

9.42 Independent random samples of $n_1 = 140$ and $n_2 = 140$ observations were randomly selected from binomial populations 1 and 2, respectively. Sample 1 had 74 successes, and sample 2 had 81 successes.

a. Suppose you have no preconceived idea as to which parameter, p_1 or p_2, is the larger, but you want to detect only a difference between the two parameters if one exists. What should you choose as the alternative hypothesis for a statistical test? The null hypothesis?

b. Calculate the standard error of the difference in the two sample proportions, $(\hat{p}_1 - \hat{p}_2)$. Make sure to use the pooled estimate for the common value of p.

c. Calculate the test statistic that you would use for the test in part a. Based on your knowledge of the standard normal distribution, is this a likely or unlikely observation, assuming that H_0 is true and the two population proportions are the same?

d. *p-value approach:* Find the p-value for the test. Test for a significant difference in the population proportions at the 1% significance level.

e. *Critical value approach:* Find the rejection region when $\alpha = .01$. Do the data provide sufficient evidence to indicate a difference in the population proportions?

9.43 Refer to Exercise 9.42. Suppose, for practical reasons, you know that p_1 cannot be larger than p_2.

a. Given this knowledge, what should you choose as the alternative hypothesis for your statistical test? The null hypothesis?

b. Does your alternative hypothesis in part a imply a one- or two-tailed test? Explain.

c. Conduct the test and state your conclusions. Test using $\alpha = .05$.

9.44 Independent random samples of 280 and 350 observations were selected from binomial populations 1 and 2, respectively. Sample 1 had 132 successes, and sample 2 had 178 successes. Do the data present sufficient evidence to indicate that the proportion of successes in population 1 is smaller than the proportion in population 2? Use one of the two methods of testing presented in this section, and explain your conclusions.

APPLICATIONS

9.45 Treatment versus Control An experiment was conducted to test the effect of a new drug on a viral infection. The infection was induced in 100 mice, and the mice were randomly split into two groups of 50. The first group, the *control group,* received no treatment for the infection. The second group received the drug. After a 30-day period, the proportions of survivors, $\hat{p}_1$ and $\hat{p}_2$, in the two groups were found to be .36 and .60, respectively.

a. Is there sufficient evidence to indicate that the drug is effective in treating the viral infection? Use $\alpha = .05$.

b. Use a 95% confidence interval to estimate the actual difference in the cure rates for the treated versus the control groups.

9.46 Movie Marketing Marketing to targeted age groups has become a standard method of advertising, even in movie theater advertising. Advertisers use computer software to track the demographics of moviegoers and then decide on the type of products to advertise before a particular movie.[1] One statistic that might be of interest is how frequently adults with children under 18 attend movies as compared to those without children. Suppose that a theater database is used to randomly select 1000 adult ticket purchasers. These adults are then surveyed and asked whether they were frequent moviegoers—that is, do they attend movies 12 or more times a year? The results are shown in the table:

	With Children under 18	Without Children
Sample size	440	560
Number who attend 12$^+$ times per year	123	145

a. Is there a significant difference in the population proportions of frequent moviegoers in these two demographic groups? Use $\alpha = .01$.

b. Why would a statistically significant difference in these population proportions be of *practical importance* to the advertiser?

9.47 M&Ms In Exercise 8.53, you investigated whether Mars, Inc., uses the same proportion of red M&Ms® in its plain and peanut varieties. Random samples of plain and peanut M&Ms provide the following sample data for the experiment:

	Plain	Peanut
Sample size	56	32
Number of red M&Ms	12	8

Use a test of hypothesis to determine whether there is a significant difference in the proportions of red candies for the two types of M&Ms. Let $\alpha = .05$ and compare your results with those of Exercise 8.53.

9.48 Hormone Therapy and Alzheimer's Disease In the last few years, many research studies have shown that the purported benefits of hormone replacement therapy (HRT) do not exist, and in fact, that hormone replacement therapy actually increases the risk of several serious diseases. A four-year experiment involving 4532 women, reported in *The Press Enterprise*, was conducted at 39 medical centers. Half of the women took placebos and half took *Prempro*, a widely prescribed type of hormone replacement therapy. There were 40 cases of dementia in the hormone group and 21 in the placebo group.[12] Is there sufficient evidence to indicate that the risk of dementia is higher for patients using *Prempro*? Test at the 1% level of significance.

9.49 HRT, continued Refer to Exercise 9.48. Calculate a 99% lower one-sided confidence bound for the difference in the risk of dementia for women using hormone replacement therapy versus those who do not. Would this difference be of *practical importance* to a woman considering HRT? Explain.

9.50 Clopidogrel A large study was conducted to test the effectiveness of an experimental blood thinner, clopidogrel, in warding off heart attacks and strokes.[13] The study involved 19,185 patients who had suffered heart attacks, strokes, or pain from clogged arteries. They were each randomly assigned to take either aspirin or clopidogrel for a period of 1 to 3 years. Of the patients taking aspirin, 5.3% suffered heart attacks, strokes, or death from cardiovascular disease; the corresponding percentage in the clopidogrel patients was 5.8%.

a. The article states that each patient was randomly assigned to one of the two medications. Explain how you could use the random number table to make these assignments.

b. Although the article does not give the sample sizes, assume that the randomization in part a results in 9925 aspirin and 9260 clopidogrel assignments. Are the results of the study statistically significant? Use the appropriate test of hypothesis.

c. What do the results of the study mean in terms of their *practical importance*?

9.51 Baby's Sleeping Position Does a baby's sleeping position affect the development of motor skills? In one study, published in the *Archives of Pediatric Adolescent Medicine,* 343 full-term infants were examined at their 4-month checkups for various developmental milestones, such as rolling over, grasping a rattle, reaching for an object, and so on.[14] The baby's predominant sleep position—either prone (on the stomach) or supine (on the back) or side—was determined by a telephone interview with the parent. The sample results for 320 of the 343 infants for whom information was received are shown here:

	Prone	Supine or Side
Number of infants	121	199
Number that roll over	93	119

The researcher reported that infants who slept in the side or supine position were less likely to roll over at the 4-month checkup than infants who slept primarily in the prone position ($P < .001$). Use a large-sample test of hypothesis to confirm or refute the researcher's conclusion.

SOME COMMENTS ON TESTING HYPOTHESES

9.7

A statistical test of hypothesis is a fairly clear-cut procedure that enables an experimenter to either reject or accept the null hypothesis H_0, with measured risks α and β. The experimenter can control the risk of falsely rejecting H_0 by selecting an appropriate value of α. On the other hand, the value of β depends on the sample size and the values of the parameter under test that are of practical importance to the experimenter. When this information is not available, an experimenter may decide to select an affordable sample size, in the hope that the sample will contain sufficient information to reject the null hypothesis. The chance that this decision is in error is given by α, whose value has been set in advance. If the sample does not provide sufficient evidence to reject H_0, the experimenter may wish to state the results of the test as "The data do not support the rejection of H_0" rather than accepting H_0 without knowing the chance of error β.

Some experimenters prefer to use the observed p-value of the test to evaluate the strength of the sample information in deciding to reject H_0. These values can usually be generated by computer and are often used in reports of statistical results:

- If the p-value is greater than .05, the results are reported as NS—not significant at the 5% level.

- If the p-value lies between .05 and .01, the results are reported as $P < .05$—significant at the 5% level.

- If the p-value lies between .01 and .001, the results are reported as $P < .01$—"highly significant" or significant at the 1% level.

- If the p-value is less than .001, the results are reported as $P < .001$—"very highly significant" or significant at the .1% level.

Still other researchers prefer to construct a confidence interval for a parameter and perform a test informally. If the value of the parameter specified by H_0 is included within the upper and lower limits of the confidence interval, then "H_0 is not rejected." If the value of the parameter specified by H_0 is not contained within the interval, then "H_0 is rejected." These results will agree with a two-tailed test; one-sided confidence bounds are used for one-tailed alternatives.

Finally, consider the choice between a one- and two-tailed test. In general, experimenters wish to know whether a treatment causes what could be a beneficial increase in a parameter or what might be a harmful decrease in a parameter. Therefore, most tests are two-tailed unless a one-tailed test is strongly dictated by practical considerations. For example, assume you will sustain a large financial loss if the mean μ is greater than μ_0 but not if it is less. You will then want to detect values larger than μ_0 with a high probability and thereby use a right-tailed test. In the same vein, if pollution levels higher than μ_0 cause critical health risks, then you will certainly wish to detect levels higher than μ_0 with a right-tailed test of hypothesis. In any case, the choice of a one- or two-tailed test should be dictated by the practical consequences that result from a decision to reject or not reject H_0 in favor of the alternative.

CHAPTER REVIEW

Key Concepts and Formulas

I. Parts of a Statistical Test

1. **Null hypothesis:** a contradiction of the alternative hypothesis

2. **Alternative hypothesis:** the hypothesis the researcher wants to support

3. **Test statistic** and its **p-value:** sample evidence calculated from the sample data

4. **Rejection region—critical values** and **significance levels:** values that separate rejection and nonrejection of the null hypothesis

5. **Conclusion:** Reject or do not reject the null hypothesis, stating the practical significance of your conclusion

II. Errors and Statistical Significance

1. The **significance level** α is the probability of rejecting H_0 when it is in fact true.

2. The **p-value** is the probability of observing a test statistic as extreme as or more extreme than the one observed; also, the smallest value of α for which H_0 can be rejected.

3. When the **p-value** is less than the **significance level** α, the null hypothesis is rejected. This happens when the **test statistic** exceeds the **critical value.**

4. In a **Type II error,** β is the probability of accepting H_0 when it is in fact false. The **power of the test** is $(1 - \beta)$, the probability of rejecting H_0 when it is false.

III. Large-Sample Test Statistics Using the *z* Distribution

To test one of the four population parameters when the sample sizes are large, use the following test statistics:

Parameter	Test Statistic
μ	$z = \dfrac{\bar{x} - \mu_0}{s/\sqrt{n}}$
p	$z = \dfrac{\hat{p} - p_0}{\sqrt{\dfrac{p_0 q_0}{n}}}$
$\mu_1 - \mu_2$	$z = \dfrac{(\bar{x}_1 - \bar{x}_2) - D_0}{\sqrt{\dfrac{s_1^2}{n_1} + \dfrac{s_2^2}{n_2}}}$
$p_1 - p_2$	$z = \dfrac{\hat{p}_1 - \hat{p}_2}{\sqrt{\hat{p}\hat{q}\left(\dfrac{1}{n_1} + \dfrac{1}{n_2}\right)}}$ or $z = \dfrac{(\hat{p}_1 - \hat{p}_2) - D_0}{\sqrt{\dfrac{\hat{p}_1 \hat{q}_1}{n_1} + \dfrac{\hat{p}_2 \hat{q}_2}{n_2}}}$

Supplementary Exercises

Starred (*) exercises are optional.

9.52 a. Define α and β for a statistical test of hypothesis.

b. For a fixed sample size n, if the value of α is decreased, what is the effect on β?

c. In order to decrease both α and β for a particular alternative value of μ, how must the sample size change?

9.53 What is the p-value for a test of hypothesis? How is it calculated for a large-sample test?

9.54 What conditions must be met so that the z test can be used to test a hypothesis concerning a population mean μ?

9.55 Define the power of a statistical test. As the alternative value of μ gets farther from μ_0, how is the power affected?

9.56 Acidity in Rainfall Refer to Exercise 8.31 and the collection of water samples to estimate the mean acidity (in pH) of rainfalls in the northeastern United States. As noted, the pH for pure rain falling through clean air is approximately 5.7. The sample of $n = 40$ rainfalls produced pH readings with $\bar{x} = 3.7$ and $s = .5$. Do the data provide sufficient evidence to indicate that the mean pH for rainfalls is more acidic ($H_a : \mu < 5.7$ pH) than pure rainwater? Test using $\alpha = .05$. Note that this inference is appropriate only for the area in which the rainwater specimens were collected.

9.57 Washing Machine Colors A manufacturer of automatic washers provides a particular model in one of three colors. Of the first 1000 washers sold, it is noted that 400 were of the first color. Can you conclude that more than one-third of all customers have a preference for the first color?

a. Find the p-value for the test.

b. If you plan to conduct your test using $\alpha = .05$, what will be your test conclusions?

9.58 Commercials in Space The commercialism of our space program[15] was the topic of Exercise 8.60. In a survey of 500 men and 500 women, 20% of the men and 26% of the women responded that space should remain commercial-free.

a. Is there a significant difference in the population proportions of men and women who think that space should remain commercial-free? Use $\alpha = .01$.

b. Can you think of any reason why a statistically significant difference in these population proportions might be of *practical importance* to the administrators of the space program? To the advertisers? To the politicians?

9.59 Bass Fishing The pH factor is a measure of the acidity or alkalinity of water. A reading of 7.0 is neutral; values in excess of 7.0 indicate alkalinity; those below 7.0 imply acidity. Loren Hill states that the best chance of catching bass occurs when the pH of the water is in the range 7.5 to 7.9.[16] Suppose you suspect that acid rain is lowering the pH of your favorite fishing spot and you wish to determine whether the pH is less than 7.5.

a. State the alternative and null hypotheses that you would choose for a statistical test.

b. Does the alternative hypothesis in part a imply a one- or a two-tailed test? Explain.

c. Suppose that a random sample of 30 water specimens gave pH readings with $\bar{x} = 7.3$ and $s = .2$. Just glancing at the data, do you think that the difference $\bar{x} - 7.5 = -.2$ is large enough to indicate that the mean pH of the water samples is less than 7.5? (Do *not* conduct the test.)

d. Now conduct a statistical test of the hypotheses in part a and state your conclusions. Test using $\alpha = .05$. Compare your statistically based decision with your intuitive decision in part c.

9.60 Pennsylvania Lottery A central Pennsylvania attorney reported that the Northumberland County district attorney's (DA) office trial record showed only 6 convictions in 27 trials from January to mid-July 1997. Four central Pennsylvania county DAs responded, "Don't judge us by statistics!"[17]

a. If the attorney's information is correct, would you reject a claim by the DA of a 50% or greater conviction rate?

b. The *actual* records show that there have been 455 guilty pleas and 48 cases that have gone to trial. Even assuming that the 455 guilty pleas are the *only* convictions of the 503 cases reported, what is the 95% confidence interval for p, the true proportion of convictions by this district attorney?

c. Using the results of part b, are you willing to reject a figure of 50% or greater for the true conviction rate? Explain.

9.61 White-Tailed Deer In an article entitled "A Strategy for Big Bucks," Charles Dickey discusses studies of the habits of white-tailed deer that indicate that they live and feed within very limited ranges— approximately 150 to 205 acres.[18] To determine whether there was a difference between the ranges of deer located in two different geographic areas, 40 deer were caught, tagged, and fitted with small radio transmitters. Several months later, the deer were tracked and identified, and the distance x from the release point was recorded. The mean and standard deviation of the distances from the release point were as follows:

	Location 1	Location 2
Sample size	40	40
Sample mean	2980 ft	3205 ft
Sample standard deviation	1140 ft	963 ft

a. If you have no preconceived reason for believing one population mean is larger than another, what would you choose for your alternative hypothesis? Your null hypothesis?

b. Does your alternative hypothesis in part a imply a one- or a two-tailed test? Explain.

c. Do the data provide sufficient evidence to indicate that the mean distances differ for the two geographic locations? Test using $\alpha = .05$.

9.62 Female Models In a study to assess various effects of using a female model in automobile advertising, 100 men were shown photographs of two automobiles matched for price, color, and size, but of different makes. One of the automobiles was shown with a female model to 50 of the men (group A), and both automobiles were shown without the model to the other 50 men (group B). In group A, the automobile shown with the model was judged as more expensive by 37 men; in group B, the same automobile was judged as the more expensive by 23 men. Do these results indicate that using a female model influences the perceived cost of an automobile? Use a one-tailed test with $\alpha = .05$.

9.63 Bolts Random samples of 200 bolts manufactured by a type A machine and 200 bolts manufactured by a type B machine showed 16 and 8 defective bolts, respectively. Do these data present sufficient evidence to suggest a difference in the performance of the machine types? Use $\alpha = .05$.

9.64 Biomass Exercise 7.63 reported that the biomass for tropical woodlands, thought to be about 35 kilograms per square meter (kg/m^2), may in fact be too high and that tropical biomass values vary regionally—from about 5 to 55 kg/m^2.[19] Suppose you measure the tropical biomass in 400 randomly selected square-meter plots and obtain $\bar{x} = 31.75$ and $s = 10.5$. Do the data present sufficient evidence to indicate that scientists are overestimating the mean biomass for tropical woodlands and that the mean is in fact lower than estimated?

a. State the null and alternative hypotheses to be tested.

b. Locate the rejection region for the test with $\alpha = .01$.

c. Conduct the test and state your conclusions.

9.65 The Death Penalty A researcher believes that the fraction p_1 of Republicans in favor of the death penalty is greater than the fraction p_2 of Democrats in favor of the death penalty. She acquired independent random samples of 200 Republicans and 200 Democrats and found 136 Republicans and 124 Democrats favoring the death penalty. Do these data support the researcher's belief?

a. Find the p-value for the test.

b. If you plan to conduct your test using $\alpha = .05$, what will be your test conclusions?

9.66* The Death Penalty, continued Refer to Exercise 9.65. Some thought should have been given to designing a test for which β is tolerably low when p_1 exceeds p_2 by an important amount. For example,

find a common sample size n for a test with $\alpha = .05$ and $\beta \leq .20$, when in fact p_1 exceeds p_2 by .1. [HINT: The maximum value of $p(1 - p)$ is .25.]

9.67 Losing Weight In a comparison of the mean 1-month weight losses for women aged 20–30 years, these sample data were obtained for each of two diets:

	Diet I	Diet II
Sample size n	40	40
Sample mean $\bar{x}$	10 lb	8 lb
Sample variance s^2	4.3	5.7

Do the data provide sufficient evidence to indicate that diet I produces a greater mean weight loss than diet II? Use $\alpha = .05$.

9.68 Increased Yield An agronomist has shown experimentally that a new irrigation/fertilization regimen produces an increase of 2 bushels per quadrat (significant at the 1% level) when compared with the regimen currently in use. The cost of implementing and using the new regimen will not be a factor if the increase in yield exceeds 3 bushels per quadrat. Is statistical significance the same as practical importance in this situation? Explain.

9.69 Breaking Strengths of Cables A test of the breaking strengths of two different types of cables was conducted using samples of $n_1 = n_2 = 100$ pieces of each type of cable.

Cable I	Cable II
$\bar{x}_1 = 1925$	$\bar{x}_2 = 1905$
$s_1 = 40$	$s_2 = 30$

Do the data provide sufficient evidence to indicate a difference between the mean breaking strengths of the two cables? Use $\alpha = .05$.

9.70 Put on the Brakes The braking ability was compared for two 2005 automobile models. Random samples of 64 automobiles were tested for each type. The recorded measurement was the distance (in feet) required to stop when the brakes were applied at 40 miles per hour. These are the computed sample means and variances:

Model I	Model II
$\bar{x}_1 = 118$	$\bar{x}_2 = 109$
$s_1^2 = 102$	$s_2^2 = 87$

Do the data provide sufficient evidence to indicate a difference between the mean stopping distances for the two models?

9.71 Spraying Fruit Trees A fruit grower wants to test a new spray that a manufacturer claims will *reduce* the loss due to insect damage. To test the claim, the grower sprays 200 trees with the new spray and 200 other trees with the standard spray. The following data were recorded:

	New Spray	Standard Spray
Mean yield per tree $\bar{x}$ (lb)	240	227
Variance s^2	980	820

a. Do the data provide sufficient evidence to conclude that the mean yield per tree treated with the new spray exceeds that for trees treated with the standard spray? Use $\alpha = .05$.

b. Construct a 95% confidence interval for the difference between the mean yields for the two sprays.

9.72 Actinomycin D A biologist hypothesizes that high concentrations of actinomycin D inhibit RNA synthesis in cells and hence the production of proteins as well. An experiment conducted to test this theory compared the RNA synthesis in cells treated with two concentrations of actinomycin D: .6 and .7 microgram per milliliter. Cells treated with the lower concentration (.6) of actinomycin D showed that 55 out of 70 developed normally, whereas only 23 out of 70 appeared to develop normally for the higher concentration (.7). Do these data provide sufficient evidence to indicate a difference between the rates of normal RNA synthesis for cells exposed to the two different concentrations of actinomycin D?

a. Find the p-value for the test.

b. If you plan to conduct your test using $\alpha = .05$, what will be your test conclusions?

9.73 SAT Scores How do California high school students compare to students nationwide in their college readiness, as measured by their SAT scores? The national average scores for the class of 2003 were 507 on the verbal portion and 519 on the math portion.[20] Suppose that 100 California students from the class of 2003 were randomly selected and their SAT scores recorded in the following table:

	Verbal	Math
Sample average	499	516
Sample standard deviation	98	96

a. Do the data provide sufficient evidence to indicate that the average verbal score for all California students in the class of 2003 is different from the national average? Test using $\alpha = .05$.

b. Do the data provide sufficient evidence to indicate that the average math score for all California stu-

dents in the class of 2003 is different from the national average? Test using $\alpha = .05$.

c. Could you use this data to determine if there is a difference between the average math and verbal scores for all California students in the class of 2003? Explain your answer.

9.74 A Maze Experiment In a maze running study, a rat is run in a T maze and the result of each run recorded. A reward in the form of food is always placed at the right exit. If learning is taking place, the rat will choose the right exit more often than the left. If no learning is taking place, the rat should randomly choose either exit. Suppose that the rat is given $n = 100$ runs in the maze and that he chooses the right exit $x = 64$ times. Would you conclude that learning is taking place? Use the p-value approach, and make a decision based on this p-value.

9.75 PCBs Polychlorinated biphenyls (PCBs) have been found to be dangerously high in some game birds found along the marshlands of the southeastern coast of the United States. The Federal Drug Administration (FDA) considers a concentration of PCBs higher than 5 parts per million (ppm) in these game birds to be dangerous for human consumption. A sample of 38 game birds produced an average of 7.2 ppm with a standard deviation of 6.2 ppm. Is there sufficient evidence to indicate that the mean ppm of PCBs in the population of game birds exceeds the FDA's recommended limit of 5 ppm? Use $\alpha = .01$.

9.76* PCBs, continued Refer to Exercise 9.75.

a. Calculate β and $1 - \beta$ if the true mean ppm of PCBs is 6 ppm.

b. Calculate β and $1 - \beta$ if the true mean ppm of PCBs is 7 ppm.

c. Find the power, $1 - \beta$, when $\mu = 8, 9, 10$, and 12. Use these values to construct a power curve for the test in Exercise 9.75.

d. For what values of μ does this test have power greater than or equal to .90?

9.77 The "Cheeseburger Bill" In the spring of 2004, the U.S. Congress considered a bill that would prevent Americans from suing fast-food giants like McDonald's for making them overweight. Although the fast-food industry may not be to blame, a study by *Children's Hospital* in Boston reports that about two-thirds of adult Americans and about 15% of children and adolescents are overweight.[21] To test this claim, a random sample of 100 children is selected, and 13 of them are found to be overweight. Is there sufficient evidence to indicate that the percentage reported by *Children's Hospital* is incorrect? Test at the 5% level of significance.

9.78 Heights and Gender It is a well-accepted fact that males are taller on the average than females. But how much taller? The genders of 105 biomedical students (Exercise 1.54) were also recorded and the data are summarized below:

	Males	Females
Sample size	48	77
Sample mean	69.58	64.43
Sample standard deviation	2.62	2.58

a. Perform a test of hypothesis to either confirm or refute our initial claim that males are taller on the average than females? Use $\alpha = .01$.

b. If the results of part a show that our claim was correct, construct a 99% confidence one-sided lower confidence bound for the average difference in heights between male and female college students. How much taller are males than females?

9.79 English as a Second Language The state of California is working very hard to make sure that all elementary-aged students whose native language is not English become proficient in English by the sixth grade. Their progress is monitored each year using the California English Language Development Test.[22] The results for two school districts in southern California for the 2003 school year are shown below.

District	Riverside	Palm Springs
Number of students tested	6124	5512
Percentage fluent	40	37

Does this data provide sufficient statistical evidence to indicate that the percentage of students who are fluent in English differs for these two districts? Test using $\alpha = .01$.

9.80 Breaststroke Swimmers How much training time does it take to become a world-class breaststroke swimmer? A survey published in *The American Journal of Sports Medicine* reported the number of meters per week swum by two groups of swimmers—those who competed only in breaststroke and those who competed in the individual medley (which includes breaststroke). The number of meters per week practicing the breaststroke swim was recorded and the summary statistics are shown below.[23]

	Breaststroke	Individual Medley
Sample size	130	80
Sample mean	9017	5853
Sample standard deviation	7162	1961

Is there sufficient evidence to indicate a difference in the average number of meters swum by these two groups of swimmers? Test using $\alpha = .01$.

9.81 Breaststroke, continued Refer to Exercise 9.80.

a. Construct a 99% confidence interval for the difference in the average number of meters swum by breaststroke versus individual medley swimmers.

b. How much longer do pure breaststroke swimmers practice that stroke than individual medley swimmers? What is the practical reason for this difference?

MY APPLET Exercises

9.82 Summer Jobs In Exercise 8.109, it was reported that amusement and theme parks average 100 full-time employees, but at peak season average 665 workers per site.[24] If $n = 40$ amusement and theme parks are sampled at peak season and the average number of full-time workers is found to be 652 with a standard deviation of 32, does this contradict the reported average of 665 workers per site?

a. What are the null and alternative hypotheses to be tested?

b. Use the **Large-Sample Test of a Population Mean** applet to find the observed value of the test statistic.

c. Use the **Large-Sample Test of a Population Mean** applet to find the *p*-value for this test.

d. Based on your results from part c, what conclusions can you draw about the average of 665 workers reported in Exercise 8.97?

9.83 Daily Wages The daily wages in a particular industry are normally distributed with a mean of $94 and a standard deviation of $11.88. Suppose a company in this industry employs 40 workers and pays them $91.50 per week on the average. Can these workers be viewed as a random sample from among all workers in the industry?

a. What are the null and alternative hypotheses to be tested?

b. Use the **Large-Sample Test of a Population Mean** applet to find the observed value of the test statistic.

c. Use the **Large-Sample Test of a Population Mean** applet to find the p-value for this test.

d. If you planned to conduct your test using $\alpha = .01$, what would be your test conclusions?

e. Was it necessary to know that the daily wages are normally distributed? Explain your answer.

9.84 Refer to Example 9.8. Use the **Power of a z-Test** applet to verify the power of the test of

$$H_0: \mu = 880 \quad \text{versus} \quad H_a: \mu \neq 880$$

for values of μ equal to 870, 875, 880, 885 and 890. Check your answers against the values shown in Table 9.2.

9.85 Refer to Example 9.8.

a. Use the method given in Example 9.8 to calculate the power of the test of

$$H_0: \mu = 880 \quad \text{versus} \quad H_a: \mu \neq 880$$

when $n = 30$ and the true value of μ is 870 tons.

b. Repeat part a using $n = 70$ and $\mu = 870$ tons.

c. Use the **Power of a z-Test** applet to verify your hand-calculated results in parts a and b.

d. What is the effect of increasing the sample size on the power of the test?

9.86 Use the appropriate slider on the **Power of a z-Test** applet to answer the following questions. Write a sentence for each part, describing what you see using the applet.

a. What effect does increasing the sample size have on the power of the test?

b. What effect does increasing the distance between the true value of μ and the hypothesized value, $\mu = 880$, have on the power of the test?

c. What effect does decreasing the significance level α have on the power of the test?

CASE STUDY **An Aspirin a Day . . . ?**

On Wednesday, January 27, 1988, the front page of the *New York Times* read "Heart attack risk found to be cut by taking aspirin: Lifesaving effects seen." A very large study of U.S. physicians showed that a single aspirin tablet taken every other day reduced by one-half the risk of heart attack in men.[25] Three days later, a headline in the *Times* read "Value of daily aspirin disputed in British study of heart attacks." How could two seemingly similar studies, both involving doctors as participants, reach such opposite conclusions?

The U.S. physicians' health study consisted of two randomized clinical trials in one. The first tested the hypothesis that 325 milligrams (mg) of aspirin taken every other day reduces mortality from cardiovascular disease. The second tested whether 50 mg of β-carotene taken on alternate days decreases the incidence of cancer. From names on an American Medical Association computer tape, 261,248 male physicians between the ages of 40 and 84 were invited to participate in the trial. Of those who responded, 59,285 were willing to participate. After the exclusion of those physicians who had a history of medical disorders, or who were currently taking aspirin or had negative reactions to aspirin, 22,071 physicians were randomized into one of four treatment groups: (1) buffered aspirin and β-carotene, (2) buffered aspirin and a β-carotene placebo, (3) aspirin placebo and β-carotene, and (4) aspirin placebo and β-carotene placebo. Thus, half were assigned to receive aspirin and half to receive β-carotene.

The study was conducted as a double-blind study, in which neither the participants nor the investigators responsible for following the participants knew to which group a participant belonged. The results of the American study concerning myocardial infarctions (the technical name for heart attacks) are given in the following table:

	American Study	
	Aspirin ($n = 11{,}037$)	Placebo ($n = 11{,}034$)
Myocardial infarction		
Fatal	5	18
Nonfatal	99	171
Total	104	189

The objective of the British study was to determine whether 500 mg of aspirin taken daily would reduce the incidence of and mortality from cardiovascular disease. In 1978 all male physicians in the United Kingdom were invited to participate. After the usual exclusions, 5139 doctors were randomly allocated to take aspirin, unless some problem developed, and one-third were randomly allocated to *avoid* aspirin. Placebo tablets were not used, so the study was not blind! The results of the British study are given here:

	British Study	
	Aspirin ($n = 3429$)	Control ($n = 1710$)
Myocardial infarction		
Fatal	89 (47.3)	47 (49.6)
Nonfatal	80 (42.5)	41 (43.3)
Total	169 (89.8)	88 (92.9)

To account for unequal sample sizes, the British study reported rates per 10,000 subject-years alive (given in parentheses).

1. Test whether the American study does in fact indicate that the rate of heart attacks for physicians taking 325 mg of aspirin every other day is significantly different from the rate for those on the placebo. Is the American claim justified?

2. Repeat the analysis using the data from the British study in which one group took 500 mg of aspirin every day and the control group took none. Based on their data, is the British claim justified?

3. Can you think of some possible reasons the results of these two studies, which were alike in some respects, produced such different conclusions?

Inference from Small Samples

© CORBIS/Sygma

MY PERSONAL TRAINER

How Do I Decide Which Test to Use?

Would You Like a Four-Day Workweek?

Will a flexible workweek schedule result in positive benefits for both employer and employee? Four obvious benefits are (1) less time traveling from field positions to the office, (2) fewer employees parked in the parking lot, (3) reduced travel expenses, and (4) allowance for employees to have another day off. But does the flexible workweek make employees more efficient and cause them to take fewer sick and personal days? The answers to some of these questions are posed in the case study at the end of this chapter.

INTRODUCTION

Suppose you need to run an experiment to estimate a population mean or the difference between two means. The process of collecting the data may be very expensive or very time-consuming. If you cannot collect a *large sample,* the estimation and test procedures of Chapters 8 and 9 are of no use to you.

This chapter introduces some equivalent statistical procedures that can be used when the *sample size is small.* The estimation and testing procedures involve these familiar parameters:

- A single population mean, μ
- The difference between two population means, $(\mu_1 - \mu_2)$
- A single population variance, σ^2
- The comparison of two population variances, σ_1^2 and σ_2^2

Small-sample tests and confidence intervals for binomial proportions will be omitted from our discussion.[†]

STUDENT'S *t* DISTRIBUTION

In conducting an experiment to evaluate a new but very costly process for producing synthetic diamonds, you are able to study only six diamonds generated by the process. How can you use these six measurements to make inferences about the average weight μ of diamonds from this process?

In discussing the sampling distribution of $\bar{x}$ in Chapter 7, we made these points:

- When the original sampled population is normal, $\bar{x}$ and $z = (\bar{x} - \mu)/(\sigma/\sqrt{n})$ both have normal distributions, *for any sample size.*
- When the original sampled population is *not* normal, $\bar{x}$, $z = (\bar{x} - \mu)/(\sigma/\sqrt{n})$, and $z \approx (\bar{x} - \mu)/(\sigma/\sqrt{n})$ all have approximately normal distributions, if the sample size is *large.*

MY TIP

When $n < 30$, the Central Limit Theorem will not guarantee that

$$\frac{\bar{x} - \mu}{s/\sqrt{n}}$$

is approximately normal.

Unfortunately, when the sample size n is small, the statistic $(\bar{x} - \mu)/(s/\sqrt{n})$ *does not* have a normal distribution. Therefore, all the critical values of z that you used in Chapters 8 and 9 are no longer correct. For example, you *cannot say* that $\bar{x}$ will lie within 1.96 standard errors of μ 95% of the time.

This problem is not new; it was studied by statisticians and experimenters in the early 1900s. To find the sampling distribution of this statistic, there are two ways to proceed:

- Use an empirical approach. Draw repeated samples and compute $(\bar{x} - \mu)/(s/\sqrt{n})$ for each sample. The relative frequency distribution that you construct using these values will approximate the shape and location of the sampling distribution.
- Use a mathematical approach to derive the actual density function or curve that describes the sampling distribution.

[†]A small-sample test for the binomial parameter p will be presented in Chapter 15.

This second approach was used by an Englishman named W.S. Gosset in 1908. He derived a complicated formula for the density function of

$$t = \frac{\bar{x} - \mu}{s/\sqrt{n}}$$

for random samples of size n from a normal population, and he published his results under the pen name "Student." Ever since, the statistic has been known as **Student's t.** It has the following characteristics:

- It is mound-shaped and symmetric about $t = 0$, just like z.

- It is more variable than z, with "heavier tails"; that is, the t curve does not approach the horizontal axis as quickly as z does. This is because the t statistic involves two random quantities, $\bar{x}$ and s, whereas the z statistic involves only the sample mean, $\bar{x}$. You can see this phenomenon in Figure 10.1.

- The shape of the t distribution depends on the sample size n. As n increases, the variability of t decreases because the estimate s of σ is based on more and more information. Eventually, when n is infinitely large, the t and z distributions are identical!

FIGURE 10.1

Standard normal z and the t distribution with 5 degrees of freedom

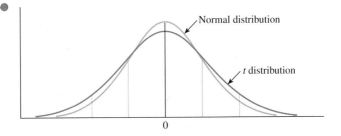

The divisor $(n - 1)$ in the formula for the sample variance s^2 is called the **number of degrees of freedom (df) associated with s^2.** It determines the *shape* of the t distribution. The origin of the term *degrees of freedom* is theoretical and refers to the number of independent squared deviations in s^2 that are available for estimating σ^2. These degrees of freedom may change for different applications and, since they specify the correct t distribution to use, you need to remember to calculate the correct degrees of freedom for each application.

The table of probabilities for the standard normal z distribution is no longer useful in calculating critical values or p-values for the t statistic. Instead, you will use Table 4 in Appendix I, which is partially reproduced in Table 10.1. When you index a particular number of degrees of freedom, the table records t_α, a value of t that has tail area a to its right, as shown in Figure 10.2.

 TIP

For a one-sample t, *df* = *n* − 1.

FIGURE 10.2

Tabulated values of Student's t

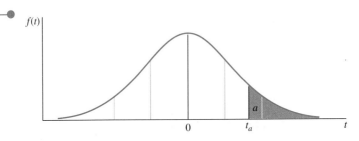

TABLE 10.1 ●   **Format of the Student's *t* Table from Table 4 in Appendix I**

df	$t_{.100}$	$t_{.050}$	$t_{.025}$	$t_{.010}$	$t_{.005}$	df
1	3.078	6.314	12.706	31.821	63.657	1
2	1.886	2.920	4.303	6.965	9.925	2
3	1.638	2.353	3.182	4.541	5.841	3
4	1.533	2.132	2.776	3.747	4.604	4
5	1.476	2.015	2.571	3.365	4.032	5
6	1.440	1.943	2.447	3.143	3.707	6
7	1.415	1.895	2.365	2.998	3.499	7
8	1.397	1.860	2.306	2.896	3.355	8
9	1.383	1.833	2.262	2.821	3.250	9
.	.	.	.	.	.	.
.	.	.	.	.	.	.
.	.	.	.	.	.	.
26	1.315	1.706	2.056	2.479	2.779	26
27	1.314	1.703	2.052	2.473	2.771	27
28	1.313	1.701	2.048	2.467	2.763	28
29	1.311	1.699	2.045	2.462	2.756	29
inf.	1.282	1.645	1.960	2.326	2.576	inf.

EXAMPLE **10.1** ── For a *t* distribution with 5 degrees of freedom, the value of *t* that has area .05 to its right is found in row 5 in the column marked $t_{.050}$. For this particular *t* distribution, the area to the right of $t = 2.015$ is .05; only 5% of all values of the *t* statistic will exceed this value.

MY APPLET

You can use the **Student's *t* Probabilities** applet to find the *t*-value described in Example 10.1. The first applet, shown in Figure 10.3, provides *t*-values and their two-tailed probabilities, while the second applet provides *t*-values and one-tailed probabilities. Use the slider on the right side of the applet to select the proper degrees of freedom. For Example 10.1, you should choose *df* = 5 and type .10 in the box marked "prob:" at the bottom of the first applet. The applet will provide the value of *t* that puts .05 in one tail of the *t* distribution. The second applet will show the identical *t* for a one-tailed area of .05. The applet in Figure 10.3 shows $t = 2.02$ which is correct to two decimal places. We will use this applet for the *MyApplet Exercises* at the end of the chapter.

FIGURE 10.3 ●

Student's *t* Probabilities applet

EXAMPLE 10.2

Suppose you have a sample of size $n = 10$ from a normal distribution. Find a value of t such that only 1% of all values of t will be smaller.

Solution The degrees of freedom that specify the correct t distribution are $df = n - 1 = 9$, and the necessary t-value must be in the lower portion of the distribution, with area .01 to its left, as shown in Figure 10.4. Since the t distribution is symmetric about 0, this value is simply the negative of the value on the right-hand side with area .01 to its right, or $-t_{.01} = -2.821$.

FIGURE 10.4

t Distribution for Example 10.2

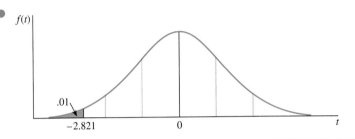

MY APPLET

Comparing the *t* and *z* Distributions

Look at one of the columns in Table 10.1. As the degrees of freedom increase, the critical value of t decreases until, when $df = $ inf., the critical t-value is the same as the critical z-value for the same tail area. You can use the **Comparing *t* and *z*** applet to visualize this concept. Look at the three applets in Figure 10.5, which show the critical values for $t_{.025}$ compared with $z_{.025}$ for $df = 8$, 29 and 100. (The slider on the right side of the applet allows you to change the df.) The red curve (black in Fig. 10.5) is the standard normal distribution, with $z_{.025} = 1.96$.

FIGURE 10.5

Comparing *t* and *z* applet

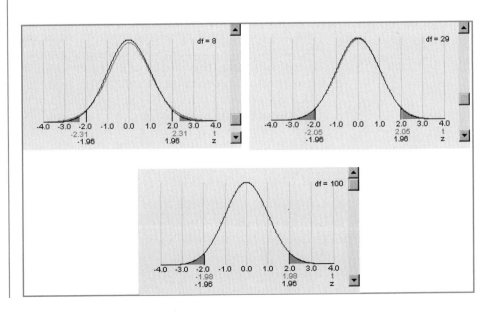

The blue curve is the *t* distribution. With 8 *df*, you can clearly see a difference in the *t* and *z* curves, especially in the critical values that cut off an area of .025 in the tails. As the degrees of freedom increase, the difference in the shapes of *t* and *z* becomes very similar, as do their critical values, until at *df* = 100, there is almost no difference. This helps to explain why we use *n* = 30 as the somewhat arbitrary dividing line between large and small samples. When *n* = 30 (*df* = 29), the critical values of *t* are quite close to their normal counterparts. Rather than produce a *t* table with rows for many more degrees of freedom, the critical values of *z* are sufficient when the sample size reaches *n* = 30.

Assumptions behind Student's *t* Distribution

The critical values of *t* allow you to make reliable inferences *only if* you follow all the rules; that is, your sample must meet these requirements specified by the *t* distribution:

Assumptions for one-sample *t*:
*   Random sample
*   Normal distribution

*   The sample must be randomly selected.
*   The population from which you are sampling must be normally distributed.

These requirements may seem quite restrictive. How can you possibly know the shape of the probability distribution for the entire population if you have only a sample? If this were a serious problem, however, the *t* statistic could be used in only very limited situations. Fortunately, the shape of the *t* distribution is not affected very much as long as the sampled population has an *approximately mound-shaped* distribution. Statisticians say that the *t* statistic is **robust,** meaning that the distribution of the statistic does not change significantly when the normality assumption is violated.

How can you tell whether your sample is from a normal population? Although there are statistical procedures designed for this purpose, the easiest and quickest way to check for normality is to use the graphical techniques of Chapter 2: Draw a dotplot or construct a stem and leaf plot. As long as your plot tends to "mound up" in the center, you can be fairly safe in using the *t* statistic for making inferences.

The random sampling requirement, on the other hand, is quite critical if you want to produce reliable inferences. If the sample is not random, or if it does not at *least behave as* a random sample, then your sample results may be affected by some unknown factor and your conclusions may be incorrect. When you design an experiment or read about experiments conducted by others, look critically at the way the data have been collected!

SMALL-SAMPLE INFERENCES CONCERNING A POPULATION MEAN

As with large-sample inference, small-sample inference can involve either **estimation** or **hypothesis testing,** depending on the preference of the experimenter. We explained the basics of these two types of inference in the earlier chapters, and we use them again now, with a different sample statistic, $t = (\bar{x} - \mu)/(s/\sqrt{n})$, and a different sampling distribution, the Student's *t*, with $(n - 1)$ degrees of freedom.

SMALL-SAMPLE HYPOTHESIS TEST FOR μ

1. Null hypothesis: $H_0 : \mu = \mu_0$
2. Alternative hypothesis:

   One-Tailed Test   **Two-Tailed Test**

   $H_a : \mu > \mu_0$ $H_a : \mu \neq \mu_0$
   (or, $H_a : \mu < \mu_0$)

3. Test statistic: $t = \dfrac{\bar{x} - \mu_0}{s/\sqrt{n}}$

4. Rejection region: Reject H_0 when

   One-Tailed Test   **Two-Tailed Test**

   $t > t_\alpha$ $t > t_{\alpha/2}$ or $t < -t_{\alpha/2}$
   (or $t < -t_\alpha$ when the
   alternative hypothesis
   is $H_a : \mu < \mu_0$)

   or when p-value $< \alpha$

The critical values of t, t_α, and $t_{\alpha/2}$ are based on $(n - 1)$ degrees of freedom. These tabulated values can be found using Table 4 of Appendix I or the **Student's t Probabilities** applet.
Assumption: The sample is randomly selected from a normally distributed population.

SMALL-SAMPLE $(1 - \alpha)100\%$ CONFIDENCE INTERVAL FOR μ

$$\bar{x} \pm t_{\alpha/2} \frac{s}{\sqrt{n}}$$

where $s/\sqrt{n}$ is the estimated standard error of $\bar{x}$, often referred to as the **standard error of the mean.**

EXAMPLE 10.3

A new process for producing synthetic diamonds can be operated at a profitable level only if the average weight of the diamonds is greater than .5 karat. To evaluate the profitability of the process, six diamonds are generated, with recorded weights .46, .61, .52, .48, .57, and .54 karat. Do the six measurements present sufficient evidence to indicate that the average weight of the diamonds produced by the process is in excess of .5 karat?

Solution   The population of diamond weights produced by this new process has mean μ, and you can set out the formal test of hypothesis in steps, as you did in Chapter 9:

1–2   **Null and alternative hypotheses:**

$$H_0: \mu = .5 \quad \text{versus} \quad H_a: \mu > .5$$

3   **Test statistic:**   You can use your calculator to verify that the mean and standard deviation for the six diamond weights are .53 and .0559, respectively. The test statistic is a t statistic, calculated as

$$t = \frac{\bar{x} - \mu_0}{s/\sqrt{n}} = \frac{.53 - .5}{.0559/\sqrt{6}} = 1.32$$

As with the large-sample tests, the test statistic provides evidence for either rejecting or accepting H_0 depending on how far from the center of the t distribution it lies.

4   **Rejection region:**   If you choose a 5% level of significance ($\alpha = .05$), the right-tailed rejection region is found using the critical values of t from Table 4 of Appendix I. With $df = n - 1 = 5$, you can reject H_0 if $t > t_{.05} = 2.015$, as shown in Figure 10.6.

5   **Conclusion:**   Since the calculated value of the test statistic, 1.32, does not fall in the rejection region, you cannot reject H_0. The data do not present sufficient evidence to indicate that the mean diamond weight exceeds .5 karat.

FIGURE 10.6 ●

Rejection region for Example 10.3

 TIP

A 95% confidence interval tells you that, if you were to construct many of these intervals (all of which would have slightly different endpoints), 95% of them would enclose the population mean.

As in Chapter 9, the conclusion to *accept* H_0 would require the difficult calculation of β, the probability of a Type II error. To avoid this problem, we choose to *not re-ject* H_0. We can then calculate the lower bound for μ using a small-sample lower one-sided confidence bound. This bound is similar to the large-sample one-sided confidence bound, except that the critical z_α is replaced by a critical t_α from Table 4. For this example, a 95% lower one-sided confidence bound for μ is:

$$\bar{x} - t_\alpha \frac{s}{\sqrt{n}}$$

$$.53 - 2.015 \frac{.0559}{\sqrt{6}}$$

$$.53 - .046$$

The 95% lower bound for μ is $\mu > .484$. The range of possible values includes mean diamond weights both smaller and greater than .5; this confirms the failure of our test to show that μ exceeds .5.

Remember from Chapter 9 that there are two ways to conduct a test of hypothesis:

- **The critical value approach:** Set up a rejection region based on the critical values of the statistic's sampling distribution. If the test statistic falls in the rejection region, you can reject H_0.
- **The p-value approach:** Calculate the p-value based on the observed value of the test statistic. If the p-value is smaller than the significance level, α, you can reject H_0. If there is no *preset* significance level, use the guidelines in Section 9.3 to judge the statistical significance of your sample results.

We used the first approach in the solution to Example 10.3. We use the second approach to solve Example 10.4.

EXAMPLE 10.4

Labels on 1-gallon cans of paint usually indicate the drying time and the area that can be covered in one coat. Most brands of paint indicate that, in one coat, a gallon will cover between 250 and 500 square feet, depending on the texture of the surface to be painted. One manufacturer, however, claims that a gallon of its paint will cover 400 square feet of surface area. To test this claim, a random sample of ten 1-gallon cans of white paint were used to paint ten identical areas using the same kind of equipment. The actual areas (in square feet) covered by these 10 gallons of paint are given here:

310	311	412	368	447
376	303	410	365	350

Do the data present sufficient evidence to indicate that the average coverage differs from 400 square feet? Find the p-value for the test, and use it to evaluate the statistical significance of the results.

Solution   To test the claim, the hypotheses to be tested are

$$H_0 : \mu = 400 \quad \text{versus} \quad H_a : \mu \neq 400$$

The sample mean and standard deviation for the recorded data are

$$\bar{x} = 365.2 \qquad s = 48.417$$

and the test statistic is

$$t = \frac{\bar{x} - \mu_0}{s/\sqrt{n}} = \frac{365.2 - 400}{48.417/\sqrt{10}} = -2.27$$

The p-value for this test is the probability of observing a value of the t statistic as contradictory to the null hypothesis as the one observed for this set of data—namely, $t = -2.27$. Since this is a two-tailed test, the p-value is the probability that either $t \leq -2.27$ or $t \geq 2.27$.

Unlike the z-table, the table for t gives the values of t corresponding to upper-tail areas equal to .100, .050, .025, .010, and .005. Consequently, you can only approximate the upper-tail area that corresponds to the probability that $t > 2.27$. Since the t statistic for this test is based on 9 df, we refer to the row corresponding to $df = 9$ in Table 4. The five critical values for various tail areas are shown in Figure 10.7, an enlargement of the tail of the t distribution with 9 degrees of freedom. The value $t = 2.27$ falls between $t_{.025} = 2.262$ and $t_{.010} = 2.821$. Therefore, the right-tail area corresponding to the probability that $t > 2.27$ lies between .01 and .025. Since this area represents only half of the p-value, you can write

$$.01 < \frac{1}{2}(p\text{-value}) < .025 \quad \text{or} \quad .02 < p\text{-value} < .05$$

FIGURE 10.7

Calculating the p-value for Example 10.4 (shaded area = $\frac{1}{2}$ p-value)

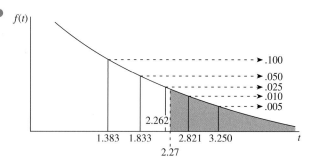

What does this tell you about the significance of the statistical results? For you to reject H_0, the p-value must be less than the specified significance level, α. Hence, you could reject H_0 at the 5% level, but not at the 2% or 1% level. Therefore, the p-value for this test would typically be reported by the experimenter as

p-value < .05   (or sometimes $P < .05$)

For this test of hypothesis, H_0 is rejected at the 5% significance level. There is sufficient evidence to indicate that the average coverage differs from 400 square feet.

Within what limits does this average coverage *really* fall? A 95% confidence interval gives the upper and lower limits for μ as:

$$\bar{x} \pm t_{\alpha/2}\left(\frac{s}{\sqrt{n}}\right)$$

$$365.2 \pm 2.262\left(\frac{48.417}{\sqrt{10}}\right)$$

$$365.2 \pm 34.63$$

Thus, you can estimate that the average area covered by 1 gallon of this brand of paint lies in the interval 330.6 to 399.8. A more precise interval estimate (a shorter interval) can generally be obtained by increasing the sample size. Notice that the upper limit of this interval is very close to the value of 400 square feet, the coverage claimed on the label. This coincides with the fact that the observed value of $t = -2.27$ is just slightly less than the left-tail critical value of $t_{.025} = -2.262$, making the p-value just slightly less than .05.

Most statistical computing packages contain programs that will implement the Student's t test or construct a confidence interval for μ when the data are properly entered into the computer's database. Most of these programs will calculate and report the *exact p-value* of the test, allowing you to quickly and accurately draw conclusions about the statistical significance of the results. The results of the *MINITAB* one-sample t test and confidence interval procedures are given in Figure 10.8. Besides the observed value of $t = -2.27$ and the confidence interval (330.6, 399.8), the output gives the sample mean, the sample standard deviation, the standard error of the mean (SE Mean $= s/\sqrt{n}$), and the exact p-value of the test (P $= .049$). This is consistent with the range for the p-value that we found using Table 4 in Appendix I:

.02 < p-value < .05

FIGURE 10.8

MINITAB output for
Example 10.4

One Sample T: Area
Test of mu = 400 vs not = 400

Variable	N	Mean	StDev	SE Mean
Area	10	365.200	48.417	15.311

Variable	95% CI	T	P
Area	(330.565, 399.835)	-2.27	0.049

MY APPLET

You can use the **Small Sample Test of a Population Mean** applet to visualize the
p-values for either one- or two-tailed tests of the population mean μ. The procedure follows the same pattern as with previous applets. You enter the values of
$\bar{x}$, n, and s and press "Enter" after each entry; the applet will calculate t and give
you the option of choosing one- or two-tailed p-values (as well as a *cumulative* and
a *middle* area that you will not need.)

FIGURE 10.9

Small Sample Test of a
Population Mean applet

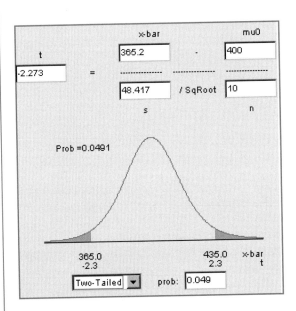

For the data of Example 10.4, the p-value is the two-tailed area to the right of $t =$
2.273 and to the left of $t = -2.273$. Can you find this same p-value in the *MINITAB*
printout shown in Figure 10.8?

You can see the value of using the computer output or the Java applet to evaluate
statistical results:

- The exact p-value eliminates the need for tables and critical values.
- All of the numerical calculations are done for you.

The most important job—which is left for the experimenter—is to *interpret* the results in terms of their practical significance!

10.3 EXERCISES

BASIC TECHNIQUES

10.1 Find the following t-values in Table 4 of Appendix I:

a. $t_{.05}$ for 5 df
b. $t_{.025}$ for 8 df
c. $t_{.10}$ for 18 df
d. $t_{.025}$ for 30 df

10.2 Find the critical value(s) of t that specify the rejection region in these situations:

a. A two-tailed test with $\alpha = .01$ and 12 df
b. A right-tailed test with $\alpha = .05$ and 16 df
c. A two-tailed test with $\alpha = .05$ and 25 df
d. A left-tailed test with $\alpha = .01$ and 7 df

10.3 Use Table 4 in Appendix I to approximate the p-value for the t statistic in each situation:

a. A two-tailed test with $t = 2.43$ and 12 df
b. A right-tailed test with $t = 3.21$ and 16 df
c. A two-tailed test with $t = -1.19$ and 25 df
d. A left-tailed test with $t = -8.77$ and 7 df

 10.4 **Test Scores** The test scores on a 100-point test were recorded for 20 students:

EX1004

71	93	91	86	75
73	86	82	76	57
84	89	67	62	72
77	68	65	75	84

a. Can you reasonably assume that these test scores have been selected from a normal population? Use a stem and leaf plot to justify your answer.
b. Calculate the mean and standard deviation of the scores.
c. If these students can be considered a random sample from the population of all students, find a 95% confidence interval for the average test score in the population.

10.5 The following $n = 10$ observations are a sample from a normal population:

7.4   7.1   6.5   7.5   7.6   6.3   6.9   7.7   6.5   7.0

a. Find the mean and standard deviation of these data.
b. Find a 99% upper one-sided confidence bound for the population mean μ.
c. Test $H_0 : \mu = 7.5$ versus $H_a : \mu < 7.5$. Use $\alpha = .01$.
d. Do the results of part b support your conclusion in part c?

APPLICATIONS

10.6 **Tuna Fish** Is there a difference in the prices of tuna, depending on the method of packaging? *Consumer Reports* gives the estimated average price for a 6-ounce can or a 7.06-ounce pouch of tuna, based on prices paid nationally in supermarkets.[1] These prices are recorded for a variety of different brands of tuna.

EX1006

Light Tuna in Water		White Tuna in Oil	White Tuna in Water	Light Tuna in Oil	
.99	.53	1.27	1.49	2.56	.62
1.92	1.41	1.22	1.29	1.92	.66
1.23	1.12	1.19	1.27	1.30	62
.85	.63	1.22	1.35	1.79	.65
.65	.67		1.29	1.23	.60
.69	.60		1.00		.67
.60	.66		1.27		
			1.28		

Source: Case Study "Pricing of Tuna" Copyright 2001 by Consumers Union of U.S., Inc., Yonkers, NY 10703-1057, a nonprofit organization. Reprinted with permission from the June 2001 issue of Consumer Reports® for educational purposes only. No commercial use or reproduction permitted. www.ConsumerReports.org.

Assume that the tuna brands included in this survey represent a random sample of all tuna brands available in the United States.

a. Find a 95% confidence interval for the average price for light tuna in water. Interpret this interval. That is, what does the "95%" refer to?
b. Find a 95% confidence interval for the average price for white tuna in oil. How does the width of this interval compare to the width of the interval in part a? Can you explain why?
c. Find 95% confidence intervals for the other two samples (white tuna in water and light tuna in oil.) Plot the four treatment means and their standard errors in a two-dimensional plot similar to Figure 8.5. What kind of broad comparisons can you make about the four treatments? (We will discuss the procedure for comparing more than two population means in Chapter 11.)

10.7 **Dissolved O$_2$ Content** Industrial wastes and sewage dumped into our rivers and streams absorb oxygen and thereby reduce the amount of dissolved oxygen available for fish and other forms of aquatic life. One state agency requires a minimum of 5 parts per million (ppm) of dissolved oxygen in order for the oxygen content to be sufficient to support aquatic life. Six water specimens taken from a river at a specific location during the low-water season (July) gave readings of 4.9, 5.1, 4.9, 5.0, 5.0, and 4.7 ppm of dissolved oxygen. Do the data provide sufficient evidence to indicate that the dissolved oxygen content is less than 5 ppm? Test using $\alpha = .05$.

10.8 Lobsters In a study of the infestation of the *Thenus orientalis* lobster by two types of barnacles, *Octolasmis tridens* and *O. lowei*, the carapace lengths (in millimeters) of 10 randomly selected lobsters caught in the seas near Singapore are measured[2]:

78	66	65	63	60	60	58	56	52	50

Find a 95% confidence interval for the mean carapace length of the *T. orientalis* lobsters.

(MY) CD EX1009 **10.9 Smoking and Lung Capacity** It is recognized that cigarette smoking has a deleterious effect on lung function. In a study of the effect of cigarette smoking on the carbon monoxide diffusing capacity (DL) of the lung, researchers found that current smokers had DL readings significantly lower than those of either exsmokers or nonsmokers. The carbon monoxide diffusing capacities for a random sample of $n = 20$ current smokers are listed here:

103.768	88.602	73.003	123.086	91.052
92.295	61.675	90.677	84.023	76.014
100.615	88.017	71.210	82.115	89.222
102.754	108.579	73.154	106.755	90.479

a. Do these data indicate that the mean DL reading for current smokers is significantly lower than 100 DL, the average for nonsmokers? Use $\alpha = .01$.

b. Find a 99% upper one-sided confidence bound for the mean DL reading for current smokers. Does this bound confirm your conclusions in part a?

(MY) CD EX1010 **10.10 Brett Favre** In Exercise 2.36 (EX0236), the number of passes completed by Brett Favre, quarterback for the Green Bay Packers, was recorded for each of the 16 regular season and 2 postseason games in the fall of 2003 (ESPN.com)[3]:

25	15	23	21	19	25
23	18	14	13	10	23
22	23	22	12	26	15

a. A stem and leaf plot of the $n = 18$ observations is shown below:

Stem-and-Leaf Display: Brett
```
Stem-and-leaf of Brett  N = 18
Leaf Unit = 1.0
 1   1   0
 3   1   23
 6   1   455
 6   1
 8   1   89
 9   2   1
 9   2   223333
 3   2   55
 1   2   6
```

Based on this plot, is it reasonable to assume that the underlying population is approximately normal, as required for the one-sample t test? Explain.

b. Calculate the mean and standard deviation for Brett Favre's per game pass completions.

c. Construct a 95% confidence interval to estimate the per game pass completions per game for Brett Favre.

10.11 Purifying Organic Compounds Organic chemists often purify organic compounds by a method known as fractional crystallization. An experimenter wanted to prepare and purify 4.85 grams (g) of aniline. Ten 4.85-g quantities of aniline were individually prepared and purified to acetanilide. The following dry yields were recorded:

3.85	3.80	3.88	3.85	3.90
3.36	3.62	4.01	3.72	3.82

Estimate the mean grams of acetanilide that can be recovered from an initial amount of 4.85 g of aniline. Use a 95% confidence interval.

10.12 Organic Compounds, continued Refer to Exercise 10.11. Approximately how many 4.85-g specimens of aniline are required if you wish to estimate the mean number of grams of acetanilide correct to within .06 g with probability equal to .95?

10.13 Bulimia Although there are many treatments for bulimia nervosa, some subjects fail to benefit from treatment. In a study to determine which factors predict who will benefit from treatment, an article in the *British Journal of Clinical Psychology* indicates that self-esteem was one of these important predictors.[4] The table gives the mean and standard deviation of self-esteem scores prior to treatment, at posttreatment, and during a follow-up:

	Pretreatment	Posttreatment	Follow-up
Sample mean $\bar{x}$	20.3	26.6	27.7
Standard deviation s	5.0	7.4	8.2
Sample size n	21	21	20

a. Use a test of hypothesis to determine whether there is sufficient evidence to conclude that the true pretreatment mean is less than 25.

b. Construct a 95% confidence interval for the true posttreatment mean.

c. In Section 10.4, we will introduce small-sample techniques for making inferences about the difference between two population means. Without the formality of a statistical test, what are you willing to conclude about the differences among the three sampled population means represented by the results in the table?

MY CD **10.14 RBC Counts** Here are the red blood
EX1014   cell counts (in 10^6 cells per microliter) of a
healthy person measured on each of 15 days:

5.4	5.2	5.0	5.2	5.5
5.3	5.4	5.2	5.1	5.3
5.3	4.9	5.4	5.2	5.2

Find a 95% confidence interval estimate of μ, the true
mean red blood cell count for this person during the
period of testing.

MY CD **10.15 Hamburger Meat** These data are the
EX1015   weights (in pounds) of 27 packages of ground
beef in a supermarket meat display:

1.08	.99	.97	1.18	1.41	1.28	.83
1.06	1.14	1.38	.75	.96	1.08	.87
.89	.89	.96	1.12	1.12	.93	1.24
.89	.98	1.14	.92	1.18	1.17	

a. Interpret the accompanying *MINITAB* printouts for
the one-sample test and estimation procedures.

MINITAB output for Exercise 10.15

One-Sample T: Weights
```
Test of mu = 1 vs not = 1

Variable     N        Mean    StDev   SE Mean
Weights      27    1.05222  0.16565   0.03188

Variable           95% CI             T       P
Weights    (0.98669, 1.11775)      1.64   0.113
```

b. Verify the calculated values of t and the upper and
lower confidence limits.

MY CD **10.16 Cholesterol** The serum cholesterol
EX1016   levels of 50 subjects randomly selected from the
L.A. Heart Data, data from an epidemiological heart
disease study on Los Angeles County employees,[5]
follow.

148	304	300	240	368	139	203	249	265	229
303	315	174	209	253	169	170	254	212	255
262	284	275	229	261	239	254	222	273	299
278	227	220	260	221	247	178	204	250	256
305	225	306	184	242	282	311	271	276	248

a. Construct a histogram for the data. Are the data
approximately mound-shaped?

b. Use a t-distribution to construct a 95% confidence
interval for the average serum cholesterol levels for
L.A. County employees.

10.17 Cholesterol, continued Refer to Exercise 10.16. Since $n > 30$, use the methods of Chapter 8
to create a large-sample 95% confidence interval for
the average serum cholesterol level for L.A. County
employees. Compare the two intervals. (**Hint:** The two
intervals should be quite similar. This is the reason we
choose to approximate the sample distribution of
$\dfrac{\bar{x} - \mu}{s/\sqrt{n}}$ with a z-distribution when $n > 30$.)

SMALL-SAMPLE INFERENCES FOR THE DIFFERENCE BETWEEN TWO POPULATION MEANS: INDEPENDENT RANDOM SAMPLES

10.4

The physical setting for the problem considered in this section is the same as the one
in Section 8.6, except that the sample sizes are no longer large. Independent random
samples of n_1 and n_2 measurements are drawn from two populations, with means
and variances μ_1, σ_1^2, μ_2, and σ_2^2, and your objective is to make inferences about
$(\mu_1 - \mu_2)$, the difference between the two population means.

When the sample sizes are small, you can no longer rely on the Central Limit Theorem to ensure that the sample means will be normal. If the original populations *are*
normal, however, then the sampling distribution of the difference in the sample means,
$(\bar{x}_1 - \bar{x}_2)$, will be normal (even for small samples) with mean $(\mu_1 - \mu_2)$ and standard error

$$\sqrt{\frac{\sigma_1^2}{n_1} + \frac{\sigma_2^2}{n_2}}$$

Assumptions for the two-sample (independent) t-test:
- Random independent samples
- Normal distributions
- $\sigma_1 = \sigma_2$

In Chapters 7 and 8, you used the sample variances, s_1^2 and s_2^2, to calculate an *estimate* of the standard error, which was then used to form a large-sample confidence interval or a test of hypothesis based on the large-sample z statistic:

$$z \approx \frac{(\bar{x}_1 - \bar{x}_2) - (\mu_1 - \mu_2)}{\sqrt{\dfrac{s_1^2}{n_1} + \dfrac{s_2^2}{n_2}}}$$

Unfortunately, when the sample sizes are small, this statistic does not have an approximately normal distribution—nor does it have a Student's t distribution. In order to form a statistic with a sampling distribution that can be derived theoretically, you must make one more assumption.

Suppose that the variability of the measurements in the two normal populations is the same and can be measured by a common variance σ^2. That is, *both populations have exactly the same shape, and $\sigma_1^2 = \sigma_2^2 = \sigma^2$.* Then the standard error of the difference in the two sample means is

$$\sqrt{\frac{\sigma_1^2}{n_1} + \frac{\sigma_2^2}{n_2}} = \sqrt{\sigma^2\left(\frac{1}{n_1} + \frac{1}{n_2}\right)}$$

It can be proven mathematically that, if you use the appropriate sample estimate s^2 for the population variance σ^2, then the resulting test statistic,

$$t = \frac{(\bar{x}_1 - \bar{x}_2) - (\mu_1 - \mu_2)}{\sqrt{s^2\left(\dfrac{1}{n_1} + \dfrac{1}{n_2}\right)}}$$

has a *Student's t distribution.* The only remaining problem is to find the sample estimate s^2 and the appropriate number of *degrees of freedom* for the t statistic.

Remember that the population variance σ^2 describes the shape of the normal distributions from which your samples come, so that either s_1^2 or s_2^2 would give you an estimate of σ^2. But why use just one when information is provided by both? A better procedure is to combine the information in both sample variances using a *weighted average,* in which the weights are determined by the relative amount of information (the number of measurements) in each sample. For example, if the first sample contained twice as many measurements as the second, you might consider giving the first sample variance twice as much weight. To achieve this result, use this formula:

$$s^2 = \frac{(n_1 - 1)s_1^2 + (n_2 - 1)s_2^2}{n_1 + n_2 - 2}$$

Remember from Section 10.3 that the degrees of freedom for the one-sample t statistic are $(n - 1)$, the denominator of the sample estimate s^2. Since s_1^2 has $(n_1 - 1)$ df and s_2^2 has $(n_2 - 1)$ df, the total number of degrees of freedom is the sum:

$$(n_1 - 1) + (n_2 - 1) = n_1 + n_2 - 2$$

shown in the denominator of the formula for s^2.

CALCULATION OF s^2

- If you have a scientific calculator, calculate each of the two sample standard deviations s_1 and s_2 separately, using the data entry procedure for your particular calculator. These values are squared and used in this formula:

$$s^2 = \frac{(n_1 - 1)s_1^2 + (n_2 - 1)s_2^2}{n_1 + n_2 - 2}$$

It can be shown that s^2 is an unbiased estimator of the common population variance σ^2. If s^2 is used to estimate σ^2 and if the samples have been randomly and independently drawn from normal populations with a common variance, then the statistic

MY TIP

For the two-sample (independent) *t*-test, $df = n_1 + n_2 - 2$

$$t = \frac{(\bar{x}_1 - \bar{x}_2) - (\mu_1 - \mu_2)}{\sqrt{s^2\left(\dfrac{1}{n_1} + \dfrac{1}{n_2}\right)}}$$

has a Student's t distribution with $(n_1 + n_2 - 2)$ degrees of freedom. The small-sample estimation and test procedures for the difference between two means are given next.

TEST OF HYPOTHESIS CONCERNING THE DIFFERENCE BETWEEN TWO MEANS: INDEPENDENT RANDOM SAMPLES

1. Null hypothesis: $H_0 : (\mu_1 - \mu_2) = D_0$, where D_0 is some specified difference that you wish to test. For many tests, you will hypothesize that there is no difference between μ_1 and μ_2; that is, $D_0 = 0$.
2. Alternative hypothesis:

One-Tailed Test	Two-Tailed Test
$H_a : (\mu_1 - \mu_2) > D_0$ [or $H_a : (\mu_1 - \mu_2) < D_0$]	$H_a : (\mu_1 - \mu_2) \neq D_0$

3. Test statistic: $t = \dfrac{(\bar{x}_1 - \bar{x}_2) - D_0}{\sqrt{s^2\left(\dfrac{1}{n_1} + \dfrac{1}{n_2}\right)}}$ where

$$s^2 = \frac{(n_1 - 1)s_1^2 + (n_2 - 1)s_2^2}{n_1 + n_2 - 2}$$

4. Rejection region: Reject H_0 when

One-Tailed Test	Two-Tailed Test
$t > t_\alpha$ [or $t < -t_\alpha$ when the alternative hypothesis is $H_a : (\mu_1 - \mu_2) < D_0$]	$t > t_{\alpha/2}$ or $t < -t_{\alpha/2}$

or when p-value $< \alpha$

(continued)

TEST OF HYPOTHESIS CONCERNING THE DIFFERENCE BETWEEN TWO MEANS: INDEPENDENT RANDOM SAMPLES *(continued)*

The critical values of t, t_α, and $t_{\alpha/2}$ are based on $(n_1 + n_2 - 2)$ df. The tabulated values can be found using Table 4 of Appendix I or the **Student's t Probabilities** applet.

Assumptions: The samples are randomly and independently selected from normally distributed populations. The variances of the populations σ_1^2 and σ_2^2 are equal.

SMALL-SAMPLE $(1 - \alpha)100\%$ CONFIDENCE INTERVAL FOR $(\mu_1 - \mu_2)$ BASED ON INDEPENDENT RANDOM SAMPLES

$$(\bar{x}_1 - \bar{x}_2) \pm t_{\alpha/2}\sqrt{s^2\left(\frac{1}{n_1} + \frac{1}{n_2}\right)}$$

where s^2 is the pooled estimate of σ^2.

EXAMPLE 10.5 — An assembly operation in a manufacturing plant requires approximately a 1-month training period for a new employee to reach maximum efficiency in assembling a device. A new method of training was suggested, and a test was conducted to compare the new method with the standard procedure. Two groups of nine new employees were trained for a period of 3 weeks, one group using the new method and the other following the standard training procedure. The length of time (in minutes) required for each employee to assemble the device was recorded at the end of the 3-week period. These measurements appear in Table 10.2. Do the data present sufficient evidence to indicate that the mean time to assemble at the end of a 3-week training period is less for the new training procedure?

TABLE 10.2 ● **Assembly Times After Two Training Procedures**

Standard Procedure	New Procedure
32	35
37	31
35	29
28	25
41	34
44	40
35	27
31	32
34	31

Solution Let μ_1 and μ_2 be the mean time to assemble after the standard and the new training procedures, respectively. Then, since you seek evidence to support the theory that $\mu_1 > \mu_2$, you can test the null hypothesis

$$H_0 : \mu_1 = \mu_2 \qquad [\text{or } H_0 : (\mu_1 - \mu_2) = 0]$$

versus the alternative hypothesis

$$H_a : \mu_1 > \mu_2 \qquad [\text{or } H_a : (\mu_1 - \mu_2) > 0]$$

To conduct the t test for these two independent samples, you must assume that the sampled populations are both normal and have the same variance σ^2. Is this reasonable? Stem and leaf plots of the data in Figure 10.10 show at least a "mounding" pattern, so that the assumption of normality is not unreasonable.

FIGURE 10.10

Stem and leaf plots for Example 10.5

Standard		New	
2	8	2	579
3	124	3	1124
3	557	3	5
4	14	4	0

Furthermore, the standard deviations of the two samples, calculated as

$$s_1 = 4.9441 \quad \text{and} \quad s_2 = 4.4752$$

 MY TIP

Stem and leaf plots can help you decide if the normality assumption is reasonable.

are not different enough for us to doubt that the two distributions may have the same shape. If you make these two assumptions and calculate (using full accuracy) the pooled estimate of the common variance as

$$s^2 = \frac{(n_1 - 1)s_1^2 + (n_2 - 1)s_2^2}{n_1 + n_2 - 2} = \frac{8(4.9441)^2 + 8(4.4752)^2}{9 + 9 - 2} = 22.2361$$

you can then calculate the test statistic,

$$t = \frac{\bar{x}_1 - \bar{x}_2}{\sqrt{s^2\left(\dfrac{1}{n_1} + \dfrac{1}{n_2}\right)}} = \frac{35.22 - 31.56}{\sqrt{22.2361\left(\dfrac{1}{9} + \dfrac{1}{9}\right)}} = 1.65$$

MY TIP

If you are using a calculator, don't round off until the final step!

The alternative hypothesis $H_a : \mu_1 > \mu_2$ or, equivalently, $H_a : (\mu_1 - \mu_2) > 0$ implies that you should use a one-tailed test in the upper tail of the t distribution with $(n_1 + n_2 - 2) = 16$ degrees of freedom. You can find the appropriate critical value for a rejection region with $\alpha = .05$ in Table 4 of Appendix I, and H_0 will be rejected if $t > 1.746$. Comparing the observed value of the test statistic $t = 1.65$ with the critical value $t_{.05} = 1.746$, you cannot reject the null hypothesis (see Figure 10.11). There is insufficient evidence to indicate that the new training procedure is superior at the 5% level of significance.

FIGURE 10.11

Rejection region for Example 10.5

EXAMPLE 10.6   Find the p-value that would be reported for the statistical test in Example 10.5.

Solution   The observed value of t for this one-tailed test is $t = 1.65$. Therefore,

$$p\text{-value} = P(t > 1.65)$$

for a t statistic with 16 degrees of freedom. Remember that you cannot obtain this probability directly from Table 4 in Appendix I; you can only *bound* the p-value using the critical values in the table. Since the observed value, $t = 1.65$, lies between $t_{.100} = 1.337$ and $t_{.050} = 1.746$, the tail area to the right of 1.65 is between .05 and .10. The p-value for this test would be reported as

$$.05 < p\text{-value} < .10$$

Because the p-value is greater than .05, most researchers would report the results as *not significant*.

(MY) **APPLET**

You can use the **Two-Sample t Test: Independent Samples** applet to visualize the p-values for either one- or two-tailed tests of the difference between two population means. The procedure follows the same pattern as with previous applets. You need to enter summary statistics—the values of $\bar{x}_1$, $\bar{x}_2$, n_1, n_2, s_1, and s_2 and press "Enter" after each entry; the applet will calculate t (assuming equal variances) and give you the option of choosing one- or two-tailed p-values, (as well as a *cumulative* and a *middle* area that you will not need.)

FIGURE 10.12

Two-Sample t Test:
Independent Samples
applet

For the data of Example 10.5, the p-value is the one-tailed area to the right of $t = 1.65$. Does the p-value confirm the conclusions for the test in Example 10.5?

EXAMPLE 10.7

Use a 95% confidence interval to estimate the difference $(\mu_1 - \mu_2)$ in Example 10.5. Does the confidence interval indicate that there is a significant difference in the means?

Solution The confidence interval formula takes a familiar form—the point estimator $(\bar{x}_1 - \bar{x}_2)$ plus or minus an amount equal to $t_{\alpha/2}$ times the standard error of the estimator. Substituting into the formula, you can calculate the 95% confidence interval:

$$(\bar{x}_1 - \bar{x}_2) \pm t_{\alpha/2}\sqrt{s^2\left(\frac{1}{n_1} + \frac{1}{n_2}\right)}$$

$$(35.22 - 31.56) \pm 2.120\sqrt{22.2361\left(\frac{1}{9} + \frac{1}{9}\right)}$$

$$3.66 \pm 4.71$$

or $-1.05 < (\mu_1 - \mu_2) < 8.37$. Since the value $(\mu_1 - \mu_2) = 0$ is included in the confidence interval, it is possible that the two means are equal. There is insufficient evidence to indicate a significant difference in the means.

Notice that the confidence interval is quite wide; if this width is not acceptable to you, you might choose to increase the size of the samples and reestimate the difference in the means using this additional information.

The two-sample procedure that uses a pooled estimate of the common variance σ^2 relies on four important assumptions:

- The samples must be *randomly selected.* Samples not randomly selected may introduce bias into the experiment and thus alter the significance levels you are reporting.

- The samples must be *independent.* If not, this is not the appropriate statistical procedure. We discuss another procedure for dependent samples in Section 10.5.

- The populations from which you sample must be *normal.* However, moderate departures from normality do not seriously affect the distribution of the test statistic, especially if the sample sizes are nearly the same.

- The population *variances should be equal* or nearly equal to ensure that the procedures are valid.

If the population variances are far from equal, there is an alternative procedure for estimation and testing that has an *approximate t* distribution in repeated sampling. As a rule of thumb, you should use this procedure if the ratio of the two sample variances,

$$\frac{\text{Larger } s^2}{\text{Smaller } s^2} > 3$$

Since the population variances are not equal, the pooled estimator s^2 is no longer appropriate, and each population variance must be estimated by its corresponding sample variance. The resulting test statistic is

$$\frac{(\bar{x}_1 - \bar{x}_2) - D_0}{\sqrt{\dfrac{s_1^2}{n_1} + \dfrac{s_2^2}{n_2}}}$$

When the sample sizes are *small,* critical values for this statistic are found using degrees of freedom approximated by the formula

$$df \approx \frac{\left(\dfrac{s_1^2}{n_1} + \dfrac{s_2^2}{n_2}\right)^2}{\dfrac{(s_1^2/n_1)^2}{(n_1 - 1)} + \dfrac{(s_2^2/n_2)^2}{(n_2 - 1)}}$$

The degrees of freedom are taken to be the integer part of this result.

Computer packages such as *MINITAB* can be used to implement this procedure, sometimes called *Satterthwaite's approximation,* as well as the *pooled method* described earlier. In fact, some experimenters choose to analyze their data using *both* methods. As long as both analyses lead to the same conclusions, you need not concern yourself with the equality or inequality of variances.

The *MINITAB* output resulting from the pooled method of analysis for the data of Example 10.5 is shown in Figure 10.13. Notice that the ratio of the two sample variances, $(4.94/4.48)^2 = 1.22$, is less than 3, which makes the pooled method appropriate. The calculated value of $t = 1.65$ and the exact p-value $= .059$ with 16 degrees of freedom are shown in the last line of the output. The exact p-value makes it quite easy for you to determine the significance or nonsignificance of the sample results. You will find instructions for generating this *MINITAB* output in the section "My *MINITAB*" at the end of this chapter.

FIGURE 10.13

MINITAB output for Example 10.5

●

Two-Sample T-Test and CI: Standard, New

```
Two-sample T for Standard vs New
                N      Mean      StDev     SE Mean
Standard   9       35.22      4.94        1.6
New        9       31.56      4.48        1.5

Difference = mu (Standard) - mu (New)
Estimate for difference: 3.66667
95% lower bound for difference: -0.21429
T-Test of difference = 0 (vs >): T-Value = 1.65  P-Value = 0.059  DF = 16
Both use Pooled StDev = 4.7155
```

If there is reason to believe that the normality assumptions have been violated, you can test for a shift in location of two population distributions using the nonparametric Wilcoxon rank sum test of Chapter 15. This test procedure, which requires fewer assumptions concerning the nature of the population probability distributions, is almost as sensitive in detecting a difference in population means when the conditions necessary for the t test are satisfied. It may be more sensitive when the normality assumption is not satisfied.

10.4 EXERCISES

BASIC TECHNIQUES

10.18 Give the number of degrees of freedom for s^2, the pooled estimator of σ^2, in these cases:

a. $n_1 = 16, n_2 = 8$

b. $n_1 = 10, n_2 = 12$

c. $n_1 = 15, n_2 = 3$

10.19 Calculate s^2, the pooled estimator for σ^2, in these cases:

a. $n_1 = 10$, $n_2 = 4$, $s_1^2 = 3.4$, $s_2^2 = 4.9$

b. $n_1 = 12$, $n_2 = 21$, $s_1^2 = 18$, $s_2^2 = 23$

10.20 Two independent random samples of sizes $n_1 = 4$ and $n_2 = 5$ are selected from each of two normal populations:

Population 1	12	3	8	5	
Population 2	14	7	7	9	6

a. Calculate s^2, the pooled estimator of σ^2.

b. Find a 90% confidence interval for $(\mu_1 - \mu_2)$, the difference between the two population means.

c. Test $H_0 : (\mu_1 - \mu_2) = 0$ against $H_a : (\mu_1 - \mu_2) < 0$ for $\alpha = .05$. State your conclusions.

10.21 Independent random samples of $n_1 = 16$ and $n_2 = 13$ observations were selected from two normal populations with equal variances:

	Population	
	1	2
Sample size	16	13
Sample mean	34.6	32.2
Sample variance	4.8	5.9

a. Suppose you wish to detect a difference between the population means. State the null and alternative hypotheses for the test.

b. Find the rejection region for the test in part a for $\alpha = .01$.

c. Find the value of the test statistic.

d. Find the approximate p-value for the test.

e. Conduct the test and state your conclusions.

10.22 Refer to Exercise 10.21. Find a 99% confidence interval for $(\mu_1 - \mu_2)$.

10.23 The *MINITAB* printout shows a test for the difference in two population means.

MINITAB output for Exercise 10.23

Two-Sample T-Test and CI: Sample 1, Sample 2

```
Two-sample T for Sample 1 vs Sample 2
          N      Mean    StDev   SE Mean
Sample 1  6     29.00     4.00      1.6
Sample 2  7     28.86     4.67      1.8

Difference = mu (Sample 1) - mu (Sample 2)
Estimate for difference: 0.14
95% CI for difference: (-5.2, 5.5)
T-Test of difference = 0 (vs not =):
T-Value = 0.06  P-Value = 0.95  DF = 11
Both use Pooled StDev = 4.38
```

a. Do the two sample standard deviations indicate that the assumption of a common population variance is reasonable?

b. What is the observed value of the test statistic? What is the p-value associated with this test?

c. What is the pooled estimate s^2 of the population variance?

d. Use the answers to part b to draw conclusions about the difference in the two population means.

e. Find the 95% confidence interval for the difference in the population means. Does this interval confirm your conclusions in part d?

APPLICATIONS

10.24 Healthy Teeth Jan Lindhe conducted a study on the effect of an oral antiplaque rinse on plaque buildup on teeth.[6] Fourteen people whose teeth were thoroughly cleaned and polished were randomly assigned to two groups of seven subjects each. Both groups were assigned to use oral rinses (no brushing) for a 2-week period. Group 1 used a rinse that contained an antiplaque agent. Group 2, the control group, received a similar rinse except that, unknown to the subjects, the rinse contained no antiplaque agent. A plaque index x, a measure of plaque buildup, was recorded at 4, 7, and 14 days. The mean and standard deviation for the 14-day plaque measurements are shown in the table for the two groups.

	Control Group	Antiplaque Group
Sample size	7	7
Mean	1.26	.78
Standard deviation	.32	.32

a. State the null and alternative hypotheses that should be used to test the effectiveness of the antiplaque oral rinse.

b. Do the data provide sufficient evidence to indicate that the oral antiplaque rinse is effective? Test using $\alpha = .05$.

c. Find the approximate p-value for the test.

MY CD **10.25 Tuna, again** In Exercise 10.6 we pre-
EX1025 sented data on the estimated average price for a 6-ounce can or a 7.06-ounce pouch of tuna, based on prices paid nationally in supermarkets. A portion of the data is reproduced in the table below. Use the *MINITAB* printout to answer the questions.

Light Tuna in Water		Light Tuna in Oil	
.99	.53	2.56	.62
1.92	1.41	1.92	.66
1.23	1.12	1.30	.62
.85	.63	1.79	.65
.65	.67	1.23	.60
.69	.60		.67
.60	.66		

MINITAB output for Exercise 10.25

Two-Sample T-Test and CI: Water, Oil

```
Two-sample T for Water vs Oil

          N      Mean    StDev   SE Mean
Water    14     0.896    0.400     0.11
Oil      11     1.147    0.679     0.20

Difference = mu (Water) - mu (Oil)
Estimate for difference: -0.250844
95% CI for difference: (-0.700004, 0.198316)
T-Test of difference = 0 (vs not =):
T-Value = -1.16  P-Value = 0.260  DF = 23
Both use Pooled StDev = 0.5389
```

a. Do the data in the table present sufficient evidence to indicate a difference in the average prices of light tuna in water versus oil? Test using $\alpha = .05$.

b. What is the p-value for the test?

c. The *MINITAB* analysis uses the pooled estimate of σ^2. Is the assumption of equal variances reasonable? Why or why not?

10.26 Runners and Cyclists Chronic anterior compartment syndrome is a condition characterized by exercise-induced pain in the lower leg. Swelling and impaired nerve and muscle function also accompany this pain, which is relieved by rest. Susan Beckham and colleagues conducted an experiment involving ten healthy runners and ten healthy cyclists to determine whether there are significant differences in pressure measurements within the anterior muscle compartment for runners and cyclists.[7] The data summary—compartment pressure in millimeters of mercury (Hg)—is as follows:

Condition	Runners Mean	Runners Standard Deviation	Cyclists Mean	Cyclists Standard Deviation
Rest	14.5	3.92	11.1	3.98
80% maximal O_2 consumption	12.2	3.49	11.5	4.95
Maximal O_2 consumption	19.1	16.9	12.2	4.47

a. Test for a significant difference in compartment pressure between runners and cyclists under the resting condition. Use $\alpha = .05$.

b. Construct a 95% confidence interval estimate of the difference in means for runners and cyclists under the condition of exercising at 80% of maximal oxygen consumption.

c. To test for a significant difference in compartment pressure at maximal oxygen consumption, should you use the pooled or unpooled t test? Explain.

10.27 Disinfectants An experiment published in *The American Biology Teacher* studied the efficacy of using 95% ethanol or 20% bleach as a disinfectant in removing bacterial and fungal contamination when culturing plant tissues. The experiment was repeated 15 times with each disinfectant, using eggplant as the plant tissue being cultured.[8] Five cuttings per plant were placed on a petri dish for each disinfectant and stored at 25°C for 4 weeks. The observation reported was the number of uncontaminated eggplant cuttings after the 4-week storage.

Disinfectant	95% Ethanol	20% Bleach
Mean	3.73	4.80
Variance	2.78095	.17143
n	15	15
Pooled variance 1.47619		

a. Are you willing to assume that the underlying variances are equal?

b. Using the information from part a, are you willing to conclude that there is a significant difference in the mean numbers of uncontaminated eggplants for the two disinfectants tested?

10.28 Titanium A geologist collected 20 different ore samples, all of the same weight, and randomly divided them into two groups. The titanium contents of the samples, found using two different methods, are listed in the table:

Method 1					Method 2				
.011	.013	.013	.015	.014	.011	.016	.013	.012	.015
.013	.010	.013	.011	.012	.012	.017	.013	.014	.015

a. Use an appropriate method to test for a significant difference in the average titanium contents using the two different methods.

b. Determine a 95% confidence interval estimate for $(\mu_1 - \mu_2)$. Does your interval estimate substantiate your conclusion in part a? Explain.

10.29 Raisins The numbers of raisins in each of 14 miniboxes (1/2-ounce size) were counted for a generic brand and for Sunmaid brand raisins:

Generic Brand				Sunmaid			
25	26	25	28	25	29	24	24
26	28	28	27	28	24	28	22
26	27	24	25	25	28	30	27
26	26			28	24		

a. Although counts cannot have a normal distribution, do these data have approximately normal distributions? (HINT: Use a histogram or stem and leaf plot.)

b. Are you willing to assume that the underlying population variances are equal? Why?

c. Use the p-value approach to determine whether there is a significant difference in the mean numbers of raisins per minibox. What are the implications of your conclusion?

10.30 Dissolved O$_2$ Content, continued Refer to Exercise 10.7, in which we measured the dissolved oxygen content in river water to determine whether a stream had sufficient oxygen to support aquatic life. A pollution control inspector suspected that a river community was releasing amounts of semitreated sewage into a river. To check his theory, he drew five randomly selected specimens of river water at a location above the town, and another five below. The dissolved oxygen readings (in parts per million) are as follows:

Above Town	4.8	5.2	5.0	4.9	5.1
Below Town	5.0	4.7	4.9	4.8	4.9

a. Do the data provide sufficient evidence to indicate that the mean oxygen content below the town is less than the mean oxygen content above? Test using $\alpha = .05$.

b. Suppose you prefer estimation as a method of inference. Estimate the difference in the mean dissolved oxygen contents for locations above and below the town. Use a 95% confidence interval.

MY CD 10.31 Freestyle Swimmers In an effort to
EX1031 compare the average swimming times for two swimmers, each swimmer was asked to swim freestyle for a distance of 100 yards at randomly selected times. The swimmers were thoroughly rested between laps and did not race against each other, so that each sample of times was an independent random sample. The times for each of ten trials are shown for the two swimmers.

Swimmer 1		Swimmer 2	
59.62	59.74	59.81	59.41
59.48	59.43	59.32	59.63
59.65	59.72	59.76	59.50
59.50	59.63	59.64	59.83
60.01	59.68	59.86	59.51

Suppose that swimmer 2 was last year's winner when the two swimmers raced. Does it appear that the average time for swimmer 2 is still faster than the average time for swimmer 1 in the 100-yard freestyle? Find the approximate p-value for the test and interpret the results.

10.32 Freestyle Swimmers, continued Refer to Exercise 10.31. Construct a lower 95% one-sided confidence bound for the difference in the average times for the two swimmers. Does this interval confirm your conclusions in Exercise 10.31?

MY CD 10.33 Comparing NFL Quarterbacks How
EX1033 does Brett Favre, quarterback for the Green Bay Packers, compare with Donovan McNabb, quarterback for the Philadelphia Eagles? The table below shows the number of completed passes for each athlete during the 2003 NFL football season (ESPN.com)[3]:

Brett Favre			Donovan McNabb		
25	23	22	19	17	18
15	18	23	18	21	15
23	14	22	18	15	27
21	13	12	16	24	23
19	10	26	11	16	21
25	23	15	9	18	10

a. Does the data indicate that there is a difference in the average number of completed passes for the two quarterbacks? Test using $\alpha = .05$.

b. Construct a 95% confidence interval for the difference in the average number of completed passes for the two quarterbacks. Does the confidence interval confirm your conclusion in part a? Explain.

MY CD 10.34 An Archeological Find An article in
EX1034 *Archaeometry* involved an analysis of 26 samples of Romano-British pottery, found at four different kiln sites in the United Kingdom.[9] The samples were analyzed to determine their chemical composition and the percentage of aluminum oxide in each of 10 samples at two sites is shown below.

Island Thorns	Ashley Rails
18.3	17.7
15.8	18.3
18.0	16.7
18.0	14.8
20.8	19.1

Does the data provide sufficient information to indicate that there is a difference in the average percentage of aluminum oxide at the two sites? Test at the 5% level of significance.

SMALL-SAMPLE INFERENCES FOR THE DIFFERENCE BETWEEN TWO MEANS: A PAIRED-DIFFERENCE TEST

10.5

To compare the wearing qualities of two types of automobile tires, A and B, a tire of type A and one of type B are randomly assigned and mounted on the rear wheels of each of five automobiles. The automobiles are then operated for a specified number of miles, and the amount of wear is recorded for each tire. These measurements appear in Table 10.3. Do the data present sufficient evidence to indicate a difference in the average wear for the two tire types?

TABLE 10.3 ● **Average Wear for Two Types of Tires**

Automobile	Tire A	Tire B
1	10.6	10.2
2	9.8	9.4
3	12.3	11.8
4	9.7	9.1
5	8.8	8.3
	$\bar{x}_1 = 10.24$	$\bar{x}_2 = 9.76$
	$s_1 = 1.316$	$s_2 = 1.328$

Table 10.3 shows a difference of $(\bar{x}_1 - \bar{x}_2) = (10.24 - 9.76) = .48$ between the two sample means, while the standard deviations of both samples are approximately 1.3. Given the variability of the data and the small number of measurements, this is a rather small difference, and you would probably not suspect a difference in the average wear for the two types of tires. Let's check your suspicions using the methods of Section 10.4.

Look at the *MINITAB* analysis in Figure 10.14. The two-sample *pooled t* test is used for testing the difference in the means based on two independent random samples. The calculated value of t used to test the null hypothesis $H_0 : \mu_1 = \mu_2$ is $t = .57$ with p-value $= .582$, a value that is not nearly small enough to indicate a significant difference in the two population means. The corresponding 95% confidence interval, given as

$$-1.448 < (\mu_1 - \mu_2) < 2.408$$

is quite wide and also does not indicate a significant difference in the population means.

FIGURE 10.14

MINITAB output using *t* test for independent samples for the tire data

```
Two-Sample T-Test and CI: Tire A, Tire B

Two-sample T for Tire A vs Tire B
            N     Mean    StDev   SE Mean
Tire A   5    10.24     1.32      0.59
Tire B   5     9.76     1.33      0.59

Difference = mu (Tire A) - mu (Tire B)
Estimate for difference: 0.480000
95% CI for difference: (-1.448239, 2.408239)
T-Test of difference = 0 (vs not =): T-Value = 0.57  P-Value = 0.582  DF = 8
Both use Pooled StDev = 1.3221
```

Take a second look at the data and you will notice that the wear measurement for type A is greater than the corresponding value for type B for *each* of the five automobiles. Wouldn't this be unlikely, if there's really no difference between the two tire types?

Consider a simple intuitive test, based on the binomial distribution of Chapter 5. If there is no difference in the mean tire wear for the two types of tires, then it is just as likely as not that tire A shows more wear than tire B. The five automobiles then correspond to five binomial trials with $p = P(\text{tire A shows more wear than tire B}) = .5$. Is the observed value of $x = 5$ positive differences unusual? The probability of observing $x = 5$ or the equally unlikely value $x = 0$ can be found in Table 1 in Appendix I to be $2(.031) = .062$, which is quite small compared to the likelihood of the more powerful t-test, which had a p-value of .58. Isn't it peculiar that the t test, which uses more information (the actual sample measurements) than the binomial test, fails to supply sufficient information for rejecting the null hypothesis?

TABLE 10.4

Differences in Tire Wear, Using the Data of Table 10.3

Automobile	A	B	$d = A - B$
1	10.6	10.2	.4
2	9.8	9.4	.4
3	12.3	11.8	.5
4	9.7	9.1	.6
5	8.8	8.3	.5
			$\bar{d} = .48$

There is an explanation for this inconsistency. The t test described in Section 10.4 is *not* the proper statistical test to be used for our example. The statistical test procedure of Section 10.4 requires that the two samples be *independent and random*. Certainly, the independence requirement is violated by the manner in which the experiment was conducted. The (pair of) measurements, an A and a B tire, for a particular automobile are definitely related. A glance at the data shows that the readings have approximately the same magnitude for a particular automobile but vary markedly from one automobile to another. This, of course, is exactly what you might expect. Tire wear is largely determined by driver habits, the balance of the wheels, and the road surface. Since each automobile has a different driver, you would expect a large amount of variability in the data from one automobile to another.

In designing the tire wear experiment, the experimenter realized that the measurements would vary greatly from automobile to automobile. If the tires (five of type A and five of type B) were randomly assigned to the ten wheels, resulting in *independent* random samples, this variability would result in a large standard error and make it difficult to detect a difference in the means. Instead, he chose to "pair" the measurements, comparing the wear for type A and type B tires on each of the five automobiles. This experimental design, sometimes called a **paired-difference** or **matched pairs** design, allows us to eliminate the car-to-car variability by looking at only the five difference measurements shown in Table 10.4. These five differences form a single random sample of size $n = 5$.

Notice that in Table 10.4 the sample mean of the differences, $d = A - B$, is calculated as

$$\bar{d} = \frac{\Sigma d_i}{n} = .48$$

and is exactly the same as the difference of the sample means: $(\bar{x}_1 - \bar{x}_2) = (10.24 - 9.76) = .48$. It should not surprise you that this can be proven to be true in general, and also that the same relationship holds for the population means. That is, the average of the population differences is

$$\mu_d = (\mu_1 - \mu_2)$$

Because of this fact, you can use the sample differences to test for a significant difference in the two population means, $(\mu_1 - \mu_2) = \mu_d$. The test is a single-sample t test of the difference measurements to test the null hypothesis

$$H_0 : \mu_d = 0 \qquad [\text{or } H_0 : (\mu_1 - \mu_2) = 0]$$

versus the alternative hypothesis

$$H_a : \mu_d \neq 0 \qquad [\text{or } H_a : (\mu_1 - \mu_2) \neq 0]$$

The test procedures take the same form as the procedures used in Section 10.3 and are described next.

PAIRED-DIFFERENCE TEST OF HYPOTHESIS FOR $(\mu_1 - \mu_2) = \mu_d$: DEPENDENT SAMPLES

1. Null hypothesis: $H_0 : \mu_d = 0$
2. Alternative hypothesis:

One-Tailed Test	**Two-Tailed Test**
$H_a : \mu_d > 0$	$H_a : \mu_d \neq 0$
(or $H_a : \mu_d < 0$)	

3. Test statistic: $t = \dfrac{\bar{d} - 0}{s_d/\sqrt{n}} = \dfrac{\bar{d}}{s_d/\sqrt{n}}$

   where n = Number of paired differences
   $\bar{d}$ = Mean of the sample differences
   s_d = Standard deviation of the sample differences

4. Rejection region: Reject H_0 when

One-Tailed Test	**Two-Tailed Test**
$t > t_\alpha$	$t > t_{\alpha/2}$ or $t < -t_{\alpha/2}$
(or $t < -t_\alpha$ when the alternative hypothesis is $H_a : \mu_d < 0$)	

   or when p-value $< \alpha$

The critical values of t, t_α, and $t_{\alpha/2}$ are based on $(n - 1)$ df. These tabulated values can be found using Table 4 or the **Student's t Probabilities** applet.

> ## $(1 - \alpha)100\%$ SMALL-SAMPLE CONFIDENCE INTERVAL FOR $(\mu_1 - \mu_2) = \mu_d$, BASED ON A PAIRED-DIFFERENCE EXPERIMENT
>
> $$\bar{d} \pm t_{\alpha/2}\left(\frac{s_d}{\sqrt{n}}\right)$$
>
> **Assumptions:** The experiment is designed as a paired-difference test so that the n differences represent a random sample from a normal population.

EXAMPLE 10.8

Do the data in Table 10.3 provide sufficient evidence to indicate a difference in the mean wear for tire types A and B? Test using $\alpha = .05$.

Solution   You can verify using your calculator that the average and standard deviation of the five difference measurements are

$$\bar{d} = .48 \quad \text{and} \quad s_d = .0837$$

Then

$$H_0 : \mu_d = 0 \quad \text{and} \quad H_a : \mu_d \neq 0$$

and

$$t = \frac{\bar{d} - 0}{s_d/\sqrt{n}} = \frac{.48}{.0837/\sqrt{5}} = 12.8$$

The critical value of t for a two-tailed statistical test, $\alpha = .05$ and 4 df, is 2.776. Certainly, the observed value of $t = 12.8$ is extremely large and highly significant. Hence, you can conclude that there is a difference in the mean wear for tire types A and B.

EXAMPLE 10.9

Find a 95% confidence interval for $(\mu_1 - \mu_2) = \mu_d$ using the data in Table 10.3.

Solution   A 95% confidence interval for the difference between the mean wears is

$$\bar{d} \pm t_{\alpha/2}\left(\frac{s_d}{\sqrt{n}}\right)$$

$$.48 \pm 2.776\left(\frac{.0837}{\sqrt{5}}\right)$$

$$.48 \pm .10$$

TIP

Confidence intervals are always interpreted in the same way! In repeated sampling, intervals constructed in this way enclose the true value of the parameter $100(1 - \alpha)\%$ of the time.

or $.38 < (\mu_1 - \mu_2) < .58$. How does the width of this interval compare with the width of an interval you might have constructed *if* you had designed the experiment in an unpaired manner? It probably would have been of the same magnitude as the interval calculated in Figure 10.14, where the observed data were *incorrectly* analyzed using the unpaired analysis. This interval, $-1.45 < (\mu_1 - \mu_2) < 2.41$, is much wider than the paired interval, which indicates that the paired difference design increased the accuracy of our estimate, and we have gained valuable information by using this design.

The *paired-difference test* or *matched pairs design* used in the tire wear experiment is a simple example of an experimental design called a **randomized block design.**

When there is a great deal of variability among the experimental units, even before any experimental procedures are implemented, the effect of this variability can be minimized by **blocking**—that is, comparing the different procedures within groups of relatively similar experimental units called **blocks.** In this way, the "noise" caused by the large variability does not mask the true differences between the procedures. We will discuss randomized block designs in more detail in Chapter 11.

 MY TIP

Paired difference test:

$df = n - 1$

It is important for you to remember that the *pairing* or *blocking* occurs when the experiment is planned, and not after the data are collected. An experimenter may choose to use pairs of identical twins to compare two learning methods. A physician may record a patient's blood pressure before and after a particular medication is given. Once you have used a paired design for an experiment, you no longer have the option of using the unpaired analysis of Section 10.4. The independence assumption has been purposely violated, and your only choice is to use the paired analysis described here!

Although pairing was very beneficial in the tire wear experiment, this may not always be the case. In the paired analysis, the degrees of freedom for the t test are cut in half—from $(n + n - 2) = 2(n - 1)$ to $(n - 1)$. This reduction *increases* the critical value of t for rejecting H_0 and also increases the width of the confidence interval for the difference in the two means. If pairing is not effective, this increase is not offset by a *decrease* in the variability, and you may in fact lose rather than gain information by pairing. This, of course, did not happen in the tire experiment—the large reduction in the standard error more than compensated for the loss in degrees of freedom.

Except for notation, the paired-difference analysis is the same as the single-sample analysis presented in Section 10.3. However, *MINITAB* provides a single procedure called **Paired** t to analyze the differences, as shown in Figure 10.15. The p-value for the paired analysis, .000, indicates a *highly significant* difference in the means. You will find instructions for generating this *MINITAB* output in the section "My *MINITAB*" at the end of this chapter.

FIGURE 10.15

MINITAB output for paired-difference analysis of tire wear data

Paired T-Test and CI: Tire A, Tire B

```
Paired T for Tire A - Tire B
                 N      Mean       StDev    SE Mean
  Tire A         5    10.2400      1.3164     0.5887
  Tire B         5     9.7600      1.3278     0.5938
  Difference     5     0.480000    0.083666   0.037417

  95% CI for mean difference: (0.376115, 0.583885)
  T-Test of mean difference = 0 (vs not = 0): T-Value = 12.83  P-Value = 0.000
```

10.5 EXERCISES

BASIC TECHNIQUES

10.35 A paired-difference experiment was conducted using $n = 10$ pairs of observations.

a. Test the null hypothesis $H_0 : (\mu_1 - \mu_2) = 0$ against $H_a : (\mu_1 - \mu_2) \neq 0$ for $\alpha = .05$, $\bar{d} = .3$, and $s_d^2 = .16$. Give the approximate p-value for the test.

b. Find a 95% confidence interval for $(\mu_1 - \mu_2)$.

c. How many pairs of observations do you need if you want to estimate $(\mu_1 - \mu_2)$ correct to within .1 with probability equal to .95?

10.36 A paired-difference experiment consists of $n = 18$ pairs, $\bar{d} = 5.7$, and $s_d^2 = 256$. Suppose you wish to detect $\mu_d > 0$.

a. Give the null and alternative hypotheses for the test.

b. Conduct the test and state your conclusions.

10.37 A paired-difference experiment was conducted to compare the means of two populations:

	Pairs				
Population	1	2	3	4	5
1	1.3	1.6	1.1	1.4	1.7
2	1.2	1.5	1.1	1.2	1.8

a. Do the data provide sufficient evidence to indicate that μ_1 differs from μ_2? Test using $\alpha = .05$.

b. Find the approximate p-value for the test and interpret its value.

c. Find a 95% confidence interval for $(\mu_1 - \mu_2)$. Compare your interpretation of the confidence interval with your test results in part a.

d. What assumptions must you make for your inferences to be valid?

APPLICATIONS

(MY) CD **10.38** Auto Insurance The cost of automobile insurance has become a sore subject in California because the rates are dependent on so many variables, such as the city in which you live, the number of cars you insure, and the company with which you are insured. Here are the 6-month premiums in 2004 for a single male, licensed for 6 to 8 years, who drives about 15,000 miles per year and has no violations or accidents.[10]

EX1038

City	Allstate	21st Century
Long Beach	$884	$996
Pomona	730	940
San Bernardino	725	836
Moreno Valley	801	762

Source: www.insurance.ca.gov

a. Why would you expect these pairs of observations to be dependent?

b. Do the data provide sufficient evidence to indicate that there is a difference in the average 6-month premiums between Allstate and 21st Century insurance? Test using $\alpha = .01$.

c. Find the approximate p-value for the test and interpret its value.

d. Find a 99% confidence interval for the difference in the average 6-month premiums for Allstate and 21st Century insurance.

e. Can we use the information in the table to make valid comparisons between Allstate and 21st Century insurance throughout the United States? Why or why not?

10.39 Runners and Cyclists II Refer to Exercise 10.26. In addition to the compartment pressures, the level of creatine phosphokinase (CPK) in blood samples, a measure of muscle damage, was determined for each of 10 runners and 10 cyclists before and after exercise.[7] The data summary—CPK values in units/liter—is as follows:

Condition	Runners Mean	Runners Standard Deviation	Cyclists Mean	Cyclists Standard Deviation
Before exercise	255.63	115.48	173.8	60.69
After exercise	284.75	132.64	177.1	64.53
Difference	29.13	21.01	3.3	6.85

a. Test for a significant difference in mean CPK values for runners and cyclists before exercise under the assumption that $\sigma_1^2 \neq \sigma_2^2$; use $\alpha = .05$. Find a 95% confidence interval estimate for the corresponding difference in means.

b. Test for a significant difference in mean CPK values for runners and cyclists after exercise under the assumption that $\sigma_1^2 \neq \sigma_2^2$; use $\alpha = .05$. Find a 95% confidence interval estimate for the corresponding difference in means.

c. Test for a significant difference in mean CPK values for runners before and after exercise.

d. Find a 95% confidence interval estimate for the difference in mean CPK values for cyclists before and after exercise. Does your estimate indicate that there is no significant difference in mean CPK levels for cyclists before and after exercise?

(MY) CD **10.40** America's Market Basket An advertisement for Albertsons, a supermarket chain in the western United States, claims that Albertsons has had consistently lower prices than four other full-service supermarkets. As part of a survey conducted by an "independent market basket price-checking company," the average weekly total, based on the prices of approximately 95 items, is given for two different supermarket chains recorded during 4 consecutive weeks in a particular month.

EX1040

Week	Albertsons	Ralphs
1	254.26	256.03
2	240.62	255.65
3	231.90	255.12
4	234.13	261.18

a. Is there a significant difference in the average prices for these two different supermarket chains?

b. What is the approximate p-value for the test conducted in part a?

c. Construct a 99% confidence interval for the difference in the average prices for the two supermarket chains. Interpret this interval.

MY CD **10.41 No Left Turn** An experiment was conducted to compare the mean reaction times to two types of traffic signs: prohibitive (No Left Turn) and permissive (Left Turn Only). Ten drivers were included in the experiment. Each driver was presented with 40 traffic signs, 20 prohibitive and 20 permissive, in random order. The mean time to reaction and the number of correct actions were recorded for each driver. The mean reaction times (in milliseconds) to the 20 prohibitive and 20 permissive traffic signs are shown here for each of the ten drivers:

Driver	Prohibitive	Permissive
1	824	702
2	866	725
3	841	744
4	770	663
5	829	792
6	764	708
7	857	747
8	831	685
9	846	742
10	759	610

a. Explain why this is a paired-difference experiment and give reasons why the pairing should be useful in increasing information on the difference between the mean reaction times to prohibitive and permissive traffic signs.

b. Do the data present sufficient evidence to indicate a difference in mean reaction times to prohibitive and permissive traffic signs? Use the p-value approach.

c. Find a 95% confidence interval for the difference in mean reaction times to prohibitive and permissive traffic signs.

10.42 Healthy Teeth II Exercise 10.24 describes a dental experiment conducted to investigate the effectiveness of an oral rinse used to inhibit the growth of plaque on teeth. Subjects were divided into two groups: One group used a rinse with an antiplaque ingredient, and the control group used a rinse containing inactive ingredients. Suppose that the plaque growth on each person's teeth was measured after using the rinse after 4 hours and then again after 8 hours. If you wish to estimate the difference in plaque growth from 4 to 8 hours, should you use a confidence interval based on a paired or an unpaired analysis? Explain.

10.43 Ground or Air? The earth's temperature (which affects seed germination, crop survival in bad weather, and many other aspects of agricultural production) can be measured using either ground-based sensors or infrared-sensing devices mounted in aircraft or space satellites. Ground-based sensoring is tedious, requiring many replications to obtain an accurate estimate of ground temperature. On the other hand, airplane or satellite sensoring of infrared waves appears to introduce a bias in the temperature readings. To determine the bias, readings were obtained at five different locations using both ground- and air-based temperature sensors. The readings (in degrees Celsius) are listed here:

Location	Ground	Air
1	46.9	47.3
2	45.4	48.1
3	36.3	37.9
4	31.0	32.7
5	24.7	26.2

a. Do the data present sufficient evidence to indicate a bias in the air-based temperature readings? Explain.

b. Estimate the difference in mean temperatures between ground- and air-based sensors using a 95% confidence interval.

c. How many paired observations are required to estimate the difference between mean temperatures for ground- versus air-based sensors correct to within .2°C, with probability approximately equal to .95?

MY CD **10.44 Red Dye** To test the comparative brightness of two red dyes, nine samples of cloth were taken from a production line and each sample was divided into two pieces. One of the two pieces in each sample was randomly chosen and red dye 1 applied; red dye 2 was applied to the remaining piece. The following data represent a "brightness score" for each piece. Is there sufficient evidence to indicate a difference in mean brightness scores for the two dyes? Use $\alpha = .05$.

Sample	1	2	3	4	5	6	7	8	9
Dye 1	10	12	9	8	15	12	9	10	15
Dye 2	8	11	10	6	12	13	9	8	13

MY CD **10.45 Tax Assessors** In response to a complaint that a particular tax assessor (A) was biased, an experiment was conducted to compare the assessor named in the complaint with another tax assessor (B) from the same office. Eight properties were selected, and each was assessed by both assessors. The assessments (in thousands of dollars) are shown in the table:

Property	Assessor A	Assessor B
1	76.3	75.1
2	88.4	86.8
3	80.2	77.3
4	94.7	90.6
5	68.7	69.1
6	82.8	81.0
7	76.1	75.3
8	79.0	79.1

Use the *MINITAB* printout to answer the questions.

MINITAB output for Exercise 10.45

Paired T-Test and CI: Assessor A, Assessor B

```
Paired T for Assessor A - Assessor B
                N      Mean     StDev    SE Mean
Assessor A   8   80.7750    7.9939     2.8263
Assessor B   8   79.2875    6.8510     2.4222
Difference   8    1.48750   1.49134    0.52727

95% lower bound for mean difference: 0.48855
T-Test of mean difference = 0 (vs > 0):
T-Value = 2.82  P-value = 0.013
```

a. Do the data provide sufficient evidence to indicate that assessor A tends to give higher assessments than assessor B?

b. Estimate the difference in mean assessments for the two assessors.

c. What assumptions must you make in order for the inferences in parts a and b to be valid?

d. Suppose that assessor A had been compared with a more stable standard—say, the average $\bar{x}$ of the assessments given by four assessors selected from the tax office. Thus, each property would be assessed by A and also by each of the four other assessors and $(x_A - \bar{x})$ would be calculated. If the test in part a is valid, can you use the paired-difference t test to test the hypothesis that the bias, the mean difference between A's assessments and the mean of the assessments of the four assessors, is equal to 0? Explain.

10.46 Memory Experiments A psychology EX1046 class performed an experiment to compare whether a recall score in which instructions to form images of 25 words were given is better than an initial recall score for which no imagery instructions were given. Twenty students participated in the experiment with the following results:

Student	With Imagery	Without Imagery	Student	With Imagery	Without Imagery
1	20	5	11	17	8
2	24	9	12	20	16
3	20	5	13	20	10
4	18	9	14	16	12
5	22	6	15	24	7
6	19	11	16	22	9
7	20	8	17	25	21
8	19	11	18	21	14
9	17	7	19	19	12
10	21	9	20	23	13

Does it appear that the average recall score is higher when imagery is used?

10.47 Music in the Workplace Before con-EX1047 tracting to have stereo music piped into each of his suites of offices, an executive had his office manager randomly select seven offices in which to have the system installed. The average time (in minutes) spent outside these offices per excursion among the employees involved was recorded before and after the music system was installed with the following results.

Office number	1	2	3	4	5	6	7
No music	8	9	5	6	5	10	7
Music	5	6	7	5	6	7	8

Would you suggest that the executive proceed with the installation? Conduct an appropriate test of hypothesis. Find the approximate *p*-value and interpret your results.

INFERENCES CONCERNING A POPULATION VARIANCE

10.6

You have seen in the preceding sections that an estimate of the population variance σ^2 is usually needed before you can make inferences about population means. Sometimes, however, the population variance σ^2 is the primary objective in an experimental investigation. It may be *more* important to the experimenter than the population mean! Consider these examples:

- Scientific measuring instruments must provide unbiased readings with a very small error of measurement. An aircraft altimeter that measures the correct altitude on the *average* is fairly useless if the measurements are in error by as much as 1000 feet above or below the correct altitude.
- Machined parts in a manufacturing process must be produced with minimum variability in order to reduce out-of-size and hence defective parts.
- Aptitude tests must be designed so that scores *will* exhibit a reasonable amount of variability. For example, an 800-point test is not very discriminatory if all students score between 601 and 605.

In previous chapters, you have used

$$s^2 = \frac{\Sigma(x_i - \bar{x})^2}{n - 1}$$

as an unbiased estimator of the population variance σ^2. This means that, in repeated sampling, the average of all your sample estimates will equal the target parameter, σ^2. But how close or far from the target is your estimator s^2 likely to be? To answer this question, we use the sampling distribution of s^2, which describes its behavior in repeated sampling.

Consider the distribution of s^2 based on repeated *random* sampling from a *normal* distribution with a specified mean and variance. We can show theoretically that the distribution begins at $s^2 = 0$ (since the variance cannot be negative) with a mean equal to σ^2. Its shape is *nonsymmetric* and changes with each different sample size and each different value of σ^2. Finding critical values for the sampling distribution of s^2 would be quite difficult and would require separate tables for each population variance. Fortunately, we can simplify the problem by *standardizing*, as we did with the z distribution.

Definition The standardized statistic

$$\chi^2 = \frac{(n - 1)s^2}{\sigma^2}$$

is called a **chi-square variable** and has a sampling distribution called the **chi-square probability distribution**, with $n - 1$ degrees of freedom.

The equation of the density function for this statistic is quite complicated to look at, but it traces the curve shown in Figure 10.16.

FIGURE 10.16

A chi-square distribution

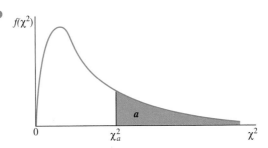

Certain critical values of the chi-square statistic, which are used for making inferences about the population variance, have been tabulated by statisticians and appear in Table 5 of Appendix I. Since the shape of the distribution varies with the sample size n or, more precisely, the degrees of freedom, $n - 1$, associated with s^2, Table 5, partially reproduced in Table 10.5, is constructed in exactly the same way as the t table, with the degrees of freedom in the first and last columns. The symbol χ_a^2 indicates that the tabulated χ^2-value has an area a to its right (see Figure 10.16).

TABLE 10.5 ● **Format of the Chi-Square Table from Table 5 in Appendix I**

df	$\chi^2_{.995}$	$\cdots$	$\chi^2_{.950}$	$\chi^2_{.900}$	$\chi^2_{.100}$	$\chi^2_{.050}$	$\cdots$	$\chi^2_{.005}$	df
1	.0000393		.0039321	.0157908	2.70554	3.84146		7.87944	1
2	.0100251		.102587	.210720	4.60517	5.99147		10.5966	2
3	.0717212		.351846	.584375	6.25139	7.81473		12.8381	3
4	.206990		.710721	1.063623	7.77944	9.48773		14.8602	4
5	.411740		1.145476	1.610310	9.23635	11.0705		16.7496	5
6	.0675727		1.63539	2.204130	10.6446	12.5916		18.5476	6
.	.		.	.	.	.		.	.
.	.		.	.	.	.		.	.
.	.		.	.	.	.		.	.
15	4.60094		7.26094	8.54675	22.3072	24.9958		32.8013	15
16	5.14224		7.96164	9.31223	23.5418	26.2962		34.2672	16
17	5.69724		8.67176	10.0852	24.7690	27.5871		35.7185	17
18	6.26481		9.39046	10.8649	25.9894	28.8693		37.1564	18
19	6.84398		10.1170	11.6509	27.2036	30.1435		38.5822	19
.	.		.	.	.	.		.	.
.	.		.	.	.	.		.	.
.	.		.	.	.	.		.	.

MY TIP

Testing one variance:
$df = n - 1$

You can see in Table 10.5 that, because the distribution is nonsymmetric and starts at 0, both upper and lower tail areas must be tabulated for the chi-square statistic. For example, the value $\chi^2_{.95}$ is the value that has 95% of the area under the curve to its right and 5% of the area to its left. This value cuts off an area equal to .05 in the lower tail of the chi-square distribution.

EXAMPLE 10.10

Check your ability to use Table 5 in Appendix I by verifying the following statements:

1. The probability that χ^2, based on $n = 16$ measurements ($df = 15$), exceeds 24.9958 is .05.

2. For a sample of $n = 6$ measurements, 95% of the area under the χ^2 distribution lies to the right of 1.145476.

These values are shaded in Table 10.5.

MY APPLET

You can use the **Chi-Square Probabilities** applet to find the χ^2-value described in Example 10.10. Since the applet provides χ^2-values and their one-tailed probabilities for the degrees of freedom that you select using the slider on the right side of the applet, you should choose $df = 5$ and type .95 in the box marked "prob:" at the bottom of the applet. The applet will provide the value of χ^2 that puts .95 in the right tail of the χ^2 distribution and hence .05 in the left tail. The applet in Figure 10.17 shows $\chi^2 = 1.14$, which differs only slightly from the value in Example 10.10. We will use this applet for the MyApplet Exercises at the end of the chapter.

FIGURE 10.17 ●

Chi-Square Probabilities
applet

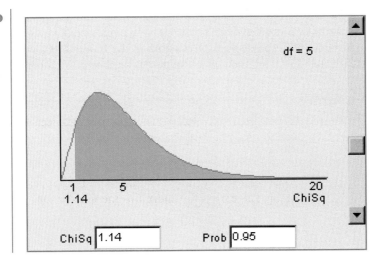

The statistical test of a null hypothesis concerning a population variance

$$H_0 : \sigma^2 = \sigma_0^2$$

uses the test statistic

$$\chi^2 = \frac{(n - 1)s^2}{\sigma_0^2}$$

Notice that when H_0 is true, s^2/σ_0^2 should be near 1, so χ^2 should be close to $(n - 1)$, the degrees of freedom. If σ^2 is really greater than the hypothesized value σ_0^2, the test statistic will tend to be larger than $(n - 1)$ and will probably fall toward the upper tail of the distribution. If $\sigma^2 < \sigma_0^2$, the test statistic will tend to be smaller than $(n - 1)$ and will probably fall toward the lower tail of the chi-square distribution. As in other testing situations, you may use either a one- or a two-tailed statistical test, depending on the alternative hypothesis. This test of hypothesis and the $(1 - \alpha)100\%$ confidence interval for σ^2 are both based on the chi-square distribution and are described next.

TEST OF HYPOTHESIS CONCERNING A POPULATION VARIANCE

1. Null hypothesis: $H_0 : \sigma^2 = \sigma_0^2$
2. Alternative hypothesis:

One-Tailed Test	**Two-Tailed Test**
$H_a : \sigma^2 > \sigma_0^2$ (or $H_a : \sigma^2 < \sigma_0^2$)	$H_a : \sigma^2 \neq \sigma_0^2$

3. Test statistic: $\chi^2 = \dfrac{(n - 1)s^2}{\sigma_0^2}$

4. Rejection region: Reject H_0 when

One-Tailed Test

$\chi^2 > \chi^2_\alpha$
(or $\chi^2 < \chi^2_{(1-\alpha)}$ when the alternative hypothesis is $H_a : \sigma^2 < \sigma_0^2$), where χ^2_α and $\chi^2_{(1-\alpha)}$ are, respectively, the upper- and lower-tail values of χ^2 that place α in the tail areas

or when p-value $< \alpha$

Two-Tailed Test

$\chi^2 > \chi^2_{\alpha/2}$ or $\chi^2 < \chi^2_{(1-\alpha/2)}$, where $\chi^2_{\alpha/2}$ and $\chi^2_{(1-\alpha/2)}$ are, respectively, the upper- and lower-tail values of χ^2 that place $\alpha/2$ in the tail areas

The critical values of χ^2 are based on $(n - 1)$ df. These tabulated values can be found using Table 5 of Appendix I or the **Chi-Square Probabilities** applet.

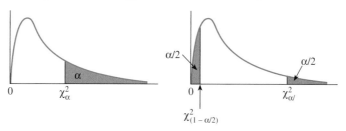

$(1 - \alpha)100\%$ CONFIDENCE INTERVAL FOR σ^2

$$\frac{(n - 1)s^2}{\chi^2_{\alpha/2}} < \sigma^2 < \frac{(n - 1)s^2}{\chi^2_{(1-\alpha/2)}}$$

where $\chi^2_{\alpha/2}$ and $\chi^2_{(1-\alpha/2)}$ are the upper and lower χ^2-values, which locate one-half of α in each tail of the chi-square distribution.
Assumption: The sample is randomly selected from a normal population.

EXAMPLE 10.11

A cement manufacturer claims that concrete prepared from his product has a relatively stable compressive strength and that the strength measured in kilograms per square centimeter (kg/cm^2) lies within a range of 40 kg/cm^2. A sample of $n = 10$ measurements produced a mean and variance equal to, respectively,

$$\bar{x} = 312 \quad \text{and} \quad s^2 = 195$$

Do these data present sufficient evidence to reject the manufacturer's claim?

Solution In Section 2.5, you learned that the range of a set of measurements should be approximately four standard deviations. The manufacturer's claim that the range of the strength measurements is within 40 kg/cm^2 must mean that the standard deviation of the measurements is roughly 10 kg/cm^2 or less. To test his claim, the appropriate hypotheses are

$$H_0 : \sigma^2 = 10^2 = 100 \quad \text{versus} \quad H_a : \sigma^2 > 100$$

If the sample variance is much larger than the hypothesized value of 100, then the test statistic

$$\chi^2 = \frac{(n - 1)s^2}{\sigma_0^2} = \frac{1755}{100} = 17.55$$

will be unusually large, favoring rejection of H_0 and acceptance of H_a. There are two ways to use the test statistic to make a decision for this test.

- **The critical value approach:** The appropriate test requires a one-tailed rejection region in the right tail of the χ^2 distribution. The critical value for $\alpha = .05$ and $(n - 1) = 9$ df is $\chi^2_{.05} = 16.9190$ from Table 5 in Appendix I. Figure 10.18 shows the rejection region; you can reject H_0 if the test statistic exceeds 16.9190. Since the observed value of the test statistic is $\chi^2 = 17.55$, you can conclude that the null hypothesis is false and that the range of concrete strength measurements exceeds the manufacturer's claim.

FIGURE 10.18

Rejection region and *p*-value (shaded) for Example 10.11

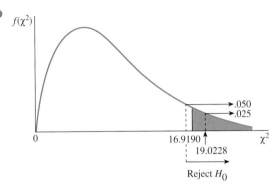

- **The *p*-value approach:** The *p*-value for a statistical test is the smallest value of α for which H_0 can be rejected. It is calculated, as in other one-tailed tests, as the area in the tail of the χ^2 distribution to the right of the observed value, $\chi^2 = 17.55$. Although computer packages allow you to calculate this area exactly, Table 5 in Appendix I allows you only to bound the *p*-value. Since the value 17.55 lies between $\chi^2_{.050} = 16.9190$ and $\chi^2_{.025} = 19.0228$, the *p*-value lies between .025 and .05. Most researchers would reject H_0 and report these results as significant at the 5% level, or $P < .05$. Again, you can reject H_0 and conclude that the range of measurements exceeds the manufacturer's claim.

EXAMPLE 10.12 ─ An experimenter is convinced that her measuring instrument had a variability measured by standard deviation $\sigma = 2$. During an experiment, she recorded the measurements 4.1, 5.2, and 10.2. Do these data confirm or disprove her assertion? Test the appropriate hypothesis, and construct a 90% confidence interval to estimate the true value of the population variance.

Solution Since there is no preset level of significance, you should choose to use the *p*-value approach in testing these hypotheses:

$$H_0 : \sigma^2 = 4 \quad \text{versus} \quad H_a : \sigma^2 \neq 4$$

Use your scientific calculator to verify that the sample variance is $s^2 = 10.57$ and the test statistic is

$$\chi^2 = \frac{(n - 1)s^2}{\sigma_0^2} = \frac{2(10.57)}{4} = 5.29$$

Since this is a two-tailed test, the rejection region is divided into two parts, half in each tail of the χ^2 distribution. If you approximate the area to the right of the observed test statistic, $\chi^2 = 5.29$, you will have only *half* of the *p*-value for the test. Since an equally unlikely value of χ^2 might occur in the lower tail of the distribution, with equal probability, you must *double* the upper area to obtain the *p*-value. With 2 *df,* the observed value, 5.29, falls between $\chi^2_{.10}$ and $\chi^2_{.05}$ so that

$$.05 < \frac{1}{2}(p\text{-value}) < .10 \quad \text{or} \quad .10 < p\text{-value} < .20$$

Since the *p*-value is greater than .10, the results are not statistically significant. There is insufficient evidence to reject the null hypothesis $H_0 : \sigma^2 = 4$.

The corresponding 90% confidence interval is

$$\frac{(n-1)s^2}{\chi^2_{\alpha/2}} < \sigma^2 < \frac{(n-1)s^2}{\chi^2_{(1-\alpha/2)}}$$

The values of $\chi^2_{(1-\alpha/2)}$ and $\chi^2_{\alpha/2}$ are

$$\chi^2_{(1-\alpha/2)} = \chi^2_{.95} = .102587$$

$$\chi^2_{\alpha/2} = \chi^2_{.05} = 5.99147$$

Substituting these values into the formula for the interval estimate, you get

$$\frac{2(10.57)}{5.99147} < \sigma^2 < \frac{2(10.57)}{.102587} \quad \text{or} \quad 3.53 < \sigma^2 < 206.07$$

Thus, you can estimate the population variance to fall into the interval 3.53 to 206.07. This very wide confidence interval indicates how little information on the population variance is obtained from a sample of only three measurements. Consequently, it is not surprising that there is insufficient evidence to reject the null hypothesis $\sigma^2 = 4$. To obtain more information on σ^2, the experimenter needs to increase the sample size.

 10.6 EXERCISES

BASIC TECHNIQUES

10.48 A random sample of $n = 25$ observations from a normal population produced a sample variance equal to 21.4. Do these data provide sufficient evidence to indicate that $\sigma^2 > 15$? Test using $\alpha = .05$.

10.49 A random sample of $n = 15$ observations was selected from a normal population. The sample mean and variance were $\bar{x} = 3.91$ and $s^2 = .3214$. Find a 90% confidence interval for the population variance σ^2.

10.50 A random sample of size $n = 7$ from a normal population produced these measurements: 1.4, 3.6, 1.7, 2.0, 3.3, 2.8, 2.9.

a. Calculate the sample variance, s^2.

b. Construct a 95% confidence interval for the population variance, σ^2.

c. Test $H_0 : \sigma^2 = .8$ versus $H_a : \sigma^2 \neq .8$ using $\alpha = .05$. State your conclusions.

d. What is the approximate *p*-value for the test in part c?

APPLICATIONS

10.51 Instrument Precision A precision instrument is guaranteed to read accurately to within 2 units. A sample of four instrument readings on the same object yielded the measurements 353, 351, 351, and 355. Test the null hypothesis that $\sigma = .7$ against the alternative $\sigma > .7$. Use $\alpha = .05$.

10.52 Instrument Precision, continued Find a 90% confidence interval for the population variance in Exercise 10.51.

10.53 Drug Potency To properly treat patients, drugs prescribed by physicians must have a potency that is accurately defined. Consequently, not only must the distribution of potency values for shipments of a drug have a mean value as specified on the drug's container, but also the variation in potency must be small. Otherwise, pharmacists would be distributing drug prescriptions that could be harmfully potent or have a low potency and be ineffective. A drug manufacturer claims that his drug is marketed with a potency of

5 ± .1 milligram per cubic centimeter (mg/cc). A random sample of four containers gave potency readings equal to 4.94, 5.09, 5.03, and 4.90 mg/cc.

a. Do the data present sufficient evidence to indicate that the mean potency differs from 5 mg/cc?

b. Do the data present sufficient evidence to indicate that the variation in potency differs from the error limits specified by the manufacturer? [HINT: It is sometimes difficult to determine exactly what is meant by limits on potency as specified by a manufacturer. Since he implies that the potency values will fall into the interval 5.0 ± .1 mg/cc with very high probability—the implication is *always*—let us assume that the range .2; or (4.9 to 5.1), represents 6σ, as suggested by the Empirical Rule. Note that letting the range equal 6σ rather than 4σ places a stringent interpretation on the manufacturer's claim. We want the potency to fall into the interval 5.0 ± .1 with very high probability.]

10.54 Drug Potency, continued Refer to Exercise 10.53. Testing of 60 additional randomly selected containers of the drug gave a sample mean and variance equal to 5.04 and .0063 (for the total of $n = 64$ containers). Using a 95% confidence interval, estimate the variance of the manufacturer's potency measurements.

10.55 Hard Hats A manufacturer of hard safety hats for construction workers is concerned about the mean and the variation of the forces helmets transmit to wearers when subjected to a standard external force. The manufacturer desires the mean force transmitted by helmets to be 800 pounds (or less), well under the legal 1000-pound limit, and σ to be less than 40. A random sample of $n = 40$ helmets was tested, and the sample mean and variance were found to be equal to 825 pounds and 2350 pounds^2, respectively.

a. If $\mu = 800$ and $\sigma = 40$, is it likely that any helmet, subjected to the standard external force, will transmit a force to a wearer in excess of 1000 pounds? Explain.

b. Do the data provide sufficient evidence to indicate that when the helmets are subjected to the standard external force, the mean force transmitted by the helmets exceeds 800 pounds?

10.56 Hard Hats, continued Refer to Exercise 10.55. Do the data provide sufficient evidence to indicate that σ exceeds 40?

(MY) CD **10.57 Light Bulbs** A manufacturer of indus-
EX1057 trial light bulbs likes its bulbs to have a mean life that is acceptable to its customers and a variation in life that is relatively small. If some bulbs fail too early in their life, customers become annoyed and shift to competitive products. Large variations above the mean reduce replacement sales, and variation in general disrupts customers' replacement schedules. A sample of 20 bulbs tested produced the following lengths of life (in hours):

| 2100 | 2302 | 1951 | 2067 | 2415 | 1883 | 2101 | 2146 | 2278 | 2019 |
| 1924 | 2183 | 2077 | 2392 | 2286 | 2501 | 1946 | 2161 | 2253 | 1827 |

The manufacturer wishes to control the variability in length of life so that σ is less than 150 hours. Do the data provide sufficient evidence to indicate that the manufacturer is achieving this goal? Test using $\alpha = .01$.

COMPARING TWO POPULATION VARIANCES

10.7

Just as a single population variance is sometimes important to an experimenter, you might also need to compare two population variances. You might need to compare the precision of one measuring device with that of another, the stability of one manufacturing process with that of another, or even the variability in the grading procedure of one college professor with that of another.

One way to compare two population variances, σ_1^2 and σ_2^2, is to use the ratio of the sample variances, s_1^2/s_2^2. If s_1^2/s_2^2 is nearly equal to 1, you will find little evidence to indicate that σ_1^2 and σ_2^2 are unequal. On the other hand, a very large or very small value for s_1^2/s_2^2 provides evidence of a difference in the population variances.

How large or small must s_1^2/s_2^2 be for sufficient evidence to exist to reject the following null hypothesis?

$$H_0 : \sigma_1^2 = \sigma_2^2$$

The answer to this question may be found by studying the distribution of s_1^2/s_2^2 in repeated sampling.

When independent random samples are drawn from two *normal* populations with *equal variances*—that is, $\sigma_1^2 = \sigma_2^2$—then s_1^2/s_2^2 has a probability distribution in repeated sampling that is known to statisticians as an **F distribution**, shown in Figure 10.19.

FIGURE 10.19

An *F* distribution with $df_1 = 10$ and $df_2 = 10$

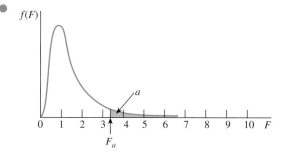

ASSUMPTIONS FOR s_1^2/s_2^2 TO HAVE AN *F* DISTRIBUTION

- Random and independent samples are drawn from each of two normal populations.
- The variability of the measurements in the two populations is the same and can be measured by a common variance, σ^2; that is, $\sigma_1^2 = \sigma_2^2 = \sigma^2$.

It is not important for you to know the complex equation of the density function for *F*. For your purposes, you need only to use the well-tabulated critical values of *F* given in Table 6 in Appendix I.

Critical values of *F* and *p*-values for significance tests can also be found using the **F Probabilities** applet shown in Figure 10.20.

 TIP

Testing two variances: $df_1 = n_1 - 1$ and $df_2 = n_2 - 1$

Like the χ^2 distribution, the shape of the *F* distribution is nonsymmetric and depends on the number of degrees of freedom associated with s_1^2 and s_2^2, represented as $df_1 = (n_1 - 1)$ and $df_2 = (n_2 - 1)$, respectively. This complicates the tabulation of critical values of the *F* distribution because a table is needed for each different combination of df_1, df_2, and *a*.

In Table 6 in Appendix I, critical values of *F* for right-tailed areas corresponding to $a = .100, .050, .025, .010,$ and $.005$ are tabulated for various combinations of df_1 numerator degrees of freedom and df_2 denominator degrees of freedom. A portion of Table 6 is reproduced in Table 10.6. The numerator degrees of freedom df_1 are listed across the top margin, and the denominator degrees of freedom df_2 are listed along the side margin. The values of *a* are listed in the second column. For a fixed combination of df_1 and df_2, the appropriate critical values of *F* are found in the line indexed by the value of *a* required.

FIGURE 10.20

F Probabilities applet

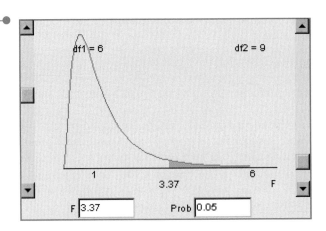

EXAMPLE 10.13

Check your ability to use Table 6 in Appendix I by verifying the following statements:

1. The value of F with area .05 to its right for $df_1 = 6$ and $df_2 = 9$ is 3.37.
2. The value of F with area .05 to its right for $df_1 = 5$ and $df_2 = 10$ is 3.33.
3. The value of F with area .01 to its right for $df_1 = 6$ and $df_2 = 9$ is 5.80.

These values are shaded in Table 10.6.

TABLE 10.6 ● **Format of the *F* Table from Table 6 in Appendix I**

df_2	a	1	2	3	4	5	6
				df_1			
1	.100	39.86	49.50	53.59	55.83	57.24	58.20
	.050	161.4	199.5	215.7	224.6	230.2	234.0
	.025	647.8	799.5	864.2	899.6	921.8	937.1
	.010	4052	4999.5	5403	5625	5764	5859
	.005	16211	20000	21615	22500	23056	23437
2	.100	8.53	9.00	9.16	9.24	9.29	9.33
	.050	18.51	19.00	19.16	19.25	19.30	19.33
	.025	38.51	39.00	39.17	39.25	39.30	39.33
	.010	98.50	99.00	99.17	99.25	99.30	99.33
	.005	198.5	199.0	199.2	199.2	199.3	199.3
3	.100	5.54	5.46	5.39	5.34	5.31	5.28
	.050	10.13	9.55	9.28	9.12	9.01	8.94
	.025	17.44	16.04	15.44	15.10	14.88	14.73
	.010	34.12	30.82	29.46	28.71	28.24	27.91
	.005	55.55	49.80	47.47	46.19	45.39	44.84
.	.			.			.
.	.			.			.
.	.			.			.
9	.100	3.36	3.01	2.81	2.69	2.61	2.55
	.050	5.12	4.26	3.86	3.63	3.48	3.37
	.025	7.21	5.71	5.08	4.72	4.48	4.32
	.010	10.56	8.02	6.99	6.42	6.06	5.80
	.005	13.61	10.11	8.72	7.96	7.47	7.13
10	.100	3.29	2.92	2.73	2.61	2.52	2.46
	.050	4.96	4.10	3.71	3.48	3.33	3.22
	.025	6.94	5.46	4.83	4.47	4.24	4.07
	.010	10.04	7.56	6.55	5.99	5.64	5.39
	.005	12.83	9.43	8.08	7.34	6.87	6.54

The statistical test of the null hypothesis

$$H_0 : \sigma_1^2 = \sigma_2^2$$

uses the test statistic

$$F = \frac{s_1^2}{s_2^2}$$

When the alternative hypothesis implies a one-tailed test—that is,

$$H_a : \sigma_1^2 > \sigma_2^2$$

you can find the right-tailed critical value for rejecting H_0 directly from Table 6 in Appendix I. However, when the alternative hypothesis requires a two-tailed test—that is,

$$H_0 : \sigma_1^2 \neq \sigma_2^2$$

the rejection region is divided between the upper and lower tails of the F distribution. These left-tailed critical values are *not given* in Table 6 for the following reason: You are free to decide which of the two populations you want to call "Population 1." If you always choose to call the population with the *larger* sample variance "Population 1," then the observed value of your test statistic will always be in the right tail of the F distribution. Even though half of the rejection region, the area $\alpha/2$ to its left, will be in the lower tail of the distribution, you will never need to use it! Remember these points, though, for a two-tailed test:

- The area in the right tail of the rejection region is only $\alpha/2$.
- The area to the right of the observed test statistic is only $1/2(p\text{-value})$.

The formal procedures for a test of hypothesis and a $(1 - \alpha)100\%$ confidence interval for two population variances are shown next.

TEST OF HYPOTHESIS CONCERNING THE EQUALITY OF TWO POPULATION VARIANCES

1. Null hypothesis: $H_0 : \sigma_1^2 = \sigma_2^2$
2. Alternative hypothesis:

One-Tailed Test	**Two-Tailed Test**
$H_a : \sigma_1^2 > \sigma_2^2$ (or $H_a : \sigma_1^2 < \sigma_2^2$)	$H_a : \sigma_1^2 \neq \sigma_2^2$

3. Test statistic:

One-Tailed Test	**Two-Tailed Test**
$F = \dfrac{s_1^2}{s_2^2}$	$F = \dfrac{s_1^2}{s_2^2}$

   where s_1^2 is the larger sample variance

4. Rejection region: Reject H_0 when

One-Tailed Test	**Two-Tailed Test**
$F > F_\alpha$	$F > F_{\alpha/2}$

   or when p-value $< \alpha$

(continued)

TEST OF HYPOTHESIS CONCERNING THE EQUALITY OF TWO POPULATION VARIANCES

(continued)

The critical values of F_α and $F_{\alpha/2}$ are based on $df_1 = (n_1 - 1)$ and $df_2 = (n_2 - 1)$. These tabulated values, for $a = .100, .050, .025, .010,$ and $.005,$ can be found using Table 6 in Appendix I, or the **F Probabilities** applet.

 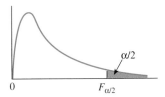

Assumptions: The samples are randomly and independently selected from normally distributed populations.

CONFIDENCE INTERVAL FOR σ_1^2/σ_2^2

$$\left(\frac{s_1^2}{s_2^2}\right)\frac{1}{F_{df_1,df_2}} < \frac{\sigma_1^2}{\sigma_2^2} < \left(\frac{s_1^2}{s_2^2}\right)F_{df_2,df_1}$$

where $df_1 = (n_1 - 1)$ and $df_2 = (n_2 - 1)$. F_{df_1,df_2} is the tabulated critical value of F corresponding to df_1 and df_2 degrees of freedom in the numerator and denominator of F, respectively, with area $\alpha/2$ to its right.

Assumptions: The samples are randomly and independently selected from normally distributed populations.

EXAMPLE 10.14 — An experimenter is concerned that the variability of responses using two different experimental procedures may not be the same. Before conducting his research, he conducts a prestudy with random samples of 10 and 8 responses and gets $s_1^2 = 7.14$ and $s_2^2 = 3.21$, respectively. Do the sample variances present sufficient evidence to indicate that the population variances are unequal?

Solution   Assume that the populations have probability distributions that are reasonably mound-shaped and hence satisfy, for all practical purposes, the assumption that the populations are normal. You wish to test these hypotheses:

$$H_0 : \sigma_1^2 = \sigma_2^2 \quad \text{versus} \quad H_a : \sigma_1^2 \neq \sigma_2^2$$

Using Table 6 in Appendix I for $\alpha/2 = .025$, you can reject H_0 when $F > 4.82$ with $\alpha = .05$. The calculated value of the test statistic is

$$F = \frac{s_1^2}{s_2^2} = \frac{7.14}{3.21} = 2.22$$

Because the test statistic does not fall into the rejection region, you cannot reject $H_0 : \sigma_1^2 = \sigma_2^2$. Thus, there is insufficient evidence to indicate a difference in the population variances.

EXAMPLE 10.15

Refer to Example 10.14 and find a 90% confidence interval for σ_1^2/σ_2^2.

Solution   The 90% confidence interval for σ_1^2/σ_2^2 is

$$\left(\frac{s_1^2}{s_2^2}\right)\frac{1}{F_{df_1,df_2}} < \frac{\sigma_1^2}{\sigma_2^2} < \left(\frac{s_1^2}{s_2^2}\right)F_{df_2,df_1}$$

where

$$
\begin{array}{ll}
s_1^2 = 7.14 & s_2^2 = 3.21 \\
df_1 = (n_1 - 1) = 9 & df_2 = (n_2 - 1) = 7 \\
F_{9,7} = 3.68 & F_{7,9} = 3.29
\end{array}
$$

Substituting these values into the formula for the confidence interval, you get

$$\left(\frac{7.14}{3.21}\right)\frac{1}{3.68} < \frac{\sigma_1^2}{\sigma_2^2} < \left(\frac{7.14}{3.21}\right)3.29 \quad \text{or} \quad .60 < \frac{\sigma_1^2}{\sigma_2^2} < 7.32$$

The calculated interval estimate .60 to 7.32 includes 1.0, the value hypothesized in H_0. This indicates that it is quite possible that $\sigma_1^2 = \sigma_2^2$ and therefore agrees with the test conclusions. Do not reject $H_0 : \sigma_1^2 = \sigma_2^2$.

The *MINITAB* command **Stat → Basic Statistics → 2 Variances** allows you to enter either raw data or summary statistics to perform the F-test for the equality of variances and calculates confidence intervals for the two individual standard deviations (which we have not discussed). The relevant printout, containing the F statistic and its p-value, is shaded in Figure 10.21.

FIGURE 10.21

MINITAB output for Example 10.14

```
Test for Equal Variances

95% Bonferroni confidence intervals for standard deviations
Sample   N     Lower      StDev      Upper
     1  10    1.74787    2.67208    5.38064
     2   8    1.12088    1.79165    4.10374


F-Test (normal distribution)
Test statistic = 2.22, p-value = 0.304
```

EXAMPLE 10.16

The variability in the amount of impurities present in a batch of chemical used for a particular process depends on the length of time the process is in operation. A manufacturer using two production lines 1 and 2 has made a slight adjustment to line 2, hoping to reduce the variability as well as the average amount of impurities in the chemical. Samples of $n_1 = 25$ and $n_2 = 25$ measurements from the two batches yield these means and variances:

$$
\begin{array}{ll}
\bar{x}_1 = 3.2 & s_1^2 = 1.04 \\
\bar{x}_2 = 3.0 & s_2^2 = .51
\end{array}
$$

Do the data present sufficient evidence to indicate that the process variability is less for line 2?

Solution   The experimenter believes that the average levels of impurities are the same for the two production lines but that her adjustment may have decreased the variability of the levels for line 2, as illustrated in Figure 10.22. This adjustment would be good for the company because it would decrease the probability of producing shipments of the chemical with unacceptably high levels of impurities.

FIGURE 10.22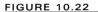

Distributions of impurity measurements for two production lines

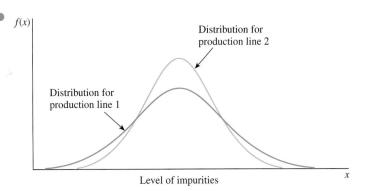

To test for a decrease in variability, the test of hypothesis is

$$H_0 : \sigma_1^2 = \sigma_2^2 \quad \text{versus} \quad H_a : \sigma_1^2 > \sigma_2^2$$

and the observed value of the test statistic is

$$F = \frac{s_1^2}{s_2^2} = \frac{1.04}{.51} = 2.04$$

Using the p-value approach, you can bound the one-tailed p-value using Table 6 in Appendix I with $df_1 = df_2 = (25 - 1) = 24$. The observed value of F falls between $F_{.050} = 1.98$ and $F_{.025} = 2.27$, so that $.025 < p\text{-value} < .05$. The results are judged significant at the 5% level, and H_0 is rejected. You can conclude that the variability of line 2 is less than that of line 1.

The F test for the difference in two population variances completes the battery of tests you have learned in this chapter for making inferences about population parameters under these conditions:

- The sample sizes are small.
- The sample or samples are drawn from normal populations.

You will find that the F and χ^2 distributions, as well as the Student's t distribution, are very important in other applications in the chapters that follow. They will be used for different estimators designed to answer different types of inferential questions, but the basic techniques for making inferences remain the same.

In the next section, we review the assumptions required for all of these inference tools, and discuss options that are available when the assumptions do not seem to be reasonably correct.

10.7 EXERCISES

BASIC TECHNIQUES

10.58 Independent random samples from two normal populations produced the variances listed here:

Sample Size	Sample Variance
16	55.7
20	31.4

a. Do the data provide sufficient evidence to indicate that σ_1^2 differs from σ_2^2? Test using $\alpha = .05$.

b. Find the approximate p-value for the test and interpret its value.

10.59 Refer to Exercise 10.58 and find a 95% confidence interval for σ_1^2/σ_2^2.

10.60 Independent random samples from two normal populations produced the given variances:

Sample Size	Sample Variance
13	18.3
13	7.9

a. Do the data provide sufficient evidence to indicate that $\sigma_1^2 > \sigma_2^2$? Test using $\alpha = .05$.

b. Find the approximate p-value for the test and interpret its value.

APPLICATIONS

10.61 SAT Scores The average total SAT scores (verbal plus math) were recorded for two groups of students: one group planning to major in engineering and one group planning to major in language/literature.

Engineering	Language/Literature
$\bar{x} = 994$	$\bar{x} = 1051$
$s = 71$	$s = 69$
$n = 15$	$n = 15$

To use the two-sample t test with a pooled estimate of σ^2, you must assume that the two population variances are equal. Test this assumption using the F test for equality of variances. What is the approximate p-value for the test?

10.62 Product Quality The stability of measurements on a manufactured product is important in maintaining product quality. In fact, it is sometimes better to have small variation in the measured value of some important characteristic of a product and have the process mean be slightly off target than to suffer wide variation with a mean value that perfectly fits requirements. The latter situation may produce a higher percentage of defective products than the former. A manufacturer of light bulbs suspected that one of her production lines was producing bulbs with a wide variation in length of life. To test this theory, she compared the lengths of life for $n = 50$ bulbs randomly sampled from the suspect line and $n = 50$ from a line that seemed to be "in control." The sample means and variances for the two samples were as follows:

"Suspect Line"	Line "in Control"
$\bar{x}_1 = 1520$	$\bar{x}_2 = 1476$
$s_1^2 = 92,000$	$s_2^2 = 37,000$

a. Do the data provide sufficient evidence to indicate that bulbs produced by the "suspect line" have a larger variance in length of life than those produced by the line that is assumed to be in control? Test using $\alpha = .05$.

b. Find the approximate p-value for the test and interpret its value.

10.63 Construct a 90% confidence interval for the variance ratio in Exercise 10.62.

10.64 Tuna III In Exercise 10.25 and dataset EX1025, you conducted a test to detect a difference in the average prices of light tuna in water versus light tuna in oil.

a. What assumption had to be made concerning the population variances so that the test would be valid?

b. Do the data present sufficient evidence to indicate that the variances violate the assumption in part a? Test using $\alpha = .05$.

10.65 Runners and Cyclists III Refer to Exercise 10.26. Susan Beckham and colleagues conducted an experiment involving 10 healthy runners and 10 healthy cyclists to determine if there are significant differences in pressure measurements within the anterior muscle compartment for runners and cyclists.[7] The data—compartment pressure, in millimeters of mercury (Hg)—are reproduced here:

	Runners		Cyclists	
Condition	Mean	Standard Deviation	Mean	Standard Deviation
Rest	14.5	3.92	11.1	3.98
80% maximal				
O$_2$ consumption	12.2	3.49	11.5	4.95
Maximal O$_2$ consumption	19.1	16.9	12.2	4.47

For each of the three variables measured in this experiment, test to see whether there is a significant difference in the variances for runners versus cyclists. Find the approximate p-values for each of these tests. Will a two-sample t test with a pooled estimate of σ^2 be appropriate for all three of these variables? Explain.

10.66 Impurities A pharmaceutical manufacturer purchases a particular material from two different suppliers. The mean level of impurities in the raw material is approximately the same for both suppliers, but the manufacturer is concerned about the variability of the impurities from shipment to shipment. If the level of impurities tends to vary excessively for one source of supply, it could affect the quality of the pharmaceutical product. To compare the variation in percentage impurities for the two suppliers, the manufacturer selects 10 shipments from each of the two suppliers and measures the percentage of impurities in the raw material for each shipment. The sample means and variances are shown in the table:

Supplier A	Supplier B
$\bar{x}_1 = 1.89$	$\bar{x}_2 = 1.85$
$s_1^2 = .273$	$s_2^2 = .094$
$n_1 = 10$	$n_2 = 10$

a. Do the data provide sufficient evidence to indicate a difference in the variability of the shipment impu-

rity levels for the two suppliers? Test using $\alpha = .01$. Based on the results of your test, what recommendation would you make to the pharmaceutical manufacturer?

b. Find a 99% confidence interval for σ_2^2 and interpret your results.

 PERSONAL TRAINER

How Do I Decide Which Test to Use?

Are you interested in testing means? If the design involves:

a. One random sample, use the one-sample t statistic.
b. Two independent random samples, are the population variances equal?
    i.   If equal, use the two-sample t statistic with pooled s^2.
    ii. If unequal, use the unpooled t with estimated df.
c. Two paired samples with random pairs, use a one-sample t for analyzing differences.

Are you interested in testing variances? If the design involves:

a. One random sample, use the χ^2 test for a single variance.
b. Two independent random samples, use the F test to compare two variances.

REVISITING THE SMALL-SAMPLE ASSUMPTIONS

 10.8

All of the tests and estimation procedures discussed in this chapter require that the data satisfy certain conditions in order that the error probabilities (for the tests) and the confidence coefficients (for the confidence intervals) be equal to the values you have specified. For example, if you construct what you believe to be a 95% confidence interval, you want to be certain that, in repeated sampling, 95% (and not 85% or 75% or less) of all such intervals will contain the parameter of interest. These conditions are summarized in these assumptions:

ASSUMPTIONS

1. For all tests and confidence intervals described in this chapter, it is assumed that **samples are randomly selected from normally distributed populations.**

2. When two samples are selected, it is assumed that they are **selected in an independent manner** except in the case of the paired-difference experiment.

3. For tests or confidence intervals concerning the difference between two population means μ_1 and μ_2 based on independent random samples, it is assumed that $\sigma_1^2 = \sigma_2^2$.

In reality, you will never know everything about the sampled population. If you did, there would be no need for sampling or statistics. It is also highly unlikely that a population will *exactly* satisfy the assumptions given in the box. Fortunately, the procedures presented in this chapter give good inferences even when the data exhibit moderate departures from the necessary conditions.

A statistical procedure that is not sensitive to departures from the conditions on which it is based is said to be **robust.** The Student's *t* tests are quite robust for moderate departures from normality. Also, as long as the sample sizes are nearly equal, there is not much difference between the pooled and unpooled *t* statistics for the difference in two population means. However, if the sample sizes are not clearly equal, and if the population variances are unequal, the pooled *t* statistic provides inaccurate conclusions.

If you are concerned that your data do not satisfy the assumptions, other options are available:

- If you can select relatively large samples, you can use one of the large-sample procedures of Chapters 8 and 9, which do not rely on the normality or equal variance assumptions.

- You may be able to use a *nonparametric test* to answer your inferential questions. These tests have been developed specifically so that few or no distributional assumptions are required for their use. Tests that can be used to compare the locations or variability of two populations are presented in Chapter 15.

CHAPTER REVIEW

Key Concepts and Formulas

I. Experimental Designs for Small Samples

1. **Single random sample:** The sampled population must be normal.

2. **Two independent random samples:** Both sampled populations must be normal.

   a. Populations have a common variance σ^2.

   b. Populations have different variances: σ_1^2 and σ_2^2.

3. **Paired-difference** or **matched pairs** design: The samples are not independent.

II. Statistical Tests of Significance

1. Based on the t, F, and χ^2 distributions

2. Use the same procedure as in Chapter 9

3. **Rejection region—critical values** and **significance levels:** based on the t, F, or χ^2 distributions with the appropriate degrees of freedom

4. **Tests of population parameters:** a single mean, the difference between two means, a single variance, and the ratio of two variances

III. Small-Sample Test Statistics

To test one of the population parameters when the sample sizes are small, use the following test statistics:

Parameter	Test Statistic	Degrees of Freedom
μ	$t = \dfrac{\bar{x} - \mu_0}{s/\sqrt{n}}$	$n - 1$
$\mu_1 - \mu_2$ (equal variances)	$t = \dfrac{(\bar{x}_1 - \bar{x}_2) - (\mu_1 - \mu_2)}{\sqrt{s^2\left(\dfrac{1}{n_1} + \dfrac{1}{n_2}\right)}}$	$n_1 + n_2 - 2$
$\mu_1 - \mu_2$ (unequal variances)	$t \approx \dfrac{(\bar{x}_1 - \bar{x}_2) - (\mu_1 - \mu_2)}{\sqrt{\dfrac{s_1^2}{n_1} + \dfrac{s_2^2}{n_2}}}$	Satterthwaite's approximation
$\mu_1 - \mu_2$ (paired samples)	$t = \dfrac{\bar{d} - \mu_d}{s_d/\sqrt{n}}$	$n - 1$
σ^2	$\chi^2 = \dfrac{(n - 1)s^2}{\sigma_0^2}$	$n - 1$
σ_1^2/σ_2^2	$F = s_1^2/s_2^2$	$n_1 - 1$ and $n_2 - 1$

MINITAB

Small-Sample Testing and Estimation

The tests and confidence intervals for population means based on the Student's t distribution are found in a *MINITAB* submenu by choosing **Stat → Basic Statistics.** You will see choices for **1-Sample t, 2-Sample t,** and **Paired t,** which will generate Dialog boxes for the procedures in Sections 10.3, 10.4, and 10.5, respectively. You must choose the columns in which the data are stored and the null and alternative hypotheses to be tested (or the confidence coefficient for a confidence interval). In the case of the two-sample t test, you must indicate whether the population variances are assumed equal or unequal, so that *MINITAB* can perform the correct test. We will display some of the Dialog boxes and Session window outputs for the examples in this chapter, beginning with the one-sample t test of Example 10.3.

First, enter the six recorded weights—.46, .61, .52, .48, .57, .54—in column C1 and name them "Weights." Use **Stat → Basic Statistics → 1-Sample t** to generate the Dialog box in Figure 10.23. To test $H_0 : \mu = .5$ versus $H_a : \mu > .5$, use the list on the left to select "Weights" for the box marked "Samples in Columns." Place your cursor in the box marked "Test mean:" and enter **.5** as the test value. Finally, use **Options** and the drop-down menu marked "Alternative" to select "greater than." Click **OK** twice to obtain the output in Figure 10.24. Notice that *MINITAB* produces a one- or a two-sided confidence interval for the single population mean; you can change the confidence coefficient from the default of **.95** in the **Options** box. Also, the **Graphs** option will produce a histogram, a box plot, or an individual value plot of the data in column C1.

Data for a two-sample t test with independent samples can be entered into the worksheet in one of two ways:

FIGURE 10.23 ●

1-Sample t (Test and Confidence Interval)

C1   Weights

● **Samples in columns:**

> Weights

○ **Summarized data**

Sample size:

Mean:

Standard deviation:

Test mean: | .5 | [required for test]

Graphs...   Options...

OK   Cancel

1-Sample t - Options

Confidence level: | 95.0 |

Alternative: | not equal ▼ |
         less than
         not equal
         greater than

Help   Cancel

FIGURE 10.24

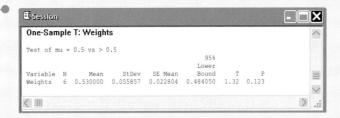

- Enter measurements from both samples into a single column and enter numbers (1 or 2) in a second column to identify the sample from which the measurement comes.
- Enter the samples in two separate columns.

If you do not have the raw data, but rather have summary statistics—the sample mean, standard deviation, and sample size—*MINITAB* 14 will allow you to use these values by selecting the radio button marked "Summarized data" and entering the appropriate values in the boxes.

Use the second method and enter the data from Example 10.5 into columns C2 and C3. Then use **Stat → Basic Statistics → 2-Sample t** to generate the Dialog box in Figure 10.25. Check "Samples in different columns," selecting C2 and C3 from the box on the left. Select the proper alternative hypothesis in the Options box, and check the "Assume equal variances" box. (Otherwise, *MINITAB* will perform Satterthwaite's approximation for unequal variances.) The two-sample output when you click **OK** twice automatically contains a 95% one- or two-sided confidence interval as well as the test statistic and *p*-value (you can change the confidence coefficient if you like). The output for Example 10.5 is shown in Figure 10.13 on page 406.

For a paired-difference test, the two samples are entered into separate columns, which we did with the tire wear data in Table 10.3. Use **Stat → Basic Statistics →**

FIGURE 10.25

Paired t to generate the Dialog box in Figure 10.26. If you have only summary statistics—the sample mean and standard deviation of the differences and sample size—*MINITAB* 14 will allow you to use these values by selecting the radio button marked "Summarized data" and entering the appropriate values in the boxes. Select C4 and C5 from the box on the left, and use **Options** to pick the proper alternative hypothesis. You may change the confidence coefficient or the test value (the default value is zero). When you click **OK** twice, you will obtain the output shown in Figure 10.15 on page 414.

The *MINITAB* command **Stat → Basic Statistics → 2 Variances** allows you to enter either raw data or summary statistics to perform the *F* test for the equality of variances, as shown in Figure 10.27. Although you cannot use *MINITAB* to perform the χ^2 test in this chapter, you can use **Calc → Probability Distributions → Chi-square** selecting "Cumulative probability" to calculate the appropriate *p*-value for the test.

FIGURE 10.26

FIGURE 10.27

Supplementary Exercises

10.67 What assumptions are made when Student's t test is used to test a hypothesis concerning a population mean?

10.68 What assumptions are made about the populations from which random samples are obtained when the t distribution is used in making small-sample inferences concerning the difference in population means?

10.69 Why use paired observations to estimate the difference between two population means rather than estimation based on independent random samples selected from the two populations? Is a paired experiment always preferable? Explain.

10.70 Impurities II A manufacturer can tolerate a small amount [.05 milligrams per liter (mg/l)] of impurities in a raw material needed for manufacturing its product. Because the laboratory test for the impurities is subject to experimental error, the manufacturer tests each batch ten times. Assume that the mean value of the experimental error is 0 and hence that the mean value of the ten test readings is an unbiased estimate of the true amount of the impurities in the batch. For a particular batch of the raw material, the mean of the ten test readings is .058 mg/l, with a standard deviation of .012 mg/l. Do the data provide sufficient evidence to indicate that the amount of impurities in the batch exceeds .05 mg/l? Find the p-value for the test and interpret its value.

10.71 Red Pine The main stem growth measured for a sample of seventeen 4-year-old red pine trees produced a mean and standard deviation equal to 11.3 and 3.4 inches, respectively. Find a 90% confidence interval for the mean growth of a population of 4-year-old red pine trees subjected to similar environmental conditions.

10.72 Sodium Hydroxide The object of a general chemistry experiment is to determine the amount (in milliliters) of sodium hydroxide (NaOH) solution needed to neutralize 1 gram of a specified acid. This will be an exact amount, but when the experiment is run in the laboratory, variation will occur as the result of experimental error. Three titrations are made using phenolphthalein as an indicator of the neutrality of the solution (pH equals 7 for a neutral solution). The three volumes of NaOH required to attain a pH of 7 in each of the three titrations are as follows: 82.10, 75.75, and

75.44 milliliters. Use a 99% confidence interval to estimate the mean number of milliliters required to neutralize 1 gram of the acid.

10.73 Sodium Chloride Measurements of water intake, obtained from a sample of 17 rats that had been injected with a sodium chloride solution, produced a mean and standard deviation of 31.0 and 6.2 cubic centimeters (cm^3), respectively. Given that the average water intake for noninjected rats observed over a comparable period of time is 22.0 cm^3, do the data indicate that injected rats drink more water than noninjected rats? Test at the 5% level of significance. Find a 90% confidence interval for the mean water intake for injected rats.

10.74 Sea Urchins An experimenter was interested in determining the mean thickness of the cortex of the sea urchin egg. The thickness was measured for $n = 10$ sea urchin eggs. These measurements were obtained:

4.5	6.1	3.2	3.9	4.7
5.2	2.6	3.7	4.6	4.1

Estimate the mean thickness of the cortex using a 95% confidence interval.

10.75 Fabricating Systems A production plant has two extremely complex fabricating systems; one system is twice as old as the other. Both systems are checked, lubricated, and maintained once every 2 weeks. The number of finished products fabricated daily by each of the systems is recorded for 30 working days. The results are given in the table. Do these data present sufficient evidence to conclude that the variability in daily production warrants increased maintenance of the older fabricating system? Use the p-value approach.

New System	Old System
$\bar{x}_1 = 246$	$\bar{x}_2 = 240$
$s_1 = 15.6$	$s_2 = 28.2$

(MY) (CD) **10.76** Fossils The data in the table are the
EX1076 diameters and heights of ten fossil specimens of a species of small shellfish, *Rotularia (Annelida) fallax,* that were unearthed in a mapping expedition near the Antarctic Peninsula.[11] The table gives an identification symbol for the fossil specimen, the fossil's diameter and height in millimeters, and the ratio of diameter to height.

Specimen	Diameter	Height	D/H
OSU 36651	185	78	2.37
OSU 36652	194	65	2.98
OSU 36653	173	77	2.25
OSU 36654	200	76	2.63
OSU 36655	179	72	2.49
OSU 36656	213	76	2.80
OSU 36657	134	75	1.79
OSU 36658	191	77	2.48
OSU 36659	177	69	2.57
OSU 36660	199	65	3.06
$\bar{x}$:	184.5	73	2.54
s:	21.5	5	.37

a. Find a 95% confidence interval for the mean diameter of the species.

b. Find a 95% confidence interval for the mean height of the species.

c. Find a 95% confidence interval for the mean ratio of diameter to height.

d. Compare the three intervals constructed in parts a, b, and c. Is the average of the ratios the same as the ratio of the average diameter to average height?

10.77 Fossils, continued Refer to Exercise 10.76 and data set EX1076. Suppose you want to estimate the mean diameter of the fossil specimens correct to within 5 millimeters with probability equal to .95. How many fossils do you have to include in your sample?

(MY)(CD) **10.78 Alcohol and Reaction Times** To test
EX1078 the effect of alcohol in increasing the reaction time to respond to a given stimulus, the reaction times of seven people were measured. After consuming 3 ounces of 40% alcohol, the reaction time for each of the seven people was measured again. Do the following data indicate that the mean reaction time after consuming alcohol was greater than the mean reaction time before consuming alcohol? Use $\alpha = .05$.

Person	1	2	3	4	5	6	7
Before	4	5	5	4	3	6	2
After	7	8	3	5	4	5	5

(MY)(CD) **10.79 Cheese, Please** Here are the prices per
EX1079 ounce of $n = 13$ different brands of individually wrapped cheese slices:

29.0	24.1	23.7	19.6	27.5
28.7	28.0	23.8	18.9	23.9
21.6	25.9	27.4		

Construct a 95% confidence interval estimate of the underlying average price per ounce of individually wrapped cheese slices.

10.80 Drug Absorption An experiment was conducted to compare the mean lengths of time required for the bodily absorption of two drugs A and B. Ten people were randomly selected and assigned to receive one of the drugs. The length of time (in minutes) for the drug to reach a specified level in the blood was recorded, and the data summary is given in the table:

Drug A	Drug B
$\bar{x}_1 = 27.2$	$\bar{x}_2 = 33.5$
$s_1^2 = 16.36$	$s_2^2 = 18.92$

a. Do the data provide sufficient evidence to indicate a difference in mean times to absorption for the two drugs? Test using $\alpha = .05$.

b. Find the approximate p-value for the test. Does this value confirm your conclusions?

c. Find a 95% confidence interval for the difference in mean times to absorption. Does the interval confirm your conclusions in part a?

10.81 Drug Absorption, continued Refer to Exercise 10.80. Suppose you wish to estimate the difference in mean times to absorption correct to within 1 minute with probability approximately equal to .95.

a. Approximately how large a sample is required for each drug (assume that the sample sizes are equal)?

b. If conducting the experiment using the sample sizes of part a will require a large amount of time and money, can anything be done to reduce the sample sizes and still achieve the 1-minute margin of error for estimation?

(MY)(CD) **10.82 Ring-Necked Pheasants** The weights
EX1082 in grams of 10 males and 10 female juvenile ring-necked pheasants are given below.

Males		Females	
1384	1672	1073	1058
1286	1370	1053	1123
1503	1659	1038	1089
1627	1725	1018	1034
1450	1394	1146	1281

a. Use a statistical test to determine if the population variance of the weights of the male birds differs from that of the females.

b. Test whether the average weight of juvenile male ring-necked pheasants exceeds that of the females by more than 300 grams. (**Hint:** The procedure that you use should take into account the results of the analysis in part a.)

MY CD **10.83 Bees** Insects hovering in flight expend
EX1083 enormous amounts of energy for their size and
weight. The data shown here were taken from a much
larger body of data collected by T. M. Casey and colleagues.[12] They show the wing stroke frequencies
(in hertz) for two different species of bees, $n_1 = 4$
Euglossa mandibularis Friese and $n_2 = 6$ *Euglossa
imperialis* Cockerell.

E. mandibularis Friese	E. imperialis Cockerell
235	180
225	169
190	180
188	185
	178
	182

a. Based on the observed ranges, do you think that a
difference exists between the two population variances?

b. Use an appropriate test to determine whether a difference exists.

c. Explain why a Student's t test with a pooled estimator s^2 is unsuitable for comparing the mean
wing stroke frequencies for the two species of
bees.

MY CD **10.84 Calcium** The calcium (Ca) content of
EX1084 a powdered mineral substance was analyzed
10 times with the following percent compositions
recorded:

.0271	.0282	.0279	.0281	.0268
.0271	.0281	.0269	.0275	.0276

a. Find a 99% confidence interval for the true calcium content of this substance.

b. What does the phrase "99% confident" mean?

c. What assumptions must you make about the sampling procedure so that this confidence interval will
be valid? What does this mean to the chemist who
is performing the analysis?

10.85 Sun or Shade? Karl Niklas and T.G. Owens
examined the differences in a particular plant, *Plantago Major L.,* when grown in full sunlight versus
shade conditions.[13] In this study, shaded plants
received direct sunlight for less than 2 hours each day,
whereas full-sun plants were never shaded. A partial
summary of the data based on $n_1 = 16$ full-sun plants
and $n_2 = 15$ shade plants is shown here:

	Full Sun		Shade	
	$\bar{x}$	s	$\bar{x}$	s
Leaf area (cm²)	128.00	43.00	78.70	41.70
Overlap area (cm²)	46.80	2.21	8.10	1.26
Leaf number	9.75	2.27	6.93	1.49
Thickness (mm)	.90	.03	.50	.02
Length (cm)	8.70	1.64	8.91	1.23
Width (cm)	5.24	.98	3.41	.61

a. What assumptions are required in order to use the
small-sample procedures given in this chapter to
compare full-sun versus shade plants? From the
summary presented, do you think that any of these
assumptions have been violated?

b. Do the data present sufficient evidence to indicate
a difference in mean leaf area for full-sun versus
shade plants?

c. Do the data present sufficient evidence to indicate
a difference in mean overlap area for full-sun versus shade plants?

10.86 Orange Juice A comparison of the precisions
of two machines developed for extracting juice from
oranges is to be made using the following data:

Machine A	Machine B
$s^2 = 3.1$ ounces²	$s^2 = 1.4$ ounces²
$n = 25$	$n = 25$

a. Is there sufficient evidence to indicate that there is
a difference in the precision of the two machines at
the 5% level of significance?

b. Find a 95% confidence interval for the ratio of the
two population variances. Does this interval confirm your conclusion from part a? Explain.

10.87 At Home or at School? Four sets of identical twins (pairs A, B, C, and D) were selected at random from a computer database of identical twins. One
child was selected at random from each pair to form an
"experimental group." These four children were sent to
school. The other four children were kept at home as a
control group. At the end of the school year, the following IQ scores were obtained:

Pair	Experimental Group	Control Group
A	110	111
B	125	120
C	139	128
D	142	135

Does this evidence justify the conclusion that lack of
school experience has a depressing effect on IQ
scores? Use the p-value approach.

(MY) CD **10.88 Dieting** Eight obese persons were
EX1088 placed on a diet for 1 month, and their weights, at the beginning and at the end of the month, were recorded:

Subjects	Weights	
	Initial	Final
1	310	263
2	295	251
3	287	249
4	305	259
5	270	233
6	323	267
7	277	242
8	299	265

Estimate the mean weight loss for obese persons when placed on the diet for a 1-month period. Use a 95% confidence interval and interpret your results. What assumptions must you make so that your inference is valid?

(MY) CD **10.89 Repair Costs** Car manufacturers try
EX1089 to design the bumpers of their automobiles to prevent costly damage in parking-lot type accidents. To compare two models of automobiles, the cars were purposely subject to a series of four front and rear impacts at 5 mph, and the repair costs were recorded.[14]

Impact Type	Honda Civic	Hyundai Elantra
Front into barrier	$403	$247
Rear into barrier	447	0
Front into angle barrier	404	407
Rear into pole	227	185

Do the data provide sufficient evidence to indicate a difference in the average cost of repair for the Honda Civic and the Hyundai Elantra? Test using $\alpha = .05$.

(MY) CD **10.90 Breathing Patterns** Research psychol-
EX1090 ogists measured the baseline breathing patterns—the total ventilation (in liters of air per minute) adjusted for body size—for each of $n = 30$ patients, so that they could estimate the average total ventilation for patients before any experimentation was done. The data, along with some *MINITAB* output, are presented here:

5.23	5.72	5.77	4.99	5.12	4.82
5.54	4.79	5.16	5.84	4.51	5.14
5.92	6.04	5.83	5.32	6.19	5.70
4.72	5.38	5.48	5.37	4.96	5.58
4.67	5.17	6.34	6.58	4.35	5.63

MINITAB output for Exercise 10.90

Stem-and-Leaf Display: Ltrs/min
```
Stem-and-leaf of Ltrs/min   N = 30
Leaf Unit = 0.10

  1      4  3
  2      4  5
  5      4  677
  8      4  899
 12      5  1111
 (4)     5  2333
 14      5  455
 11      5  6777
  7      5  889
  4      6  01
  2      6  3
  1      6  5
```

Descriptive Statistics: Ltrs/min
Variable	N	N*	Mean	SE Mean	StDev
Ltrs/min	30	0	5.3953	0.0997	0.5462

Minimum	Q1	Median	Q3	Maximum
4.3500	4.9825	5.3750	5.7850	6.5800

a. What information does the stem and leaf plot give you about the data? Why is this important?

b. Use the *MINITAB* output to construct a 99% confidence interval for the average total ventilation for patients.

10.91 Reaction Times A comparison of reaction times (in seconds) for two different stimuli in a psychological word-association experiment produced the following results when applied to a random sample of 16 people:

Stimulus 1	1	3	2	1	2	1	3	2
Stimulus 2	4	2	3	3	1	2	3	3

Do the data present sufficient evidence to indicate a difference in mean reaction times for the two stimuli? Test using $\alpha = .05$.

10.92 Reaction Times II Refer to Exercise 10.91. Suppose that the word-association experiment is conducted using eight people as blocks and making a comparison of reaction times within each person; that is, each person is subjected to both stimuli in a random order. The reaction times (in seconds) for the experiment are as follows:

Person	Stimulus 1	Stimulus 2
1	3	4
2	1	2
3	1	3
4	2	1
5	1	2
6	2	3
7	3	3
8	2	3

Do the data present sufficient evidence to indicate a difference in mean reaction times for the two stimuli? Test using $\alpha = .05$.

10.93 Refer to Exercises 10.91 and 10.92. Calculate a 95% confidence interval for the difference in the two population means for each of these experimental designs. Does it appear that blocking increased the amount of information available in the experiment?

(MY) CD **10.94 Impact Strength** The following data EX1094 are readings (in foot-pounds) of the impact strengths of two kinds of packaging material:

A	B
1.25	.89
1.16	1.01
1.33	.97
1.15	.95
1.23	.94
1.20	1.02
1.32	.98
1.28	1.06
1.21	.98

MINITAB output for Exercise 10.94

Two-Sample T-Test and CI: A, B
```
Two-sample T for A vs B
    N    Mean    StDev   SE Mean
A   9   1.2367   0.0644    0.021
B   9   0.9778   0.0494    0.016

Difference = mu (A) - mu (B)
Estimate for difference: 0.258889
95% CI for difference: (0.201506, 0.316272)
T-Test of difference = 0 (vs not =):
T-Value = 9.56  P-Value = 0.000  DF = 16
Both use Pooled StDev = 0.0574
```

a. Use the *MINITAB* printout to determine whether there is evidence of a difference in the mean strengths for the two kinds of material.

b. Are there practical implications to your results?

10.95 Cake Mixes An experiment was conducted to compare the densities (in ounces per cubic inch) of cakes prepared from two different cake mixes. Six cake pans were filled with batter A, and six were filled with batter B. Expecting a variation in oven temperature, the experimenter placed a pan filled with batter A and another with batter B *side by side* at six different locations in the oven. The six paired observations of densities are as follows:

Batter A	.135	.102	.098	.141	.131	.144
Batter B	.129	.120	.112	.152	.135	.163

a. Do the data present sufficient evidence to indicate a difference between the average densities of cakes prepared using the two types of batter?

b. Construct a 95% confidence interval for the difference between the average densities for the two mixes.

10.96 Under what assumptions can the F distribution be used in making inferences about the ratio of population variances?

10.97 Got Milk? A dairy is in the market for a new container-filling machine and is considering two models, manufactured by company A and company B. Ruggedness, cost, and convenience are comparable in the two models, so the deciding factor is the variability of fills. The model that produces fills with the smaller variance is preferred. If you obtain samples of fills for each of the two models, an F test can be used to test for the equality of population variances. Which type of rejection region would be most favored by each of these individuals?

a. The manager of the dairy—Why?

b. A sales representative for company A—Why?

c. A sales representative for company B—Why?

10.98 Got Milk II Refer to Exercise 10.97. Wishing to demonstrate that the variability of fills is less for her model than for her competitor's, a sales representative for company A acquired a sample of 30 fills from her company's model and a sample of 10 fills from her competitor's model. The sample variances were $s_A^2 = .027$ and $s_B^2 = .065$, respectively. Does this result provide statistical support at the .05 level of significance for the sales representative's claim?

10.99 Chemical Purity A chemical manufacturer claims that the purity of his product never varies by more than 2%. Five batches were tested and given purity readings of 98.2, 97.1, 98.9, 97.7, and 97.9%.

a. Do the data provide sufficient evidence to contradict the manufacturer's claim? (HINT: To be generous, let a range of 2% equal 4σ.)

b. Find a 90% confidence interval for σ^2.

10.100 16 Ounce Cans? A cannery prints "weight 16 ounces" on its label. The quality control supervisor selects nine cans at random and weighs them. She finds $\bar{x} = 15.7$ and $s = .5$. Do the data present sufficient evidence to indicate that the mean weight is less than that claimed on the label?

10.101 Reaction Time III A psychologist wishes to verify that a certain drug increases the reaction time to a given stimulus. The following reaction times (in tenths of a second) were recorded before and after injection of the drug for each of four subjects:

	Reaction Time	
Subject	Before	After
1	7	13
2	2	3
3	12	18
4	12	13

Test at the 5% level of significance to determine whether the drug significantly increases reaction time.

(MY) CD 10.102 Food Production At a time when EX10102 energy conservation is so important, some scientists think closer scrutiny should be given to the cost (in energy) of producing various forms of food. Suppose you wish to compare the mean amount of oil required to produce 1 acre of corn versus 1 acre of cauliflower. The readings (in barrels of oil per acre), based on 20-acre plots, seven for each crop, are shown in the table. Use these data to find a 90% confidence interval for the difference between the mean amounts of oil required to produce these two crops.

Corn	Cauliflower
5.6	15.9
7.1	13.4
4.5	17.6
6.0	16.8
7.9	15.8
4.8	16.3
5.7	17.1

10.103 Alcohol and Altitude The effect of alcohol consumption on the body appears to be much greater at high altitudes than at sea level. To test this theory, a scientist randomly selects 12 subjects and randomly divides them into two groups of six each. One group is put into a chamber that simulates conditions at an altitude of 12,000 feet, and each subject ingests a drink containing 100 cubic centimeters (cc) of alcohol. The second group receives the same drink in a chamber that simulates conditions at sea level. After 2 hours, the amount of alcohol in the blood (grams per 100 cc) for each subject is measured. The data are shown in the table. Do the data provide sufficient evidence to support the theory that retention of alcohol in the blood is greater at high altitudes?

Sea Level	12,000 Feet
.07	.13
.10	.17
.09	.15
.12	.14
.09	.10
.13	.14

10.104 Stock Risks The closing prices of two common stocks were recorded for a period of 15 days. The means and variances are

$$\bar{x}_1 = 40.33 \quad \bar{x}_2 = 42.54$$
$$s_1^2 = 1.54 \quad s_2^2 = 2.96$$

a. Do these data present sufficient evidence to indicate a difference between the variabilities of the closing prices of the two stocks for the populations associated with the two samples? Give the p-value for the test and interpret its value.

b. Construct a 99% confidence interval for the ratio of the two population variances.

10.105 Auto Design An experiment is conducted to compare two new automobile designs. Twenty people are randomly selected, and each person is asked to rate each design on a scale of 1 (poor) to 10 (excellent). The resulting ratings will be used to test the null hypothesis that the mean level of approval is the same for both designs against the alternative hypothesis that one of the automobile designs is preferred. Do these data satisfy the assumptions required for the Student's t test of Section 10.4? Explain.

10.106 Safety Programs The data shown here were collected on lost-time accidents (the figures given are mean work-hours lost per month over a period of 1 year) before and after an industrial safety program was put into effect. Data were recorded for six industrial plants. Do the data provide sufficient evidence to indicate whether the safety program was effective in reducing lost-time accidents? Test using $\alpha = .01$.

	Plant Number					
	1	2	3	4	5	6
Before program	38	64	42	70	58	30
After program	31	58	43	65	52	29

(MY) CD 10.107 Two Different Entrees To compare EX10107 the demand for two different entrees, the manager of a cafeteria recorded the number of purchases of each entree on seven consecutive days. The data are shown in the table. Do the data provide sufficient evidence to indicate a greater mean demand for one of the entrees? Use the *MINITAB* printout.

Day	A	B
Monday	420	391
Tuesday	374	343
Wednesday	434	469
Thursday	395	412
Friday	637	538
Saturday	594	521
Sunday	679	625

MINITAB output for Exercise 10.107

Paired T-Test and CI: A, B

```
Paired T for A - B
              N      Mean      StDev    SE Mean
A             7    504.714   127.245     48.094
B             7    471.286    97.445     36.831
Difference    7     33.4286   47.5039    17.9548

95% CI for mean difference:
(-10.5052, 77.3623)
T-Test of mean difference = 0 (vs not = 0):
T-Value = 1.86  P-Value = 0.112
```

10.108 Pollution Control The EPA limit on the allowable discharge of suspended solids into rivers and streams is 60 milligrams per liter (mg/l) per day. A study of water samples selected from the discharge at a phosphate mine shows that over a long period, the mean daily discharge of suspended solids is 48 mg/l, but day-to-day discharge readings are variable. State inspectors measured the discharge rates of suspended solids for $n = 20$ days and found $s^2 = 39$ (mg/l)^2. Find a 90% confidence interval for σ^2. Interpret your results.

10.109 Enzymes Two methods were used to measure the specific activity (in units of enzyme activity per milligram of protein) of an enzyme. One unit of enzyme activity is the amount that catalyzes the formation of 1 micromole of product per minute under specified conditions. Use an appropriate test or estimation procedure to compare the two methods of measurement. Comment on the validity of any assumptions you need to make.

Method 1	125	137	130	151	142
Method 2	137	143	151	156	149

10.110 Connector Rods A producer of machine parts claimed that the diameters of the connector rods produced by his plant had a variance of at most .03 inch^2. A random sample of 15 connector rods from his plant produced a sample mean and variance of .55 inch and .053 inch^2, respectively.

a. Is there sufficient evidence to reject his claim at the $\alpha = .05$ level of significance?

b. Find a 95% confidence interval for the variance of the rod diameters.

10.111 Sleep and the College Student How much sleep do you get on a typical school night? A group of 10 college students were asked to report the number of hours that they slept on the previous night with the following results:

7, 6, 7.25, 7, 8.5, 5, 8, 7, 6.75, 6

a. Find a 99% confidence interval for the average number of hours that college students sleep.

b. What assumptions are required in order for this confidence interval to be valid?

(MY CD) **10.112 Arranging Objects** The following
EX10112 data are the response times in seconds for $n = 25$ first graders to arrange three objects by size.

5.2	3.8	5.7	3.9	3.7
4.2	4.1	4.3	4.7	4.3
3.1	2.5	3.0	4.4	4.8
3.6	3.9	4.8	5.3	4.2
4.7	3.3	4.2	3.8	5.4

Find a 95% confidence interval for the average response time for first graders to arrange three objects by size. Interpret this interval.

10.113 Finger-Lickin' Good! Maybe too good, according to tests performed by the consumer testing division of *Good Housekeeping*. Nutritional information provided by Kentucky Fried Chicken claims that each small bag of Potato Wedges contains 4.8 ounces of food, for a total of 280 calories. A sample of 10 orders from KFC restaurants in New York and New Jersey averaged 358 calories.[15] If the standard deviation of this sample was $s = 54$, is there sufficient evidence to indicate that the average number of calories in small bags of KFC Potato Wedges is greater than advertised? Test at the 1% level of significance.

10.114 Mall Rats An article in *American Demographics* investigated consumer habits at the mall. We tend to spend the most money shopping on the weekends, and, in particular, on Sundays from 4 to 6 P.M. Wednesday morning shoppers spend the least![16] Suppose that a random sample of 20 weekend shoppers and a random sample of 20 weekday shoppers were selected, and the amount spent per trip to the mall was recorded.

	Weekends	Weekdays
Sample size	20	20
Sample mean	$78	$67
Sample standard deviation	$22	$20

a. Is it reasonable to assume that the two population variances are equal? Use the F test to test this hypothesis with $\alpha = .05$.

b. Based on the results of part a, use the appropriate test to determine whether there is a difference in the average amount spent per trip on weekends versus weekdays. Use $\alpha = .05$.

 10.115 Border Wars As the costs of prescription drugs escalate, more and more senior citizens are ordering prescriptions from Canada, or actually crossing the border to buy prescription drugs. The price of a typical prescription for 9 best-selling drugs

EX10115

was recorded at randomly selected stores in both the United States and in Canada.[17]

Drug	U.S.	Canada
Lipitor	$290	$179
Zocor	412	211
Prilosec	117	72
Norvasc	139	125
Zyprexa	571	396
Paxil	276	171
Prevacid	484	196
Celebrex	161	67
Zoloft	235	156

a. Is there sufficient evidence to indicate that the average cost of prescription drugs in the United States is different from the average cost in Canada? Use $\alpha = .01$.

b. What is the approximate the p-value for this test? Does this confirm your conclusions in part a?

MY APPLET Exercises

10.116 Use the **Student's t Probabilities** applet to find the following probabilities:

a. $P(t > 1.2)$ with 5 df

b. $P(t > 2) + P(t < -2)$ with 10 df

c. $P(t < -3.3)$ with 8 df

d. $P(t > .6)$ with 12 df

10.117 Use the **Student's t Probabilities** applet to find the following critical values:

a. an upper one-tailed rejection region with $\alpha = .05$ and 11 df

b. a two-tailed rejection region with $\alpha = .05$ and 7 df

c. a lower one-tailed rejection region with $\alpha = .01$ and 15 df

10.118 Refer to the **Interpreting Confidence Intervals** applet.

a. Suppose that you have a random sample of size $n = 10$ from a population with unknown mean μ. What formula would you use to construct a 95% confidence interval for the unknown population mean?

b. Use the [One Sample] button in the first applet to create a single 95% confidence interval for μ. Use the formula in part a and the information given in

the applet to verify the confidence limits provided. (The applet rounds to the nearest integer.) Did this confidence interval enclose the true value, $\mu = 100$?

10.119 Refer to the **Interpreting Confidence Intervals** applet.

a. Use the [10 Samples] button in the first applet to create ten 95% confidence intervals for μ.

b. Are the widths of these intervals all the same? Explain why or why not.

c. How many of the intervals work properly and enclose the true value of μ?

d. Try this simulation again by clicking the [10 Samples] button a few more times and counting the number of intervals that work correctly. Is it close to our 95% confidence level?

e. Use the [10 Samples] button in the second applet to create ten 99% confidence intervals for μ. How many of these intervals work properly?

10.120 Refer to the **Interpreting Confidence Intervals** applet.

a. Use the [100 Samples] button to create one hundred 95% confidence intervals for μ. How many of the intervals work properly and enclose the true value of μ?

b. Repeat the instructions of part a to construct 99% confidence intervals. How many of the intervals work properly and enclose the true value of μ?

c. Try this simulation again by clicking the 100 Samples button a few more times and counting the number of intervals that work correctly. Use both the 95% and 99% confidence intervals. Do the percentage of intervals that work come close to our 95% and 99% confidence levels?

10.121 A random sample of $n = 12$ observations from a normal population produced $\bar{x} = 47.1$ and $s^2 = 4.7$. Test the hypothesis H_0: $\mu = 48$ against H_a: $\mu \neq 48$. Use the **Small Test of a Population Mean** applet and a 5% significance level.

10.122 **SAT Scores** In Exercise 9.73, we reported that the national average SAT scores for the class of 2003 were 507 on the verbal portion and 519 on the math portion. Suppose that we only have a small random sample of 15 California students from the class of 2003; their SAT scores are recorded in the following table.

	Verbal	Math
Sample average	499	516
Sample standard deviation	98	96

a. Use the **Small Sample Test of a Population Mean** applet. Do the data provide sufficient evidence to indicate that the average verbal score for all California students in the class of 2003 is different from the national average? Test using $\alpha = .05$.

b. Use the **Small Sample Test of a Population Mean** applet. Do the data provide sufficient evidence to indicate that the average math score for all California students in the class of 2003 is different from the national average? Test using $\alpha = .05$.

10.123 **Surgery Recovery Times** The length of time to recovery was recorded for patients randomly assigned and subjected to two different surgical procedures. The data (recorded in days) are as follows:

	Procedure I	Procedure II
Sample average	7.3	8.9
Sample variance	1.23	1.49
Sample size	11	13

Do the data present sufficient evidence to indicate a difference between the mean recovery times for the two surgical procedures? Perform the test of hypothesis, calculating the test statistic and the approximate p-value by hand. Then check your results using the **Two Sample t test: Independent Samples** applet.

10.124 **Stock Prices** Refer to Exercise 10.104 in which we reported the closing prices of two common stocks, recorded over a period of 15 days.

$\bar{x}_1 = 40.33 \quad \bar{x}_2 = 42.54$
$s_1^2 = 1.54 \quad s_2^2 = 2.96$

Use the **Two Sample t test: Independent Samples** applet. Do the data provide sufficient evidence to indicate that the average prices of the two common stocks are different? Use the p-value to access the significance of the test.

CASE STUDY

How Would You Like a Four-Day Workweek?

Will a flexible workweek schedule result in positive benefits for both employer and employee? Is a more rested employee, who spends less time commuting to and from work, likely to be more efficient and take less time off for sick leave and personal leave? A report on the benefits of flexible work schedules that appeared in *Environmental Health* looked at the records of $n = 11$ employees who worked in a satellite office in a county health department in Illinois under a 4-day workweek schedule.[18] Employees worked a conventional workweek in year 1 and a 4-day workweek in year 2. Some statistics for these employees are shown in the following table:

	Personal Leave		Sick Leave	
Employee	Year 2	Year 1	Year 2	Year 1
1	26	33	30	37
2	18	37	61	45
3	24	20	59	56
4	19	26	2	9
5	17	1	79	92
6	34	2	63	65
7	19	13	71	21
8	18	22	83	62
9	9	22	35	26
10	36	13	81	73
11	26	18	79	21

1. A 4-day workweek ensures that employees will have one more day that need not be spent at work. One possible result is a reduction in the average number of personal-leave days taken by employees on a 4-day work schedule. Do the data indicate that this is the case? Use the p-value approach to testing to reach your conclusion.

2. A 4-day workweek schedule might also have an effect on the average number of sick-leave days an employee takes. Should a directional alternative be used in this case? Why or why not?

3. Construct a 95% confidence interval to estimate the average difference in days taken for sick leave between these 2 years. What do you conclude about the difference between the average number of sick-leave days for these two work schedules?

4. Based on the analysis of these two variables, what can you conclude about the advantages of a 4-day workweek schedule?

Case Study from "Four-Day Work Week Improves Environment," by C. S. Catlin, *Environmental Health,* Vol. 59, No. 7, March 1997. Copyright 1997 National Environmental Health Association. Reprinted by permission.

$n = 10$ in each of four professorial ranks, for both male and female professors

Gender	Assistant Professor		Associate Professor		Full Professor	
Male	$56.9	$57.4	$61.0	$65.4	$93.4	$94.5
	56.9	55.2	68.7	68.2	95.3	88.4
	57.8	57.2	69.1	67.3	96.5	90.3
	61.3	57.9	67.0	69.8	95.6	90.9
	60.5	60.5	61.1	64.1	102.3	93.9
Female	49.6	52.0	56.4	57.3	95.3	85.9
	50.6	51.6	62.9	65.6	82.0	87.8
	46.6	54.9	56.9	64.0	76.5	87.0
	57.4	55.9	58.9	60.4	88.5	81.7
	55.6	52.8	64.6	62.0	80.6	82.8

Source: Based on "Average Salary for Men and Women Faculty, by Category, Affiliation, and Academic Rank 2000–2001," *Academe,* March–April 2001, p. 39. Copyright © 2001 American Association of University Professors.

a. Identify the design used in this survey.

b. Use the appropriate analysis of variance for these data.

c. Do the data indicate that the salary at the different ranks vary by gender?

d. If there is no interaction, determine whether there are differences in salaries by rank, and whether there are differences by gender. Discuss your results.

e. Plot the average salaries using an interaction plot. If the main effect of ranks is significant, use Tukey's method of pairwise comparisons to determine if there are significant differences among the ranks.

CASE STUDY   "A Fine Mess"

Do you risk a parking ticket by parking where you shouldn't or forgetting how much time you have left on the parking meter? Do the fines associated with various parking infractions vary depending on the city in which you receive a parking ticket? To look at this issue, the fines imposed for overtime parking, parking in a red zone, and parking next to a fire hydrant were recorded for 13 cities in southern California.[11]

City	Overtime Parking	Red Zone	Fire Hydrant
Long Beach	$17	$30	$30
Bakersfield	17	33	33
Orange	22	30	32
San Bernardino	20	30	78
Riverside	21	30	30
San Luis Obispo	8	20	75
Beverly Hills	23	38	30
Palm Springs	22	28	46
Laguna Beach	22	22	32
Del Mar	25	40	55
Los Angeles	20	55	30
San Diego	35	60	60
Newport Beach	32	42	30

Source: From "A Fine Mess," by R. McGarvey, *Avenues,* July/August 1994. Reprinted by permission of the author.

1. Identify the design used for the data collection in this case study.

2. Analyze the data using the appropriate analysis. What can you say about the variation among the cities in this study? Among fines for the three types of violations? Can Tukey's procedure be of use in further delineating any significant differences you may find? Would confidence interval estimates be useful in your analysis?

3. Summarize the results of your analysis of these data.

Linear Regression and Correlation

© Getty Images

GENERAL OBJECTIVES

In this chapter, we consider the situation in which the mean value of a random variable y is related to another variable x. By measuring both y and x for each experimental unit, thereby generating bivariate data, you can use the information provided by x to estimate the average value of y and to predict values of y for preassigned values of x.

CHAPTER INDEX

 PERSONAL TRAINER

How Do I Make Sure That My Calculations Are Correct?

Is Your Car "Made in the U.S.A."?

The phrase "made in the U.S.A." has become a battle cry in the past few years as American workers try to protect their jobs from overseas competition. In the case study at the end of this chapter, we explore the changing attitudes of American consumers toward automobiles made outside the United States, using a simple linear regression analysis.

INTRODUCTION

12.1

High school seniors, freshmen entering college, their parents, and a university administration are concerned about the academic achievement of a student after he or she has enrolled in a university. Can you estimate or predict a student's grade point average (GPA) at the end of the freshman year before the student enrolls in the university? At first glance this might seem like a difficult problem. However, you would expect highly motivated students who have graduated with a high class rank from a high school with superior academic standards to achieve a high GPA at the end of the college freshman year. On the other hand, students who lack motivation or who have achieved only moderate success in high school are not expected to do so well. You would expect the college achievement of a student to be a function of several variables:

- Rank in high school class
- High school's overall rating
- High school GPA
- SAT scores

This problem is of a fairly general nature. You are interested in a random variable *y* (college GPA) that is related to a number of independent variables. The objective is to create a *prediction equation* that expresses *y* as a function of these independent variables. Then, if you can measure the independent variables, you can substitute these values into the prediction equation and obtain the prediction for *y*—the student's college GPA in our example. But which variables should you use as predictors? How strong is their relationship to *y*? How do you construct a good prediction equation for *y* as a function of the selected predictor variables? We will answer these questions in the next two chapters.

In this chapter, we restrict our attention to the simple problem of predicting *y* as a linear function of a single predictor variable *x*. This problem was originally addressed in Chapter 3 in the discussion of *bivariate data*. Remember that we used the equation of a straight line to describe the relationship between *x* and *y* and we described the strength of the relationship using the correlation coefficient *r*. We rely on some of these results as we revisit the subject of linear regression and correlation.

A SIMPLE LINEAR PROBABILISTIC MODEL

12.2

Consider the problem of trying to predict the value of a response *y* based on the value of an independent variable *x*. The best-fitting line of Chapter 3,

$$y = a + bx$$

was based on a *sample* of *n* bivariate observations drawn from a larger *population* of measurements. The line that describes the relationship between *y* and *x* in the *population* is similar to, but not the same as, the best-fitting line from the *sample*. How can you construct a **population model** to describe the relationship between a random variable *y* and a related independent variable *x*?

You begin by assuming that the variable of interest, *y*, is *linearly* related to an independent variable *x*. To describe the linear relationship, you can use the **deterministic model**

$$y = \alpha + \beta x$$

where α is the y-intercept—the value of y when $x = 0$—and β is the slope of the line, defined as the change in y for a one-unit change in x, as shown in Figure 12.1. This model describes a deterministic relationship between the variable of interest y, sometimes called the **response variable,** and the independent variable x, often called the **predictor variable.** That is, the linear equation determines an exact value of y when the value of x is given. Is this a realistic model for an experimental situation? Consider the following example.

FIGURE 12.1

The y-intercept and slope for a line

 TIP

slope = change in y for a 1-unit change in x

y-intercept = value of y when $x = 0$

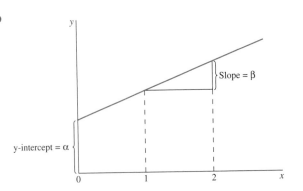

Table 12.1 displays the mathematics achievement test scores for a random sample of $n = 10$ college freshmen, along with their final calculus grades. A bivariate plot of these scores and grades is given in Figure 12.2. You can use the **Building a Scatterplot** applet to refresh your memory as to how this plot is drawn. Notice that the points *do not lie exactly on a line* but rather seem to be deviations about an underlying line. A simple way to modify the deterministic model is to add a **random error component** to explain the deviations of the points about the line. A particular response y is described using the **probabilistic model**

$$y = \alpha + \beta x + \epsilon$$

TABLE 12.1

Mathematics Achievement Test Scores and Final Calculus Grades for College Freshmen

Student	Mathematics Achievement Test Score	Final Calculus Grade
1	39	65
2	43	78
3	21	52
4	64	82
5	57	92
6	47	89
7	28	73
8	75	98
9	34	56
10	52	75

FIGURE 12.2

Scatterplot of the data in
Table 12.1

The first part of the equation, $\alpha + \beta x$—called the **line of means**—describes the average value of y for a given value of x. The error component ϵ allows each individual response y to deviate from the line of means by a small amount.

In order to use this *probabilistic model* for making inferences, you need to be more specific about this "small amount," ϵ.

ASSUMPTIONS ABOUT THE RANDOM ERROR ϵ

Assume that the values of ϵ satisfy these conditions:

*   Are independent in the probabilistic sense
*   Have a mean of 0 and a common variance equal to σ^2
*   Have a normal probability distribution

These assumptions about the random error ϵ are shown in Figure 12.3 for three fixed values of x—say, x_1, x_2, and x_3. Notice the similarity between these assumptions and the assumptions necessary for the tests in Chapters 10 and 11. We will revisit these assumptions later in this chapter and provide some diagnostic tools for you to use in checking their validity.

FIGURE 12.3

Linear probabilistic model

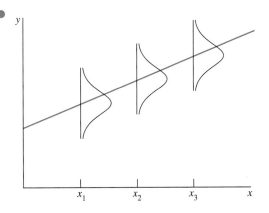

Remember that this model is created for a population of measurements that is generally unknown to you. However, you can use sample information to estimate the values of α and β, which are the coefficients of the line of means, $E(y) = \alpha + \beta x$. These estimates are used to form the best-fitting line for a given set of data, called the **least squares line** or **regression line.** We review how to calculate the intercept and the slope of this line in the next section.

THE METHOD OF LEAST SQUARES

12.3

The statistical procedure for finding the best-fitting line for a set of bivariate data does mathematically what you do visually when you move a ruler until you think you have minimized the vertical distances, or deviations, from the ruler to a set of points. The formula for the best-fitting line is

$$\hat{y} = a + bx$$

MY TIP

slope = coefficient of x

y-intercept = constant term

where a and b are the estimates of the intercept and slope parameters α and β, respectively. The fitted line for the data in Table 12.1 is shown in the **Method of Least Squares** applet, Figure 12.4. The red vertical lines (light teal in Figure 12.4) drawn from the prediction line to each point (x_i, y_i) represent the deviations of the points from the line.

FIGURE 12.4

Method of Least Squares applet

To minimize the distances from the points to the fitted line, you can use the **principle of least squares.**

PRINCIPLE OF LEAST SQUARES

The line that minimizes the sum of squares of the deviations of the observed values of y from those predicted is the **best-fitting line.** The sum of squared deviations is commonly called the **sum of squares for error** (SSE) and defined as

$$SSE = \Sigma(y_i - \hat{y}_i)^2 = \Sigma(y_i - a - bx_i)^2$$

Look at the regression line and the data points in Figure 12.4. SSE is the sum of the squared distances represented by the area of the yellow squares (light teal in Figure 12.4).

Finding the values of a and b, the estimates of α and β, uses differential calculus, which is beyond the scope of this text. Rather than derive their values, we will simply present formulas for calculating the values of a and b—called the **least-squares estimators** of α and β. We will use notation that is based on the **sums of squares** for the variables in the regression problem, which are similar in form to the sums of squares used in Chapter 11. These formulas look different from the formulas presented in Chapter 3, but they are in fact algebraically identical!

You should use the data entry method for your scientific calculator to enter the sample data.

- If your calculator has only a one-variable statistics function, you can still save some time in finding the necessary sums and sums of squares.

- If your calculator has a two-variable statistics function, or if you have a graphing calculator, the calculator will automatically store all of the sums and sums of squares as well as the values of a, b, and the correlation coefficient r.

- Make sure you consult your calculator manual to find the easiest way to obtain the least squares estimators.

LEAST-SQUARES ESTIMATORS OF α AND β

$$b = \frac{S_{xy}}{S_{xx}} \quad \text{and} \quad a = \bar{y} - b\bar{x}$$

where the quantities S_{xy} and S_{xx} are defined as

$$S_{xy} = \Sigma(x_i - \bar{x})(y_i - \bar{y}) = \Sigma x_i y_i - \frac{(\Sigma x_i)(\Sigma y_i)}{n}$$

and

$$S_{xx} = \Sigma(x_i - \bar{x})^2 = \Sigma x_i^2 - \frac{(\Sigma x_i)^2}{n}$$

Notice that the sum of squares of the x-values is found using the computing formula given in Section 2.3 and the sum of the cross-products is the numerator of the *covariance* defined in Section 3.4.

EXAMPLE 12.1   Find the least-squares prediction line for the calculus grade data in Table 12.1.

Solution   Use the data in Table 12.2 and the data entry method in your scientific calculator to find the following sums of squares:

$$S_{xx} = \Sigma x_i^2 - \frac{(\Sigma x_i)^2}{n} = 23{,}634 - \frac{(460)^2}{10} = 2474$$

$$S_{xy} = \Sigma x_i y_i - \frac{(\Sigma x_i)(\Sigma y_i)}{n} = 36{,}854 - \frac{(460)(760)}{10} = 1894$$

$$\bar{y} = \frac{\Sigma y_i}{n} = \frac{760}{10} = 76 \qquad \bar{x} = \frac{\Sigma x_i}{n} = \frac{460}{10} = 46$$

TABLE 12.2 ● **Calculations for the Data in Table 12.1**

y_i	x_i	x_i^2	$x_i y_i$	y_i^2
65	39	1521	2535	4225
78	43	1849	3354	6084
52	21	441	1092	2704
82	64	4096	5248	6724
92	57	3249	5244	8464
89	47	2209	4183	7921
73	28	784	2044	5329
98	75	5625	7350	9604
56	34	1156	1904	3136
75	52	2704	3900	5625
Sum 760	460	23,634	36,854	59,816

Then

 TIP

You can predict y for a given value of x by substituting x into the equation to find $\hat{y}$.

$$b = \frac{S_{xy}}{S_{xx}} = \frac{1894}{2474} = .76556 \quad \text{and} \quad a = \bar{y} - b\bar{x} = 76 - (.76556)(46) = 40.78424$$

The least-squares regression line is then

$$\hat{y} = a + bx = 40.78424 + .76556x$$

The graph of this line is shown in Figure 12.4. It can now be used to predict y for a given value of x—either by referring to Figure 12.4 or by substituting the proper value of x into the equation. For example, if a freshman scored $x = 50$ on the achievement test, the student's predicted calculus grade is (using full decimal accuracy)

$$\hat{y} = a + b(50) = 40.78424 + (.76556)(50) = 79.06$$

MY PERSONAL TRAINER

How Do I Make Sure That My Calculations Are Correct?

- Be careful of rounding errors. Carry at least six significant figures, and round off only in reporting the end result.
- Use a scientific or graphing calculator to do all the work for you. Most of these calculators will calculate the values for a and b if you enter the data properly.
- Use a computer software program if you have access to one.
- Always plot the data and graph the line. If the line does not fit through the points, you have probably made a mistake!

MY APPLET

You can use the **Method of Least Squares** applet to find the values of a and b that determine the *best fitting line*, $\hat{y} = a + bx$. The horizontal line that you see is the line $y = \bar{y}$. Use your mouse to drag the line and watch the yellow squares change size. The object is to make SSE—the total area of the yellow squares—as small as possible. The value of SSE is the red portion of the bar on the left of the applet (teal in Figure 12.4) marked SSE = _____. When you think that you have minimized SSE, click the [Find Best Model] button and see how well you did!

12.4

AN ANALYSIS OF VARIANCE FOR LINEAR REGRESSION

In Chapter 11, you used the analysis of variance procedures to divide the total variation in the experiment into portions attributed to various factors of interest to the experimenter. In a regression analysis, the response y is related to the independent variable x. Hence, the total variation in the response variable y, given by

$$\text{Total SS} = S_{yy} = \Sigma(y_i - \bar{y})^2 = \Sigma y_i^2 - \frac{(\Sigma y_i)^2}{n}$$

is divided into two portions:

- SSR (sum of squares for regression) measures the amount of variation explained by using the regression line with one independent variable x

- SSE (sum of squares for error) measures the "residual" variation in the data that is not explained by the independent variable x

so that

Total SS = SSR + SSE

For a particular value of the response y_i, you can visualize this breakdown in the variation using the vertical distances illustrated in Figure 12.5. You can see that SSR is the sum of the squared deviations of the differences between the estimated response without using x ($\bar{y}$) and the estimated response using x (the regression line, $\hat{y}$); SSE is the sum of the squared differences between the regression line ($\hat{y}$) and the point y.

FIGURE 12.5

Deviations from the fitted line

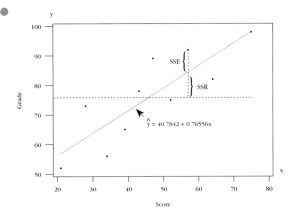

It is not too hard to show algebraically that

$$\text{SSR} = \Sigma(\hat{y}_i - \bar{y}_i)^2 = \Sigma(a + bx_i - \bar{y})^2 = \Sigma(\bar{y} - b\bar{x} + bx_i - \bar{y})^2 = b^2\Sigma(x_i - \bar{x})^2$$

$$= \left(\frac{S_{xy}}{S_{xx}}\right)^2 S_{xx} = \frac{(S_{xy})^2}{S_{xx}}$$

Since Total SS = SSR + SSE, you can complete the partition by calculating

$$\text{SSE} = \text{Total SS} - \text{SSR} = S_{yy} - \frac{(S_{xy})^2}{S_{xx}}$$

Remember from Chapter 11 that each of the various sources of variation, when divided by the appropriate **degrees of freedom,** provides an estimate of the variation in the experiment. These estimates are called **mean squares**—MS = SS/*df*—and are displayed in an ANOVA table.

In examining the degrees of freedom associated with each of these sums of squares, notice that the total degrees of freedom for n measurements is $(n - 1)$. Since estimating the regression line, $\hat{y} = a + bx_i = \bar{y} - b\bar{x} + bx_i$, involves estimating *one additional* parameter β, there is *one* degree of freedom associated with SSR, leaving $(n - 2)$ degrees of freedom with SSE.

As with all ANOVA tables we have discussed, the mean square for error,

$$MSE = s^2 = \frac{SSE}{n - 2}$$

is an unbiased estimator of the underlying variance σ^2. The analysis of variance table is shown in Table 12.3.

TABLE 12.3 ● **Analysis of Variance for Linear Regression**

Source	df	SS	MS
Regression	1	$\dfrac{(S_{xy})^2}{S_{xx}}$	MSR
Error	$n - 2$	$S_{yy} - \dfrac{(S_{xy})^2}{S_{xx}}$	MSE
Total	$n - 1$	S_{yy}	

For the data in Table 12.1, you can calculate

$$\text{Total SS} = S_{yy} = \Sigma y_i^2 - \frac{(\Sigma y_i)^2}{n} = 59,816 - \frac{(760)^2}{10} = 2056$$

$$SSR = \frac{(S_{xy})^2}{S_{xx}} = \frac{(1894)^2}{2474} = 1449.9741$$

so that

$$SSE = \text{Total SS} - SSR = 2056 - 1449.9741 = 606.0259$$

and

$$MSE = \frac{SSE}{n - 2} = \frac{606.0259}{8} = 75.7532$$

The analysis of variance table, part of the *linear regression output* generated by *MINITAB*, is the lower shaded section in the printout in Figure 12.6. The first two lines give the equation of the least-squares line, $\hat{y} = 40.8 + .766x$. The least-squares estimates a and b are given with greater accuracy in the column labeled "Coef." You can find instructions for generating this output in the section "My *MINITAB*" at the end of this chapter.

FIGURE 12.6

MINITAB output for the data of Table 12.1

Regression Analysis: y versus x

```
The regression equation is
y = 40.8 + 0.766 x

Predictor       Coef     SE Coef          T        P
Constant      40.784       8.507       4.79    0.001
x             0.7656      0.1750       4.38    0.002

S = 8.70363       R-Sq = 70.5%     R-Sq(adj) = 66.8%

Analysis of Variance
Source          DF          SS          MS        F        P
Regression       1      1450.0      1450.0    19.14    0.002
Residual Error   8       606.0        75.8
Total            9      2056.0
```

MY TIP

Look for *a* and *b* in the column called "Coef."

The *MINITAB* output also gives some information about the variation in the experiment. Each of the least-squares estimates, *a* and *b*, has an associated standard error, labeled "SE Coef" in Figure 12.6. In the middle of the printout, you will find the best unbiased estimate of σ—$S = \sqrt{MSE} = \sqrt{75.7532} = 8.70363$, which measures the **residual error,** the unexplained or "leftover" variation in the experiment. It will not surprise you to know that the t and F statistics and their p-values found in the printout are used to test statistical hypotheses. We explain these entries in the next section.

12.4 EXERCISES

BASIC TECHNIQUES

12.1 Graph the line corresponding to the equation $y = 2x + 1$ by graphing the points corresponding to $x = 0, 1,$ and 2. Give the y-intercept and slope for the line.

12.2 Graph the line corresponding to the equation $y = -2x + 1$ by graphing the points corresponding to $x = 0, 1,$ and 2. Give the y-intercept and slope for the line. How is this line related to the line $y = 2x + 1$ of Exercise 12.1?

12.3 Give the equation and graph for a line with y-intercept equal to 3 and slope equal to -1.

12.4 Give the equation and graph for a line with y-intercept equal to -3 and slope equal to 1.

12.5 What is the difference between deterministic and probabilistic mathematical models?

12.6 You are given five points with these coordinates:

x	-2	-1	0	1	2
y	1	1	3	5	5

a. Use the data entry method on your scientific or graphing calculator to enter the $n = 5$ observations. Find the sums of squares and cross-products, $S_{xx}, S_{xy},$ and S_{yy}.

b. Find the least-squares line for the data.

c. Plot the five points and graph the line in part b. Does the line appear to provide a good fit to the data points?

d. Construct the ANOVA table for the linear regression.

12.7 Six points have these coordinates:

x	1	2	3	4	5	6
y	5.6	4.6	4.5	3.7	3.2	2.7

a. Find the least-squares line for the data.

b. Plot the six points and graph the line. Does the line appear to provide a good fit to the data points?

c. Use the least-squares line to predict the value of y when $x = 3.5$.

d. Fill in the missing entries in the *MINITAB* analysis of variance table.

MINITAB ANOVA table for Exercise 12.7

```
Analysis of Variance

Source           DF        SS        MS
Regression       *         ***       5.4321
Residual Error   *         0.1429    ***
Total            *         5.5750
```

APPLICATIONS

12.8 Professor Asimov Professor Isaac Asimov was one of the most prolific writers of all time. Prior to his death, he wrote nearly 500 books during a 40-year career. In fact, as his career progressed, he became even more productive in terms of the number of books written within a given period of time.[1] The data give the time in months required to write his books in increments of 100:

Number of Books, x	100	200	300	400	490
Time in Months, y	237	350	419	465	507

a. Assume that the number of books x and the time in months y are linearly related. Find the least-squares line relating y to x.

b. Plot the time as a function of the number of books written using a scatterplot, and graph the least-squares line on the same paper. Does it seem to provide a good fit to the data points?

c. Construct the ANOVA table for the linear regression.

 12.9 A Chemical Experiment Using a chem-
EX1209 ical procedure called *differential pulse polarography*, a chemist measured the peak current generated (in microamperes) when a solution containing a given amount of nickel (in parts per billion) is added to a buffer[2]:

x = Ni (ppb)	y = Peak Current (mA)
19.1	.095
38.2	.174
57.3	.256
76.2	.348
95	.429
114	.500
131	.580
150	.651
170	.722

a. Use the data entry method for your calculator to calculate the preliminary sums of squares and cross-products, S_{xx}, S_{yy}, and S_{xy}.

b. Calculate the least-squares regression line.

c. Plot the points and the fitted line. Does the assumption of a linear relationship appear to be reasonable?

d. Use the regression line to predict the peak current generated when a solution containing 100 ppb of nickel is added to the buffer.

e. Construct the ANOVA table for the linear regression.

 12.10 Sleep Deprivation A study was conducted to determine the effects of sleep deprivation on people's ability to solve problems without sleep. A total of 10 subjects participated in the study, two at each of five sleep deprivation levels—8, 12, 16, 20, and 24 hours. After his or her specified sleep deprivation period, each subject was administered a set of simple addition problems, and the number of errors was recorded. These results were obtained:

Number of Errors, y	8, 6	6, 10	8, 14
Number of Hours without Sleep, x	8	12	16

Number of Errors, y	14, 12	16, 12
Number of Hours without Sleep, x	20	24

a. How many pairs of observations are in the experiment?

b. What are the total number of degrees of freedom?

c. Complete the *MINITAB* printout.

MINITAB output for Exercise 12.10

```
Regression Analysis: y versus x

The regression equation is
y = 3.00 + 0.475 x

Predictor        Coef      SE Coef         T        P
Constant         3.000       2.127      1.41    0.196
x                ***        0.1253      3.79    0.005

S = 2.24165     R-Sq = 64.2%      R-Sq(adj) = 59.8%

Analysis of Variance
Source           DF        SS        MS        F        P
Regression       **        72.200    72.200    14.37   0.005
Residual Error   **        ***       5.025
Total            **        ***
```

d. What is the least-squares prediction equation?

e. Use the prediction equation to predict the number of errors for a person who has not slept for 10 hours.

 12.11 Achievement Tests The Academic
EX1211 Performance Index (API) is a measure of school

achievement based on the results of the Stanford 9 Achievement test. Scores range from 200 to 1000, with 800 considered a long-range goal for schools. The following table shows the API for eight elementary schools in Riverside County, California, along with the percent of students at that school who are considered English Language Learners (ELL).[3]

School	1	2	3	4	5	6	7	8
API	588	659	710	657	669	641	557	743
ELL	58	22	14	30	11	26	39	6

a. Which of the two variables is the independent variable and which is the dependent variable? Explain your choice.

b. Use a scatterplot to plot the data. Is the assumption of a linear relationship between x and y reasonable?

c. Assuming that x and y are linearly related, calculate the least-squares regression line.

d. Plot the line on the scatterplot in part b. Does the line fit through the data points?

(MY) CD **12.12 How Long Is It?** How good are you at
EX1212 estimating? To test a subject's ability to estimate sizes, he was shown 10 different objects and asked to estimate their length or diameter. The object was then measured, and the results were recorded in the table below.

Object	Estimated (inches)	Actual (inches)
Pencil	7.00	6.00
Dinner plate	9.50	10.25
Book 1	7.50	6.75
Cell phone	4.00	4.25
Photograph	14.50	15.75
Toy	3.75	5.00
Belt	42.00	41.50
Clothespin	2.75	3.75
Book 2	10.00	9.25
Calculator	3.50	4.75

a. Find the least-squares regression line for predicting the actual measurement as a function of the estimated measurement.

b. Plot the points and the fitted line. Does the assumption of a linear relationship appear to be reasonable?

(MY) CD **12.13 Test Interviews** Of two personnel
EX1213 evaluation techniques available, the first requires a two-hour test interview while the second can be completed in less than an hour. The scores for each of the 15 individuals who took both tests are given in the next table.

Applicant	Test 1 (x)	Test 2(y)
1	75	38
2	89	56
3	60	35
4	71	45
5	92	59
6	105	70
7	55	31
8	87	52
9	73	48
10	77	41
11	84	51
12	91	58
13	75	45
14	82	49
15	76	47

a. Construct a scatterplot for the data. Does the assumption of linearity appear to be reasonable?

b. Find the least-squares line for the data.

c. Use the regression line to predict the score on the second test for an applicant who scored 85 on Test 1.

12.14 Test Interviews, continued Refer to Exercise 12.13. Construct the ANOVA table for the linear regression relating y, the score on Test 2, to x, the score on Test 1.

(MY) CD **12.15 Armspan and Height** Leonardo
EX1215 daVinci (1452–1519) drew a sketch of a man, indicating that a person's armspan (measuring across the back with your arms outstretched to make a "t") is roughly equal to the person's height. To test this claim, we measured eight people with the following results:

Person	1	2	3	4
Armspan (inches)	68	62.25	65	69.5
Height (inches)	69	62	65	70

Person	5	6	7	8
Armspan (inches)	68	69	62	60.25
Height (inches)	67	67	63	62

a. Draw a scatterplot for armspan and height. Use the same scale on both the horizontal and vertical axes. Describe the relationship between the two variables.

b. If da Vinci is correct, and a person's armspan is roughly the same as the person's height, what should the slope of the regression line be?

c. Calculate the regression line for predicting height based on a person's armspan. Does the value of the slope b confirm your conclusions in part b?

d. If a person has an armspan of 62 inches, what would you predict the person's height to be?

(MY) CD **12.16 Strawberries** The following data were
EX1216 obtained in an experiment relating the dependent variable, y (texture of strawberries), with x (coded storage temperature).

x	−2	−2	0	2	2
y	4.0	3.5	2.0	0.5	0.0

a. Find the least-squares line for the data.

b. Plot the data points and graph the least-squares line as a check on your calculations.

c. Construct the ANOVA table.

12.5 TESTING THE USEFULNESS OF THE LINEAR REGRESSION MODEL

In considering linear regression, you may ask two questions:

- Is the independent variable x useful in predicting the response variable y?
- If so, how well does it work?

This section examines several statistical tests and measures that will help you reach some answers. Once you have determined that the model is working, you can then use the model for predicting the response y for a given value of x.

Inferences Concerning β, the Slope of the Line of Means

Is the least-squares regression line useful? That is, is the regression equation that uses information provided by x substantially better than the simple predictor $\bar{y}$ that does not rely on x? If the independent variable x is *not useful* in the population model $y = \alpha + \beta x + \epsilon$, then the value of y does not change for different values of x. The only way that this happens for all values of x is when the slope β of the line of means equals 0. This would indicate that the relationship between y and x is not linear, so that the initial question about the usefulness of the independent variable x can be restated as: Is there a linear relationship between x and y?

You can answer this question by using either a test of hypothesis or a confidence interval for β. Both of these procedures are based on the sampling distribution of b, the sample estimator of the slope β. It can be shown that, if the assumptions about the random error ϵ are valid, then the estimator b has a normal distribution in repeated sampling with mean

$$E(b) = \beta$$

and standard error given by

$$SE = \sqrt{\frac{\sigma^2}{S_{xx}}}$$

where σ^2 is the variance of the random error ϵ. Since the value of σ^2 is estimated with $s^2 = MSE$, you can base inferences on the statistic given by

$$t = \frac{b - \beta}{\sqrt{MSE/S_{xx}}}$$

which has a t distribution with $df = (n - 2)$, the degrees of freedom associated with MSE.

TEST OF HYPOTHESIS CONCERNING THE SLOPE OF A LINE

1. Null hypothesis: $H_0 : \beta = \beta_0$
2. Alternative hypothesis:

One-Tailed Test	**Two-Tailed Test**
$H_a : \beta > \beta_0$	$H_a : \beta \neq \beta_0$
(or $\beta < \beta_0$)	

3. Test statistic: $t = \dfrac{b - \beta_0}{\sqrt{MSE/S_{xx}}}$

   When the assumptions given in Section 12.2 are satisfied, the test statistic will have a Student's t distribution with $(n - 2)$ degrees of freedom.

4. Rejection region: Reject H_0 when

One-Tailed Test	**Two-Tailed Test**
$t > t_\alpha$	$t > t_{\alpha/2}$ or $t < -t_{\alpha/2}$
(or $t < -t_\alpha$ when the alternative hypothesis is $H_a : \beta < \beta_0$)	

or when p-value $< \alpha$

The values of t_α and $t_{\alpha/2}$ can be found using Table 4 in Appendix I or the **t Probabilities** applet. Use the values of t corresponding to $(n - 2)$ degrees of freedom.

EXAMPLE (12.2) — Determine whether there is a significant linear relationship between the calculus grades and test scores listed in Table 12.1. Test at the 5% level of significance.

Solution The hypotheses to be tested are

$$H_0 : \beta = 0 \quad \text{versus} \quad H_a : \beta \neq 0$$

and the observed value of the test statistic is calculated as

$$t = \frac{b - 0}{\sqrt{\text{MSE}/S_{xx}}} = \frac{.7656 - 0}{\sqrt{75.7532/2474}} = 4.38$$

with $(n - 2) = 8$ degrees of freedom. With $\alpha = .05$, you can reject H_0 when $t > 2.306$ or $t < -2.306$. Since the observed value of the test statistic falls into the rejection region, H_0 is rejected and you can conclude that there is a significant linear relationship between the calculus grades and the test scores for the population of college freshmen.

MY APPLET

You can use the *t* **Test for the Slope** applet shown in Figure 12.7 to find *p*-values or rejection regions for this test. You must first calculate the standard error $\text{SE} = \sqrt{\text{MSE}/S_{xx}}$, type its value into the box marked "Std Error," and press "Enter."

FIGURE 12.7

t Test for the Slope applet

- If you enter the value of *b* into the formula at the top of the applet and press "Enter," the applet will calculate the test statistic and its one- or two-tailed *p*-value.

- If you enter the significance level α in the box marked "prob:" and select the "One-Tailed" or "Two-Tailed" option in the drop-down list, the applet will calculate the positive value of *t* necessary for rejecting H_0. (You could also use the **Student's *t* Probabilities** applet to find the critical values.)

What is the *p*-value for the test performed in Example 12.2? Does this *p*-value confirm our conclusions?

Another way to make inferences about the value of β is to construct a confidence interval for β and examine the range of possible values for β.

A $(1 - \alpha)100\%$ CONFIDENCE INTERVAL FOR β

$$b \pm t_{\alpha/2}(\text{SE})$$

where $t_{\alpha/2}$ is based on $(n - 2)$ degrees of freedom and

$$\text{SE} = \sqrt{\frac{s^2}{S_{xx}}} = \sqrt{\frac{\text{MSE}}{S_{xx}}}$$

EXAMPLE 12.3

Find a 95% confidence interval estimate of the slope β for the calculus grade data in Table 12.1.

Solution Substituting previously calculated values into

$$b \pm t_{.025}\sqrt{\frac{\text{MSE}}{S_{xx}}}$$

you have

$$.766 \pm 2.306\sqrt{\frac{75.7532}{2474}}$$

$$.766 \pm .404$$

The resulting 95% confidence interval is .362 to 1.170. Since the interval does not contain 0, you can conclude that the true value of β is not 0, and you can reject the null hypothesis $H_0 : \beta = 0$ in favor of $H_a : \beta \neq 0$, a conclusion that agrees with the findings in Example 12.2. Furthermore, the confidence interval estimate indicates that there is an increase from as little as .4 to as much as 1.2 points in a calculus test score for each 1-point increase in the achievement test score.

If you are using computer software to perform the regression analysis, you will find the t statistic and its p-value on the printout. Look at the *MINITAB* regression analysis printout reproduced in Figure 12.8. In the second portion of the printout, you will find the least-squares estimates a ("Constant") and b ("x") in the column marked "Coef," their standard errors ("SE Coef"), the calculated value of the t statistic ("T") used for testing the hypothesis that the parameter equals 0, and its p-value ("P"). The t test for significant regression, $H_0 : \beta = 0$, has a p-value of $P = .002$, and the null hypothesis is rejected, as in Example 12.2. Does this agree with the p-value found using the *t* **Test for the Slope** applet in Figure 12.7? In any event, there is a significant linear relationship between x and y.

FIGURE 12.8

MINITAB output for the calculus grade data

 TIP

Look for the standard error of b in the column marked "SE Coef."

Regression Analysis: y versus x

```
The regression equation is
y = 40.8 + 0.766 x

Predictor        Coef      SE Coef         T        P
Constant       40.784        8.507      4.79    0.001
x              0.7656       0.1750      4.38    0.002

S = 8.70363     R-Sq = 70.5%      R-Sq(adj) = 66.8%

Analysis of Variance
Source          DF           SS           MS        F        P
Regression       1       1450.0       1450.0    19.14    0.002
Residual Error   8        606.0         75.8
Total            9       2056.0
```

The Analysis of Variance *F* Test

The analysis of variance portion of the printout in Figure 12.8 shows an F statistic given by

 TIP

ANOVA *F* tests are always one-tailed (upper-tail).

$$F = \frac{\text{MSR}}{\text{MSE}} = 19.14$$

with 1 numerator degree of freedom and $(n - 2) = 8$ denominator degrees of freedom. This is an *equivalent test statistic* that can also be used for testing the hypothesis $H_0 : \beta = 0$. Notice that, within rounding error, the value of F is equal to t^2 with the identical *p*-value. In this case, if you use five-decimal-place accuracy prior to rounding, you find that $t^2 = (.76556/.17498)^2 = (4.37513)^2 = 19.14175 \approx 19.14 = F$ as given in the printout. This is no accident and results from the fact that the square of a t statistic with *df* degrees of freedom has the same distribution as an F statistic with 1 numerator and *df* denominator degrees of freedom. The F test is a more general test of the usefulness of the model and can be used when the model has more than one independent variable.

Measuring the Strength of the Relationship: The Coefficient of Determination

How well does the regression model fit? To answer this question, you can use a measure related to the *correlation coefficient r*, introduced in Chapter 3. Remember that

$$r = \frac{S_{xy}}{S_x S_y} = \frac{S_{xy}}{\sqrt{S_{xx} S_{yy}}} \qquad \text{for } -1 \le r \le 1$$

where s_{xy}, s_x, and s_y were defined in Chapter 3 and the various sums of squares were defined in Section 12.4.

The sum of squares for regression, SSR, in the analysis of variance measures the portion of the total variation, Total SS $= S_{yy}$, that can be explained by the regression of y on x. The remaining portion, SSE, is the "unexplained" variation attributed to random error. One way to measure the strength of the relationship between the response variable y and the predictor variable x is to calculate the **coefficient of determination**—the proportion of the total variation that is explained by the linear regression of y on x. For the calculus grade data, this proportion is equal to

 TIP

On computer printouts, r^2 is often given as a **percentage** rather than a proportion.

$$\frac{\text{SSR}}{\text{Total SS}} = \frac{1450}{2056} = .705 \quad \text{or} \quad 70.5\%$$

Since Total SS $= S_{yy}$ and SSR $= \dfrac{(S_{xy})^2}{S_{xx}}$, you can write

$$\frac{\text{SSR}}{\text{Total SS}} = \frac{(S_{xy})^2}{S_{xx} S_{yy}} = \left(\frac{S_{xy}}{\sqrt{S_{xx} S_{yy}}}\right)^2 = r^2$$

Therefore, the coefficient of determination, which was calculated as SSR/Total SS, is simply the square of the correlation coefficient r. It is the entry labeled "R-Sq" in Figure 12.8.

Remember that the analysis of variance table isolates the variation due to regression (SSR) from the total variation in the experiment. Doing so reduces the amount of *random variation* in the experiment, now measured by SSE rather than Total SS. In this context, the **coefficient of determination, r^2,** can be defined as follows:

Definition   The **coefficient of determination** r^2 can be interpreted as the percent reduction in the total variation in the experiment obtained by using the regression line $\hat{y} = a + bx$, instead of ignoring x and using the sample mean $\bar{y}$ to predict the response variable y.

For the calculus grade data, a reduction of $r^2 = .705$ or 70.5% is substantial. The regression model is working very well!

Interpreting the Results of a Significant Regression

Once you have performed the t or F test to determine the significance of the linear regression, you must interpret your results carefully. The slope β of the line of means is estimated based on data from only a particular region of observation. Even if you do not reject the null hypothesis that the slope of the line equals 0, it does not necessarily mean that y and x are unrelated. It may be that you have committed a Type II error—falsely declaring that the slope is 0 and that x and y are unrelated.

Fitting the Wrong Model

It may happen that y and x are perfectly related in a nonlinear way, as shown in Figure 12.9. Here are three possibilities:

FIGURE 12.9

Curvilinear relationship

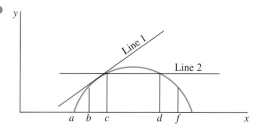

- If observations were taken only within the interval $b < x < c$, the relationship would appear to be linear with a positive slope.
- If observations were taken only within the interval $d < x < f$, the relationship would appear to be linear with a negative slope.
- If the observations were taken over the interval $c < x < d$, the line would be fitted with a slope close to 0, indicating no linear relationship between y and x.

For the example shown in Figure 12.9, no straight line accurately describes the true relationship between x and y, which is really a *curvilinear relationship*. In this case, we have chosen the *wrong model* to describe the relationship. Sometimes this type of mistake can be detected using residual plots, the subject of Section 12.7.

Extrapolation

One serious problem is to apply the results of a linear regression analysis to values of x that are *not included* within the range of the fitted data. This is called **extrapolation** and can lead to serious errors in prediction, as shown for line 1 in Figure 12.9.

Prediction results would be good over the interval $b < x < c$ but would seriously overestimate the values of y for $x > c$.

Causality

When there is a significant regression of y and x, it is tempting to conclude that x *causes* y. However, it is possible that one or more unknown variables that you have not even measured and that are not included in the analysis may be causing the observed relationship. In general, the statistician reports the results of an analysis but leaves conclusions concerning causality to scientists and investigators who are experts in these areas. These experts are better prepared to make such decisions!

12.5 EXERCISES

BASIC TECHNIQUES

12.17 Refer to Exercise 12.6. The data are reproduced below.

x	−2	−1	0	1	2
y	1	1	3	5	5

a. Do the data present sufficient evidence to indicate that y and x are linearly related? Test the hypothesis that $\beta = 0$ at the 5% level of significance.

b. Use the ANOVA table from Exercise 12.6 to calculate $F = MSR/MSE$. Verify that the square of the t statistic used in part a is equal to F.

c. Compare the two-tailed critical value for the t test in part a with the critical value for F with $\alpha = .05$. What is the relationship between the critical values?

12.18 Refer to Exercise 12.17. Find a 95% confidence interval for the slope of the line. What does the phrase "95% confident" mean?

12.19 Refer to Exercise 12.7. The data, along with the *MINITAB* analysis of variance table are reproduced below.

x	1	2	3	4	5	6
y	5.6	4.6	4.5	3.7	3.2	2.7

MINITAB ANOVA table for Exercise 12.19

Regression Analysis: y versus x

```
Analysis of Variance
Source           DF       SS        MS       F       P
Regression        1   5.4321   5.4321   152.10   0.000
Residual Error    4   0.1429   0.0357
Total             5   5.5750
```

a. Do the data provide sufficient evidence to indicate that y and x are linearly related? Use the information in the *MINITAB* printout to answer this question at the 1% level of significance.

b. Calculate the coefficient of determination r^2. What information does this value give about the usefulness of the linear model?

APPLICATIONS

(MY) CD **12.20 Air Pollution** An experiment was
EX1220 designed to compare several different types of air pollution monitors.[4] The monitor was set up, and then exposed to different concentrations of ozone, ranging between 15 and 230 parts per million (ppm) for periods of 8–72 hours. Filters on the monitor were then analyzed, and the amount (in micrograms) of sodium nitrate (NO_3) recorded by the monitor was measured. The results for one type of monitor are given in the table.

Ozone, x (ppm/hr)	.8	1.3	1.7	2.2	2.7	2.9
NO_3, y (μg)	2.44	5.21	6.07	8.98	10.82	12.16

a. Find the least-squares regression line relating the monitor's response to the ozone concentration.

b. Do the data provide sufficient evidence to indicate that there is a linear relationship between the ozone concentration and the amount of sodium nitrate detected?

c. Calculate r^2. What does this value tell you about the effectiveness of the linear regression analysis?

(MY) CD **12.21 The Cost of Flying** How is the cost of
EX1221 a plane flight related to the length of the trip? The table shows the average round-trip coach airfare paid by customers of American Airlines on each of 18 heavily traveled U.S. air routes.[5]

Route	Distance (miles)	Cost
Dallas–Austin	178	$125
Houston–Dallas	232	123
Chicago–Detroit	238	148
Chicago–St. Louis	262	136
Chicago–Cleveland	301	129
Chicago–Atlanta	593	162
New York–Miami	1092	224
New York–San Juan	1608	264
New York–Chicago	714	287
Chicago–Denver	901	256
Dallas–Salt Lake	1005	365
New York–Dallas	1374	459
Chicago–Seattle	1736	424
Los Angeles–Chicago	1757	361
Los Angeles–Atlanta	1946	309
New York–Los Angeles	2463	444
Los Angeles–Honolulu	2556	323
New York–San Francisco	2574	513

a. If you want to estimate the cost of a flight based on the distance traveled, which variable is the response variable and which is the independent predictor variable?

b. Assume that there is a linear relationship between cost and distance. Calculate the least-squares regression line describing cost as a linear function of distance.

c. Plot the data points and the regression line. Does it appear that the line fits the data?

d. Use the appropriate statistical tests and measures to explain the usefulness of the regression model for predicting cost.

12.22 Professor Asimov, continued Refer to the data in Exercise 12.8, relating x, the number of books written by Professor Isaac Asimov, to y, the number of months he took to write his books (in increments of 100). The data are reproduced below.

Number of Books, x	100	200	300	400	490
Time in Months, y	237	350	419	465	507

a. Do the data support the hypothesis that $\beta = 0$? Use the p-value approach, bounding the p-value using Table 4 of Appendix I or finding the exact p-value using the t **Test for the Slope** applet. Explain your conclusions in practical terms.

b. Use the ANOVA table in Exercise 12.8, part c, to calculate the coefficient of determination r^2. What percentage reduction in the total variation is achieved by using the linear regression model?

c. Plot the data or refer to the plot in Exercise 12.8, part b. Do the results of parts a and b indicate that

the model provides a good fit for the data? Are there any assumptions that may have been violated in fitting the linear model?

12.23 Refer to the sleep deprivation experiment described in Exercise 12.10 and data set EX1210. The data and the *MINITAB* printout are reproduced here.

Number of Errors, y	8, 6	6, 10	8, 14
Number of Hours without Sleep, x	8	12	16

Number of Errors, y	14, 12	16, 12
Number of Hours without Sleep, x	20	24

MINITAB output for Exercise 12.23

Regression Analysis: y versus x

```
The regression equation is
y = 3.00 + 0.475 x

Predictor     Coef    SE Coef      T       P
Constant     3.000      2.127    1.41   0.196
x           0.4750      0.1253    3.79   0.005

S = 2.24165    R-Sq = 64.2%    R-Sq(adj) = 59.8%

Analysis of Variance

Source          DF       SS       MS       F       P
Regression       1    72.200   72.200   14.37   0.005
Residual Error   8    40.200    5.025
Total            9   112.400
```

a. Do the data present sufficient evidence to indicate that the number of errors is linearly related to the number of hours without sleep? Identify the two test statistics in the printout that can be used to answer this question.

b. Would you expect the relationship between y and x to be linear if x varied over a wider range (say, $x = 4$ to $x = 48$)?

c. How do you describe the strength of the relationship between y and x?

d. What is the best estimate of the common population variance σ^2?

e. Find a 95% confidence interval for the slope of the line.

12.24 Strawberries II The following data (Exercise 12.16 and data set EX1216) were obtained in an experiment relating the dependent variable, y (texture of strawberries), with x (coded storage temperature). Use the information from Exercise 12.16 to answer the following questions:

x	−2	−2	0	2	2
y	4.0	3.5	2.0	0.5	0.0

a. What is the best estimate of σ^2, the variance of the random error ε?

b. Do the data indicate that texture and storage temperature are linearly related? Use $\alpha = .05$.

c. Calculate the coefficient of determination, r^2.

d. Of what value is the *linear* model in increasing the accuracy of prediction as compared to the predictor, $\bar{y}$?

 12.25 Laptops and Learning In Exercise
EX1225 1.61 we described an informal experiment conducted at McNair Academic High School in Jersey City, New Jersey. Two freshman algebra classes were studied, one of which used laptop computers at school and at home, while the other class did not. In each class, students were given a survey at the beginning and end of the semester, measuring his or her technological level. The scores were recorded for the end of semester survey (x) and the final examination (y) for the laptop group.[6] The data and the *MINITAB* printout are shown here.

Student	Posttest	Final Exam	Student	Posttest	Final Exam
1	100	98	11	88	84
2	96	97	12	92	93
3	88	88	13	68	57
4	100	100	14	84	84
5	100	100	15	84	81
6	96	78	16	88	83
7	80	68	17	72	84
8	68	47	18	88	93
9	92	90	19	72	57
10	96	94	20	88	83

Regression Analysis: y versus x

```
The regression equation is
y = -26.8 + 1.26 x

Predictor     Coef    SE Coef       T       P
Constant    -26.82      14.76   -1.82   0.086
x           1.2617     0.1685    7.49   0.000

S = 7.61912     R-Sq = 75.7%    R-Sq(adj) = 74.3%

Analysis of Variance
Source          DF        SS       MS       F       P
Regression       1    3254.0   3254.0   56.05   0.000
Residual Error  18    1044.9     58.1
Total           19    4299.0
```

a. Construct a scatterplot for the data. Does the assumption of linearity appear to be reasonable?

b. What is the equation of the regression line used for predicting final exam score as a function of the posttest score?

c. Do the data present sufficient evidence to indicate that final exam score is linearly related to the posttest score? Use $\alpha = .01$.

d. Find a 99% confidence interval for the slope of the regression line.

12.26 Laptops and Learning, continued Refer to Exercise 12.25.

a. Use the *MINITAB* printout to find the value of the coefficient of determination, r^2. Show that $r^2 = \text{SSR/Total SS}$.

b. What percentage reduction in the total variation is achieved by using the linear regression model?

12.27 Armspan and Height II In Exercise 12.15 (data set EX1215), we measured the armspan and height of eight people with the following results:

Person	1	2	3	4
Armspan (inches)	68	62.25	65	69.5
Height (inches)	69	62	65	70

Person	5	6	7	8
Armspan (inches)	68	69	62	60.25
Height (inches)	67	67	63	62

a. Does the data provide sufficient evidence to indicate that there is a linear relationship between armspan and height? Test at the 5% level of significance.

b. Construct a 95% confidence interval for the slope of the line of means, β.

c. If Leonardo da Vinci is correct, and a person's armspan is roughly the same as the person's height, the slope of the regression line is approximately equal to 1. Is this supposition confirmed by the confidence interval constructed in part b? Explain.

DIAGNOSTIC TOOLS FOR CHECKING THE REGRESSION ASSUMPTIONS

12.6

Even though you have determined—using the *t*-test for the slope (or the ANOVA *F*-test) and the value of r^2—that *x* is useful in predicting the value of *y*, the results of a regression analysis are valid only when the data satisfy the necessary regression assumptions.

REGRESSION ASSUMPTIONS

- The relationship between y and x must be linear, given by the model

$$y = \alpha + \beta x + \epsilon$$

- The values of the random error term ϵ (1) are independent, (2) have a mean of 0 and a common variance σ^2, independent of x, and (3) are normally distributed.

Since these assumptions are quite similar to those presented in Chapter 11 for an analysis of variance, it should not surprise you to find that the **diagnostic tools** for checking these assumptions are the same as those we used in that chapter. These tools involve the analysis of the **residual error,** the unexplained variation in each observation once the variation explained by the regression model has been removed.

Dependent Error Terms

The error terms are often dependent when the observations are collected at regular time intervals. When this is the case, the observations make up a **time series** whose error terms are correlated. This in turn causes bias in the estimates of model parameters. Time series data should be analyzed using time series methods. You will find a discussion of time-series analysis in the text *Statistics for Management and Economics,* 7th edition, by Mendenhall, Beaver and Beaver.

Residual Plots

The other regression assumptions can be checked using **residual plots,** which are fairly complicated to construct by hand but easy to use once a computer has graphed them for you!

In simple linear regression, you can use the **plot of residuals versus fit** to check for a constant variance as well as to make sure that the linear model is in fact adequate. This plot should be free of any patterns. It should appear as a random scatter of points about 0 on the vertical axis with approximately the same vertical spread for all values of $\hat{y}$. One property of the residuals is that they sum to 0 and therefore have a sample mean of 0. The plot of the residuals versus fit for the calculus grade example is shown in Figure 12.10. There are no apparent patterns in this residual plot, which indicates that the model is adequate for these data.

FIGURE 12.10

Plot of the residuals versus $\hat{y}$ for Example 12.1

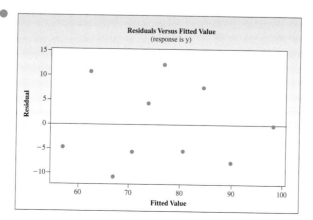

Recall from Chapter 11 that the **normal probability plot** is a graph that plots the residuals against the expected value of that residual if it had come from a normal distribution. When the residuals are normally distributed or approximately so, the plot should appear as a straight line, sloping upward. If the residuals have a standard normal distribution, then the line slopes upward at a 45° angle. The normal probability plot for the residuals in Example 12.1 is given in Figure 12.11. With the exception of the fourth and fifth plotted points, the remaining points appear to lie approximately on a straight line. This plot is not unusual and does not indicate underlying nonnormality. The most serious violations of the normality assumption usually appear in the tails of the distribution because this is where the normal distribution differs most from other types of distributions with a similar mean and measure of spread. Hence, curvature in either or both of the two ends of the normal probability plot is indicative of nonnormality.

MY **TIP**

Residuals vs. fits ⇔ random scatter

Normal plot ⇔ straight line, sloping up

FIGURE 12.11

Normal probability plot of residuals for Example 12.1

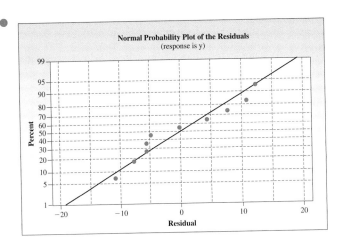

12.6 **EXERCISES**

BASIC TECHNIQUES

12.28 What diagnostic plot can you use to determine whether the data satisfy the normality assumption? What should the plot look like for normal residuals?

12.29 What diagnostic plot can you use to determine whether the incorrect model has been used? What should the plot look like if the correct model has been used?

12.30 What diagnostic plot can you use to determine whether the assumption of equal variance has been violated? What should the plot look like when the variances are equal for all values of x?

12.31 Refer to the data in Exercise 12.7. The normal probability plot and the residuals versus fitted values plots generated by *MINITAB* are shown here. Does it

appear that any regression assumptions have been violated? Explain.

MINITAB output for Exercise 12.31

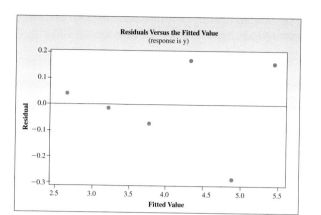

APPLICATIONS

12.32 Air Pollution Refer to Exercise 12.20, in which an air pollution monitor's response to ozone was recorded for several different concentrations of ozone. Use the *MINITAB* residual plots to comment on the validity of the regression assumptions.

MINITAB output for Exercise 12.32

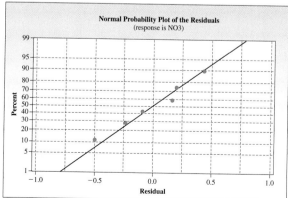

12.33 Professor Asimov Refer to Exercise 12.8, in which the number of books x written by Isaac Asimov

are related to the number of months y he took to write them. A plot of the data is shown.

a. Can you see any pattern other than a linear relationship in the original plot?

b. The value of r^2 for these data is .959. What does this tell you about the fit of the regression line?

c. Look at the accompanying diagnostic plots for these data. Do you see any pattern in the residuals? Does this suggest that the relationship between number of months and number of books written is something other than linear?

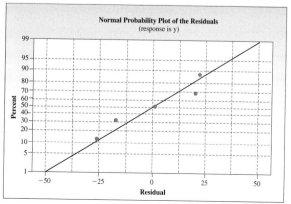

12.34 Laptops and Learning, again Refer to the data given in Exercise 12.25. The *MINITAB* printout is reproduced here.

```
Regression Analysis: y versus x

The regression equation is
y = -26.8 + 1.26 x

Predictor      Coef    SE Coef       T        P
Constant     -26.82      14.76    -1.82    0.086
x            1.2617      0.1685     7.49    0.000

S = 7.61912    R-Sq = 75.7%    R-Sq(adj) = 74.3%

Analysis of Variance
Source           DF      SS      MS       F       P
Regression        1   3254.0  3254.0   56.05   0.000
Residual Error   18   1044.9    58.1
Total            19   4299.0
```

a. What assumptions must be made about the distribution of the random error, ε?

b. What is the best estimate of σ^2, the variance of the random error, ε?

c. Use the diagnostic plots for these data to comment on the validity of the regression assumptions.

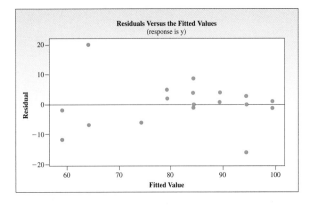

12.35 LCD TVs In Exercise 3.19, *Consumer Reports* gave the price (*y*) for the top 10 LCD standard definition TVs in the 14 to 20 inch category along with the screen size (*x*) in inches[7]:

Brand	Price	Size
Sharp LC-20E1U	$1200	20
Sony KLV-15SR1	800	15
Panasonic TC-20LA1	1050	20
Panasonic TC-17LA1	750	17
Gateway GTW-L18M103	700	18
Panasonic TC-14LA1	500	14
Gateway GTW-L17M103	600	17
Toshiba 14VL43U	670	14
Toshiba 20VL43U	1200	20
Sharp LC-15E1U	650	15

Does the price of an LCD TV depend on the size of the screen? Suppose we assume that the relationship between *x* and *y* is linear, and run a linear regression, resulting in a value $r^2 = .708$.

a. What does the value of r^2 tell you about the strength of the relationship between price and screen size?

b. The residual plot for this data, generated by *MINITAB*, is shown below. Does this plot suggest that the relationship between price and screen size might be something other than linear?

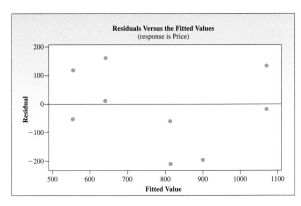

c. Plot the values of *x* and *y* using a scatterplot. Does this plot confirm your suspicions in part b? Explain.

ESTIMATION AND PREDICTION USING THE FITTED LINE

12.7

Now that you have

- tested the fitted regression line, $\hat{y} = a + bx$, to make sure that it is useful for prediction and
- used the diagnostic tools to make sure that none of the regression assumptions have been violated

you are ready to use the line for one of its two purposes:

- Estimating the average value of y for a given value of x
- Predicting a particular value of y for a given value of x

The sample of n pairs of observations have been chosen from a population in which the *average* value of y is related to the value of the predictor variable x by the **line of means,**

$$E(y) = \alpha + \beta x$$

an unknown line, shown as a broken line in Figure 12.12. Remember that for a fixed value of x—say, x_0—the *particular* values of y deviate from the line of means. These values of y have a normal distribution with mean equal to $\alpha + \beta x_0$ and variance σ^2, as shown in Figure 12.12.

FIGURE 12.12

Distribution of y for $x = x_0$

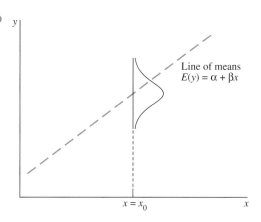

Line of means
$E(y) = \alpha + \beta x$

$x = x_0$

Since the computed values of a and b vary from sample to sample, each new sample produces a different regression line $\hat{y} = a + bx$, which can be used either to estimate the line of means or to predict a particular value of y. Figure 12.13 shows one of the possible configurations of the fitted line (teal), the unknown line of means (gray), and a particular value of y (the teal dot).

FIGURE 12.13

Error in estimating $E(y)$ and in predicting y

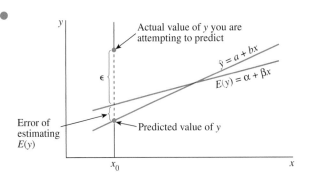

How far will our estimator $\hat{y} = a + bx_0$ be from the quantity to be estimated or predicted? This depends, as always, on the variability in our estimator, measured by its **standard error.** It can be shown that

$$\hat{y} = a + bx_0$$

the estimated value of y when $x = x_0$, is an unbiased estimator of the line of means, $\alpha + \beta x_0$, and that $\hat{y}$ is normally distributed with the standard error of $\hat{y}$ estimated by

$$SE(\hat{y}) = \sqrt{MSE\left(\frac{1}{n} + \frac{(x_0 - \bar{x})^2}{S_{xx}}\right)}$$

Estimation and testing are based on the statistic

$$t = \frac{\hat{y} - E(y)}{SE(\hat{y})}$$

which has a t distribution with $(n - 2)$ degrees of freedom.

To form a $(1 - \alpha)100\%$ confidence interval for the average value of y when $x = x_0$, measured by the line of means, $\alpha + \beta x_0$, you can use the usual form for a confidence interval based on the t distribution:

$$\hat{y} \pm t_{\alpha/2}SE(\hat{y})$$

 TIP

For a given value of x, the prediction interval is always wider than the confidence interval.

If you choose to predict a *particular* value of y when $x = x_0$, however, there is some additional error in the prediction because of the deviation of y from the line of means. If you examine Figure 12.13, you can see that the error in prediction has two components:

- The error in using the fitted line to estimate the line of means
- The error caused by the deviation of y from the line of means, measured by σ^2

The variance of the difference between y and $\hat{y}$ is the sum of these two variances and forms the basis for the standard error of $(y - \hat{y})$ used for prediction:

$$SE(y - \hat{y}) = \sqrt{MSE\left[1 + \frac{1}{n} + \frac{(x_0 - \bar{x})^2}{S_{xx}}\right]}$$

and the $(1 - \alpha)100\%$ prediction interval is formed as

$$\hat{y} \pm t_{\alpha/2}SE(y - \hat{y})$$

$(1 - \alpha)100\%$ CONFIDENCE AND PREDICTION INTERVALS

- For estimating the average value of y when $x = x_0$:

$$\hat{y} \pm t_{\alpha/2}\sqrt{MSE\left[\frac{1}{n} + \frac{(x_0 - \bar{x})^2}{S_{xx}}\right]}$$

- For predicting a particular value of y when $x = x_0$:

$$\hat{y} \pm t_{\alpha/2}\sqrt{MSE\left[1 + \frac{1}{n} + \frac{(x_0 - \bar{x})^2}{S_{xx}}\right]}$$

where $t_{\alpha/2}$ is the value of t with $(n - 2)$ degrees of freedom and area $\alpha/2$ to its right.

EXAMPLE 12.4 Use the information in Example 12.1 to estimate the average calculus grade for students whose achievement score is 50, with a 95% confidence interval.

Solution   The point estimate of $E(y|x_0 = 50)$, the average calculus grade for students whose achievement score is 50, is

$$\hat{y} = 40.78424 + .76556(50) = 79.06$$

The standard error of $\hat{y}$ is

$$\sqrt{MSE\left[\frac{1}{n} + \frac{(x_0 - \bar{x})^2}{S_{xx}}\right]} = \sqrt{75.7532\left[\frac{1}{10} + \frac{(50 - 46)^2}{2474}\right]} = 2.840$$

and the 95% confidence interval is

$$79.06 \pm 2.306(2.840)$$
$$79.06 \pm 6.55$$

Our results indicate that the average calculus grade for students who score 50 on the achievement test will lie between 72.51 and 85.61.

EXAMPLE 12.5 A student took the achievement test and scored 50 but has not yet taken the calculus test. Using the information in Example 12.1, predict the calculus grade for this student with a 95% prediction interval.

Solution   The predicted value of y is $\hat{y} = 79.06$, as in Example 12.4. However, the error in prediction is measured by $SE(y - \hat{y})$, and the 95% prediction interval is

$$79.06 \pm 2.306\sqrt{75.7532\left[1 + \frac{1}{10} + \frac{(50 - 46)^2}{2474}\right]}$$

$$79.06 \pm 2.306(9.155)$$
$$79.06 \pm 21.11$$

or from 57.95 to 100.17. The prediction interval is *wider* than the confidence interval in Example 12.4 because of the extra variability in predicting the actual value of the response y.

One particular point on the line of means is often of interest to experimenters, the **y-intercept α**—the average value of y when $x_0 = 0$.

EXAMPLE 12.6

Prior to fitting a line to the calculus grade-achievement score data, you may have thought that a score of 0 on the achievement test would predict a grade of 0 on the calculus test. This implies that we should fit a model with α equal to 0. Do the data support the hypothesis of a 0 intercept?

Solution You can answer this question by constructing a 95% confidence interval for the y-intercept α, which is the average value of y when $x = 0$. The estimate of α is

$$\hat{y} = 40.784 + .76556(0) = 40.784 = a$$

and the 95% confidence interval is

$$\hat{y} \pm t_{\alpha/2}\sqrt{MSE\left[\frac{1}{n} + \frac{(x_0 - \bar{x})^2}{S_{xx}}\right]}$$

$$40.784 \pm 2.306\sqrt{75.7532\left[\frac{1}{10} + \frac{(0 - 46)^2}{2474}\right]}$$

$$40.784 \pm 19.617$$

or from 21.167 to 60.401, an interval that does not contain the value $\alpha = 0$. Hence, it is unlikely that the y-intercept is 0. You should include a nonzero intercept in the model $y = \alpha + \beta x + \epsilon$.

For this special situation in which you are interested in testing or estimating the y-intercept α for the line of means, the inferences involve the sample estimate a. The test for a 0 intercept is given in Figure 12.14 in the shaded line labeled "Constant." The coefficient given as 40.784 is a, with standard error given in the column labeled "SE Coef" as 8.507, which agrees with the value calculated in Example 12.6. The value of $t = 4.79$ is found by dividing a by its standard error with p-value $= .001$.

FIGURE 12.14

Portion of the *MINITAB* output for Example 12.6

Predictor	Coef	SE Coef	T	P
Constant	40.784	8.507	4.79	0.001
x	0.7656	0.1750	4.38	0.002

You can see that it is quite time-consuming to calculate these estimation and prediction intervals by hand. Moreover, it is difficult to maintain accuracy in your calculations. Fortunately, computer programs can perform these calculations for you. The *MINITAB* regression command provides an option for either estimation or prediction when you specify the necessary value(s) of x. The printout in Figure 12.15 gives the values of $\hat{y} = 79.06$ labeled "Fit," the standard error of $\hat{y}$, SE($\hat{y}$), labeled "SE Fit," the *confidence interval* for the average value of y when $x = 50$, labeled "95.0% CI," and the prediction interval for y when $x = 50$, labeled "95.0% PI."

FIGURE 12.15

MINITAB option for estimation and prediction

```
Predicted Values for New Observations
New Obs     Fit     SE Fit      95% CI           95% PI
1          79.06     2.84    (72.51,  85.61)   (57.95,  100.17)

Values of Predictors for New Observations
New Obs        x
1            50.0
```

The confidence bands and prediction bands generated by *MINITAB* for the calculus grades data are shown in Figure 12.16. Notice that in general the confidence bands are narrower than the prediction bands for every value of the achievement test score x. Certainly you would expect predictions for an individual value to be much more variable than estimates of the average value. Also notice that the bands seem to get wider as the value of x_0 gets farther from the mean $\bar{x}$. This is because the standard errors used in the confidence and prediction intervals contain the term $(x_0 - \bar{x})^2$, which gets larger as the two values diverge. In practice, this means that estimation and prediction are more accurate when x_0 is near the center of the range of the x-values. You can locate the calculated confidence and prediction intervals when $x = 50$ in Figure 12.16.

FIGURE 12.16

Confidence and prediction intervals for the data in Table 12.1

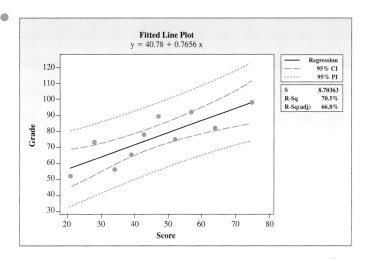

12.7 **EXERCISES**

BASIC TECHNIQUES

12.36 Refer to Exercise 12.6.

a. Estimate the average value of y when $x = 1$, using a 90% confidence interval.

b. Find a 90% prediction interval for some value of y to be observed in the future when $x = 1$.

12.37 Refer to Exercise 12.7. Portions of the *MINITAB* printout are shown here.

MINITAB output for Exercise 12.37

```
Regression Analysis: y versus x

The regression equation is
y = 6.00 - 0.557 x

Predictor      Coef    SE Coef        T        P
Constant     6.0000     0.1759    34.10    0.000
x           -0.55714    0.04518   -12.33    0.000

Predicted Values for New Observations
New Obs    Fit    SE Fit     95.0% CI          95.0% PI
1       4.8857    0.1027  (4.6006,  5.1708)  (4.2886,  5.4829)
2       1.5429    0.2174  (0.9392,  2.1466)  (0.7430,  2.3427) X
X denotes a point that is an outlier in the predictors.

Values of Predictors for New Observations
New Obs       x
1          2.00
2          8.00
```

a. Find a 95% confidence interval for the average value of y when $x = 2$.

b. Find a 95% prediction interval for some value of y to be observed in the future when $x = 2$.

c. The last line in the third section of the printout indicates a problem with one of the fitted values. What value of x corresponds to the fitted value $\hat{y} = 1.5429$? What problem has the *MINITAB* program detected?

APPLICATIONS

12.38 **What to Buy?** A marketing research experiment was conducted to study the relationship between the length of time necessary for a buyer to reach a decision and the number of alternative package designs of a product presented. Brand names were eliminated from the packages to reduce the effects of brand preferences. The buyers made their selections using the manufacturer's product descriptions on the packages as the only buying guide. The length of time necessary to reach a decision was recorded for 15 participants in the marketing research study.

Length of Decision Time, y (sec)	5, 8, 8, 7, 9	7, 9, 8, 9, 10	10, 11, 10, 12, 9
Number of Alternatives, x	2	3	4

a. Find the least-squares line appropriate for these data.

b. Plot the points and graph the line as a check on your calculations.

c. Calculate s^2.

d. Do the data present sufficient evidence to indicate that the length of decision time is linearly related to the number of alternative package designs? (Test at the $\alpha = .05$ level of significance.)

e. Find the approximate p-value for the test and interpret its value.

f. If they are available, examine the diagnostic plots to check the validity of the regression assumptions.

g. Estimate the average length of time necessary to reach a decision when three alternatives are presented, using a 95% confidence interval.

MY CD **12.39 Housing Prices** If you try to rent an
EX1239 apartment or buy a house, you find that real estate representatives establish apartment rents and house prices on the basis of square footage of heated floor space. The data in the table give the square footages and sales prices of $n = 12$ houses randomly selected from those sold in a small city. Use the *MINITAB* printout to answer the questions.

Square Feet, x	Price, y	Square Feet, x	Price, y
1,460	$188,700	1,977	$205,400
2,108	209,300	1,610	197,000
1,743	201,400	1,530	192,400
1,499	191,100	1,759	198,200
1,864	202,400	1,821	204,300
2,391	214,900	2,216	211,700

Plot of data for Exercise 12.39

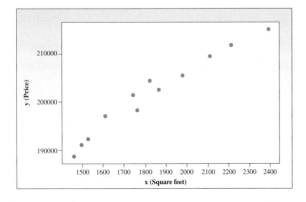

MINITAB output for Exercise 12.39

Regression Analysis: y versus x

```
The regression equation is
y = 151206 + 27.4 x

Predictor        Coef     SE Coef         T        P
Constant       151206        3389     44.62    0.000
x              27.406       1.828     14.99    0.000

S = 1792.72      R-Sq = 95.7%      R-Sq(adj) = 95.3%

Predicted Values for New Observations
New Obs    Fit    SE Fit       95.0% CI              95.0% PI
1       199989       526   (198817,  201161)   (195826,  204151)
2       206018       602   (204676,  207360)   (201804,  210232)

Values of Predictors for New Observations
New Obs        x
1           1780
2           2000
```

a. Can you see any pattern other than a linear relationship in the original plot?

b. The value of r^2 for these data is .957. What does this tell you about the fit of the regression line?

c. Look at the accompanying diagnostic plots for these data. Do you see any pattern in the residuals? Does this suggest that the relationship between price and square feet is something other than linear?

MINITAB output for Exercise 12.39

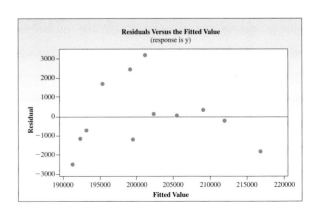

12.40 Housing Prices II Refer to Exercise 12.39 and data set EX1239.

a. Estimate the average increase in the price for an increase of 1 square foot for houses sold in the city. Use a 99% confidence interval. Interpret your estimate.

b. A real estate salesperson needs to estimate the average sales price of houses with a total of 2,000 square feet of heated space. Use a 95% confidence interval and interpret your estimate.

c. Calculate the price per square foot for each house and then calculate the sample mean. Why is this estimate of the average cost per square foot not equal to the answer in part a? Should it be? Explain.

d. Suppose that a house with 1780 square feet of heated floor space is offered for sale. Construct a 95% prediction interval for the price at which the house will sell.

12.41 Strawberries III The following data (Exercises 12.16 and 12.24) were obtained in an experiment relating the dependent variable, y (texture of strawberries), with x (coded storage temperature).

x	-2	-2	0	2	2
y	4.0	3.5	2.0	0.5	0.0

a. Estimate the expected strawberry texture for a coded storage temperature of $x = -1$. Use a 99% confidence interval.

b. Predict the particular value of y when $x = 1$ with a 99% prediction interval.

c. At what value of x will the width of the prediction interval for a particular value of y be a minimum, assuming n remains fixed?

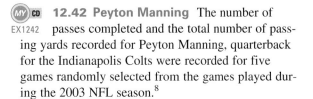

12.42 Peyton Manning The number of passes completed and the total number of passing yards recorded for Peyton Manning, quarterback for the Indianapolis Colts were recorded for five games randomly selected from the games played during the 2003 NFL season.[8]

Week	Completions	Total Yards
8	22	269
15	25	290
2	14	173
10	28	347
11	27	401

a. What is the least-squares line relating the total passing yards to the number of pass completions for Peyton Manning?

b. What proportion of the total variation is explained by the regression of total passing yards (y) on the number of pass completions (x)?

c. If they are available, examine the diagnostic plots to check the validity of the regression assumptions.

12.43 Peyton Manning, continued Refer to Exercise 12.42.

a. Estimate the average number of passing yards for games in which Peyton throws 21 completed passes using a 95% confidence interval.

b. Predict the actual number of passing yards for games in which Peyton throws 21 completed passes using a 95% prediction interval.

c. Would it be advisable to use the least-squares line from Exercise 12.42 to predict Peyton's total number of passing yards for a game in which he threw only five completed passes? Explain.

CORRELATION ANALYSIS

12.8

In Chapter 3, we introduced the *correlation coefficient* as a measure of the strength of the linear relationship between two variables. The correlation coefficient, r—formally called the **Pearson product moment sample coefficient of correlation**—is defined next.

PEARSON PRODUCT MOMENT COEFFICIENT OF CORRELATION

$$r = \frac{S_{xy}}{s_x s_y} = \frac{S_{xy}}{\sqrt{S_{xx}S_{yy}}} \qquad \text{for } -1 \le r \le 1$$

The variances and covariance can be found by direct calculation, by using a calculator with a two-variable statistics capacity, or by using a statistical package such as *MINITAB*. The variances and covariance are calculated as

 TIP

r is always between
-1 and $+1$

$$s_{xy} = \frac{S_{xy}}{n-1} \qquad s_x^2 = \frac{S_{xx}}{n-1} \qquad s_y^2 = \frac{S_{yy}}{n-1}$$

and use S_{xy}, S_{xx}, and S_{yy}, the same quantities used in regression analysis earlier in this chapter. In general, when a sample of n individuals or experimental units is selected and two variables are measured on each individual or unit so that *both variables are random,* the correlation coefficient r is the appropriate measure of linearity for use in this situation.

EXAMPLE 12.7 The heights and weights of $n = 10$ offensive backfield football players are randomly selected from a county's football all-stars. Calculate the correlation coefficient for the heights (in inches) and weights (in pounds) given in Table 12.4.

TABLE 12.4 ● **Heights and Weights of $n = 10$ Backfield All-Stars**

Player	Height, x	Weight, y
1	73	185
2	71	175
3	75	200
4	72	210
5	72	190
6	75	195
7	67	150
8	69	170
9	71	180
10	69	175

Solution You should use the appropriate data entry method of your scientific calculator to verify the calculations for the sums of squares and cross-products:

$$S_{xy} = 328 \qquad S_{xx} = 60.4 \qquad S_{yy} = 2610$$

using the calculational formulas given earlier in this chapter. Then

$$r = \frac{328}{\sqrt{(60.4)(2610)}} = .8261$$

or $r = .83$. This value of r is fairly close to 1, the largest possible value of r, which indicates a fairly strong positive linear relationship between height and weight.

There is a direct relationship between the calculational formulas for the correlation coefficient r and the slope of the regression line b. Since the numerator of both quantities is S_{xy}, both r and b have the same sign. Therefore, the correlation coefficient has these general properties:

 TIP

The sign of r is always
the **same** as the sign of
the slope b.

- When $r = 0$, the slope is 0, and there is no linear relationship between x and y.

- When r is positive, so is b, and there is a positive linear relationship between x and y.

- When r is negative, so is b, and there is a negative linear relationship between x and y.

In Section 12.5, we showed that

$$r^2 = \frac{SSR}{\text{Total SS}} = \frac{\text{Total SS} - SSE}{\text{Total SS}}$$

In this form, you can see that r^2 can never be greater than 1, so that $-1 \leq r \leq 1$. Moreover, you can see the relationship between the random variation (measured by SSE) and r^2.

- If there is no random variation and all the points fall on the regression line, then $SSE = 0$ and $r^2 = 1$.
- If the points are randomly scattered and there is no variation explained by regression, then $SSR = 0$ and $r^2 = 0$.

(MY) APPLET

You can use the **Exploring Correlation** applet to visualize the connection between the value of r and the pattern of points shown in the scatterplot. Use your mouse to move the slider at the bottom of the scatterplot. You will see the value of r change as the pattern of the points changes. Try to reproduce the patterns described above for $r^2 = 1$ and $r^2 = 0$.

FIGURE 12.17

Exploring Correlation applet

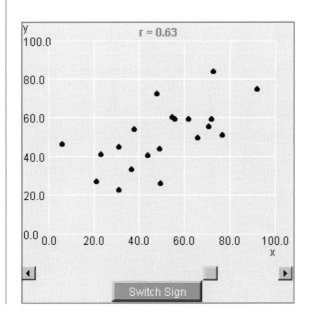

Figure 12.18 shows four typical scatterplots and their associated correlation coefficients. Notice that in scatterplot (d) there appears to be a curvilinear relationship between x and y, but r is approximately 0, which reinforces the fact that r is a measure of a *linear* (not *curvilinear*) relationship between two variables.

FIGURE 12.18 ●

Some typical scatterplots
with approximate values
of r

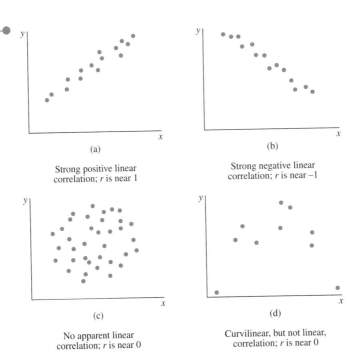

(a)

Strong positive linear
correlation; r is near 1

(b)

Strong negative linear
correlation; r is near -1

(c)

No apparent linear
correlation; r is near 0

(d)

Curvilinear, but not linear,
correlation; r is near 0

Consider a population generated by measuring two random variables on each experimental unit. In this *bivariate* population, the **population correlation coefficient** ρ (Greek letter rho) is calculated and interpreted as it is in the sample. In this situation, the experimenter can test the hypothesis that there is no correlation between the variables x and y using a test statistic that is *exactly equivalent* to the test of the slope β in Section 12.5. The test procedure is shown next.

TEST OF HYPOTHESIS CONCERNING THE CORRELATION COEFFICIENT ρ

1. Null hypothesis: $H_0 : \rho = 0$
2. Alternative hypothesis:

One-Tailed Test	Two-Tailed Test
$H_a : \rho > 0$ (or $\rho < 0$)	$H_a : \rho \neq 0$

MY TIP

You can prove that

$$t = r\sqrt{\frac{n-2}{1-r^2}}$$

$$= \frac{b-0}{\sqrt{MSE/S_{xx}}}.$$

3. Test statistic: $t = r\sqrt{\dfrac{n-2}{1-r^2}}$

When the assumptions given in Section 12.2 are satisfied, the test statistic will have a Student's t distribution with $(n-2)$ degrees of freedom.

4. Rejection region: Reject H_0 when

One-Tailed Test	Two-Tailed Test
$t > t_\alpha$ (or $t < -t_\alpha$ when the alternative hypothesis is $H_a : \rho < 0$) or when p-value $< \alpha$	$t > t_{\alpha/2}$ or $t < -t_{\alpha/2}$

The values of t_α and $t_{\alpha/2}$ can be found using Table 4 in Appendix I or the ***t*-Probabilities** applet. Use the values of t corresponding to $(n-2)$ degrees of freedom.

EXAMPLE 12.8

Refer to the height and weight data in Example 12.7. The correlation of height and weight was calculated to be $r = .8261$. Is this correlation significantly different from 0?

Solution To test the hypotheses

$$H_0 : \rho = 0 \quad \text{versus} \quad H_a : \rho \neq 0$$

the value of the test statistic is

$$t = r\sqrt{\frac{n-2}{1-r^2}} = .8261\sqrt{\frac{10-2}{1-(.8261)^2}} = 4.15$$

MY TIP

The *t* value and *p*-value for testing $H_0 : \rho = 0$ will be identical to the *t* and *p*-value for testing $H_0 : \beta = 0$.

which for $n = 10$ has a t distribution with 8 degrees of freedom. Since this value is greater than $t_{.005} = 3.355$, the two-tailed p-value is less than $2(.005) = .01$, and the correlation is declared significant at the 1% level $(P < .01)$. The value $r^2 = .8261^2 = .6824$ means that about 68% of the variation in one of the variables is explained by the other. The *MINITAB* printout in Figure 12.19 displays the correlation r and the exact p-value for testing its significance.

FIGURE 12.19 ●

MINITAB output for Example 12.8

```
Correlations: x, y

Pearson correlation of x and y = 0.826
P-Value = 0.003
```

If the linear coefficients of correlation between y and each of two variables x_1 and x_2 are calculated to be .4 and .5, respectively, it does not follow that a predictor using both variables will account for $[(.4)^2 + (.5)^2] = .41$, or a 41% reduction in the sum of squares of deviations. Actually, x_1 and x_2 might be highly correlated and therefore contribute virtually the same information for the prediction of y.

Finally, remember that r is a measure of **linear correlation** and that x and y could be perfectly related by some **curvilinear** function when the observed value of r is equal to 0. The problem of estimating or predicting y using information given by several independent variables, $x_1, x_2, \ldots, x_k$, is the subject of Chapter 13.

 12.8 EXERCISES

BASIC TECHNIQUES

12.44 How does the coefficient of correlation measure the strength of the linear relationship between two variables y and x?

12.45 Describe the significance of the algebraic sign and the magnitude of r.

12.46 What value does r assume if all the data points fall on the same straight line in these cases?

a. The line has positive slope.

b. The line has negative slope.

12.47 You are given these data:

x	−2	−1	0	1	2
y	2	2	3	4	4

a. Plot the data points. Based on your graph, what will be the sign of the sample correlation coefficient?

b. Calculate r and r^2 and interpret their values.

12.48 You are given these data:

x	1	2	3	4	5	6
y	7	5	5	3	2	0

a. Plot the six points on graph paper.

b. Calculate the sample coefficient of correlation r and interpret.

c. By what percentage was the sum of squares of deviations reduced by using the least-squares predictor $\hat{y} = a + bx$ rather than $\bar{y}$ as a predictor of y?

12.49 Reverse the slope of the line in Exercise 12.48 by reordering the y observations, as follows:

x	1	2	3	4	5	6
y	0	2	3	5	5	7

Repeat the steps of Exercise 12.48. Notice the change in the sign of r and the relationship between the values of r^2 of Exercise 12.48 and this exercise.

APPLICATIONS

12.50 Lobster The table gives the numbers of *Octolasmis tridens* and *O. lowei* barnacles on each of 10 lobsters.[9] Does it appear that the barnacles compete for space on the surface of a lobster?

Lobster Field Number	O. tridens	O. lowei
A061	645	6
A062	320	23
A066	401	40
A070	364	9
A067	327	24
A069	73	5
A064	20	86
A068	221	0
A065	3	109
A063	5	350

a. If they do compete, do you expect the number x of *O. tridens* and the number y of *O. lowei* barnacles to be positively or negatively correlated? Explain.

b. If you want to test the theory that the two types of barnacles compete for space by conducting a test of the null hypothesis "the population correlation coefficient ρ equals 0," what is your alternative hypothesis?

c. Conduct the test in part b and state your conclusions.

12.51 Social Skills Training A social skills training program was implemented with seven mildly challenged students in a study to determine whether the program caused improvement in pre/post measures and behavior ratings. For one such test, the pre- and posttest scores for the seven students are given in the table.[10]

Subject	Pretest	Posttest
Earl	101	113
Ned	89	89
Jasper	112	121
Charlie	105	99
Tom	90	104
Susie	91	94
Lori	89	99

a. What type of correlation, if any, do you expect to see between the pre- and posttest scores? Plot the data. Does the correlation appear to be positive or negative?

b. Calculate the correlation coefficient, r. Is there a significant positive correlation?

12.52 Hockey G. W. Marino investigated the variables related to a hockey player's ability to make a fast start from a stopped position.[11] In the experiment, each skater started from a stopped position and attempted to move as rapidly as possible over a 6-meter distance. The correlation coefficient r between a skater's stride rate (number of strides per second) and the length of time to cover the 6-meter distance for the sample of 69 skaters was $-.37$.

a. Do the data provide sufficient evidence to indicate a correlation between stride rate and time to cover the distance? Test using $\alpha = .05$.

b. Find the approximate p-value for the test.

c. What are the practical implications of the test in part a?

12.53 Hockey II Refer to Exercise 12.52. Marino calculated the sample correlation coefficient r for the stride rate and the average acceleration rate for the 69 skaters to be .36. Do the data provide sufficient evidence to indicate a correlation between stride rate and average acceleration for the skaters? Use the p-value approach.

12.54 Geothermal Power Geothermal
EX1254   power is an important source of energy. Since
the amount of energy contained in 1 pound of water is
a function of its temperature, you might wonder
whether water obtained from deeper wells contains
more energy per pound. The data in the table are
reproduced from an article on geothermal systems by
A.J. Ellis.[12]

Location of Well	Average (max.) Drill Hole Depth (m)	Average (max.) Temperature (°C)
El Tateo, Chile	650	230
Ahuachapan, El Salvador	1,000	230
Namafjall, Iceland	1,000	250
Larderello (region), Italy	600	200
Matsukawa, Japan	1,000	220
Cerro Prieto, Mexico	800	300
Wairakei, New Zealand	800	230
Kizildere, Turkey	700	190
The Geysers, United States	1,500	250

Is there a significant positive correlation between average maximum drill hole depth and average maximum
temperature?

12.55 Cheese, Please! The demand for healthy
foods that are low in fat and calories has resulted in a
large number of "low-fat" or "fat-free" products. The
table shows the number of calories and the amount of
sodium (in milligrams) per slice for five different
brands of fat-free American cheese.

Brand	Sodium (mg)	Calories
Kraft Fat Free Singles	300	30
Ralphs Fat Free Singles	300	30
Borden Fat Free	320	30
Healthy Choice Fat Free	290	30
Smart Beat American	180	25

a. Should you use the methods of linear regression
analysis or correlation analysis to analyze the data?
Explain.

b. Analyze the data to determine the nature of the relationship between sodium and calories in fat-free
American cheese. Use any statistical tests that are
appropriate.

12.56 Body Temperature and Heart Rate
EX1256   Is there any relationship between these two variables? To find out, we randomly selected 12 people
from a data set constructed by Allen Shoemaker

(*Journal of Statistics Education*) and recorded their
body temperature and heart rate.[13]

Person	1	2	3	4	5	6
Temperature (degrees)	96.3	97.4	98.9	99.0	99.0	96.8
Heart Rate (beats per minute)	70	68	80	75	79	75

Person	7	8	9	10	11	12
Temperature (degrees)	98.4	98.4	98.8	98.8	99.2	99.3
Heart Rate (beats per minute)	74	84	73	84	66	68

a. Find the correlation coefficient r, relating body
temperature to heart rate.

b. Is there sufficient evidence to indicate that there is a
correlation between these two variables? Test at the
5% level of significance.

12.57 Baseball Stats Does a team's batting
EX1257   average depend in any way on the number of
home runs hit by the team? The data in the table show
the number of team home runs and the overall team
batting average for eight randomly selected major
league baseball teams in the very early stages of the
2004 baseball season.[14]

Team	Total Home Runs	Team Batting Average
Anaheim	26	.279
Atlanta	21	.271
Baltimore	21	.282
Boston	26	.257
Chicago	38	.276
Houston	25	.290
Philadelphia	26	.245
San Diego	17	.271

Source: espn.com.

a. Plot the points using a scatterplot. Does it appear
that there is any relationship between total home
runs and team batting average?

b. Is there a significant positive correlation between
total home runs and team batting average? Test at
the 5% level of significance.

c. Do you think that the relationship between these
two variables would be different if we had looked
at an entire baseball season, rather than just the
very early part of the season?

CHAPTER REVIEW

Key Concepts and Formulas

I. A Linear Probabilistic Model

1. When the data exhibit a linear relationship, the appropriate model is $y = \alpha + \beta x + \epsilon$.

2. The random error ϵ has a normal distribution with mean 0 and variance σ^2.

II. Method of Least Squares

1. Estimates a and b, for a and b, are chosen to minimize SSE, the sum of squared deviations about the regression line, $\hat{y} = a + bx$.

2. The least-squares estimates are $b = S_{xy}/S_{xx}$ and $a = \bar{y} - b\bar{x}$.

III. Analysis of Variance

1. Total SS = SSR + SSE, where Total SS = S_{yy} and SSR = $(S_{xy})^2/S_{xx}$.

2. The best estimate of σ^2 is MSE = SSE/$(n - 2)$.

IV. Testing, Estimation, and Prediction

1. A test for the significance of the linear regression—$H_0 : \beta = 0$—can be implemented using one of two test statistics:

$$t = \frac{b}{\sqrt{\text{MSE}/S_{xx}}} \quad \text{or} \quad F = \frac{\text{MSR}}{\text{MSE}}$$

2. The strength of the relationship between x and y can be measured using

$$r^2 = \frac{\text{SSR}}{\text{Total SS}}$$

which gets closer to 1 as the relationship gets stronger.

3. Use residual plots to check for nonnormality, inequality of variances, or an incorrectly fit model.

4. Confidence intervals can be constructed to estimate the intercept α and slope β of the regression line and to estimate the average value of y, $E(y)$, for a given value of x.

5. Prediction intervals can be constructed to predict a particular observation, y, for a given value of x. For a given x, prediction intervals are always wider than confidence intervals.

V. Correlation Analysis

1. Use the correlation coefficient to measure the relationship between x and y when both variables are random:

$$r = \frac{S_{xy}}{\sqrt{S_{xx}S_{yy}}}$$

2. The sign of r indicates the direction of the relationship; r near 0 indicates no linear relationship, and r near 1 or -1 indicates a strong linear relationship.

3. A test of the significance of the correlation coefficient uses the statistic

$$t = r\sqrt{\frac{n - 2}{1 - r^2}}$$

and is identical to the test of the slope β.

 MINITAB

Linear Regression Procedures

In Chapter 3, we used some of the linear regression procedures available in *MINITAB* to obtain a graph of the best-fitting least-squares regression line and to calculate the correlation coefficient r for a bivariate data set. Now that you have studied the testing and estimation techniques for a simple linear regression analysis, more *MINITAB* options are available to you.

Consider the relationship between $x =$ mathematics achievement test score and $y =$ final calculus grade, which was used as an example throughout this chapter. Enter the data into the first two columns of a *MINITAB* worksheet. If you use **Graph → Scatterplot → Simple,** you can generate the scatterplot for the data, as shown in Figure 12.2. However, the main inferential tools for linear regression analysis are generated using **Stat → Regression → Regression.** (You will use this same sequence of commands in Chapter 13 when you study *multiple regression analysis.*) The Dialog box for the Regression command is shown in Figure 12.20.

Select **y** for the "Response" variable and **x** for the "Predictor" variable. You may now choose to generate some residual plots to check the validity of your regression assumptions before you use the model for estimation or prediction. Choose **Graphs** to display the Dialog box in Figure 12.21.

FIGURE 12.20

FIGURE 12.21

We have used **Regular** residual plots, checking the boxes for "Normal plot of residuals" and "Residuals versus fits." Click **OK** to return to the main Dialog box. If you now choose **Options,** you can obtain confidence and prediction intervals for either of these cases:

- A single value of x (typed in the box marked "Prediction intervals for new observations")
- Several values of x stored in a column (say, C3) of the worksheet

Enter the value $x = 50$ in Figure 12.22 to match the output given in Figure 12.15. When you click **OK** twice, the regression output is generated as shown in Figure 12.23. The two diagnostic plots will appear in separate graphics windows.

FIGURE 12.22

FIGURE 12.23

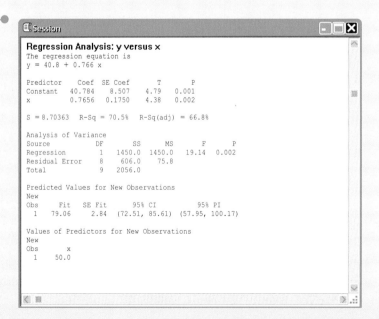

If you wish, you can now plot the data points, the regression line, and the upper and lower confidence and prediction limits (see Figure 12.16) using **Stat → Regression → Fitted line plot.** Select y and x for the response and predictor variables and click "Display confidence interval" and "Display prediction interval" in the **Options** Dialog box. Make sure that **Linear** is selected as the "Type of Regression Model," so that you will obtain a linear fit to the data.

Recall that in Chapter 3, we used the command **Stat → Basic Statistics → Correlation** to obtain the value of the correlation coefficient r. Make sure that the box marked "Display p-values" is checked. The output for this command (using the test/grade data) is shown in Figure 12.24. Notice that the p-value for the test of $H_0 : \rho = 0$ is identical to the p-value for the test of $H_0 : \beta = 0$ because the tests are exactly equivalent!

FIGURE 12.24

```
Session                                    ⊟□✕

Correlations: x, y

Pearson correlation of x and y = 0.840
P-Value = 0.002
```

Supplementary Exercises

12.58 Potency of an Antibiotic An experiment was conducted to observe the effect of an increase in temperature on the potency of an antibiotic. Three 1-ounce portions of the antibiotic were stored for equal lengths of time at each of these temperatures: 30°, 50°, 70°, and 90°. The potency readings observed at each temperature of the experimental period are listed here:

Potency Readings, y	38, 43, 29	32, 26, 33	19, 27, 23	14, 19, 21
Temperature, x	30°	50°	70°	90°

Use an appropriate computer program to answer these questions:

a. Find the least-squares line appropriate for these data.

b. Plot the points and graph the line as a check on your calculations.

c. Construct the ANOVA table for linear regression.

d. If they are available, examine the diagnostic plots to check the validity of the regression assumptions.

e. Estimate the change in potency for a 1-unit change in temperature. Use a 95% confidence interval.

f. Estimate the average potency corresponding to a temperature of 50°. Use a 95% confidence interval.

g. Suppose that a batch of the antibiotic was stored at 50° for the same length of time as the experimental period. Predict the potency of the batch at the end of the storage period. Use a 95% prediction interval.

12.59 Plant Science An experiment was conducted to determine the effect of soil applications of various levels of phosphorus on the inorganic phosphorus levels in a particular plant. The data in the table represent the levels of inorganic phosphorus in micromoles (μmol) per gram dry weight of sudan grass roots grown in the greenhouse for 28 days, in the absence of zinc. Use the *MINITAB* output to answer the questions.

Phosphorus Applied, x	Phosphorus in Plant, y
.5 μmol	204
	195
	247
	245
.25 μmol	159
	127
	95
	144
.10 μmol	128
	192
	84
	71

a. Plot the data. Do the data appear to exhibit a linear relationship?

b. Find the least-squares line relating the plant phosphorus levels y to the amount of phosphorus applied to the soil x. Graph the least-squares line as a check on your answer.

c. Do the data provide sufficient evidence to indicate that the amount of phosphorus present in the plant is linearly related to the amount of phosphorus applied to the soil?

d. Estimate the mean amount of phosphorus in the plant if .20 μmol of phosphorus is applied to the soil, in the absence of zinc. Use a 90% confidence interval.

MINITAB output for Exercise 12.59

```
Regression Analysis: y versus x

The regression equation is
y = 80.9 + 271 x

Predictor      Coef    SE Coef       T        P
Constant      80.85      22.40    3.61    0.005
x            270.82      68.31    3.96    0.003

S = 39.0419    R-Sq = 61.1%    R-Sq(adj) = 57.2%

Predicted Values for New Observations
New Obs    Fit   SE Fit      95.0% CI          95.0% PI
1        135.0    12.6   (112.1,  157.9)  (60.6,  209.4)

Values of Predictors for New Observations
New Obs        x
1          0.200
```

MY CD **12.60 Track Stats!** An experiment was conducted to investigate the effect of a training program on the length of time for a typical male college student to complete the 100-yard dash. Nine students were placed in the program. The reduction y in time to complete the 100-yard dash was measured for three students at the end of 2 weeks, for three at the end of 4 weeks, and for three at the end of 6 weeks of training. The data are given in the table.
EX1260

Reduction in Time, y (sec)	1.6, .8, 1.0	2.1, 1.6, 2.5	3.8, 2.7, 3.1
Length of Training, x (wk)	2	4	6

Use an appropriate computer software package to analyze these data. State any conclusions you can draw.

MY CD **12.61 Nematodes** Some varieties of nematodes, roundworms that live in the soil and frequently are so small as to be invisible to the naked eye, feed on the roots of lawn grasses and other plants. This pest, which is particularly troublesome in warm climates, can be treated by the application of nematicides. Data collected on the percent kill of nematodes for various rates of application (dosages given in pounds per acre of active ingredient) are as follows:
EX1261

Rate of Application, x	2	3	4	5
Percent Kill, y	50, 56, 48	63, 69, 71	86, 82, 76	94, 99, 97

MINITAB diagnostic plots for Exercise 12.61

Use an appropriate computer printout to answer these questions:

a. Calculate the coefficient of correlation r between rates of application x and percent kill y.

b. Calculate the coefficient of determination r^2 and interpret.

c. Fit a least-squares line to the data.

d. Suppose you wish to estimate the mean percent kill for an application of 4 pounds of the nematicide per acre. What do the diagnostic plots generated by

MINITAB tell you about the validity of the regression assumptions? Which assumptions may have been violated? Can you explain why?

12.62 Knee Injuries Athletes and others suffering the same type of injury to the knee often require anterior and posterior ligament reconstruction. In order to determine the proper length of bone-patellar tendon-bone grafts, experiments were done using three imaging techniques to determine the required length of the grafts, and these results were compared to the actual length required. A summary of the results of a simple linear regression analysis for each of these three methods is given in the following table.[15]

Imaging Technique	Coefficient of Determination, r^2	Intercept	Slope	p-value
Radiographs	0.80	−3.75	1.031	<0.0001
Standard MRI	0.43	20.29	0.497	0.011
3-Dimensional MRI	0.65	1.80	0.977	<0.0001

a. What can you say about the significance of each of the three regression analyses?

b. How would you rank the effectiveness of the three regression analyses? What is the basis of your decision?

c. How do the values of r^2 and the p-values compare in determining the best predictor of actual graft lengths of ligament required?

(MY) CD EX1263 **12.63 Achievement Tests II** Refer to Exercise 12.11 and data set EX1211 regarding the relationship between the Academic Performance Index (API), a measure of school achievement based on the results of the Stanford 9 Achievement test, and the percent of students who are considered English Language Learners (ELL). The following table shows the API for eight elementary schools in Riverside County, California, along with the percent of students at that school who are considered English Language Learners.[3]

School	1	2	3	4	5	6	7	8
API	588	659	710	657	669	641	557	743
ELL	58	22	14	30	11	26	39	6

a. Use an appropriate program to analyze the relationship between API and ELL.

b. Explain all pertinent details of your analysis.

12.64 How Long Is It? Refer to Exercise 12.12 and data set EX1212 regarding a subject's ability to estimate sizes. The table that follows gives the actual and estimated lengths of the specified objects.

Object	Estimated (inches)	Actual (inches)
Pencil	7.00	6.00
Dinner plate	9.50	10.25
Book 1	7.50	6.75
Cell phone	4.00	4.25
Photograph	14.50	15.75
Toy	3.75	5.00
Belt	42.00	41.50
Clothespin	2.75	3.75
Book 2	10.00	9.25
Calculator	3.50	4.75

a. Use an appropriate program to analyze the relationship between the actual and estimated lengths of the listed objects.

b. Explain all pertinent details of your analysis.

(MY) CD EX1265 **12.65 Tennis, Anyone?** If you play tennis, you know that tennis racquets vary in their physical characteristics. The data in the accompanying table give measures of bending stiffness and twisting stiffness as measured by engineering tests for 12 tennis racquets:

Racquet	Bending Stiffness, x	Twisting Stiffness, y
1	419	227
2	407	231
3	363	200
4	360	211
5	257	182
6	622	304
7	424	384
8	359	194
9	346	158
10	556	225
11	474	305
12	441	235

a. If a racquet has bending stiffness, is it also likely to have twisting stiffness? Do the data provide evidence that x and y are positively correlated?

b. Calculate the coefficient of determination r^2 and interpret its value.

(MY) CD EX1266 **12.66 Avocado Research** Movement of avocados into the United States from certain areas is prohibited because of the possibility of bringing fruit flies into the country with the avocado shipments. However, certain avocado varieties supposedly are resistant to fruit fly infestation before they soften as a result of ripening. The data in the table resulted from an experiment in which avocados ranging from 1 to 9 days after harvest were exposed to Mediterranean fruit flies. Penetrability of the avocados was measured on the day of exposure, and the percentage of the avocado fruit infested was assessed.

Days after Harvest	Penetrability	Percentage Infected
1	.91	30
2	.81	40
4	.95	45
5	1.04	57
6	1.22	60
7	1.38	75
9	1.77	100

Use the *MINITAB* printout of the regression of percentage infected (y) on days after harvest (x) to analyze the relationship between these two variables. Explain all pertinent parts of the printout and interpret the results of any tests.

MINITAB output for Exercise 12.66

Regression Analysis: Percent versus x

```
The regression equation is
Percent = 18.4 + 8.18 x

Predictor       Coef     SE Coef         T        P
Constant      18.427       5.110      3.61    0.015
x             8.1768      0.9285      8.81    0.000

S = 6.35552     R-Sq = 93.9%      R-Sq(adj) = 92.7%

Analysis of Variance
Source           DF          SS        MS       F        P
Regression        1      3132.9    3132.9   77.56    0.000
Residual Error    5       202.0      40.4
Total             6      3334.9
```

12.67 Avocados II Refer to Exercise 12.66. Suppose the experimenter wants to examine the relationship between the penetrability and the number of days after harvest. Does the method of linear regression discussed in this chapter provide an appropriate method of analysis? If not, what assumptions have been violated? Use the *MINITAB* diagnostic plots provided.

MINITAB diagnostic plots for Exercise 12.67

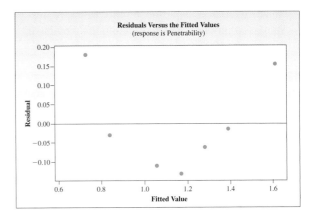

12.68 Metabolism and Weight Gain Why is it that one person may tend to gain weight, even if he eats no more and exercises no less than a slim friend? Recent studies suggest that the factors that control metabolism may depend on your genetic makeup. One study involved 11 pairs of identical twins fed about 1000 calories per day more than needed to maintain initial weight. Activities were kept constant, and exercise was minimal. At the end of 100 days, the changes in body weight (in kilograms) were recorded for the 22 twins.[16] Is there a significant positive correlation between the changes in body weight for the twins? Can you conclude that this similarity is caused by genetic similarities? Explain.

Pair	Twin A	Twin B
1	4.2	7.3
2	5.5	6.5
3	7.1	5.7
4	7.0	7.2
5	7.8	7.9
6	8.2	6.4
7	8.2	6.5
8	9.1	8.2
9	11.5	6.0
10	11.2	13.7
11	13.0	11.0

12.69 Movie Reviews How many weeks can a movie run and still make a reasonable profit? The data that follow show the number of weeks in release (x) and the average per-site gross (y) for the top 20 movies during the last weekend in January 2004.[17]

Movie	Per-site Average	Weeks in Release
1. The Butterfly Effect	$6551	1
2. Along Came Polly	5460	2
3. Win a Date with Tad Hamilton	2700	1
4. Big Fish	2916	7
5. Lord of the Rings: The Return of the King	2653	6
6. Cheaper by the Dozen	2269	5
7. Cold Mountain	1786	5
8. Torque	1824	2
9. Something's Gotta Give	1875	7
10. Mystic River	2538	16
11. Calendar Girls	2470	6
12. The Last Samurai	1953	8
13. Monster	6219	5
14. My Baby's Daddy	1658	3
15. Disney's Teacher's Pet	704	2
16. Peter Pan	1090	5
17. Paycheck	1062	5
18. The Cooler	2238	9
19. Mona Lisa Smile	882	6
20. 21 Grams	2130	10

Source: Entertainment Weekly

a. Plot the points in a scatterplot. Does it appear that the relationship between x and y is linear? How would you describe the direction and strength of the relationship?

b. Calculate the value of r^2. What percentage of the overall variation is explained by using the linear model rather than $\bar{y}$ to predict the response variable y?

c. What is the regression equation? Do the data provide evidence to indicate that x and y are linearly related? Test using a 5% significance level.

d. Given the results of parts b and c, is it appropriate to use the regression line for estimation and prediction? Explain your answer.

12.70 In addition to increasingly large bounds on error, why should an experimenter refrain from predicting y for values of x outside the experimental region?

12.71 If the experimenter stays within the experimental region, when will the error in predicting a particular value of y be maximum?

12.72 Oatmeal, Anyone? An agricultural experimenter, investigating the effect of the amount of nitrogen x applied in 100 pounds per acre on the yield of oats y measured in bushels per acre, collected the following data:

x	1	2	3	4
y	22	38	57	68
	19	41	54	65

a. Find the least-squares line for the data.

b. Construct the ANOVA table.

c. Is there sufficient evidence to indicate that the yield of oats is linearly related to the amount of nitrogen applied? Use $\alpha = .05$.

d. Predict the expected yield of oats with 95% confidence if 250 pounds of nitrogen per acre are applied.

e. Estimate the average increase in yield for an increase of 100 pounds of nitrogen per acre with 99% confidence.

f. Calculate r^2 and explain its significance in terms of predicting y, the yield of oats.

12.73 Fresh Roses A horticulturalist devised a scale to measure the freshness of roses that were packaged and stored for varying periods of time before transplanting. The freshness measurement y and the length of time in days that the rose is packaged and stored before transplanting x are given below.

x	5	10	15	20	25
y	15.3	13.6	9.8	5.5	1.8
	16.8	13.8	8.7	4.7	1.0

a. Fit a least-squares line to the data.

b. Construct the ANOVA table.

c. Is there sufficient evidence to indicate that freshness is linearly related to storage time? Use $\alpha = .05$.

d. Estimate the mean rate of change in freshness for a 1-day increase in storage time usig a 98% confidence interval.

e. Estimate the expected freshness measurement for a storage time of 14 days with a 95% confidence interval.

f. Of what value is the linear model in reference to $\bar{y}$ in predicting freshness?

12.74 Lexus, Inc. The makers of the *Lexus* automobile have steadily increased their sales since their U.S. launch in 1989. However, the rate of increase changed in 1996 when *Lexus* introduced a line of trucks. The sales of *Lexus* from 1996 to 2003 are shown in the table[8]:

Year	1996	1997	1998	1999	2000	2001	2002	2003
Sales ($000)	80	100	155	180	210	225	230	260

Source: Adapted from Automotive News, 26 January 2004.

a. Plot the data using a scatterplot. How would you describe the relationship between year and sales of *Lexus*?

b. Find the least-squares regression line relating the sales of *Lexus* to the year being measured?

c. Is there sufficient evidence to indicate that sales are linearly related to year? Use $\alpha = .05$.

d. Predict the sales of *Lexus* for the year 2004 using a 95% prediction interval.

e. If they are available, examine the diagnostic plots to check the validity of the regression assumptions.

f. If you were to predict the sales of *Lexus* in the year 2015, what problems might arise with your prediction?

(MY) CD **12.75 Starbucks** Here is some nutritional
EX1275 data for a sampling of Starbuck's products (16 fluid ounces), taken from the company website, www.starbucks.com.[19] The complete data set (star-bucks.mtp) can be found with the other data sets on the CD-ROM accompanying your textbook.

Data Display

Row	Product	Calories	Fat Calories
1	CaffèMocha-nowhip	300	110
2	CaramelFrappuccino® BlendedCoffee-nowhip	280	30
3	ChocolateBrownie Frappuccino® BlendedCoffee-nowhip	370	80
4	ChocolateMalt Frappuccino® BlendedCrème-whip	610	200
5	EggnogLatte-nowhip	410	180
6	HotChocolate-nowhip	340	140
7	IcedCaffèMocha-whip	350	180
8	IcedWhiteChocolate Mocha-whip	490	210
9	MochaFrappuccino® BlendedCoffee-whip	420	150
10	PeppermintMocha-nowhip	370	110
11	Tazo®ChaiCrème Frappuccino® BlendedTea-nowhip	370	40
12	ToffeeNutCrème-whip	460	220
13	ToffeNutLatte-nowhip	330	120

Data Display *(continued)*

Row	Product	Calories	Fat Calories
14	VanillaFrappuccino® BlendedCrème-whip	480	150
15	WhiteHotChocolate-whip	580	250

Row	Total Fat (g)	Saturated Fat (g)	Cholesterol (mg)
1	12.0	7	40
2	3.5	2	15
3	9.0	6	15
4	22.0	11	65
5	20.0	12	115
6	15.0	8	50
7	20.0	12	75
8	24.0	16	75
9	16.0	10	65
10	12.0	6	40
11	4.5	1	<5
12	24.0	15	100
13	13.0	8	50
14	17.0	9	55
15	28.0	19	95

Row	Sodium (mg)	Total Carbs (g)	Fiber (g)	Sugar (g)	Protein (g)
1	150	41	2	31	13
2	250	57	0	48	5
3	310	69	2	56	7
4	430	90	2	72	15
5	240	41	0	38	17
6	190	42	2	35	15
7	105	37	2	27	9
8	220	58	0	54	11
9	260	61	0	51	6
10	150	59	2	49	13
11	370	69	0	64	15
12	380	45	0	44	14
13	340	41	0	38	13
14	380	66	0	62	15
15	310	65	0	64	17

Use the appropriate statistical methods to analyze the relationships between some of the nutritional variables given in the table. Write a summary report explaining any conclusions that you can draw from your analysis.

MY APPLET Exercises

You can refresh your memory about regression lines and the correlation coefficient by doing the MyApplet Exercises at the end of Chapter 3.

12.76 a. Graph the line corresponding to the equation $y = 0.5x + 3$ by graphing the points corresponding to $x = 0$, 1, and 2. Give the y-intercept and slope for the line.

b. Check your graph using the **How a Line Works** applet.

12.77 a. Graph the line corresponding to the equation $y = -0.5x + 3$ by graphing the points corresponding to $x = 0$, 1, and 2. Give the y-intercept and slope for the line.

b. Check your graph using the **How a Line Works** applet.

c. How is this line related to the line $y = 0.5x + 3$ of Exercise 12.76?

12.78 The *MINITAB* printout for the data in Table 12.1 is shown below.

MINITAB output for Exercise 12.78.

Regression Analysis: y versus x

```
The regression equation is
y = 40.8 + 0.766 x

Predictor      Coef     SE Coef        T        P
Constant     40.784       8.507     4.79    0.001
x            0.7656      0.1750     4.38    0.002

S = 8.70363     R-Sq = 70.5%     R-Sq(adj) = 66.8%

Analysis of Variance
Source          DF       SS        MS       F       P
Regression       1   1450.0    1450.0   19.14   0.002
Residual Error   8    606.0      75.8
Total            9   2056.0
```

a. Use the **Method of Least Squares** applet to find the values of a and b that determine the *best fitting line*, $\hat{y} = a + bx$. When you think that you have minimized SSE, click the [Find Best Model] button and see how well you did. What is the equation of the line? Does it match the regression equation given in the *MINITAB* printout?

b. Find the values of SSE and r^2 on the **Method of Least Squares** applet. Find these values on the *MINITAB* printout and confirm that they are the same.

c. Use the values of b and its standard error SE(b) from the *MINITAB* printout along with the t **Test for the Slope** applet to verify the value of the t statistic and its p-value, given in the printout.

12.79 Use the first applet in **Building a Scatterplot** to create a scatterplot for the data in Table 12.1. Verify your plot using Figure 12.2.

12.80 Is your overall satisfaction with a hotel room correlated with the cost of the room? Satisfaction scores and median room prices were recorded for nine "budget" hotels with the following results[20]:

Hotel	Median Rate	Score
Sleep Inn	55	80
Microtel	50	75
Super 8 Motel	50	70
Red Roof Inn	50	69
Motel 6	40	68
Days Inn	55	64
Econo Lodge	50	63
Travelodge	55	62
Knights Inn	45	57

EX1280

Source: Case Study "Hotel Room Cost and Satisfaction" Copyright 2001 by Consumers Union of U.S., Inc., Yonkers, NY 10703-1057, a nonprofit organization. Reprinted with permission from the July 2001 issue of Consumer Reports® for educational purposes only. No commercial use or reproduction permitted. www.ConsumerReports.org.

a. Calculate the correlation coefficient r between price and overall score. How would you describe the relationship between price and overall score?

b. Use the applet called **Correlation and the Scatterplot** to plot the nine data points. What is the correlation coefficient shown on the applet? Compare with the value you calculated in part a.

c. Describe the pattern that you see in the scatterplot. Are there any outliers? If so, how would you explain them?

CASE STUDY

 Foreign Cars

Is Your Car "Made in the U.S.A."?

The phrase "made in the U.S.A." has become a battle cry in the past few years as U.S. workers try to protect their jobs from overseas competition. For the past few decades, a major trade imbalance in the United States has been caused by a flood of imported goods that enter the country and are sold at lower cost than comparable American-made goods. One prime concern is the automotive industry, in which the number of imported cars steadily increased during the 1970s and 1980s. The U.S. automobile industry has been besieged with complaints about product quality, worker layoffs, and high prices and has spent billions in advertising and research to produce an American-made car that will satisfy consumer demands. Have they been successful in stopping the flood of imported cars purchased by American consumers? The data in the table represent the numbers of imported cars y sold in the United States (in millions) for the years 1969–2003.[21] To simplify the analysis, we have coded the year using the coded variable $x =$ Year $-$ 1969.

Year	(Year − 1969), x	Number of Imported Cars, y	Year	(Year − 1969), x	Number of Imported Cars, y
1969	0	1.1	1987	18	3.1
1970	1	1.3	1988	19	3.1
1971	2	1.6	1989	20	2.8
1972	3	1.6	1990	21	2.5
1973	4	1.8	1991	22	2.1
1974	5	1.4	1992	23	2.0
1975	6	1.6	1993	24	1.8
1976	7	1.5	1994	25	1.8
1977	8	2.1	1995	26	1.6
1978	9	2.0	1996	27	1.4
1979	10	2.3	1997	28	1.4
1980	11	2.4	1998	29	1.4
1981	12	2.3	1999	30	1.8
1982	13	2.2	2000	31	2.1
1983	14	2.4	2001	32	2.2
1984	15	2.4	2002	33	2.3
1985	16	2.8	2003	34	2.2
1986	17	3.2			

1. Using a scatterplot, plot the data for the years 1969–1988. Does there appear to be a linear relationship between the number of imported cars and the year?

2. Use a computer software package to find the least-squares line for predicting the number of imported cars as a function of year for the years 1969–1988.

3. Is there a significant linear relationship between the number of imported cars and the year?

4. Use the computer program to predict the number of cars that will be imported using 95% prediction intervals for each of the years 2001, 2002, and 2003.

5. Now look at the actual data points for the years 2001–2003. Do the predictions obtained in step 4 provide accurate estimates of the *actual* values observed in these years? Explain.

6. Add the data for 1989–2003 to your database, and recalculate the regression line. What effect have the new data points had on the slope? What is the effect on SSE?

7. Given the form of the scatterplot for the years 1969–2003, does it appear that a straight line provides an accurate model for the data? What other type of model might be more appropriate? (Use residual plots to help answer this question.)

Appendix I
Tables

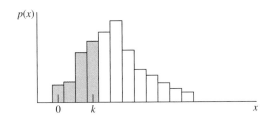

TABLE 1   Cumulative Binomial Probabilities

Tabulated values are $P(x \leq k) = p(0) + p(1) + \cdots + p(k)$
(Computations are rounded at the third decimal place.)

$n = 2$

							p							
k	.01	.05	.10	.20	.30	.40	.50	.60	.70	.80	.90	.95	.99	k
0	.980	.902	.810	.640	.490	.360	.250	.160	.090	.040	.010	.002	.000	0
1	1.000	.998	.990	.960	.910	.840	.750	.640	.510	.360	.190	.098	.020	1
2	1.000	1.000	1.000	1.000	1.000	1.000	1.000	1.000	1.000	1.000	1.000	1.000	1.000	2

$n = 3$

							p							
k	.01	.05	.10	.20	.30	.40	.50	.60	.70	.80	.90	.95	.99	k
0	.970	.857	.729	.512	.343	.216	.125	.064	.027	.008	.001	.000	.000	0
1	1.000	.993	.972	.896	.784	.648	.500	.352	.216	.104	.028	.007	.000	1
2	1.000	1.000	.999	.992	.973	.936	.875	.784	.657	.488	.271	.143	.030	2
3	1.000	1.000	1.000	1.000	1.000	1.000	1.000	1.000	1.000	1.000	1.000	1.000	1.000	3

$n = 4$

							p							
k	.01	.05	.10	.20	.30	.40	.50	.60	.70	.80	.90	.95	.99	k
0	.961	.815	.656	.410	.240	.130	.062	.026	.008	.002	.000	.000	.000	0
1	.999	.986	.948	.819	.652	.475	.312	.179	.084	.027	.004	.000	.000	1
2	1.000	1.000	.996	.973	.916	.821	.688	.525	.348	.181	.052	.014	.001	2
3	1.000	1.000	1.000	.998	.992	.974	.938	.870	.760	.590	.344	.185	.039	3
4	1.000	1.000	1.000	1.000	1.000	1.000	1.000	1.000	1.000	1.000	1.000	1.000	1.000	4

TABLE 1 *(continued)*

$n = 5$

							p							
k	.01	.05	.10	.20	.30	.40	.50	.60	.70	.80	.90	.95	.99	k
0	.951	.774	.590	.328	.168	.078	.031	.010	.002	.000	.000	.000	.000	0
1	.999	.977	.919	.737	.528	.337	.188	.087	.031	.007	.000	.000	.000	1
2	1.000	.999	.991	.942	.837	.683	.500	.317	.163	.058	.009	.001	.000	2
3	1.000	1.000	1.000	.993	.969	.913	.812	.663	.472	.263	.081	.023	.001	3
4	1.000	1.000	1.000	1.000	.998	.990	.969	.922	.832	.672	.410	.226	.049	4
5	1.000	1.000	1.000	1.000	1.000	1.000	1.000	1.000	1.000	1.000	1.000	1.000	1.000	5

$n = 6$

							p							
k	.01	.05	.10	.20	.30	.40	.50	.60	.70	.80	.90	.95	.99	k
0	.941	.735	.531	.262	.118	.047	.016	.004	.001	.000	.000	.000	.000	0
1	.999	.967	.886	.655	.420	.233	.109	.041	.011	.002	.000	.000	.000	1
2	1.000	.998	.984	.901	.744	.544	.344	.179	.070	.017	.001	.000	.000	2
3	1.000	1.000	.999	.983	.930	.821	.656	.456	.256	.099	.016	.002	.000	3
4	1.000	1.000	1.000	.998	.989	.959	.891	.767	.580	.345	.114	.033	.001	4
5	1.000	1.000	1.000	1.000	.999	.996	.984	.953	.882	.738	.469	.265	.059	5
6	1.000	1.000	1.000	1.000	1.000	1.000	1.000	1.000	1.000	1.000	1.000	1.000	1.000	6

$n = 7$

							p							
k	.01	.05	.10	.20	.30	.40	.50	.60	.70	.80	.90	.95	.99	k
0	.932	.698	.478	.210	.082	.028	.008	.002	.000	.000	.000	.000	.000	0
1	.998	.956	.850	.577	.329	.159	.062	.019	.004	.000	.000	.000	.000	1
2	1.000	.996	.974	.852	.647	.420	.227	.096	.029	.005	.000	.000	.000	2
3	1.000	1.000	.997	.967	.874	.710	.500	.290	.126	.033	.003	.000	.000	3
4	1.000	1.000	1.000	.995	.971	.904	.773	.580	.353	.148	.026	.004	.000	4
5	1.000	1.000	1.000	1.000	.996	.981	.938	.841	.671	.423	.150	.044	.002	5
6	1.000	1.000	1.000	1.000	1.000	.998	.992	.972	.918	.790	.522	.302	.068	6
7	1.000	1.000	1.000	1.000	1.000	1.000	1.000	1.000	1.000	1.000	1.000	1.000	1.000	7

$n = 8$

							p							
k	.01	.05	.10	.20	.30	.40	.50	.60	.70	.80	.90	.95	.99	k
0	.923	.663	.430	.168	.058	.017	.004	.001	.000	.000	.000	.000	.000	0
1	.997	.943	.813	.503	.255	.106	.035	.009	.001	.000	.000	.000	.000	1
2	1.000	.994	.962	.797	.552	.315	.145	.050	.011	.001	.000	.000	.000	2
3	1.000	1.000	.995	.944	.806	.594	.363	.174	.058	.010	.000	.000	.000	3
4	1.000	1.000	1.000	.990	.942	.826	.637	.406	.194	.056	.005	.000	.000	4
5	1.000	1.000	1.000	.999	.989	.950	.855	.685	.448	.203	.038	.006	.000	5
6	1.000	1.000	1.000	1.000	.999	.991	.965	.894	.745	.497	.187	.057	.003	6
7	1.000	1.000	1.000	1.000	1.000	.999	.996	.983	.942	.832	.570	.337	.077	7
8	1.000	1.000	1.000	1.000	1.000	1.000	1.000	1.000	1.000	1.000	1.000	1.000	1.000	8

TABLE 1 *(continued)*

n = 9

							p							
k	.01	.05	.10	.20	.30	.40	.50	.60	.70	.80	.90	.95	.99	*k*
0	.914	.630	.387	.134	.040	.010	.002	.000	.000	.000	.000	.000	.000	0
1	.997	.929	.775	.436	.196	.071	.020	.004	.000	.000	.000	.000	.000	1
2	1.000	.992	.947	.738	.463	.232	.090	.025	.004	.000	.000	.000	.000	2
3	1.000	.999	.992	.914	.730	.483	.254	.099	.025	.003	.000	.000	.000	3
4	1.000	1.000	.999	.980	.901	.733	.500	.267	.099	.020	.001	.000	.000	4
5	1.000	1.000	1.000	.997	.975	.901	.746	.517	.270	.086	.008	.001	.000	5
6	1.000	1.000	1.000	1.000	.996	.975	.910	.768	.537	.262	.053	.008	.000	6
7	1.000	1.000	1.000	1.000	1.000	.996	.980	.929	.804	.564	.225	.071	.003	7
8	1.000	1.000	1.000	1.000	1.000	1.000	.998	.990	.960	.866	.613	.370	.086	8
9	1.000	1.000	1.000	1.000	1.000	1.000	1.000	1.000	1.000	1.000	1.000	1.000	1.000	9

n = 10

							p							
k	.01	.05	.10	.20	.30	.40	.50	.60	.70	.80	.90	.95	.99	*k*
0	.904	.599	.349	.107	.028	.006	.001	.000	.000	.000	.000	.000	.000	0
1	.996	.914	.736	.376	.149	.046	.011	.002	.000	.000	.000	.000	.000	1
2	1.000	.988	.930	.678	.383	.167	.055	.012	.002	.000	.000	.000	.000	2
3	1.000	.999	.987	.879	.650	.382	.172	.055	.011	.001	.000	.000	.000	3
4	1.000	1.000	.998	.967	.850	.633	.377	.166	.047	.006	.000	.000	.000	4
5	1.000	1.000	1.000	.994	.953	.834	.623	.367	.150	.033	.002	.000	.000	5
6	1.000	1.000	1.000	.999	.989	.945	.828	.618	.350	.121	.013	.001	.000	6
7	1.000	1.000	1.000	1.000	.998	.988	.945	.833	.617	.322	.070	.012	.000	7
8	1.000	1.000	1.000	1.000	1.000	.998	.989	.954	.851	.624	.264	.086	.004	8
9	1.000	1.000	1.000	1.000	1.000	1.000	.999	.994	.972	.893	.651	.401	.096	9
10	1.000	1.000	1.000	1.000	1.000	1.000	1.000	1.000	1.000	1.000	1.000	1.000	1.000	10

n = 11

							p							
k	.01	.05	.10	.20	.30	.40	.50	.60	.70	.80	.90	.95	.99	*k*
0	.895	.569	.314	.086	.020	.004	.000	.000	.000	.000	.000	.000	.000	0
1	.995	.898	.697	.322	.113	.030	.006	.001	.000	.000	.000	.000	.000	1
2	1.000	.985	.910	.617	.313	.119	.033	.006	.001	.000	.000	.000	.000	2
3	1.000	.998	.981	.839	.570	.296	.113	.029	.004	.000	.000	.000	.000	3
4	1.000	1.000	.997	.950	.790	.533	.274	.099	.022	.002	.000	.000	.000	4
5	1.000	1.000	1.000	.988	.922	.754	.500	.246	.078	.012	.000	.000	.000	5
6	1.000	1.000	1.000	.998	.978	.901	.726	.467	.210	.050	.003	.000	.000	6
7	1.000	1.000	1.000	1.000	.996	.971	.887	.704	.430	.161	.019	.002	.000	7
8	1.000	1.000	1.000	1.000	.999	.994	.967	.881	.687	.383	.090	.015	.000	8
9	1.000	1.000	1.000	1.000	1.000	.999	.994	.970	.887	.678	.303	.102	.005	9
10	1.000	1.000	1.000	1.000	1.000	1.000	1.000	.996	.980	.914	.686	.431	.105	10
11	1.000	1.000	1.000	1.000	1.000	1.000	1.000	1.000	1.000	1.000	1.000	1.000	1.000	11

TABLE 1 *(continued)*

n = 12

							p							
k	.01	.05	.10	.20	.30	.40	.50	.60	.70	.80	.90	.95	.99	*k*
0	.886	.540	.282	.069	.014	.002	.000	.000	.000	.000	.000	.000	.000	0
1	.994	.882	.659	.275	.085	.020	.003	.000	.000	.000	.000	.000	.000	1
2	1.000	.980	.889	.558	.253	.083	.019	.003	.000	.000	.000	.000	.000	2
3	1.000	.998	.974	.795	.493	.225	.073	.015	.002	.000	.000	.000	.000	3
4	1.000	1.000	.996	.927	.724	.438	.194	.057	.009	.001	.000	.000	.000	4
5	1.000	1.000	.999	.981	.882	.665	.387	.158	.039	.004	.000	.000	.000	5
6	1.000	1.000	1.000	.996	.961	.842	.613	.335	.118	.019	.001	.000	.000	6
7	1.000	1.000	1.000	.999	.991	.943	.806	.562	.276	.073	.004	.000	.000	7
8	1.000	1.000	1.000	1.000	.998	.985	.927	.775	.507	.205	.026	.002	.000	8
9	1.000	1.000	1.000	1.000	1.000	.997	.981	.917	.747	.442	.111	.020	.000	9
10	1.000	1.000	1.000	1.000	1.000	1.000	.997	.980	.915	.725	.341	.118	.006	10
11	1.000	1.000	1.000	1.000	1.000	1.000	1.000	.998	.986	.931	.718	.460	.114	11
12	1.000	1.000	1.000	1.000	1.000	1.000	1.000	1.000	1.000	1.000	1.000	1.000	1.000	12

n = 15

							p							
k	.01	.05	.10	.20	.30	.40	.50	.60	.70	.80	.90	.95	.99	*k*
0	.860	.463	.206	.035	.005	.000	.000	.000	.000	.000	.000	.000	.000	0
1	.990	.829	.549	.167	.035	.005	.000	.000	.000	.000	.000	.000	.000	1
2	1.000	.964	.816	.398	.127	.027	.004	.000	.000	.000	.000	.000	.000	2
3	1.000	.995	.944	.648	.297	.091	.018	.002	.000	.000	.000	.000	.000	3
4	1.000	.999	.987	.836	.515	.217	.059	.009	.001	.000	.000	.000	.000	4
5	1.000	1.000	.998	.939	.722	.403	.151	.034	.004	.000	.000	.000	.000	5
6	1.000	1.000	1.000	.982	.869	.610	.304	.095	.015	.001	.000	.000	.000	6
7	1.000	1.000	1.000	.996	.950	.787	.500	.213	.050	.004	.000	.000	.000	7
8	1.000	1.000	1.000	.999	.985	.905	.696	.390	.131	.018	.000	.000	.000	8
9	1.000	1.000	1.000	1.000	.996	.966	.849	.597	.278	.061	.002	.000	.000	9
10	1.000	1.000	1.000	1.000	.999	.991	.941	.783	.485	.164	.013	.001	.000	10
11	1.000	1.000	1.000	1.000	1.000	.998	.982	.909	.703	.352	.056	.005	.000	11
12	1.000	1.000	1.000	1.000	1.000	1.000	.996	.973	.873	.602	.184	.036	.000	12
13	1.000	1.000	1.000	1.000	1.000	1.000	1.000	.995	.965	.833	.451	.171	.010	13
14	1.000	1.000	1.000	1.000	1.000	1.000	1.000	1.000	.995	.965	.794	.537	.140	14
15	1.000	1.000	1.000	1.000	1.000	1.000	1.000	1.000	1.000	1.000	1.000	1.000	1.000	15

TABLE 1 *(continued)*

$n = 20$

k	.01	.05	.10	.20	.30	.40	.50	.60	.70	.80	.90	.95	.99	k
							p							
0	.818	.358	.122	.012	.001	.000	.000	.000	.000	.000	.000	.000	.000	0
1	.983	.736	.392	.069	.008	.001	.000	.000	.000	.000	.000	.000	.000	1
2	.999	.925	.677	.206	.035	.004	.000	.000	.000	.000	.000	.000	.000	2
3	1.000	.984	.867	.411	.107	.016	.001	.000	.000	.000	.000	.000	.000	3
4	1.000	.997	.957	.630	.238	.051	.006	.000	.000	.000	.000	.000	.000	4
5	1.000	1.000	.989	.804	.416	.126	.021	.002	.000	.000	.000	.000	.000	5
6	1.000	1.000	.998	.913	.608	.250	.058	.006	.000	.000	.000	.000	.000	6
7	1.000	1.000	1.000	.968	.772	.416	.132	.021	.001	.000	.000	.000	.000	7
8	1.000	1.000	1.000	.990	.887	.596	.252	.057	.005	.000	.000	.000	.000	8
9	1.000	1.000	1.000	.997	.952	.755	.412	.128	.017	.001	.000	.000	.000	9
10	1.000	1.000	1.000	.999	.983	.872	.588	.245	.048	.003	.000	.000	.000	10
11	1.000	1.000	1.000	1.000	.995	.943	.748	.404	.113	.010	.000	.000	.000	11
12	1.000	1.000	1.000	1.000	.999	.979	.868	.584	.228	.032	.000	.000	.000	12
13	1.000	1.000	1.000	1.000	1.000	.994	.942	.750	.392	.087	.002	.000	.000	13
14	1.000	1.000	1.000	1.000	1.000	.998	.979	.874	.584	.196	.011	.000	.000	14
15	1.000	1.000	1.000	1.000	1.000	1.000	.994	.949	.762	.370	.043	.003	.000	15
16	1.000	1.000	1.000	1.000	1.000	1.000	.999	.984	.893	.589	.133	.016	.000	16
17	1.000	1.000	1.000	1.000	1.000	1.000	1.000	.996	.965	.794	.323	.075	.001	17
18	1.000	1.000	1.000	1.000	1.000	1.000	1.000	.999	.992	.931	.608	.264	.017	18
19	1.000	1.000	1.000	1.000	1.000	1.000	1.000	1.000	.999	.988	.878	.642	.182	19
20	1.000	1.000	1.000	1.000	1.000	1.000	1.000	1.000	1.000	1.000	1.000	1.000	1.000	20

TABLE 1 *(continued)*

n = 25

							p							
k	.01	.05	.10	.20	.30	.40	.50	.60	.70	.80	.90	.95	.99	*k*
0	.778	.277	.072	.004	.000	.000	.000	.000	.000	.000	.000	.000	.000	0
1	.974	.642	.271	.027	.002	.000	.000	.000	.000	.000	.000	.000	.000	1
2	.998	.873	.537	.098	.009	.000	.000	.000	.000	.000	.000	.000	.000	2
3	1.000	.966	.764	.234	.033	.002	.000	.000	.000	.000	.000	.000	.000	3
4	1.000	.993	.902	.421	.090	.009	.000	.000	.000	.000	.000	.000	.000	4
5	1.000	.999	.967	.617	.193	.029	.002	.000	.000	.000	.000	.000	.000	5
6	1.000	1.000	.991	.780	.341	.074	.007	.000	.000	.000	.000	.000	.000	6
7	1.000	1.000	.998	.891	.512	.154	.022	.001	.000	.000	.000	.000	.000	7
8	1.000	1.000	1.000	.953	.677	.274	.054	.004	.000	.000	.000	.000	.000	8
9	1.000	1.000	1.000	.983	.811	.425	.115	.013	.000	.000	.000	.000	.000	9
10	1.000	1.000	1.000	.994	.902	.586	.212	.034	.002	.000	.000	.000	.000	10
11	1.000	1.000	1.000	.998	.956	.732	.345	.078	.006	.000	.000	.000	.000	11
12	1.000	1.000	1.000	1.000	.983	.846	.500	.154	.017	.000	.000	.000	.000	12
13	1.000	1.000	1.000	1.000	.994	.922	.655	.268	.044	.002	.000	.000	.000	13
14	1.000	1.000	1.000	1.000	.998	.966	.788	.414	.098	.006	.000	.000	.000	14
15	1.000	1.000	1.000	1.000	1.000	.987	.885	.575	.189	.017	.000	.000	.000	15
16	1.000	1.000	1.000	1.000	1.000	.996	.946	.726	.323	.047	.000	.000	.000	16
17	1.000	1.000	1.000	1.000	1.000	.999	.978	.846	.488	.109	.002	.000	.000	17
18	1.000	1.000	1.000	1.000	1.000	1.000	.993	.926	.659	.220	.009	.000	.000	18
19	1.000	1.000	1.000	1.000	1.000	1.000	.998	.971	.807	.383	.033	.001	.000	19
20	1.000	1.000	1.000	1.000	1.000	1.000	1.000	.991	.910	.579	.098	.007	.000	20
21	1.000	1.000	1.000	1.000	1.000	1.000	1.000	.998	.967	.766	.236	.034	.000	21
22	1.000	1.000	1.000	1.000	1.000	1.000	1.000	1.000	.991	.902	.463	.127	.002	22
23	1.000	1.000	1.000	1.000	1.000	1.000	1.000	1.000	.998	.973	.729	.358	.026	23
24	1.000	1.000	1.000	1.000	1.000	1.000	1.000	1.000	1.000	.996	.928	.723	.222	24
25	1.000	1.000	1.000	1.000	1.000	1.000	1.000	1.000	1.000	1.000	1.000	1.000	1.000	25

TABLE 2 **Cumulative Poisson Probabilities**
Tabulated values are $P(x \le k) = p(0) + p(1) + \cdots + p(k)$
(Computations are rounded at the third decimal place.)

						μ						
k	.1	.2	.3	.4	.5	.6	.7	.8	.9	1.0	1.5
0	.905	.819	.741	.670	.607	.549	.497	.449	.407	.368	.223
1	.995	.982	.963	.938	.910	.878	.844	.809	.772	.736	.558
2	1.000	.999	.996	.992	.986	.977	.966	.953	.937	.920	.809
3		1.000	1.000	.999	.998	.997	.994	.991	.987	.981	.934
4				1.000	1.000	1.000	.999	.999	.998	.996	.981
5							1.000	1.000	1.000	.999	.996
6										1.000	.999
7											1.000

						μ					
k	2.0	2.5	3.0	3.5	4.0	4.5	5.0	5.5	6.0	6.5	7.0
0	.135	.082	.055	.033	.018	.011	.007	.004	.003	.002	.001
1	.406	.287	.199	.136	.092	.061	.040	.027	.017	.011	.007
2	.677	.544	.423	.321	.238	.174	.125	.088	.062	.043	.030
3	.857	.758	.647	.537	.433	.342	.265	.202	.151	.112	.082
4	.947	.891	.815	.725	.629	.532	.440	.358	.285	.224	.173
5	.983	.958	.916	.858	.785	.703	.616	.529	.446	.369	.301
6	.995	.986	.966	.935	.889	.831	.762	.686	.606	.563	.450
7	.999	.996	.988	.973	.949	.913	.867	.809	.744	.673	.599
8	1.000	.999	.996	.990	.979	.960	.932	.894	.847	.792	.729
9		1.000	.999	.997	.992	.983	.968	.946	.916	.877	.830
10			1.000	.999	.997	.993	.986	.975	.957	.933	.901
11				1.000	.999	.998	.995	.989	.980	.966	.947
12					1.000	.999	.998	.996	.991	.984	.973
13						1.000	.999	.998	.996	.993	.987
14							1.000	.999	.999	.997	.994
15								1.000	.999	.999	.998
16									1.000	1.000	.999
17											1.000

TABLE 2 *(continued)*

k	7.5	8.0	8.5	9.0	9.5	10.0	12.0	15.0	20.0
0	.001	.000	.000	.000	.000	.000	.000	.000	.000
1	.005	.003	.002	.001	.001	.000	.000	.000	.000
2	.020	.014	.009	.006	.004	.003	.001	.000	.000
3	.059	.042	.030	.021	.015	.010	.002	.000	.000
4	.132	.100	.074	.055	.040	.029	.008	.001	.000
5	.241	.191	.150	.116	.089	.067	.020	.003	.000
6	.378	.313	.256	.207	.165	.130	.046	.008	.000
7	.525	.453	.386	.324	.269	.220	.090	.018	.001
8	.662	.593	.523	.456	.392	.333	.155	.037	.002
9	.776	.717	.653	.587	.522	.458	.242	.070	.005
10	.862	.816	.763	.706	.645	.583	.347	.118	.011
11	.921	.888	.849	.803	.752	.697	.462	.185	.021
12	.957	.936	.909	.876	.836	.792	.576	.268	.039
13	.978	.966	.949	.926	.898	.864	.682	.363	.066
14	.990	.983	.973	.959	.940	.917	.772	.466	.105
15	.995	.992	.986	.978	.967	.951	.844	.568	.157
16	.998	.996	.993	.989	.982	.973	.899	.664	.221
17	.999	.998	.997	.995	.991	.986	.937	.749	.297
18	1.000	.999	.999	.998	.996	.993	.963	.819	.381
19		1.000	.999	.999	.998	.997	.979	.875	.470
20			1.000	1.000	.999	.998	.988	.917	.559
21					1.000	.999	.994	.947	.644
22						1.000	.997	.967	.721
23							.999	.981	.787
24							.999	.989	.843
25							1.000	.994	.888
26								.997	.922
27								.998	.948
28								.999	.966
29								1.000	.978
30									.987
31									.992
32									.995
33									.997
34									.999
35									.999
36									1.000

μ (column header spanning 7.5 through 20.0)

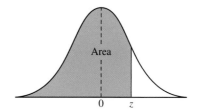

TABLE 3 **Areas under the Normal Curve**

z	.00	.01	.02	.03	.04	.05	.06	.07	.08	.09
−3.4	.0003	.0003	.0003	.0003	.0003	.0003	.0003	.0003	.0003	.0002
−3.3	.0005	.0005	.0005	.0004	.0004	.0004	.0004	.0004	.0004	.0003
−3.2	.0007	.0007	.0006	.0006	.0006	.0006	.0006	.0005	.0005	.0005
−3.1	.0010	.0009	.0009	.0009	.0008	.0008	.0008	.0008	.0007	.0007
−3.0	.0013	.0013	.0013	.0012	.0012	.0011	.0011	.0011	.0010	.0010
−2.9	.0019	.0018	.0017	.0017	.0016	.0016	.0015	.0015	.0014	.0014
−2.8	.0026	.0025	.0024	.0023	.0023	.0022	.0021	.0021	.0020	.0019
−2.7	.0035	.0034	.0033	.0032	.0031	.0030	.0029	.0028	.0027	.0026
−2.6	.0047	.0045	.0044	.0043	.0041	.0040	.0039	.0038	.0037	.0036
−2.5	.0062	.0060	.0059	.0057	.0055	.0054	.0052	.0051	.0049	.0048
−2.4	.0082	.0080	.0078	.0075	.0073	.0071	.0069	.0068	.0066	.0064
−2.3	.0107	.0104	.0102	.0099	.0096	.0094	.0091	.0089	.0087	.0084
−2.2	.0139	.0136	.0132	.0129	.0125	.0122	.0119	.0116	.0113	.0110
−2.1	.0179	.0174	.0170	.0166	.0162	.0158	.0154	.0150	.0146	.0143
−2.0	.0228	.0222	.0217	.0212	.0207	.0202	.0197	.0192	.0188	.0183
−1.9	.0287	.0281	.0274	.0268	.0262	.0256	.0250	.0244	.0239	.0233
−1.8	.0359	.0351	.0344	.0336	.0329	.0322	.0314	.0307	.0301	.0294
−1.7	.0446	.0436	.0427	.0418	.0409	.0401	.0392	.0384	.0375	.0367
−1.6	.0548	.0537	.0526	.0516	.0505	.0495	.0485	.0475	.0465	.0455
−1.5	.0668	.0655	.0643	.0630	.0618	.0606	.0594	.0582	.0571	.0559
−1.4	.0808	.0793	.0778	.0764	.0749	.0735	.0722	.0708	.0694	.0681
−1.3	.0968	.0951	.0934	.0918	.0901	.0885	.0869	.0853	.0838	.0823
−1.2	.1151	.1131	.1112	.1093	.1075	.1056	.1038	.1020	.1003	.0985
−1.1	.1357	.1335	.1314	.1292	.1271	.1251	.1230	.1210	.1190	.1170
−1.0	.1587	.1562	.1539	.1515	.1492	.1469	.1446	.1423	.1401	.1379
−0.9	.1841	.1814	.1788	.1762	.1736	.1711	.1685	.1660	.1635	.1611
−0.8	.2119	.2090	.2061	.2033	.2005	.1977	.1949	.1922	.1894	.1867
−0.7	.2420	.2389	.2358	.2327	.2296	.2266	.2236	.2206	.2177	.2148
−0.6	.2743	.2709	.2676	.2643	.2611	.2578	.2546	.2514	.2483	.2451
−0.5	.3085	.3050	.3015	.2981	.2946	.2912	.2877	.2843	.2810	.2776
−0.4	.3446	.3409	.3372	.3336	.3300	.3264	.3228	.3192	.3156	.3121
−0.3	.3821	.3783	.3745	.3707	.3669	.3632	.3594	.3557	.3520	.3483
−0.2	.4207	.4168	.4129	.4090	.4052	.4013	.3974	.3936	.3897	.3859
−0.1	.4602	.4562	.4522	.4483	.4443	.4404	.4364	.4325	.4286	.4247
−0.0	.5000	.4960	.4920	.4880	.4840	.4801	.4761	.4721	.4681	.4641

TABLE 3 *(continued)*

z	.00	.01	.02	.03	.04	.05	.06	.07	.08	.09
0.0	.5000	.5040	.5080	.5120	.5160	.5199	.5239	.5279	.5319	.5359
0.1	.5398	.5438	.5478	.5517	.5557	.5596	.5636	.5675	.5714	.5753
0.2	.5793	.5832	.5871	.5910	.5948	.5987	.6026	.6064	.6103	.6141
0.3	.6179	.6217	.6255	.6293	.6331	.6368	.6406	.6443	.6480	.6517
0.4	.6554	.6591	.6628	.6664	.6700	.6736	.6772	.6808	.6844	.6879
0.5	.6915	.6950	.6985	.7019	.7054	.7088	.7123	.7157	.7190	.7224
0.6	.7257	.7291	.7324	.7357	.7389	.7422	.7454	.7486	.7517	.7549
0.7	.7580	.7611	.7642	.7673	.7704	.7734	.7764	.7794	.7823	.7852
0.8	.7881	.7910	.7939	.7967	.7995	.8023	.8051	.8078	.8106	.8133
0.9	.8159	.8186	.8212	.8238	.8264	.8289	.8315	.8340	.8365	.8389
1.0	.8413	.8438	.8461	.8485	.8508	.8531	.8554	.8577	.8599	.8621
1.1	.8643	.8665	.8686	.8708	.8729	.8749	.8770	.8790	.8810	.8830
1.2	.8849	.8869	.8888	.8907	.8925	.8944	.8962	.8980	.8997	.9015
1.3	.9032	.9049	.9066	.9082	.9099	.9115	.9131	.9147	.9162	.9177
1.4	.9192	.9207	.9222	.9236	.9251	.9265	.9279	.9292	.9306	.9319
1.5	.9332	.9345	.9357	.9370	.9382	.9394	.9406	.9418	.9429	.9441
1.6	.9452	.9463	.9474	.9484	.9495	.9505	.9515	.9525	.9535	.9545
1.7	.9554	.9564	.9573	.9582	.9591	.9599	.9608	.9616	.9625	.9633
1.8	.9641	.9649	.9656	.9664	.9671	.9678	.9686	.9693	.9699	.9706
1.9	.9713	.9719	.9726	.9732	.9738	.9744	.9750	.9756	.9761	.9767
2.0	.9772	.9778	.9783	.9788	.9793	.9798	.9803	.9808	.9812	.9817
2.1	.9821	.9826	.9830	.9834	.9838	.9842	.9846	.9850	.9854	.9857
2.2	.9861	.9864	.9868	.9871	.9875	.9878	.9881	.9884	.9887	.9890
2.3	.9893	.9896	.9898	.9901	.9904	.9906	.9909	.9911	.9913	.9916
2.4	.9918	.9920	.9922	.9925	.9927	.9929	.9931	.9932	.9934	.9936
2.5	.9938	.9940	.9941	.9943	.9945	.9946	.9948	.9949	.9951	.9952
2.6	.9953	.9955	.9956	.9957	.9959	.9960	.9961	.9962	.9963	.9964
2.7	.9965	.9966	.9967	.9968	.9969	.9970	.9971	.9972	.9973	.9974
2.8	.9974	.9975	.9976	.9977	.9977	.9978	.9979	.9979	.9980	.9981
2.9	.9981	.9982	.9982	.9983	.9984	.9984	.9985	.9985	.9986	.9986
3.0	.9987	.9987	.9987	.9988	.9988	.9989	.9989	.9989	.9990	.9990
3.1	.9990	.9991	.9991	.9991	.9992	.9992	.9992	.9992	.9993	.9993
3.2	.9993	.9993	.9994	.9994	.9994	.9994	.9994	.9995	.9995	.9995
3.3	.9995	.9995	.9995	.9996	.9996	.9996	.9996	.9996	.9996	.9997
3.4	.9997	.9997	.9997	.9997	.9997	.9997	.9997	.9997	.9997	.9998

TABLE 4
Critical Values of t

df	$t_{.100}$	$t_{.050}$	$t_{.025}$	$t_{.010}$	$t_{.005}$	df
1	3.078	6.314	12.706	31.821	63.657	1
2	1.886	2.920	4.303	6.965	9.925	2
3	1.638	2.353	3.182	4.541	5.841	3
4	1.533	2.132	2.776	3.747	4.604	4
5	1.476	2.015	2.571	3.365	4.032	5
6	1.440	1.943	2.447	3.143	3.707	6
7	1.415	1.895	2.365	2.998	3.499	7
8	1.397	1.860	2.306	2.896	3.355	8
9	1.383	1.833	2.262	2.821	3.250	9
10	1.372	1.812	2.228	2.764	3.169	10
11	1.363	1.796	2.201	2.718	3.106	11
12	1.356	1.782	2.179	2.681	3.055	12
13	1.350	1.771	2.160	2.650	3.012	13
14	1.345	1.761	2.145	2.624	2.977	14
15	1.341	1.753	2.131	2.602	2.947	15
16	1.337	1.746	2.120	2.583	2.921	16
17	1.333	1.740	2.110	2.567	2.898	17
18	1.330	1.734	2.101	2.552	2.878	18
19	1.328	1.729	2.093	2.539	2.861	19
20	1.325	1.725	2.086	2.528	2.845	20
21	1.323	1.721	2.080	2.518	2.831	21
22	1.321	1.717	2.074	2.508	2.819	22
23	1.319	1.714	2.069	2.500	2.807	23
24	1.318	1.711	2.064	2.492	2.797	24
25	1.316	1.708	2.060	2.485	2.787	25
26	1.315	1.706	2.056	2.479	2.779	26
27	1.314	1.703	2.052	2.473	2.771	27
28	1.313	1.701	2.048	2.467	2.763	28
29	1.311	1.699	2.045	2.462	2.756	29
∞	1.282	1.645	1.960	2.326	2.576	∞

SOURCE: From "Table of Percentage Points of the *t*-Distribution," *Biometrika* 32 (1941):300. Reproduced by permission of the *Biometrika* Trustees.

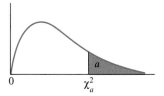

TABLE 5
Critical Values
of Chi-Square

df	$\chi^2_{.995}$	$\chi^2_{.990}$	$\chi^2_{.975}$	$\chi^2_{.950}$	$\chi^2_{.900}$
1	.0000393	.0001571	.0009821	.0039321	.0157908
2	.0100251	.0201007	.0506356	.102587	.210720
3	.0717212	.114832	.215795	.351846	.584375
4	.206990	.297110	.484419	.710721	1.063623
5	.411740	.554300	.831211	1.145476	1.61031
6	.675727	.872085	1.237347	1.63539	2.20413
7	.989265	1.239043	1.68987	2.16735	2.83311
8	1.344419	1.646482	2.17973	2.73264	3.48954
9	1.734926	2.087912	2.70039	3.32511	4.16816
10	2.15585	2.55821	3.24697	3.94030	4.86518
11	2.60321	3.05347	3.81575	4.57481	5.57779
12	3.07382	3.57056	4.40379	5.22603	6.30380
13	3.56503	4.10691	5.00874	5.89186	7.04150
14	4.07468	4.66043	5.62872	6.57063	7.78953
15	4.60094	5.22935	6.26214	7.26094	8.54675
16	5.14224	5.81221	6.90766	7.96164	9.31223
17	5.69724	6.40776	7.56418	8.67176	10.0852
18	6.26481	7.01491	8.23075	9.39046	10.8649
19	6.84398	7.63273	8.90655	10.1170	11.6509
20	7.43386	8.26040	9.59083	10.8508	12.4426
21	8.03366	8.89720	10.28293	11.5913	13.2396
22	8.64272	9.54249	10.9823	12.3380	14.0415
23	9.26042	10.19567	11.6885	13.0905	14.8479
24	9.88623	10.8564	12.4011	13.8484	15.6587
25	10.5197	11.5240	13.1197	14.6114	16.4734
26	11.1603	12.1981	13.8439	15.3791	17.2919
27	11.8076	12.8786	14.5733	16.1513	18.1138
28	12.4613	13.5648	15.3079	16.9279	18.9392
29	13.1211	14.2565	16.0471	17.7083	19.7677
30	13.7867	14.9535	16.7908	18.4926	20.5992
40	20.7065	22.1643	24.4331	26.5093	29.0505
50	27.9907	29.7067	32.3574	34.7642	37.6886
60	35.5346	37.4848	40.4817	43.1879	46.4589
70	43.2752	45.4418	48.7576	51.7393	55.3290
80	51.1720	53.5400	57.1532	60.3915	64.2778
90	59.1963	61.7541	65.6466	69.1260	73.2912
100	67.3276	70.0648	74.2219	77.9295	82.3581

TABLE 5 (continued)	$\chi^2_{.100}$	$\chi^2_{.050}$	$\chi^2_{.025}$	$\chi^2_{.010}$	$\chi^2_{.005}$	df
	2.70554	3.84146	5.02389	6.63490	7.87944	1
	4.60517	5.99147	7.37776	9.21034	10.5966	2
	6.25139	7.81473	9.34840	11.3449	12.8381	3
	7.77944	9.48773	11.1433	13.2767	14.8602	4
	9.23635	11.0705	12.8325	15.0863	16.7496	5
	10.6446	12.5916	14.4494	16.8119	18.5476	6
	12.0170	14.0671	16.0128	18.4753	20.2777	7
	13.3616	15.5073	17.5346	20.0902	21.9550	8
	14.6837	16.9190	19.0228	21.6660	23.5893	9
	15.9871	18.3070	20.4831	23.2093	25.1882	10
	17.2750	19.6751	21.9200	24.7250	26.7569	11
	18.5494	21.0261	23.3367	26.2170	28.2995	12
	19.8119	22.3621	24.7356	27.6883	29.8194	13
	21.0642	23.6848	26.1190	29.1413	31.3193	14
	22.3072	24.9958	27.4884	30.5779	32.8013	15
	23.5418	26.2962	28.8485	31.9999	34.2672	16
	24.7690	27.8571	30.1910	33.4087	35.7185	17
	25.9894	28.8693	31.5264	34.8053	37.1564	18
	27.2036	30.1435	32.8523	36.1908	38.5822	19
	28.4120	31.4104	34.1696	37.5662	39.9968	20
	29.6151	32.6705	35.4789	38.9321	41.4010	21
	30.8133	33.9244	36.7807	40.2894	42.7956	22
	32.0069	35.1725	38.0757	41.6384	44.1813	23
	33.1963	36.4151	39.3641	42.9798	45.5585	24
	34.3816	37.6525	40.6465	44.3141	46.9278	25
	35.5631	38.8852	41.9232	45.6417	48.2899	26
	36.7412	40.1133	43.1944	46.9630	49.6449	27
	37.9159	41.3372	44.4607	48.2782	50.9933	28
	39.0875	42.5569	45.7222	49.5879	52.3356	29
	40.2560	43.7729	46.9792	50.8922	53.6720	30
	51.8050	55.7585	59.3417	63.6907	66.7659	40
	63.1671	67.5048	71.4202	76.1539	79.4900	50
	74.3970	79.0819	83.2976	88.3794	91.9517	60
	85.5271	90.5312	95.0231	100.425	104.215	70
	96.5782	101.879	106.629	112.329	116.321	80
	107.565	113.145	118.136	124.116	128.299	90
	118.498	124.342	129.561	135.807	140.169	100

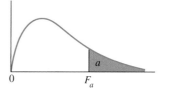

TABLE 6 Percentage Points of the *F* Distribution

df_2	a	\multicolumn{9}{c}{df_1}								
		1	2	3	4	5	6	7	8	9
1	.100	39.86	49.50	53.59	55.83	57.24	58.20	58.91	59.44	59.86
	.050	161.4	199.5	215.7	224.6	230.2	234.0	236.8	238.9	240.5
	.025	647.8	799.5	864.2	899.6	921.8	937.1	948.2	956.7	963.3
	.010	4052	4999.5	5403	5625	5764	5859	5928	5982	6022
	.005	16211	20000	21615	22500	23056	23437	23715	23925	24091
2	.100	8.53	9.00	9.16	9.24	9.29	9.33	9.35	9.37	9.38
	.050	18.51	19.00	19.16	19.25	19.30	19.33	19.35	19.37	19.38
	.025	38.51	39.00	39.17	39.25	39.30	39.33	39.36	39.37	39.39
	.010	98.50	99.00	99.17	99.25	99.30	99.33	99.36	99.37	99.39
	.005	198.5	199.0	199.2	199.2	199.3	199.3	199.4	199.4	199.4
3	.100	5.54	5.46	5.39	5.34	5.31	5.28	5.27	5.25	5.24
	.050	10.13	9.55	9.28	9.12	9.01	8.94	8.89	8.85	8.81
	.025	17.44	16.04	15.44	15.10	14.88	14.73	14.62	14.54	14.47
	.010	34.12	30.82	29.46	28.71	28.24	27.91	27.64	27.49	27.35
	.005	55.55	49.80	47.47	46.19	45.39	44.84	44.43	44.13	43.88
4	.100	4.54	4.32	4.19	4.11	4.05	4.01	3.98	3.95	3.94
	.050	7.71	6.94	6.59	6.39	6.26	6.16	6.09	6.04	6.00
	.025	12.22	10.65	9.98	9.60	9.36	9.20	9.07	8.98	8.90
	.010	21.20	18.00	16.69	15.98	15.52	15.21	14.98	14.80	14.66
	.005	31.33	26.28	24.26	23.15	22.46	21.97	21.62	21.35	21.14
5	.100	4.06	3.78	3.62	3.52	3.45	3.40	3.37	3.34	3.32
	.050	6.61	5.79	5.41	5.19	5.05	4.95	4.88	4.82	4.77
	.025	10.01	8.43	7.76	7.39	7.15	6.98	6.85	6.76	6.68
	.010	16.26	13.27	12.06	11.39	10.97	10.67	10.46	10.29	10.16
	.005	22.78	18.31	16.53	15.56	14.94	14.51	14.20	13.96	13.77
6	.100	3.78	3.46	3.29	3.18	3.11	3.05	3.01	2.98	2.96
	.050	5.99	5.14	4.76	4.53	4.39	4.28	4.21	4.15	4.10
	.025	8.81	7.26	6.60	6.23	5.99	5.82	5.70	5.60	5.52
	.010	13.75	10.92	9.78	9.15	8.75	8.47	8.26	8.10	7.98
	.005	18.63	14.54	12.92	12.03	11.46	11.07	10.79	10.57	10.39
7	.100	3.59	3.26	3.07	2.96	2.88	2.83	2.78	2.75	2.72
	.050	5.59	4.74	4.35	4.12	3.97	3.87	3.79	3.73	3.68
	.025	8.07	6.54	5.89	5.52	5.29	5.12	4.99	4.90	4.82
	.010	12.25	9.55	8.45	7.85	7.46	7.19	6.99	6.84	6.72
	.005	16.24	12.40	10.88	10.05	9.52	9.16	8.89	8.68	8.51
8	.100	3.46	3.11	2.92	2.81	2.73	2.67	2.62	2.59	2.56
	.050	5.32	4.46	4.07	3.84	3.69	3.58	3.50	3.44	3.39
	.025	7.57	6.06	5.42	5.05	4.82	4.65	4.53	4.43	4.36
	.010	11.26	8.65	7.59	7.01	6.63	6.37	6.18	6.03	5.91
	.005	14.69	11.04	9.60	8.81	8.30	7.95	7.69	7.50	7.34
9	.100	3.36	3.01	2.81	2.69	2.61	2.55	2.51	2.47	2.44
	.050	5.12	4.26	3.86	3.63	3.48	3.37	3.29	3.23	3.18
	.025	7.21	5.71	5.08	4.72	4.48	4.32	4.20	4.10	4.03
	.010	10.56	8.02	6.99	6.42	6.06	5.80	5.61	5.47	5.35
	.005	13.61	10.11	8.72	7.96	7.47	7.13	6.88	6.69	6.54

TABLE 6 *(continued)*

					df_1							
10	12	15	20	24	30	40	60	120	∞	*a*	df_2
60.19	60.71	61.22	61.74	62.00	62.26	62.53	62.79	63.06	63.33	.100	1
241.9	243.9	245.9	248.0	249.1	250.1	251.2	252.2	253.3	254.3	.050	
968.6	976.7	984.9	993.1	997.2	1001	1006	1010	1014	1018	.025	
6056	6106	6157	6209	6235	6261	6287	6313	6339	6366	.010	
24224	24426	24630	24836	24940	25044	25148	25253	25359	25465	.005	
9.39	9.41	9.42	9.44	9.45	9.46	9.47	9.47	9.48	9.49	.100	2
19.40	19.41	19.43	19.45	19.45	19.46	19.47	19.48	19.49	19.50	.050	
39.40	39.41	39.43	39.45	39.46	39.46	39.47	39.48	39.49	39.50	.025	
99.40	99.42	99.43	99.45	99.46	99.47	99.47	99.48	99.49	99.50	.010	
199.4	199.4	199.4	199.4	199.5	199.5	199.5	199.5	199.5	199.5	.005	
5.23	5.22	5.20	5.18	5.18	5.17	5.16	5.15	5.14	5.13	.100	3
8.79	8.74	8.70	8.66	8.64	8.62	8.59	8.57	8.55	8.53	.050	
14.42	14.34	14.25	14.17	14.12	14.08	14.04	13.99	13.95	13.90	.025	
27.23	27.05	26.87	26.69	26.60	26.50	26.41	26.32	26.22	26.13	.010	
43.69	43.39	43.08	42.78	42.62	42.47	42.31	42.15	41.99	41.83	.005	
3.92	3.90	3.87	3.84	3.83	3.82	3.80	3.79	3.78	3.76	.100	4
5.96	5.91	5.86	5.80	5.77	5.75	5.72	5.69	5.66	5.63	.050	
8.84	8.75	8.66	8.56	8.51	8.46	8.41	8.36	8.31	8.26	.025	
14.55	14.37	14.20	14.02	13.93	13.84	13.75	13.65	13.56	13.46	.010	
20.97	20.70	20.44	20.17	20.03	19.89	19.75	19.61	19.47	19.32	.005	
3.30	3.27	3.24	3.21	3.19	3.17	3.16	3.14	3.12	3.10	.100	5
4.74	4.68	4.62	4.56	4.53	4.50	4.46	4.43	4.40	4.36	.050	
6.62	6.52	6.43	6.33	6.28	6.23	6.18	6.12	6.07	6.02	.025	
10.05	9.89	9.72	9.55	9.47	9.38	9.29	9.20	9.11	9.02	.010	
13.62	13.38	13.15	12.90	12.78	12.66	12.53	12.40	12.27	12.14	.005	
2.94	2.90	2.87	2.84	2.82	2.80	2.78	2.76	2.74	2.72	.100	6
4.06	4.00	3.94	3.87	3.84	3.81	3.77	3.74	3.70	3.67	.050	
5.46	5.37	5.27	5.17	5.12	5.07	5.01	4.96	4.90	4.85	.025	
7.87	7.72	7.56	7.40	7.31	7.23	7.14	7.06	6.97	6.88	.010	
10.25	10.03	9.81	9.59	9.47	9.36	9.24	9.12	9.00	8.88	.005	
2.70	2.67	2.63	2.59	2.58	2.56	2.54	2.51	2.49	2.47	.100	7
3.64	3.57	3.51	3.44	3.41	3.38	3.34	3.30	3.27	3.23	.050	
4.76	4.67	4.57	4.47	4.42	4.36	4.31	4.25	4.20	4.14	.025	
6.62	6.47	6.31	6.16	6.07	5.99	5.91	5.82	5.74	5.65	.010	
8.38	8.18	7.97	7.75	7.65	7.53	7.42	7.31	7.19	7.08	.005	
2.54	2.50	2.46	2.42	2.40	2.38	2.36	2.34	2.32	2.29	.100	8
3.35	3.28	3.22	3.15	3.12	3.08	3.04	3.01	2.97	2.93	.050	
4.30	4.20	4.10	4.00	3.95	3.89	3.84	3.78	3.73	3.67	.025	
5.81	5.67	5.52	5.36	5.28	5.20	5.12	5.03	4.95	4.86	.010	
7.21	7.01	6.81	6.61	6.50	6.40	6.29	6.18	6.06	5.95	.005	
2.42	2.38	2.34	2.30	2.28	2.25	2.23	2.21	2.18	2.16	.100	9
3.14	3.07	3.01	2.94	2.90	2.86	2.83	2.79	2.75	2.71	.050	
3.96	3.87	3.77	3.67	3.61	3.56	3.51	3.45	3.39	3.33	.025	
5.26	5.11	4.96	4.81	4.73	4.65	4.57	4.48	4.40	4.31	.010	
6.42	6.23	6.03	5.83	5.73	5.62	5.52	5.41	5.30	5.19	.005	

TABLE 6 *(continued)*

df_2	a	1	2	3	4	5	6	7	8	9
						df_1				
10	.100	3.29	2.92	2.73	2.61	2.52	2.46	2.41	2.38	2.35
	.050	4.96	4.10	3.71	3.48	3.33	3.22	3.14	3.07	3.02
	.025	6.94	5.46	4.83	4.47	4.24	4.07	3.95	3.85	3.78
	.010	10.04	7.56	6.55	5.99	5.64	5.39	5.20	5.06	4.94
	.005	12.83	9.43	8.08	7.34	6.87	6.54	6.30	6.12	5.97
11	.100	3.23	2.86	2.66	2.54	2.45	2.39	2.34	2.30	2.27
	.050	4.84	3.98	3.59	3.36	3.20	3.09	3.01	2.95	2.90
	.025	6.72	5.26	4.63	4.28	4.04	3.88	3.76	3.66	3.59
	.010	9.65	7.21	6.22	5.67	5.32	5.07	4.89	4.74	4.63
	.005	12.23	8.91	7.60	6.88	6.42	6.10	5.86	5.68	5.54
12	.100	3.18	2.81	2.61	2.48	2.39	2.33	2.28	2.24	2.21
	.050	4.75	3.89	3.49	3.26	3.11	3.00	2.91	2.85	2.80
	.025	6.55	5.10	4.47	4.12	3.89	3.73	3.61	3.51	3.44
	.010	9.33	6.93	5.95	5.41	5.06	4.82	4.64	4.50	4.39
	.005	11.75	8.51	7.23	6.52	6.07	5.76	5.52	5.35	5.20
13	.100	3.14	2.76	2.56	2.43	2.35	2.28	2.23	2.20	2.16
	.050	4.67	3.81	3.41	3.18	3.03	2.92	2.83	2.77	2.71
	.025	6.41	4.97	4.35	4.00	3.77	3.60	3.48	3.39	3.31
	.010	9.07	6.70	5.74	5.21	4.86	4.62	4.44	4.30	4.19
	.005	11.37	8.19	6.93	6.23	5.79	5.48	5.25	5.08	4.94
14	.100	3.10	2.73	2.52	2.39	2.31	2.24	2.19	2.15	2.12
	.050	4.60	3.74	3.34	3.11	2.96	2.85	2.76	2.70	2.65
	.025	6.30	4.86	4.24	3.89	3.66	3.50	3.38	3.29	3.21
	.010	8.86	6.51	5.56	5.04	4.69	4.46	4.28	4.14	4.03
	.005	11.06	7.92	6.68	6.00	5.56	5.26	5.03	4.86	4.72
15	.100	3.07	2.70	2.49	2.36	2.27	2.21	2.16	2.12	2.09
	.050	4.54	3.68	3.29	3.06	2.90	2.79	2.71	2.64	2.59
	.025	6.20	4.77	4.15	3.80	3.58	3.41	3.29	3.20	3.12
	.010	8.68	6.36	5.42	4.89	4.56	4.32	4.14	4.00	3.89
	.005	10.80	7.70	6.48	5.80	5.37	5.07	4.85	4.67	4.54
16	.100	3.05	2.67	2.46	2.33	2.24	2.18	2.13	2.09	2.06
	.050	4.49	3.63	3.24	3.01	2.85	2.74	2.66	2.59	2.54
	.025	6.12	4.69	4.08	3.73	3.50	3.34	3.22	3.12	3.05
	.010	8.53	6.23	5.29	4.77	4.44	4.20	4.03	3.89	3.78
	.005	10.58	7.51	6.30	5.64	5.21	4.91	4.69	4.52	4.38
17	.100	3.03	2.64	2.44	2.31	2.22	2.15	2.10	2.06	2.03
	.050	4.45	3.59	3.20	2.96	2.81	2.70	2.61	2.55	2.49
	.025	6.04	4.62	4.01	3.66	3.44	3.28	3.16	3.06	2.98
	.010	8.40	6.11	5.18	4.67	4.34	4.10	3.93	3.79	3.68
	.005	10.38	7.35	6.16	5.50	5.07	4.78	4.56	4.39	4.25
18	.100	3.01	2.62	2.42	2.29	2.20	2.13	2.08	2.04	2.00
	.050	4.41	3.55	3.16	2.93	2.77	2.66	2.58	2.51	2.46
	.025	5.98	4.56	3.95	3.61	3.38	3.22	3.10	3.01	2.93
	.010	8.29	6.01	5.09	4.58	4.25	4.01	3.84	3.71	3.60
	.005	10.22	7.21	6.03	5.37	4.96	4.66	4.44	4.28	4.14
19	.100	2.99	2.61	2.40	2.27	2.18	2.11	2.06	2.02	1.98
	.050	4.38	3.52	3.13	2.90	2.74	2.63	2.54	2.48	2.42
	.025	5.92	4.51	3.90	3.56	3.33	3.17	3.05	2.96	2.88
	.010	8.18	5.93	5.01	4.50	4.17	3.94	3.77	3.63	3.52
	.005	10.07	7.09	5.92	5.27	4.85	4.56	4.34	4.18	4.04
20	.100	2.97	2.59	2.38	2.25	2.16	2.09	2.04	2.00	1.96
	.050	4.35	3.49	3.10	2.87	2.71	2.60	2.51	2.45	2.39
	.025	5.87	4.46	3.86	3.51	3.29	3.13	3.01	2.91	2.84
	.010	8.10	5.85	4.94	4.43	4.10	3.87	3.70	3.56	3.46
	.005	9.94	6.99	5.82	5.17	4.76	4.47	4.26	4.09	3.96

TABLE 6 *(continued)*

10	12	15	20	24	30	40	60	120	∞	a	df₂
					df_1						
2.32	2.28	2.24	2.20	2.18	2.16	2.13	2.11	2.08	2.06	.100	10
2.98	2.91	2.85	2.77	2.74	2.70	2.66	2.62	2.58	2.54	.050	
3.72	3.62	3.52	3.42	3.37	3.31	3.26	3.20	3.14	3.08	.025	
4.85	4.71	4.56	4.41	4.33	4.25	4.17	4.08	4.00	3.91	.010	
5.85	5.66	5.47	5.27	5.17	5.07	4.97	4.86	4.75	4.64	.005	
2.25	2.21	2.17	2.12	2.10	2.08	2.05	2.03	2.00	1.97	.100	11
2.85	2.79	2.72	2.65	2.61	2.57	2.53	2.49	2.45	2.40	.050	
3.53	3.43	3.33	3.23	3.17	3.12	3.06	3.00	2.94	2.88	.025	
4.54	4.40	4.25	4.10	4.02	3.94	3.86	3.78	3.69	3.60	.010	
5.42	5.24	5.05	4.86	4.76	4.65	4.55	4.44	4.34	4.23	.005	
2.19	2.15	2.10	2.06	2.04	2.01	1.99	1.96	1.93	1.90	.100	12
2.75	2.69	2.62	2.54	2.51	2.47	2.43	2.38	2.34	2.30	.050	
3.37	3.28	3.18	3.07	3.02	2.96	2.91	2.85	2.79	2.72	.025	
4.30	4.16	4.01	3.86	3.78	3.70	3.62	3.54	3.45	3.36	.010	
5.09	4.91	4.72	4.53	4.43	4.33	4.23	4.12	4.01	3.90	.005	
2.14	2.10	2.05	2.01	1.98	1.96	1.93	1.90	1.88	1.85	.100	13
2.67	2.60	2.53	2.46	2.42	2.38	2.34	2.30	2.25	2.21	.050	
3.25	3.15	3.05	2.95	2.89	2.84	2.78	2.72	2.66	2.60	.025	
4.10	3.96	3.82	3.66	3.59	3.51	3.43	3.34	3.25	3.17	.010	
4.82	4.64	4.46	4.27	4.17	4.07	3.97	3.87	3.76	3.65	.005	
2.10	2.05	2.01	1.96	1.94	1.91	1.89	1.86	1.83	1.80	.100	14
2.60	2.53	2.46	2.39	2.35	2.31	2.27	2.22	2.18	2.13	.050	
3.15	3.05	2.95	2.84	2.79	2.73	2.67	2.61	2.55	2.49	.025	
3.94	3.80	3.66	3.51	3.43	3.35	3.27	3.18	3.09	3.00	.010	
4.60	4.43	4.25	4.06	3.96	3.86	3.76	3.66	3.55	3.44	.005	
2.06	2.02	1.97	1.92	1.90	1.87	1.85	1.82	1.79	1.76	.100	15
2.54	2.48	2.40	2.33	2.29	2.25	2.20	2.16	2.11	2.07	.050	
3.06	2.96	2.86	2.76	2.70	2.64	2.59	2.52	2.46	2.40	.025	
3.80	3.67	3.52	3.37	3.29	3.21	3.13	3.05	2.96	2.87	.010	
4.42	4.25	4.07	3.88	3.79	3.69	3.58	3.48	3.37	3.26	.005	
2.03	1.99	1.94	1.89	1.87	1.84	1.81	1.78	1.75	1.72	.100	16
2.49	2.42	2.35	2.28	2.24	2.19	2.15	2.11	2.06	2.01	.050	
2.99	2.89	2.79	2.68	2.63	2.57	2.51	2.45	2.38	2.32	.025	
3.69	3.55	3.41	3.26	3.18	3.10	3.02	2.93	2.84	2.75	.010	
4.27	4.10	3.92	3.73	3.64	3.54	3.44	3.33	3.22	3.11	.005	
2.00	1.96	1.91	1.86	1.84	1.81	1.78	1.75	1.72	1.69	.100	17
2.45	2.38	2.31	2.23	2.19	2.15	2.10	2.06	2.01	1.96	.050	
2.92	2.82	2.72	2.62	2.56	2.50	2.44	2.38	2.32	2.25	.025	
3.59	3.46	3.31	3.16	3.08	3.00	2.92	2.83	2.75	2.65	.010	
4.14	3.97	3.79	3.61	3.51	3.41	3.31	3.21	3.10	2.98	.005	
1.98	1.93	1.89	1.84	1.81	1.78	1.75	1.72	1.69	1.66	.100	18
2.41	2.34	2.27	2.19	2.15	2.11	2.06	2.02	1.97	1.92	.050	
2.87	2.77	2.67	2.56	2.50	2.44	2.38	2.32	2.26	2.19	.025	
3.51	3.37	3.23	3.08	3.00	2.92	2.84	2.75	2.66	2.57	.010	
4.03	3.86	3.68	3.50	3.40	3.30	3.20	3.10	2.99	2.87	.005	
1.96	1.91	1.86	1.81	1.79	1.76	1.73	1.70	1.67	1.63	.100	19
2.38	2.31	2.23	2.16	2.11	2.07	2.03	1.98	1.93	1.88	.050	
2.82	2.72	2.62	2.51	2.45	2.39	2.33	2.27	2.20	2.13	.025	
3.43	3.30	3.15	3.00	2.92	2.84	2.76	2.67	2.58	2.49	.010	
3.93	3.76	3.59	3.40	3.31	3.21	3.11	3.00	2.89	2.78	.005	
1.94	1.89	1.84	1.79	1.77	1.74	1.71	1.68	1.64	1.61	.100	20
2.35	2.28	2.20	2.12	2.08	2.04	1.99	1.95	1.90	1.84	.050	
2.77	2.68	2.57	2.46	2.41	2.35	2.29	2.22	2.16	2.09	.025	
3.37	3.23	3.09	2.94	2.86	2.78	2.69	2.61	2.52	2.42	.010	
3.85	3.68	3.50	3.32	3.22	3.12	3.02	2.92	2.81	2.69	.005	

TABLE 6 *(continued)*

df_2	a	1	2	3	4	5	6	7	8	9
								df_1		
21	.100	2.96	2.57	2.36	2.23	2.14	2.08	2.02	1.98	1.95
	.050	4.32	3.47	3.07	2.84	2.68	2.57	2.49	2.42	2.37
	.025	5.83	4.42	3.82	3.48	3.25	3.09	2.97	2.87	2.80
	.010	8.02	5.78	4.87	4.37	4.04	3.81	3.64	3.51	3.40
	.005	9.83	6.89	5.73	5.09	4.68	4.39	4.18	4.01	3.88
22	.100	2.95	2.56	2.35	2.22	2.13	2.06	2.01	1.97	1.93
	.050	4.30	3.44	3.05	2.82	2.66	2.55	2.46	2.40	2.34
	.025	5.79	4.38	3.78	3.44	3.22	3.05	2.93	2.84	2.76
	.010	7.95	5.72	4.82	4.31	3.99	3.76	3.59	3.45	3.35
	.005	9.73	6.81	5.65	5.02	4.61	4.32	4.11	3.94	3.81
23	.100	2.94	2.55	2.34	2.21	2.11	2.05	1.99	1.95	1.92
	.050	4.28	3.42	3.03	2.80	2.64	2.53	2.44	2.37	2.32
	.025	5.75	4.35	3.75	3.41	3.18	3.02	2.90	2.81	2.73
	.010	7.88	5.66	4.76	4.26	3.94	3.71	3.54	3.41	3.30
	.005	9.63	6.73	5.58	4.95	4.54	4.26	4.05	3.88	3.75
24	.100	2.93	2.54	2.33	2.19	2.10	2.04	1.98	1.94	1.91
	.050	4.26	3.40	3.01	2.78	2.62	2.51	2.42	2.36	2.30
	.025	5.72	4.32	3.72	3.38	3.15	2.99	2.87	2.78	2.70
	.010	7.82	5.61	4.72	4.22	3.90	3.67	3.50	3.36	3.26
	.005	9.55	6.66	5.52	4.89	4.49	4.20	3.99	3.83	3.69
25	.100	2.92	2.53	2.32	2.18	2.09	2.02	1.97	1.93	1.89
	.050	4.24	3.39	2.99	2.76	2.60	2.49	2.40	2.34	2.28
	.025	5.69	4.29	3.69	3.35	3.13	2.97	2.85	2.75	2.68
	.010	7.77	5.57	4.68	4.18	3.85	3.63	3.46	3.32	3.22
	.005	9.48	6.60	5.46	4.84	4.43	4.15	3.94	3.78	3.64
26	.100	2.91	2.52	2.31	2.17	2.08	2.01	1.96	1.92	1.88
	.050	4.23	3.37	2.98	2.74	2.59	2.47	2.39	2.32	2.27
	.025	5.66	4.27	3.67	3.33	3.10	2.94	2.82	2.73	2.65
	.010	7.72	5.53	4.64	4.14	3.82	3.59	3.42	3.29	3.18
	.005	9.41	6.54	5.41	4.79	4.38	4.10	3.89	3.73	3.60
27	.100	2.90	2.51	2.30	2.17	2.07	2.00	1.95	1.91	1.87
	.050	4.21	3.35	2.96	2.73	2.57	2.46	2.37	2.31	2.25
	.025	5.63	4.24	3.65	3.31	3.08	2.92	2.80	2.71	2.63
	.010	7.68	5.49	4.60	4.11	3.78	3.56	3.39	3.26	3.15
	.005	9.34	6.49	5.36	4.74	4.34	4.06	3.85	3.69	3.56
28	.100	2.89	2.50	2.29	2.16	2.06	2.00	1.94	1.90	1.87
	.050	4.20	3.34	2.95	2.71	2.56	2.45	2.36	2.29	2.24
	.025	5.61	4.22	3.63	3.29	3.06	2.90	2.78	2.69	2.61
	.010	7.64	5.45	4.57	4.07	3.75	3.53	3.36	3.23	3.12
	.005	9.28	6.44	5.32	4.70	4.30	4.02	3.81	3.65	3.52
29	.100	2.89	2.50	2.28	2.15	2.06	1.99	1.93	1.89	1.86
	.050	4.18	3.33	2.93	2.70	2.55	2.43	2.35	2.28	2.22
	.025	5.59	4.20	3.61	3.27	3.04	2.88	2.76	2.67	2.59
	.010	7.60	5.42	4.54	4.04	3.73	3.50	3.33	3.20	3.09
	.005	9.23	6.40	5.28	4.66	4.26	3.98	3.77	3.61	3.48
30	.100	2.88	2.49	2.28	2.14	2.05	1.98	1.93	1.88	1.85
	.050	4.17	3.32	2.92	2.69	2.53	2.42	2.33	2.27	2.21
	.025	5.57	4.18	3.59	3.25	3.03	2.87	2.75	2.65	2.57
	.010	7.56	5.39	4.51	4.02	3.70	3.47	3.30	3.17	3.07
	.005	9.18	6.35	5.24	4.62	4.23	3.95	3.74	3.58	3.45

TABLE 6 *(continued)*

10	12	15	20	24	30	40	60	120	∞	a	df_2
						df_1					
1.92	1.87	1.83	1.78	1.75	1.72	1.69	1.66	1.62	1.59	.100	21
2.32	2.25	2.18	2.10	2.05	2.01	1.96	1.92	1.87	1.81	.050	
2.73	2.64	2.53	2.42	2.37	2.31	2.25	2.18	2.11	2.04	.025	
3.31	3.17	3.03	2.88	2.80	2.72	2.64	2.55	2.46	2.36	.010	
3.77	3.60	3.43	3.24	3.15	3.05	2.95	2.84	2.73	2.61	.005	
1.90	1.86	1.81	1.76	1.73	1.70	1.67	1.64	1.60	1.57	.100	22
2.30	2.23	2.15	2.07	2.03	1.98	1.94	1.89	1.84	1.78	.050	
2.70	2.60	2.50	2.39	2.33	2.27	2.21	2.14	2.08	2.00	.025	
3.26	3.12	2.98	2.83	2.75	2.67	2.58	2.50	2.40	2.31	.010	
3.70	3.54	3.36	3.18	3.08	2.98	2.88	2.77	2.66	2.55	.005	
1.89	1.84	1.80	1.74	1.72	1.69	1.66	1.62	1.59	1.55	.100	23
2.27	2.20	2.13	2.05	2.01	1.96	1.91	1.86	1.81	1.76	.050	
2.67	2.57	2.47	2.36	2.30	2.24	2.18	2.11	2.04	1.97	.025	
3.21	3.07	2.93	2.78	2.70	2.62	2.54	2.45	2.35	2.26	.010	
3.64	3.47	3.30	3.12	3.02	2.92	2.82	2.71	2.60	2.48	.005	
1.88	1.83	1.78	1.73	1.70	1.67	1.64	1.61	1.57	1.53	.100	24
2.25	2.18	2.11	2.03	1.98	1.94	1.89	1.84	1.79	1.73	.050	
2.64	2.54	2.44	2.33	2.27	2.21	2.15	2.08	2.01	1.94	.025	
3.17	3.03	2.89	2.74	2.66	2.58	2.49	2.40	2.31	2.21	.010	
3.59	3.42	3.25	3.06	2.97	2.87	2.77	2.66	2.55	2.43	.005	
1.87	1.82	1.77	1.72	1.69	1.66	1.63	1.59	1.56	1.52	.100	25
2.24	2.16	2.09	2.01	1.96	1.92	1.87	1.82	1.77	1.71	.050	
2.61	2.51	2.41	2.30	2.24	2.18	2.12	2.05	1.98	1.91	.025	
3.13	2.99	2.85	2.70	2.62	2.54	2.45	2.36	2.27	2.17	.010	
3.54	3.37	3.20	3.01	2.92	2.82	2.72	2.61	2.50	2.38	.005	
1.86	1.81	1.76	1.71	1.68	1.65	1.61	1.58	1.54	1.50	.100	26
2.22	2.15	2.07	1.99	1.95	1.90	1.85	1.80	1.75	1.69	.050	
2.59	2.49	2.39	2.28	2.22	2.16	2.09	2.03	1.95	1.88	.025	
3.09	2.96	2.81	2.66	2.58	2.50	2.42	2.33	2.23	2.13	.010	
3.49	3.33	3.15	2.97	2.87	2.77	2.67	2.56	2.45	2.33	.005	
1.85	1.80	1.75	1.70	1.67	1.64	1.60	1.57	1.53	1.49	.100	27
2.20	2.13	2.06	1.97	1.93	1.88	1.84	1.79	1.73	1.67	.050	
2.57	2.47	2.36	2.25	2.19	2.13	2.07	2.00	1.93	1.85	.025	
3.06	2.93	2.78	2.63	2.55	2.47	2.38	2.29	2.20	2.10	.010	
3.45	3.28	3.11	2.93	2.83	2.73	2.63	2.52	2.41	2.29	.005	
1.84	1.79	1.74	1.69	1.66	1.63	1.59	1.56	1.52	1.48	.100	28
2.19	2.12	2.04	1.96	1.91	1.87	1.82	1.77	1.71	1.65	.050	
2.55	2.45	2.34	2.23	2.17	2.11	2.05	1.98	1.91	1.83	.025	
3.03	2.90	2.75	2.60	2.52	2.44	2.35	2.26	2.17	2.06	.010	
3.41	3.25	3.07	2.89	2.79	2.69	2.59	2.48	2.37	2.25	.005	
1.83	1.78	1.73	1.68	1.65	1.62	1.58	1.55	1.51	1.47	.100	29
2.18	2.10	2.03	1.94	1.90	1.85	1.81	1.75	1.70	1.64	.050	
2.53	2.43	2.32	2.21	2.15	2.09	2.03	1.96	1.89	1.81	.025	
3.00	2.87	2.73	2.57	2.49	2.41	2.33	2.23	2.14	2.03	.010	
3.38	3.21	3.04	2.86	2.76	2.66	2.56	2.45	2.33	2.21	.005	
1.82	1.77	1.72	1.67	1.64	1.61	1.57	1.54	1.50	1.46	.100	30
2.16	2.09	2.01	1.93	1.89	1.84	1.79	1.74	1.68	1.62	.050	
2.51	2.41	2.31	2.20	2.14	2.07	2.01	1.94	1.87	1.79	.025	
2.98	2.84	2.70	2.55	2.47	2.39	2.30	2.21	2.11	2.01	.010	
3.34	3.18	3.01	2.82	2.73	2.63	2.52	2.42	2.30	2.18	.005	

TABLE 6 *(continued)*

df_2	a		1	2	3	4	5	6	7	8	9
							df_1				
40	.100		2.84	2.44	2.23	2.09	2.00	1.93	1.87	1.83	1.79
	.050		4.08	3.23	2.84	2.61	2.45	2.34	2.25	2.18	2.12
	.025		5.42	4.05	3.46	3.13	2.90	2.74	2.62	2.53	2.45
	.010		7.31	5.18	4.31	3.83	3.51	3.29	3.12	2.99	2.89
	.005		8.83	6.07	4.98	4.37	3.99	3.71	3.51	3.35	3.22
60	.100		2.79	2.39	2.18	2.04	1.95	1.87	1.82	1.77	1.74
	.050		4.00	3.15	2.76	2.53	2.37	2.25	2.17	2.10	2.04
	.025		5.29	3.93	3.34	3.01	2.79	2.63	2.51	2.41	2.33
	.010		7.08	4.98	4.13	3.65	3.34	3.12	2.95	2.82	2.72
	.005		8.49	5.79	4.73	4.14	3.76	3.49	3.29	3.13	3.01
120	.100		2.75	2.35	2.13	1.99	1.90	1.82	1.77	1.72	1.68
	.050		3.92	3.07	2.68	2.45	2.29	2.17	2.09	2.02	1.96
	.025		5.15	3.80	3.23	2.89	2.67	2.52	2.39	2.30	2.22
	.010		6.85	4.79	3.95	3.48	3.17	2.96	2.79	2.66	2.56
	.005		8.18	5.54	4.50	3.92	3.55	3.28	3.09	2.93	2.81
∞	.100		2.71	2.30	2.08	1.94	1.85	1.77	1.72	1.67	1.63
	.050		3.84	3.00	2.60	2.37	2.21	2.10	2.01	1.94	1.63
	.025		5.02	3.69	3.12	2.79	2.57	2.41	2.29	2.19	2.11
	.010		6.63	4.61	3.78	3.32	3.02	2.80	2.64	2.51	2.41
	.005		7.88	5.30	4.28	3.72	3.35	3.09	2.90	2.74	2.62

TABLE 6 *(continued)*

10	12	15	20	24	30	40	60	120	∞	a	df₂

df_1

10	12	15	20	24	30	40	60	120	∞	a	df_2
1.76	1.71	1.66	1.61	1.57	1.54	1.51	1.47	1.42	1.38	.100	40
2.08	2.00	1.92	1.84	1.79	1.74	1.69	1.64	1.58	1.51	.050	
2.39	2.29	2.18	2.07	2.01	1.94	1.88	1.80	1.72	1.64	.025	
2.80	2.66	2.52	2.37	2.29	2.20	2.11	2.02	1.92	1.80	.010	
3.12	2.95	2.78	2.60	2.50	2.40	2.30	2.18	2.06	1.93	.005	
1.71	1.66	1.60	1.54	1.51	1.48	1.44	1.40	1.35	1.29	.100	60
1.99	1.92	1.84	1.75	1.70	1.65	1.59	1.53	1.47	1.39	.050	
2.27	2.17	2.06	1.94	1.88	1.82	1.74	1.67	1.58	1.48	.025	
2.63	2.50	2.35	2.20	2.12	2.03	1.94	1.84	1.73	1.60	.010	
2.90	2.74	2.57	2.39	2.29	2.19	2.08	1.96	1.83	1.69	.005	
1.65	1.60	1.55	1.48	1.45	1.41	1.37	1.32	1.26	1.19	.100	120
1.91	1.83	1.75	1.66	1.61	1.55	1.50	1.43	1.35	1.25	.050	
2.16	2.05	1.94	1.82	1.76	1.69	1.61	1.53	1.43	1.31	.025	
2.47	2.34	2.19	2.03	1.95	1.86	1.76	1.66	1.53	1.38	.010	
2.71	2.54	2.37	2.19	2.09	1.98	1.87	1.75	1.61	1.43	.005	
1.60	1.55	1.49	1.42	1.38	1.34	1.30	1.24	1.17	1.00	.100	∞
1.83	1.75	1.67	1.57	1.52	1.46	1.39	1.32	1.22	1.00	.050	
2.05	1.94	1.83	1.71	1.64	1.57	1.48	1.39	1.27	1.00	.025	
2.32	2.18	2.04	1.88	1.79	1.70	1.59	1.47	1.32	1.00	.010	
2.52	2.36	2.19	2.00	1.90	1.79	1.67	1.53	1.36	1.00	.005	

TABLE 7 Critical Values of T for the Wilcoxon Rank Sum Test, $n_1 \leq n_2$

TABLE 7(a)
5% Left-Tailed Critical Values

							n_1							
n_2	2	3	4	5	6	7	8	9	10	11	12	13	14	15
3	—	6												
4	—	6	11											
5	3	7	12	19										
6	3	8	13	20	28									
7	3	8	14	21	29	39								
8	4	9	15	23	31	41	51							
9	4	10	16	24	33	43	54	66						
10	4	10	17	26	35	45	56	69	82					
11	4	11	18	27	37	47	59	72	86	100				
12	5	11	19	28	38	49	62	75	89	104	120			
13	5	12	20	30	40	52	64	78	92	108	125	142		
14	6	13	21	31	42	54	67	81	96	112	129	147	166	
15	6	13	22	33	44	56	69	84	99	116	133	152	171	192

TABLE 7(b)
2.5% Left-Tailed Critical Values

							n_1							
n_2	2	3	4	5	6	7	8	9	10	11	12	13	14	15
4	—	—	10											
5	—	6	11	17										
6	—	7	12	18	26									
7	—	7	13	20	27	36								
8	3	8	14	21	29	38	49							
9	3	8	14	22	31	40	51	62						
10	3	9	15	23	32	42	53	65	78					
11	3	9	16	24	34	44	55	68	81	96				
12	4	10	17	26	35	46	58	71	84	99	115			
13	4	10	18	27	37	48	60	73	88	103	119	136		
14	4	11	19	28	38	50	62	76	91	106	123	141	160	
15	4	11	20	29	40	52	65	79	94	110	127	145	164	184

SOURCE: Adapted from "An Extended Table of Critical Values for the Mann-Whitney (Wilcoxon) Two-Sample Statistics" by Roy C. Milton, *Journal of the American Statistical Association*, Volume 59, Number 307 (September 1964). Reproduced with the permission of the Editor, *Journal of the American Statistical Association*.

TABLE 7(c)
1% Left-Tailed Critical Values

n_2	\$n_1\$ 2	3	4	5	6	7	8	9	10	11	12	13	14	15
3	—	—												
4	—	—	—											
5	—	—	10	16										
6	—	—	11	17	24									
7	—	6	11	18	25	34								
8	—	6	12	19	27	35	45							
9	—	7	13	20	28	37	47	59						
10	—	7	13	21	29	39	49	61	74					
11	—	7	14	22	30	40	51	63	77	91				
12	—	8	15	23	32	42	53	66	79	94	109			
13	3	8	15	24	33	44	56	68	82	97	113	130		
14	3	8	16	25	34	45	58	71	85	100	116	134	152	
15	3	9	17	26	36	47	60	73	88	103	120	138	156	176

TABLE 7(d)
.5% Left-Tailed Critical Values

n_2	\$n_1\$ 3	4	5	6	7	8	9	10	11	12	13	14	15
3	—												
4	—	—											
5	—	—	15										
6	—	10	16	23									
7	—	10	16	24	32								
8	—	11	17	25	34	42							
9	6	11	18	26	35	45	56						
10	6	12	19	27	37	47	58	71					
11	6	12	20	28	38	49	61	73	87				
12	7	13	21	30	40	51	63	76	90	105			
13	7	13	22	31	41	53	65	79	93	109	125		
14	7	14	22	32	43	54	67	81	96	112	129	147	
15	8	15	23	33	44	56	69	84	99	115	133	151	171

TABLE 8
Critical Values of *T* in the Wilcoxon Signed-Rank Test, *n* = 5(1)50

One-Sided	Two-Sided	n = 5	n = 6	n = 7	n = 8	n = 9	n = 10
α = .050	α = .10	1	2	4	6	8	11
α = .025	α = .05		1	2	4	6	8
α = .010	α = .02			0	2	3	5
α = .005	α = .01				0	2	3

One-Sided	Two-Sided	n = 11	n = 12	n = 13	n = 14	n = 15	n = 16
α = .050	α = .10	14	17	21	26	30	36
α = .025	α = .05	11	14	17	21	25	30
α = .010	α = .02	7	10	13	16	20	24
α = .005	α = .01	5	7	10	13	16	19

One-Sided	Two-Sided	n = 17	n = 18	n = 19	n = 20	n = 21	n = 22
α = .050	α = .10	41	47	54	60	68	75
α = .025	α = .05	35	40	46	52	59	66
α = .010	α = .02	28	33	38	43	49	56
α = .005	α = .01	23	28	32	37	43	49

One-Sided	Two-Sided	n = 23	n = 24	n = 25	n = 26	n = 27	n = 28
α = .050	α = .10	83	92	101	110	120	130
α = .025	α = .05	73	81	90	98	107	117
α = .010	α = .02	62	69	77	85	93	102
α = .005	α = .01	55	68	68	76	84	92

One-Sided	Two-Sided	n = 29	n = 30	n = 31	n = 32	n = 33	n = 34
α = .050	α = .10	141	152	163	175	188	201
α = .025	α = .05	127	137	148	159	171	183
α = .010	α = .02	111	120	130	141	151	162
α = .005	α = .01	100	109	118	128	138	149

One-Sided	Two-Sided	n = 35	n = 36	n = 37	n = 38	n = 39	
α = .050	α = .10	214	228	242	256	271	
α = .025	α = .05	195	208	222	235	250	
α = .010	α = .02	174	186	198	211	224	
α = .005	α = .01	160	171	183	195	208	

One-Sided	Two-Sided	n = 40	n = 41	n = 42	n = 43	n = 44	n = 45
α = .050	α = .10	287	303	319	336	353	371
α = .025	α = .05	264	279	295	311	327	344
α = .010	α = .02	238	252	267	281	297	313
α = .005	α = .01	221	234	248	262	277	292

One-Sided	Two-Sided	n = 46	n = 47	n = 48	n = 49	n = 50	
α = .050	α = .10	389	408	427	446	466	
α = .025	α = .05	361	379	397	415	434	
α = .010	α = .02	329	345	362	380	398	
α = .005	α = .01	307	323	339	356	373	

SOURCE: From "Some Rapid Approximate Statistical Procedures" (1964) 28, by F. Wilcoxon and R.A. Wilcox. Reproduced with the kind permission of Lederle Laboratories, a division of American Cyanamid Company.

TABLE 9
Critical Values of Spearman's Rank Correlation Coefficient for a One-Tailed Test

n	$\alpha = .05$	$\alpha = .025$	$\alpha = .01$	$\alpha = .005$
5	.900	—	—	—
6	.829	.886	.943	—
7	.714	.786	.893	—
8	.643	.738	.833	.881
9	.600	.683	.783	.833
10	.564	.648	.745	.794
11	.523	.623	.736	.818
12	.497	.591	.703	.780
13	.475	.566	.673	.745
14	.457	.545	.646	.716
15	.441	.525	.623	.689
16	.425	.507	.601	.666
17	.412	.490	.582	.645
18	.399	.476	.564	.625
19	.388	.462	.549	.608
20	.377	.450	.534	.591
21	.368	.438	.521	.576
22	.359	.428	.508	.562
23	.351	.418	.496	.549
24	.343	.409	.485	.537
25	.336	.400	.475	.526
26	.329	.392	.465	.515
27	.323	.385	.456	.505
28	.317	.377	.448	.496
29	.311	.370	.440	.487
30	.305	.364	.432	.478

SOURCE: From "Distribution of Sums of Squares of Rank Differences for Small Samples" by E.G. Olds, *Annals of Mathematical Statistics* 9 (1938). Reproduced with the permission of the editor, *Annals of Mathematical Statistics.*

TABLE 10 **Random Numbers**

Line	1	2	3	4	5	6	7	8	9	10	11	12	13	14
							Column							
1	10480	15011	01536	02011	81647	91646	69179	14194	62590	36207	20969	99570	91291	90700
2	22368	46573	25595	85393	30995	89198	27982	53402	93965	34095	52666	19174	39615	99505
3	24130	48360	22527	97265	76393	64809	15179	24830	49340	32081	30680	19655	63348	58629
4	42167	93093	06243	61680	07856	16376	39440	53537	71341	57004	00849	74917	97758	16379
5	37570	39975	81837	16656	06121	91782	60468	81305	49684	60672	14110	06927	01263	54613
6	77921	06907	11008	42751	27756	53498	18602	70659	90655	15053	21916	81825	44394	42880
7	99562	72905	56420	69994	98872	31016	71194	18738	44013	48840	63213	21069	10634	12952
8	96301	91977	05463	07972	18876	20922	94595	56869	69014	60045	18425	84903	42508	32307
9	89579	14342	63661	10281	17453	18103	57740	84378	25331	12566	58678	44947	05585	56941
10	84575	36857	53342	53988	53060	59533	38867	62300	08158	17983	16439	11458	18593	64952
11	28918	69578	88231	33276	70997	79936	56865	05859	90106	31595	01547	85590	91610	78188
12	63553	40961	48235	03427	49626	69445	18663	72695	52180	20847	12234	90511	33703	90322
13	09429	93969	52636	92737	88974	33488	36320	17617	30015	08272	84115	27156	30613	74952
14	10365	61129	87529	85689	48237	52267	67689	93394	01511	26358	85104	20285	29975	89868
15	07119	97336	71048	08178	77233	13916	47564	81056	97735	85977	29372	74461	28551	90707
16	51085	12765	51821	51259	77452	16308	60756	92144	49442	53900	70960	63990	75601	40719
17	02368	21382	52404	60268	89368	19885	55322	44819	01188	65255	64835	44919	05944	55157
18	01011	54092	33362	94904	31273	04146	18594	29852	71585	85030	51132	01915	92747	64951
19	52162	53916	46369	58586	23216	14513	83149	98736	23495	64350	94738	17752	35156	35749
20	07056	97628	33787	09998	42698	06691	76988	13602	51851	46104	88916	19509	25625	58104
21	48663	91245	85828	14346	09172	30168	90229	04734	59193	22178	30421	61666	99904	32812
22	54164	58492	22421	74103	47070	25306	76468	26384	58151	06646	21524	15227	96909	44592
23	32639	32363	05597	24200	13363	38005	94342	28728	35806	06912	17012	64161	18296	22851
24	29334	27001	87637	87308	58731	00256	45834	15398	46557	41135	10367	07684	36188	18510
25	02488	33062	28834	07351	19731	92420	60952	61280	50001	67658	32586	86679	50720	94953
26	81525	72295	04839	96423	24878	82651	66566	14778	76797	14780	13300	87074	79666	95725
27	29676	20591	68086	26432	46901	20849	89768	81536	86645	12659	92259	57102	80428	25280
28	00742	57392	39064	66432	84673	40027	32832	61362	98947	96067	64760	64585	96096	98253
29	05366	04213	25669	26422	44407	44048	37937	63904	45766	66134	75470	66520	34693	90449
30	91921	26418	64117	94305	26766	25940	39972	22209	71500	64568	91402	42416	07844	69618
31	00582	04711	87917	77341	42206	35126	74087	99547	81817	42607	43808	76655	62028	76630
32	00725	69884	62797	56170	86324	88072	76222	36086	84637	93161	76038	65855	77919	88006
33	69011	65795	95876	55293	18988	27354	26575	08625	40801	59920	29841	80150	12777	48501
34	25976	57948	29888	88604	67917	48708	18912	82271	65424	69774	33611	54262	85963	03547
35	09763	83473	73577	12908	30883	18317	28290	35797	05998	41688	34952	37888	38917	88050
36	91567	42595	27958	30134	04024	86385	29880	99730	55536	84855	29080	09250	79656	73211
37	17955	56349	90999	49127	20044	59931	06115	20542	18059	02008	73708	83517	36103	42791
38	46503	18584	18845	49618	02304	51038	20655	58727	28168	15475	56942	53389	20562	87338
39	92157	89634	94824	78171	84610	82834	09922	25417	44137	48413	25555	21246	35509	20468
40	14577	62765	35605	81263	39667	47358	56873	56307	61607	49518	89656	20103	77490	18062
41	98427	07523	33362	64270	01638	92477	66969	98420	04880	45585	46565	04102	46880	45709
42	34914	63976	88720	82765	34476	17032	87589	40836	32427	70002	70663	88863	77775	69348
43	70060	28277	39475	46473	23219	53416	94970	25832	69975	94884	19661	72828	00102	66794
44	53976	54914	06990	67245	68350	82948	11398	42878	80287	88267	47363	46634	06541	97809
45	76072	29515	40980	07391	58745	25774	22987	80059	39911	96189	41151	14222	60697	59583
46	90725	52210	83974	29992	65831	38857	50490	83765	55657	14361	31720	57375	56228	41546
47	64364	67412	33339	31926	14883	24413	59744	92351	97473	89286	35931	04110	23726	51900
48	08962	00358	31662	25388	61642	34072	81249	35648	56891	69352	48373	45578	78547	81788
49	95012	68379	93526	70765	10592	04542	76463	54328	02349	17247	28865	14777	62730	92277
50	15664	10493	20492	38391	91132	21999	59516	81652	27195	48223	46751	22923	32261	85653

Source: Abridged from *Handbook of Tables for Probability and Statistics,* 2d ed. Edited by William H. Beyer (Cleveland: The Chemical Rubber Company, 1968). Reproduced by permission of CRC Press, Inc.

TABLE 10 *(continued)*

Line	1	2	3	4	5	6	7	8	9	10	11	12	13	14
									Column					
51	16408	81899	04153	53381	79401	21438	83035	92350	36693	31238	59649	91754	72772	02338
52	18629	81953	05520	91962	04739	13092	97662	24822	94730	06496	35090	04822	86774	98289
53	73115	35101	47498	87637	99016	71060	88824	71013	18735	20286	23153	72924	35165	43040
54	57491	16703	23167	49323	45021	33132	12544	41035	80780	45393	44812	12515	98931	91202
55	30405	83946	23792	14422	15059	45799	22716	19792	09983	74353	68668	30429	70735	25499
56	16631	35006	85900	98275	32388	52390	16815	69298	82732	38480	73817	32523	41961	44437
57	96773	20206	42559	78985	05300	22164	24369	54224	35033	19687	11052	91491	60383	19746
58	38935	64202	14349	82674	66523	44133	00697	35552	35970	19124	63318	29686	03387	59846
59	31624	76384	17403	53363	44167	64486	64758	75366	76554	31601	12614	33072	60332	92325
60	78919	19474	23632	27889	47914	02584	37680	20801	72152	39339	34806	08930	85001	87820
61	03931	33309	57047	74211	63445	17361	62825	39908	05607	91284	68833	25570	38818	46920
62	74426	33278	43972	10119	89917	15665	52872	73823	73144	88662	88970	74492	51805	99378
63	09066	00903	20795	95452	92648	45454	09552	88815	16553	51125	79375	97596	16296	66092
64	42238	12426	87025	14267	20979	04508	64535	31355	86064	29472	47689	05974	52468	16834
65	16153	08002	26504	41744	81959	65642	74240	56302	00033	67107	77510	70625	28725	34191
66	21457	40742	29820	96783	29400	21840	15035	34537	33310	06116	95240	15957	16572	06004
67	21581	57802	02050	89728	17937	37621	47075	42080	97403	48626	68995	43805	33386	21597
68	55612	78095	83197	33732	05810	24813	86902	60397	16489	03264	88525	42786	05269	92532
69	44657	66999	99324	51281	84463	60563	79312	93454	68876	25471	93911	25650	12682	73572
70	91340	84979	46949	81973	37949	61023	43997	15263	80644	43942	89203	71795	99533	50501
71	91227	21199	31935	27022	84067	05462	35216	14486	29891	68607	41867	14951	91696	85065
72	50001	38140	66321	19924	72163	09538	12151	06878	91903	18749	34405	56087	82790	70925
73	65390	05224	72958	28609	81406	39147	25549	48542	42627	45233	57202	94617	23772	07896
74	27504	96131	83944	41575	10573	08619	64482	73923	36152	05184	94142	25299	84387	34925
75	37169	94851	39117	89632	00959	16487	65536	49071	39782	17095	02330	74301	00275	48280
76	11508	70225	51111	38351	19444	66499	71945	05422	13442	78675	84081	66938	93654	59894
77	37449	30362	06694	54690	04052	53115	62757	95348	78662	11163	81651	50245	34971	52924
78	46515	70331	85922	38329	57015	15765	97161	17869	45349	61796	66345	81073	49106	79860
79	30986	81223	42416	58353	21532	30502	32305	86482	05174	07901	54339	58861	74818	46942
80	63798	64995	46583	09785	44160	78128	83991	42865	92520	83531	80377	35909	81250	54238
81	82486	84846	99254	67632	43218	50076	21361	64816	51202	88124	41870	52689	51275	83556
82	21885	32906	92431	09060	64297	51674	64126	62570	26123	05155	59194	52799	28225	85762
83	60336	98782	07408	53458	13564	59089	26445	29789	85205	41001	12535	12133	14645	23541
84	43937	46891	24010	25560	86355	33941	25786	54990	71899	15475	95434	98227	21824	19585
85	97656	63175	89303	16275	07100	92063	21942	18611	47348	20203	18534	03862	78095	50136
86	03299	01221	05418	38982	55758	92237	26759	86367	21216	98442	08303	56613	91511	75928
87	79626	06486	03574	17668	07785	76020	79924	25651	83325	88428	85076	72811	22717	50585
88	85636	68335	47539	03129	65651	11977	02510	26113	99447	68645	34327	15152	55230	93448
89	18039	14367	61337	06177	12143	46609	32989	74014	64708	00533	35398	58408	13261	47908
90	08362	15656	60627	36478	65648	16764	53412	09013	07832	41574	17639	82163	60859	75567
91	79556	29068	04142	16268	15387	12856	66227	38358	22478	73373	88732	09443	82558	05250
92	92608	82674	27072	32534	17075	27698	98204	63863	11951	34648	88022	56148	34925	57031
93	23982	25835	40055	67006	12293	02753	14827	23235	35071	99704	37543	11601	35503	85171
94	09915	96306	05908	97901	28395	14186	00821	80703	70426	75647	76310	88717	37890	40129
95	59037	33300	26695	62247	69927	76123	50842	43834	86654	70959	79725	93872	28117	19233
96	42488	78077	69882	61657	34136	79180	97526	43092	04098	73571	80799	76536	71255	64239
97	46764	86273	63003	93017	31204	36692	40202	35275	57306	55543	53203	18098	47625	88684
98	03237	45430	55417	63282	90816	17349	88298	90183	36600	78406	06216	95787	42579	90730
99	86591	81482	52667	61582	14972	90053	89534	76036	49199	43716	97548	04379	46370	28672
100	38534	01715	94964	87288	65680	43772	39560	12918	86737	62738	19636	51132	25739	56947

TABLE 11(a)
Percentage Points of the Studentized Range, $q_{.05}(k, df)$; Upper 5% Points

					k					
df	2	3	4	5	6	7	8	9	10	11
1	17.97	26.98	32.82	37.08	40.41	43.12	45.40	47.36	49.07	50.59
2	6.08	8.33	9.80	10.88	11.74	12.44	13.03	13.54	13.99	14.39
3	4.50	5.91	6.82	7.50	8.04	8.48	8.85	9.18	9.46	9.72
4	3.93	5.04	5.76	6.29	6.71	7.05	7.35	7.60	7.83	8.03
5	3.64	4.60	5.22	5.67	6.03	6.33	6.58	6.80	6.99	7.17
6	3.46	4.34	4.90	5.30	5.63	5.90	6.12	6.32	6.49	6.65
7	3.34	4.16	4.68	5.06	5.36	5.61	5.82	6.00	6.16	6.30
8	3.26	4.04	4.53	4.89	5.17	5.40	5.60	5.77	5.92	6.05
9	3.20	3.95	4.41	4.76	5.02	5.24	5.43	5.59	5.74	5.87
10	3.15	3.88	4.33	4.65	4.91	5.12	5.30	5.46	5.60	5.72
11	3.11	3.82	4.26	4.57	4.82	5.03	5.20	5.35	5.49	5.61
12	3.08	3.77	4.20	4.51	4.75	4.95	5.12	5.27	5.39	5.51
13	3.06	3.73	4.15	4.45	4.69	4.88	5.05	5.19	5.32	5.43
14	3.03	3.70	4.11	4.41	4.64	4.83	4.99	5.13	5.25	5.36
15	3.01	3.67	4.08	4.37	4.60	4.78	4.94	5.08	5.20	5.31
16	3.00	3.65	4.05	4.33	4.56	4.74	4.90	5.03	5.15	5.26
17	2.98	3.63	4.02	4.30	4.52	4.70	4.86	4.99	5.11	5.21
18	2.97	3.61	4.00	4.28	4.49	4.67	4.82	4.96	5.07	5.17
19	2.96	3.59	3.98	4.25	4.47	4.65	4.79	4.92	5.04	5.14
20	2.95	3.58	3.96	4.23	4.45	4.62	4.77	4.90	5.01	5.11
24	2.92	3.53	3.90	4.17	4.37	4.54	4.68	4.81	4.92	5.01
30	2.89	3.49	3.85	4.10	4.30	4.46	4.60	4.72	4.82	4.92
40	2.86	3.44	3.79	4.04	4.23	4.39	4.52	4.63	4.73	4.82
60	2.83	3.40	3.74	3.98	4.16	4.31	4.44	4.55	4.65	4.73
120	2.80	3.36	3.68	3.92	4.10	4.24	4.36	4.47	4.56	4.64
∞	2.77	3.31	3.63	3.86	4.03	4.17	4.29	4.39	4.47	4.55

TABLE 11(a)
(continued)

					k				
12	13	14	15	16	17	18	19	20	df
51.96	53.20	54.33	55.36	56.32	57.22	58.04	58.83	59.56	1
14.75	15.08	15.38	15.65	15.91	16.14	16.37	16.57	16.77	2
9.95	10.15	10.35	10.52	10.69	10.84	10.98	11.11	11.24	3
8.21	8.37	8.52	8.66	8.79	8.91	9.03	9.13	9.23	4
7.32	7.47	7.60	7.72	7.83	7.93	8.03	8.12	8.21	5
6.79	6.92	7.03	7.14	7.24	7.34	7.43	7.51	7.59	6
6.43	6.55	6.66	6.76	6.85	6.94	7.02	7.10	7.17	7
6.18	6.29	6.39	6.48	6.57	6.65	6.73	6.80	6.87	8
5.98	6.09	6.19	6.28	6.36	6.44	6.51	6.58	6.64	9
5.83	5.93	6.03	6.11	6.19	6.27	6.34	6.40	6.47	10
5.71	5.81	5.90	5.98	6.06	6.13	6.20	6.27	6.33	11
5.61	5.71	5.80	5.88	5.95	6.02	6.09	6.15	6.21	12
5.53	5.63	5.71	5.79	5.86	5.93	5.99	6.05	6.11	13
5.46	5.55	5.64	5.71	5.79	5.85	5.91	5.97	6.03	14
5.40	5.49	5.57	5.65	5.72	5.78	5.85	5.90	5.96	15
5.35	5.44	5.52	5.59	5.66	5.73	5.79	5.84	5.90	16
5.31	5.39	5.47	5.54	5.61	5.67	5.73	5.79	5.84	17
5.27	5.35	5.43	5.50	5.57	5.63	5.69	5.74	5.79	18
5.23	5.31	5.39	5.46	5.53	5.59	5.65	5.70	5.75	19
5.20	5.28	5.36	5.43	5.49	5.55	5.61	5.66	5.71	20
5.10	5.18	5.25	5.32	5.38	5.44	5.49	5.55	5.59	24
5.00	5.08	5.15	5.21	5.27	5.33	5.38	5.43	5.47	30
4.90	4.98	5.04	5.11	5.16	5.22	5.27	5.31	5.36	40
4.81	4.88	4.94	5.00	5.06	5.11	5.15	5.20	5.24	60
4.71	4.78	4.84	4.90	4.95	5.00	5.04	5.09	5.13	120
4.62	4.68	4.74	4.80	4.85	4.89	4.93	4.97	5.01	∞

SOURCE: From *Biometrika Tables for Statisticians,* Vol. 1, 3rd ed., edited by E.S. Pearson and H.O. Hartley (Cambridge University Press, 1966). Reproduced by permission of the *Biometrika* Trustees.

TABLE 11(b)
Percentage Points of the Studentized Range, $q_{.01}(k, df)$; Upper 1% Points

df	k									
	2	3	4	5	6	7	8	9	10	11
1	90.03	135.0	164.3	185.6	202.2	215.8	227.2	237.0	245.6	253.2
2	14.04	19.02	22.29	24.72	26.63	28.20	29.53	30.68	31.69	32.59
3	8.26	10.62	12.17	13.33	14.24	15.00	15.64	16.20	16.69	17.13
4	6.51	8.12	9.17	9.96	10.58	11.10	11.55	11.93	12.27	12.57
5	5.70	6.98	7.80	8.42	8.91	9.32	9.67	9.97	10.24	10.48
6	5.24	6.33	7.03	7.56	7.97	8.32	8.61	8.87	9.10	9.30
7	4.95	5.92	6.54	7.01	7.37	7.68	7.94	8.17	8.37	8.55
8	4.75	5.64	6.20	6.62	6.96	7.24	7.47	7.68	7.86	8.03
9	4.60	5.43	5.96	6.35	6.66	6.91	7.13	7.33	7.49	7.65
10	4.48	5.27	5.77	6.14	6.43	6.67	6.87	7.05	7.21	7.36
11	4.39	5.15	5.62	5.97	6.25	6.48	6.67	6.84	6.99	7.13
12	4.32	5.05	5.50	5.84	6.10	6.32	6.51	6.67	6.81	6.94
13	4.26	4.96	5.40	5.73	5.98	6.19	6.37	6.53	6.67	6.79
14	4.21	4.89	5.32	5.63	5.88	6.08	6.26	6.41	6.54	6.66
15	4.17	4.84	5.25	5.56	5.80	5.99	6.16	6.31	6.44	6.55
16	4.13	4.79	5.19	5.49	5.72	5.92	6.08	6.22	6.35	6.46
17	4.10	4.74	5.14	5.43	5.66	5.85	6.01	6.15	6.27	6.38
18	4.07	4.70	5.09	5.38	5.60	5.79	5.94	6.08	6.20	6.31
19	4.05	4.67	5.05	5.33	5.55	5.73	5.89	6.02	6.14	6.25
20	4.02	4.64	5.02	5.29	5.51	5.69	5.84	5.97	6.09	6.19
24	3.96	4.55	4.91	5.17	5.37	5.54	5.69	5.81	5.92	6.02
30	3.89	4.45	4.80	5.05	5.24	5.40	5.54	5.65	5.76	5.85
40	3.82	4.37	4.70	4.93	5.11	5.26	5.39	5.50	5.60	5.69
60	3.76	4.28	4.59	4.82	4.99	5.13	5.25	5.36	5.45	5.53
120	3.70	4.20	4.50	4.71	4.87	5.01	5.12	5.21	5.30	5.37
∞	3.64	4.12	4.40	4.60	4.76	4.88	4.99	5.08	5.16	5.23

TABLE 11(b)
(continued)

12	13	14	15	16	17	18	19	20	df
260.0	266.2	271.8	277.0	281.8	286.3	290.0	294.3	298.0	1
33.40	34.13	34.81	35.43	36.00	36.53	37.03	37.50	37.95	2
17.53	17.89	18.22	18.52	18.81	19.07	19.32	19.55	19.77	3
12.84	13.09	13.32	13.53	13.73	13.91	14.08	14.24	14.40	4
10.70	10.89	11.08	11.24	11.40	11.55	11.68	11.81	11.93	5
9.48	9.65	9.81	9.95	10.08	10.21	10.32	10.43	10.54	6
8.71	8.86	9.00	9.12	9.24	9.35	9.46	9.55	9.65	7
8.18	8.31	8.44	8.55	8.66	8.76	8.85	8.94	9.03	8
7.78	7.91	8.03	8.13	8.23	8.33	8.41	8.49	8.57	9
7.49	7.60	7.71	7.81	7.91	7.99	8.08	8.15	8.23	10
7.25	7.36	7.46	7.56	7.65	7.73	7.81	7.88	7.95	11
7.06	7.17	7.26	7.36	7.44	7.52	7.59	7.66	7.73	12
6.90	7.01	7.10	7.19	7.27	7.35	7.42	7.48	7.55	13
6.77	6.87	6.96	7.05	7.13	7.20	7.27	7.33	7.39	14
6.66	6.76	6.84	6.93	7.00	7.07	7.14	7.20	7.26	15
6.56	6.66	6.74	6.82	6.90	6.97	7.03	7.09	7.15	16
6.48	6.57	6.66	6.73	6.81	6.87	6.94	7.00	7.05	17
6.41	6.50	6.58	6.65	6.72	6.79	6.85	6.91	6.97	18
6.34	6.43	6.51	6.58	6.65	6.72	6.78	6.84	6.89	19
6.28	6.37	6.45	6.52	6.59	6.65	6.71	6.77	6.82	20
6.11	6.19	6.26	6.33	6.39	6.45	6.51	6.56	6.61	24
5.93	6.01	6.08	6.14	6.20	6.26	6.31	6.36	6.41	30
5.76	5.83	5.90	5.96	6.02	6.07	6.12	6.16	6.21	40
5.60	5.67	5.73	5.78	5.84	5.89	5.93	5.97	6.01	60
5.44	5.50	5.56	5.61	5.66	5.71	5.75	5.79	5.83	120
5.29	5.35	5.40	5.45	5.49	5.54	5.57	5.61	5.65	∞

The columns are headed by k.

Data Sources

Introduction

1. MSNBC News, http://www.msnbc.com/ID/3965938, 17 January 2004.
2. Fox News, http://www.foxnews.com/story/0,2933,102511,00.html, 10 February 2004.
3. "Hot News: 98.6 Not Normal," *The Press-Enterprise* (Riverside, CA), 23 September 1992.

Chapter 1

1. "White House 2004: Democrats," *Time*/CNN Poll, Harris Interactive, http://www.pollingreport.com/wh04gen.htm, 6 February 2004.
2. "Run the Country? Most Teens Would Pass," http://abcnews.go.com/images/pdf/943a1TeensandthePresidency.pdf, 22 January 2004.
3. "Who They Are," *Time*, 5 January 2004, pp. 98–99.
4. "Getting Back to Work," http://www.usatoday.com/snapshot/news/2001-07-17-backtowork.htm, 2 October 2001.
5. William A. McGeveran, Jr., ed., *The World Almanac and Book of Facts, 2004* (New York: St. Martin's Press, 2004).
6. Adapted from Alison Stein Wellner, "The Next 25 Years," *American Demographics*, April 2003, p. 24.
7. "Report to the Community," Riverside County Office of Education, 16 February 2004, p. 3.
8. Borgna Brunner, ed., *Time Almanac, 2004* (Needham, MA: Pearson Education, 2004).
9. *Entertainment Weekly* #749/750, 6 February 2004, p. 125.
10. Robert P. Wilder, D. Brennan, and D. E. Schotte, "A Standard Measure for Exercise Prescription for Aqua Running," *American Journal of Sports Medicine* 21, no. 1 (1993):45.
11. *Princeton Review*, Irvine, CA, 1993.
12. http://www.kentuckyderby.com/2004/derby_history/derby_statistics/1976_2003.htm.
13. Robyn Greenspan, "Home Is Where the Network Is," http://www.clickz.com/stats/big_picture/applications/article.php/1301_3073431, 5 September 2003.
14. "Gallup Glance," http://www.gallup.com, 18 September 2001.
15. Robyn Greenspan, "Small Biz Gets Up to Speed," http://www.clickz.com/stats/markets/smallbiz/article.php/10098_33038711, 26 January 2004.
16. GEICO Insurance Advertisement, *Time*, 10 November 2003.
17. A. Tubb, A. J. Parker, and G. Nickless, "The Analysis of Romano-British Pottery by Atomic Absorption Spectrophotometry," *Archaeometry* 22 (1980):153.
18. Nutritional information from www.bk.com, www.papajohns.com, and other package labels.
19. Lawrence E. Levine and Victorina Wasmuth, "Laptops, Technology, and Algebra 1: A Case Study of an Experiment," *Mathematics Teacher* 97, no. 2 (February 2004):136.
20. A. Azzalini and A. W. Bowman, "A Look at Some Data on the Old Faithful Geyser," *Applied Statistics* (1990):57.

21. http://www.starbucks.com/retail/locator/default.aspx, 13 March 2004.

22. P. A. Mackowiak, S. S. Wasserman, and M. M. Levine, "A Critical Appraisal of 98.6 Degrees F, the Upper Limit of the Normal Body Temperature, and Other Legacies of Carl Reinhold August Wunderlich," *Journal of the American Medical Association* (268):1578–1580.

23. Allen L. Shoemaker, "What's Normal? Temperture, Gender, and Heart Rate," *Journal of Statistics Education* (1996).

24. http://cyberatlas.internet.com/big_picture/traffic_patterns/article/0,,5931_3301321,00.html, 16 February 2004.

Chapter 2

1. "Premium Comparison Surveys," http://www.insurance.ca.gov/docs/FS-Surveys.htm, 25 February 2004.

2. Borgna Brunner, ed., *Time Almanac, 2004* (Needham, MA: Pearson Education, Inc., 2004).

3. "Birth Order and the Baby Boom," *American Demographics* (Trend Cop), March 1997, p. 10.

4. "Tuna Goes Upscale," *Consumer Reports,* June 2001, p. 19.

5. http://www.starbucks.com/retail/locator/default.aspx, 13 March 2004.

6. "LCD TVs: Thin Is In," *Consumer Reports,* March 2004, p. 27.

7. A. Tubb, A. J. Parker, and G. Nickless, "The Analysis of Romano-British Pottery by Atomic Absorption Spectrophotometry," *Archaeometry* 22 (1980):153.

8. A. Azzalini and A. W. Bowman, "A Look at Some Data on the Old Faithful Geyser," *Applied Statistics* (1990):57.

9. http://sports.espn.go.com/nfl/players/, 28 February 2004.

10. Allen L. Shoemaker, "What's Normal? Temperature, Gender, and Heart Rate," *Journal of Statistics Education* (1996).

11. "Four People Movers," *Consumer Reports,* July 1997, p. 57.

12. "Favorite Camping Activity," http://www.usatoday.com/snapshot/news/2001-05-22-camping.htm, 26 September 2001 (*Source:* Wirthlin Worldwide for Coleman Company).

13. www.mlb.com, 28 February 2004.

14. "Ideal Family Size," http://www.usatoday.com/snapshot/news/2001-07-06-familysize.htm, 26 September 2001 (*Source:* Gallup).

15. "Microwaving Leftovers," http://www.usatoday.com/snapshot/life/2001-09-06-microwaving-leftovers.htm, 26 September 2001 (*Source:* Opinion Research for Tupperware).

16. "Waiting for a Prescription," http://www.usatoday.com/snapshot/life/2001-06-10-drugwait.htm, 26 September 2001 (*Source:* Opinion Research for AmeriSource Health).

17. William A. McGeveran, Jr., ed., *The World Almanac and Book of Facts, 2004* (New York: St. Martin's Press, 2004).

Chapter 3

1. Adapted from Michael J. Weiss, "The New Summer Break," *American Demographics,* August 2001, p. 55.

2. www.bls.gov/data/home.htm, 29 February 2004.

3. William A. McGeveran, Jr., ed., *The World Almanac and Book of Facts, 2004* (New York: St. Martin's Press, 2004).

4. Gregory K. Torrey, S. F. Vasa, J. W. Maag, and J. J. Kramer, "Social Skills Interventions Across School Settings: Case Study Reviews of Students with Mild Disabilities," *Psychology in the Schools* 29 (July 1992):248.

5. Mark Rechtin, "Boss Puts 'Buzz' on U.S. Luxury Leader's To-do List," *Automotive News,* 26 January 2004, p. 4.

6. "LCD TVs: Thin Is In," *Consumer Reports,* March 2004, p. 27.

7. Stellan Ohlsson, "The Learning Curve for Writing Books: Evidence from Professor Asimov," *Psychological Science* 3, no. 6 (1992):380–382.

8. "The Armed Forces: Who Are They," *Time,* 29 December 2003 to 5 January 2004, p. 98.

9. "Summery Scoreboard," *Entertainment Weekly* #614, 14 September 2001, p. 13.

10. http://sports.espn.go.com/nfl/players/gamelog?statsld=1025, 28 February 2004.

11. A. Tubb, A. J. Parker, and G. Nickless, "The Analysis of Romano-British Pottery by Atomic Absorption Spectrophotometry," *Archaeometry* 22 (1980):153.

12. Robyn Greenspan, "Home Is Where the Network Is," http://www.clickz.com/stats/big_picture/applications/article.php/1301_3073431, 5 September 2003.

13. Bettye Wells Miller, "Faith Shows Ballot Clout," *The Press-Enterprise* (Riverside, CA) 1 March 2004, p. A7.

14. Borgna Brunner, ed., *Time Almanac, 2004* (Needham, MA: Pearson Education), 2004.

15. "Music on the Move," *Consumer Reports,* July 2001, p. 41.

16. "College-Entry Scores Show Little Change in California," *The Press-Enterprise* (Riverside, CA), 29 August 2001, p. A-10.

17. "Dishwashers," *Consumer Reports* 60, no. 8 (August 1995):536.

Chapter 4

1. C. Salmon, J. P. Cartron, and P. Rouger, *The Human Blood Groups* (New York: Masson Publishing, 1984).

2. Table adapted from http://www.pollingreport.com/defense.htm#World%20Affairs, 3 March 2004.

3. Bruce E. Morgan and Michael A. Oberlander, "An Examination of Injuries in Major League Soccer," *The American Journal of Sports Medicine,* 29(4), 2001, pp. 426–429.

4. http://sports.espn.go.com/nba/, 23 March 2004.

5. Data adapted from "Demo Memo," *American Demographics,* May 1997, p. 32; and *The World Almanac and Book of Facts, 2004,* p. 377.

6. David Van Biema, "Does Heaven Exist?" *Time,* 24 March 1997, p. 73.

7. Adapted from David L. Wheeler, "More Social Roles Means Fewer Colds," *Chronicle of Higher Education* XLIII, no. 44 (July 11, 1997):A13.

8. Andrew S. Levy, M. J. Wetzler, M. Lewars, and W. Laughlin, "Knee Injuries in Women Collegiate Rugby Players," *The American Journal of Sports Medicine* 25, no. 3 (1997):360.

9. P. D. Franklin, R. A. Lemon, and H. S. Barden, "Accuracy of Imaging the Menisci on an In-Office, Dedicated, Magnetic Resonance Imaging Extremity System," *The American Journal of Sports Medicine* 25, no. 3 (1997):382.

10. Michael Crichton, *Congo* (New York: Knopf, 1980).

Chapter 5

1. http://www.collegeboard.com/student/testing/sat/scores/understanding/average.html, Copyright © collegeboard.com, Inc., 2004.

2. http://usatoday.com/snapshot/life/2001-06-11-potter.htm. *Source:* Ipsos-NPD Book Trends 9 October 2001.

3. Christy Fisher, "The Not-So-Great American Road Trip," *American Demographics,* May 1997, p. 47.

4. http://www.hsus.org/ace/11831, 24 March 2004.

5. L. D. Williams, P. S. Hamilton, B. W. Wilson, and M. D. Estock, "An Outbreak of *Escherichia coli 01257:H7* involving Long Term Shedding and Person-to-Person Transmission in a Child Care Center," *Environmental Health,* May 1997, p. 9.

6. Darryl Haralson and Jeff Dionisa, *USA Today.* Source: State Farm. http://usatoday.com/news/snapshot.htm?section=M&label=2004-01-30-car.jpg.

7. Data adapted from *American Demographics,* May 1997, p. 32.

8. "Call It in the Air," *The Press-Enterprise* (Riverside, CA), 19 October 1992.

9. Mark A. Atkinson, "Diet, Genetics, and Diabetes," *Food Technology* 51, no. 3 (March 1997), p. 77.

10. Darryl Haralson and Jeff Dionisa, *USA Today.* Source: Caravan survey for IHOP. http://usatoday.com/news/snapshot.htm?section=M&label=2004-01-30-car.jpg.

11. John Fetto, "Twenty-One and Counting. . . ," *American Demographics,* 25, no. 7, September 2003, p. 48.

12. http://www.usatoday.com 1 January 2004.

13. Sandra Yin, Pamela Paul, and David Whelan, "What Summer Break?" *American Demographics,* July 2001, p. 64.

14. Matthew L. Wald, "Cancers Near a Reactor: A Mystery and a Debate," *New York Times,* 21 May 1987, p. A-22.

Chapter 6

1. Paul M. Sommers, "Presidential Candidates Who Measure Up," *Chance* 9, no. 3 (Summer 1996):30, and http://www.whitehouse.gov/news/releases/2001/20010804-2.html.

2. Adapted from A. M. Goodman and A. R. Ennos, "The Response of Field-grown Sunflower and Maize to Mechanical Support," *Annals of Botany* 79 (1997):703.

3. John Fetto, "Shop Around the Clock," *American Demographics,* September 2003, p. 18.

4. "Medical Encyclopedia: Pulse," *Medline Plus: Trusted Health Information for You,* http://www.nlm.nih.gov/medlineplus/ency/article/003399.htm#Normal%20Values, 2 April 2004.

5. "Favorite Flavors," *Source:* International Consumer Research for LifeSavers, http://www.usatoday.com/snapshot/life/2001-08-07-lifesaver.htm. 11 October 2001.

6. PepsiCo Annual Report, http://co2.mobular.net/ccbn/7/195/203, 25 March 2004, p. 10.

7. Sonja Steptoe, "Ready, Set, Relax!" *Time,* 27 October 2003, p. 38.

8. Philip A. Altman and D. S. Dittmer, *The Biology Data Book,* 2nd ed., Vol I. (Bethesda, MD: Federation of American Societies for Experimental Biology, 1964), p. 137.

9. Allen L. Shoemaker, "What's Normal? Temperature, Gender, and Heart Rate," *Journal of Statistics Education* (1996).

10. "What Stresses Us Out," http://www.usatoday.com/snapshot/life/2001-05-23-stress.htm. *Source:* CyberPulse for Wrigley Healthcare's Surpass, 11 October 2001.

11. "Average Salary for Men and Women Faculty," *Academe: Bulletin of the American Association of University Professors,* 189, no. 2 (March–April 2003), p. 37.

12. Daniel Seligman, "Keeping Up," *Fortune,* 27 July 1981.

Chapter 7

1. "Chlorinated Water Byproduct, Rat Cancer Linked," *The Press-Enterprise* (Riverside, CA), 18 June 1997, p. A-6.

2. Chris Gilberg, J. L. Cos, H. Kashima, and K. Eberle, "Survey Biases: When Does the Interviewer's Race Matter?" *Chance,* Fall 1996, p. 23.

3. P. D. Franklin, R. A. Lemon, and H. S. Barden, "Accuracy of Imaging the Menisci on an In-Office, Dedicated, Magnetic Resonance Imaging Extremity System," *The American Journal of Sports Medicine* 25, no. 3 (1997):382.

4. "New Drug a Bit Better Than Aspirin," *The Press-Enterprise* (Riverside, CA), 14 November 1996, p. A-14.

5. "Space Exploration," CNN/*USA Today*/Gallup Poll, http://www.pollingreport.com/science.htm#Space, 5 April 2004.

6. "ASK AMERICA: 2003 Nationwide Policy Survey," Congressional District #44, 23 June 2003.

7. "Average Salary for Men and Women Faculty," *Academe: Bulletin of the American Association of University Professors,* 189, no. 2 (March–April 2003), 36.

8. Allen L. Shoemaker, "What's Normal? Temperature, Gender, and Heart Rate," *Journal of Statistics Education* (1996).

9. Nicola Maffulli, V. Testa, G. Capasso, and A. Sullo, "Calcific Insertional Achilles Tendinopathy," *The American Journal of Sports Medicine* 32, no. 1 (January/February 2004):174.

10. A. Elizabeth Sloan, "How Sweet It is," *Food Technology* 51, no. 3 (March 1997):26.

11. "Views Mixed on Role of Internet in Education," *The Press-Enterprise* (Riverside, CA) 21 August 2001, p. A-7.

12. http://us.mms.com/us/about/products/milkchocolate, 3 March 2004.

13. Judy Holland, "'Cheeseburger Bill' on the Menu," *The Press-Enterprise* (Riverside, CA), 9 March 2004, p. E1.

14. Adam Fernandez, "Nuts About You," *American Demographics* 26, no. 1 (February 2004):14.

15. P. C. Karalekas, Jr., C. R. Ryan, and F. B. Taylor, "Control of Lead, Copper, and Iron Pipe Corrosion in Boston," *American Water Works Journal,* February, 1983.

16. *Science News* 136 (19 August 1989):124.

17. "Half in Study Were Abused As Adults," *The Press-Enterprise* (Riverside, CA), 15 July 1997, p. D-4.

18. Catherine M. Santaniello and R. E. Koning, "Are Radishes Really Allelopathic to Lettuce?" *The American Biology Teacher* 58, no. 2 (February 1996):102.

19. http://www.gallup.com/poll/indicators/indairlines.asp#RelatedAnalyses. Gallup Poll News Service, 16 October 2001.

20. J. Hackl, *Journal of Quality Technology,* April 1991.

21. Daniel Seligman, "The Road to Monte Carlo," *Fortune,* 15 April 1985.

Chapter 8

1. *Science News* 136 (19 August 1989):124.

2. Laurie Lucas, "It's Elementary, Mister," *The Press-Enterprise* (Riverside, CA), 28 May 1997, p. D-1.

3. Mike Tharp, "Ready, Set Go. Why We Love Our Games—Sports Crazy," *U.S. News & World Report,* 15 July 1997, p. 31.

4. Adapted from "Suite Dreams," *Consumer Reports,* July 2001, p. 12–16.

5. "Space Exploration," Associated Press Poll, http://www.pollingreport.com/science.htm#Space, 5 April 2004.

6. "Caught in the Middle," *American Demographics,* July 2001, p. 14–15.

7. "Movie Mania Ebbing?" Gallup Poll of 800 adults March 16–18, 2001. Web site: http://www.usatoday.com/snapshot/news/2001-06-14-moviemania.htm.

8. Alison Stein Wellner, "A New Cure for Shoppus Interuptus," *The Marketing Tool Directory,* 2002.

9. Allen L. Shoemaker, "What's Normal? Temperature, Gender, and Heart Rate," *Journal of Statistics Education* (1996).

10. Christopher Reynolds, "Rocking the Vote," *American Demographics* 26, no. 1, (February 2004):48.

11. William Leonard, Barbara Speziale, and John Pernick, "Performance Assessment of a Standards-Based High School Biology Curriculum," *The American Biology Teacher* 63, no. 5 (2001):310–316.

12. "Voters Want Welfare Recipients Put to Work, Statewide Poll Finds," *The Press-Enterprise* (Riverside, CA), 30 May 1997.

13. Mark Gillespie, "Baseball Fans Overwhelmingly Want Mandatory Steroid Testing," Gallup News Service, http://gallup.com/content/print.aspx?ci=11245, 14 February 2004.

14. David L. Wheeler, "More Social Roles Means Fewer Colds," *Chronicle of Higher Education* XLIII, no. 44 (11 July 1997):A13.

15. Adapted from "Toplines: To the Moon?" Rebecca Gardyn, ed., *American Demographics,* August 2001, p. 9.

16. Reed Abelson, "A Survey of Wall St. Finds Women Disheartened," *The New York Times on the Web,* http://www.nytimes.com, July 26, 2001.

17. Shannon Dortch, "American Weighs In," *American Demographics,* June 1997, p. 39.

18. Adapted from A. M. Goodman and A. R. Ennos, "The Responses of Field-grown Sunflower and Maize to Mechanical Support," *Annals of Botary* 79 (1997):703.

19. Adapted from "I'm a Yankee Doodle Dandy," Knowledge Networks: 2000, *American Demographics,* July 2001, p. 9.

20. G. Wayne Marino, "Selected Mechanical Factors Associated with Acceleration in Ice Skating," *Research Quarterly for Exercise and Sport* 54, no. 3 (1983).

21. Sandra Yin, Pamela Paul, and David Whelan, "What Summer Break?" *American Demographics,* July 2001, p. 64.

22. Jeffrey M. Jones, "CNN/*USA Today*/Gallup Poll—Public Expects Gas Price Increase to Be Permanent," http://gallup.com/content/print/.aspx?ci=11257, 8 April 2004.

Chapter 9

1. Adapted from Paul, Pamela, "Coming Soon: More Ads Tailored to Your Tastes," *American Demographics,* August 2001, p. 28.

2. *The Princeton Review* (Irvine, CA, 1993).

3. Allen L. Shoemaker, "What's Normal? Temperature, Gender, and Heart Rate," *Journal of Statistics Education* (1996).

4. "Hot News: 98.6 Not Normal," *The Press-Enterprise* (Riverside, CA), 23 September 1992.

5. Nicola Maffulli, V. Testa, G. Capasso, and A. Sullo, "Calcific Insertional Achilles Tendinopathy," *The American Journal of Sports Medicine* 32, no. 1 (January/February 2004):174.

6. Adapted from "Suite Dreams," *Consumer Reports,* July 2001, pp. 12–16.

7. Lance Wallace and Terrence Slonecker, "Ambient Air Concentrations of Fine ($PM_{2.5}$) Manganese in the U.S. National Parks and in California and Canadian Cities: The Possible Impact of Adding MMT to Unleaded Gasoline," *Journal of the Air and Waste Management Association* 47 (June 1997):642–651.

8. Diane Crispell, "TV Soloists," *American Demographics,* May 1997, p. 32

9. "Seeing the World Through Tinted Lenses," *Washington Post,* 16 March 1993, p. 5.

10. Christopher Reynolds, "Rocking the Vote," *American Demographics* 26, no. 1 (February 2004):48.

11. "U.S. Pet Ownership Statistics," The Humane Society of the United States, http://www.hsus.org/ace/11831, 24 March 2004.

12. Denise Grady, "Study finds Alzheimer's Danger in Hormone Therapy," *The Press-Enterprise* (Riverside, CA), 28 May 2003.

13. "New Drug a Bit Better Than Aspirin," *The Press-Enterprise* (Riverside, CA), 14 November 1996, p. A-14.

14. Jonathan W. Jantz, C. D. Blosser, and L. A. Fruechting, "A Motor Milestone Change Noted with a Change in Sleep Position," *Archives of Pediatric Adolescent Medicine* 151 (June 1997):565.

15. Adapted from "Toplines: To the Moon?" Rebecca Gardyn, ed., *American Demographics,* August 2001, p. 9.

16. Loren Hill, *Bassmaster,* September/October 1980.

17. Joe Sylvester, "Area District Attorneys: Don't Judge Us by Statistics," *The Daily Item* (Sunbury, PA), 17 August 1997, p. A-1.

18. Charles Dickey, "A Strategy for Big Bucks," *Field and Stream,* October 1990.

19. *Science News* 136 (19 August 1989):124.

20. "2003 College-Bound Seniors: A Profile of SAT Program Test Takers," The College Board Summary Reporting Service, http://www.collegeboard.com/prod_downloads/about/news_info/cbsenior/yr2003/pdf/2003_TOTALGRP_PRD.pdf, 14 April 2004.

21. Judy Holland, "'Cheeseburger Bill' on the Menu," *The Press-Enterprise* (Riverside, CA), 9 March 2004, p. E1.

22. Cadonna Peyton, "Pupils Build English Skills," *The Press-Enterprise* (Riverside, CA), 19 March 2004, p. B-1.

23. Kurt Grote, T. L. Lincoln, and J. G. Gamble, "Hip Adductor Injury in Competitive Swimmers," *The American Journal of Sports Medicine* 32, no. 1 (January/February 2004):104.

24. Sandra Yin, Pamela Paul, and David Whelan, "What Summer Break?" *American Demographics,* July 2001, p. 64.

25. Joel B. Greenhouse and Samuel W. Greenhouse, "An Aspirin a Day . . . ?" *Chance: New Directions for Statistics and Computing* 1, no. 4 (1988):24–31.

Chapter 10

1. "Tuna Goes Upscale," *Consumer Reports,* June 2001, p. 19.

2. W. B. Jeffries, H. K. Voris, and C. M. Yang, "Diversity ad Distribution of the Pedunculate Barnacles *Octolasmis* Gray, 1825 Epizoic on the Scyllarid Lobster *Thenus orientalis* (Lund, 1793)," *Crustaceana* 46, no. 3 (1984).

3. http://sports.espn.go.com/nfl/players/, 28 February 2004.

4. Wendy K. Baell and E. H. Wertheim, "Predictors of Outcome in the Treatment of Bulimia Nervosa," *British Journal of Clinical Psychology* 31 (1992):330–332.

5. "L.A. Heart Data." Adapted from data found at http://www-unix.oit.umass.edu/~statdata/statdata/data/laheart.dat.

6. Jan D. Lindhe, "Clinical Assessment of Antiplaque Agents," *Compendium of Continuing Education in Dentistry,* Suppl. 5 (1984).

7. Susan J. Beckham, W. A. Grana, P. Buckley, J. E. Breasile, and P. L. Claypool, "A Comparison of Anterior Compartment Pressures in Competitive Runners and Cyclists," *American Journal of Sports Medicine* 21, no. 1 (1992):36.

8. Michael A. Brehm, J. S. Buguliskis, D. K. Hawkins, E. S. Lee, D. Sabapathi, and R. A. Smith, "Determining Differences in Efficacy of Two Disinfectants Using *t*-tests," *The American Biology Teacher* 58, no. 2 (February 1996):111.

9. A. Tubb, A. J. Parker, and G. Nickless, "The Analysis of Romano-British Pottery by Atomic Absorption Spectrophotometry," *Archaeometry* 22 (1980):153.

10. "Premium Comparison Surveys," http://www.insurance.ca.gov/docs/FS-Surveys.htm, 25 February 2004.

11. Carlos E. Macellari, "Revision of Serpulids of the Genus *Rotularia (Annelida)* at Seymour Island (Antarctic Peninsula) and Their Value in Stratigraphy," *Journal of Paleontology* 58, no. 4 (1984).

12. T. M. Casey, M. L. May, and K. R. Morgan, "Flight Energetics of Euglossine Bees in Relation to Morphology and Wing Stroke Frequency," *Journal of Experimental Biology* 116 (1985).

13. Karl J. Niklas and T. G. Owens, "Physiological and Morphological Modifications of *Plantago Major (Plantaginaceae)* in Response to Light Conditions," *American Journal of Botany* 76, no. 3 (1989):370–382.

14. "Expensive Bumper Repair: Latest Crash Tests," *Consumers' Research,* April 2001, pp. 20–21.

15. "KFC: Too Finger-Lickin' Good?" *Good Housekeeping* Saavy Consumer Product Tests, http://magazines.ivillage.com/goodhousekeeping/print/0,,446041,00.html, 11 March 2004.

16. John Fetto, "Shop Around the Clock," *American Demographics* 25, no. 7 (September 2003):18.

17. Adapted from Donald L. Barlett and James B. Steele, "Why We Pay So Much for Drugs," *Time,* 2 February 2004, p. 44.

18. Charles S. Catlin, "Four-day Work Week Improves Environment," *Environmental Health,* March 1997, p. 12.

Chapter 11

1. "Tuna Goes Upscale," *Consumer Reports,* June 2001, p. 19.

2. "2003 Auto Insurance," California Department of Insurance, http:// cdinswww.insurance.ca.gov/pls/wu_survey_auto/apsw_get_prem$auto_mc.querylist, 23 April 2004.

3. Devona Wells, "One Stop Shopping Embraced," *The Press-Enterprise* (Riverside, CA), 20 February 2004, p. A-5.

4. H. F. Barsam and Z. M. Simutis, "Coputer-Based Graphics for Terrain Visualisation Training," *Human Factors,* no. 26, 1984. Copyright 1984 by the Human Factors Society, Inc. Reproduced by permission.

5. "How Fares Differ by Airport and Airline," *Consumer Reports,* July 1997, p. 24.

6. *The Press-Enterprise* (Riverside, CA), 11 February 1993.

7. "Expensive Bumper Repair: Latest Crash Tests," *Consumers' Research,* April 2001, pp. 20–22.

8. Adapted from "Average Salary for Men and Women Faculty, by Category, Affiliation, and Academic Rank 2002–2003," *Academe: Bulletin of the American Association of University Professors* (March–April 2003):37.

9. A. Tubb, A. J. Parker, and G. Nickless, "The Analysis of Romano-British Pottery by Atomic Absorption Spectrophotometry," *Archaeometry* 22 (1980):153.

10. Harry R. Weber, "Is the Fix In?" *The Press-Enterprise* (Riverside, CA), 19 February 2004, p. G-1.

11. Robert McGarvey, "A Fine Mess," *Avenues,* July/August 1994, pp. 19–25.

Chapter 12

1. Stellan Ohlsson, "The Learning Curve for Writing Books: Evidence from Professor Asimov," *Psychological Science* 3, no. 6 (1992):380–382.

2. Daniel C. Harris, *Quantitative Chemical Analysis,* 3rd ed. (New York: Freeman, 1991).

3. "2001 Academic Performance Index (API) Report," *The Press-Enterprise* (Riverside, CA), 16 October 2001, p. A8.

4. Adapted from J. Zhou and S. Smith, "Measurement of Ozone Concentrations in Ambient Air Using a Badge-Type Passive Monitor," *Journal of the Air & Waste Management Association* 47 (June 1997):697.

5. "Round-Trip Fares on America's Most Popular Routes," *Consumer Reports,* July 1997, p. 25.

6. Lawrence E. Levine and Victorina Wasmuth, "Laptops, Technology, and Algebra 1: A Case Study of an Experiment," *Mathematics Teacher* 97, no. 2 (February 2004):136.

7. "LCD TVs: Thin Is In," *Consumer Reports,* March 2004, p. 27.

8. "#18 Peyton Manning," http://sports.espn.go.com/nfl/players/profile?statsId=4256, 4 May 2004.

9. W. B. Jeffries, H. K. Voris, and C. M. Yang, "Diversity and Distribution of the Pedunculat Barnacles *Octolasmis* Gray, 1825 Epizoic on the Scyllarid Lobster, *Thenus orientalis* (Lund, 1793)," *Crustaceana* 46, no. 3 (1984).

10. Gregory K. Torrey, S. F. Vasa, J. W. Maag, and J. J. Kramer, "Social Skills Interventions Across School Settings: Cast Study Reviews of Students with Mild Disabilities," *Psychology in the Schools* 29 (July 1992):248.

11. G. Wayne Marino, "Selected Mechanical Factors Associated with Acceleration in Ice Skating," *Research Quarterly for Exercise and Sport* 54, no. 3 (1983).

12. A. J. Ellis, "Geothermal Systems," *American Scientist,* September/October 1975.

13. Allen L. Shoemaker, "What's Normal? Temperature, Gender, and Heart Rate," *Journal of Statistics Education* (1996).

14. http://sports.espn.go.com/mlb/teams/batting, 4 May 2004.

15. David R. McAllister et al., "A Comparison of Preoperative Imaging Techniques for Predicting Patellar Tendon Graft Length before Cruciate Ligament Reconstruction," *The American Journal of Sports Medicine,* 20(4):461–465.

16. Henry Gleitman, *Basic Psychology,* 4th ed. (New York: Norton, 1996).

17. *Entertainment Weekly,* #749/750, 6 February 2004, p. 125.

18. Mark Rechtin, "Boss Puts 'Buzz' on U.S. Luxury Leader's To-do List," *Automotive News,* 26 January 2004, p. 4.

19. "Starbucks Beverage and Food Details," http://www.starbucks.com/retail/nutrition_info.asp, 4 May 2004.

20. "Suite Dreams," *Consumer Reports,* July 2001, p. 12.

21. *Automotive News: 1997 Market Data Book,* 28 May 1997, p. 50; *2001 Market Data Book* (online), http://www.automews.com/datacenter.cms, 20 September 2001; and *2004 Market Data Book,* 24 May 2004, p. 40.

Chapter 13

1. W. S. Good, "Productivity in the Retail Grocery Trade," *Journal of Retailing* 60, no. 3 (1984).

2. "Burgers from the Garden," *Consumer Reports,* July 1997, p. 36.

3. Adapted from J. Zhou and S. Smith, "Measurement of Ozone Concentrations in Ambient Air Using a Badge-Type Passive Monitor," *Journal of the Air & Waste Management Association* (June 1997):697.

4. "LCD TVs: Thin Is In," *Consumer Reports,* March 2004, p. 27.

5. "2001 Academic Performance Index (API) report," *The Press-Enterprise* (Riverside, CA), 16 October 2001, p. A8.

6. R. Blair and R. Miser, "Biotin Bioavailability from Protein Supplements and Cereal Grains for Growing Broiler Chickens," *International Journal of Vitamin and Nutrition Research* 59 (1989):55–58.

7. "Round-Trip Fares on America's Most Popular Routes," *Consumer Reports,* July 1997, p. 25.

8. Jeff Howe, "Total Control," *American Demographics,* July 2001, p. 30.

9. "Tuna Goes Upscale," *Consumer Reports,* June 2001, p. 19.

10. *Automotive News: 1997 Market Data Book,* 28 May 1997, p. 50; *2001 Market Data Book* (online), http://www.automews.com/datacenter.cms, 20 September 2001; and *2004 Market Data Book,* 24 May 2004, p. 40.

Chapter 14

1. Daniel Q. Haney "Mondays May Be Hazardous," *The Press-Enterprise* (Riverside, CA), 17 November 1992, p. A16.

2. "What Colors Come in Your Bag?" http://us.mms.com/us/about/products/milkchocolate/, 17 May 2004.

3. "What Colors Come in Your Bag?" http://usmms.com/us/about/products/peanut/, 17 May 2004.

4. Adapted from Margaret Hornblower, "Great Xpectations," *Time,* 9 June 1997, p. 58.

5. Adapted from Linda Schmittroth, ed., *Statistical Record of Women Worldwide* (Detroit and London: Gale Research, 1991).

6. Adapted from: http://www.USATODAY.com/snapshot/life/2001-06-10-drugwiat.htm. 26 September 2001.

7. Adapted from Dana Blanton, "Poll: Most Believe 'Cover-Up' of JFK Assassination Facts," http://www.foxnews.com/story/0,2933,102511,00.html, 10 February 2004.

8. "No Shows," *American Demographics,* 25, no. 9 (November 2003):11.

9. Adapted from Tamar Lewin, "Report Looks at a Generation, and Caring for Young and Old," *The New York Times on the Web,* 11 July 2001.

10. Adapted from Rebecca Piirto Heath, "Life on Easy Street," *American Demographics,* April 1997, p. 33.

11. W. W. Menard, "Time, Chance and the Origin of Manganese Nodules," *American Scientist,* September/October, 1976.

12. Elizabeth A. Crowley, "Churchgoing Rises with Age," http://www.usatoday.com/news/snapshot.htm, 30 January 2004.

13. Adapted from Bettye Wells Miller, "Faith Shows Ballot Clout," *The Press-Enterprise* (Riverside, CA) 1 March 2004, p. A7.

14. Jonathan W. Jantz, C. D. Blosser, and L. A. Fruechting, "A Motor Milestone Change Noted with a Change in Sleep Position," *Archives of Pediatric Adolescent Medicine* 151 (June 1997):565.

15. Adapted from: "Salmonella May Taint Many Holiday Turkeys," *The Press-Enterprise* (Riverside, CA) 20 November 2001, p. A3.

16. M. Martin Costa and M. Gatz, "Determination of Authorship Credit in Published Dissertations," *Psychological Science* 3, no. 6 (1992):54.

17. Dan Smith, "Motorists Have Little Respect for Others' Skills," *The Press-Enterprise* (Riverside, CA), 15 March 1991.

18. "White Cars Still Favored, but Green Fast Approaching," *The Press-Enterprise* (Riverside, CA), 19 April 1993.

19. "Every Dad Has His Day," *Time,* 16 June 1997, p. 16.

20. Doreen Matsui, R. Lim, T. Tschen, and M. J. Rieder, "Assessment of the Palatability of β-Lactamase-Resistant Antibiotics in Children," *Archives of Pediatric Adolescent Medicine* 151 (June 1997):599.

21. Andrew S. Levy, M. J. Wetzler, M. Lewars, and W. Laughlin, "Knee Injuries in Women Collegiate Rugby Players," *The American Journal of Sports Medicine* 25, no. 3 (1997):360.

22. Adapted from David L. Wheeler, "More Social Roles Means Fewer Colds," *Chronicle of Higher Education* XLIII, no. 44 (July 11, 1997):A13.

23. Carole Day and Del Lowenthal, "The Use of Open Group Discussions in Marketing Library Services to Young Adults," *British Journal of Educational Psychology* 62 (1992):324–340.

Chapter 15

1. T. M. Casey, M. L. May, and K. R. Morgan, "Flight Energetics of Euglossine Bees in Relation to Morphology and Wing Stroke Frequency," *Journal of Experimental Biology* 116 (1985).

2. "Alzheimer's Test Set for New Memory Drug," *The Press-Enterprise* (Riverside, CA), 18 November 1997, p. A-4.

3. *Science News* 136 (August 1989):126.

4. D. Matsui et al., "Assessment of the Palatability of β-Lactamase-Resistant Antibiotics in Children," *Archives of Pediatric Adolescent Medicine* 151 (1997):559–601.

5. Scott K. Powers and M. B. Walker, "Physiological and Anatomical Characteristics of Outstanding Female Junior Tennis Players," *Research Quarterly for Exercise and Sport* 53, no. 2 (1983).

6. *Science News,* 1989, p. 116.

7. G. Merrington, L. Winder, and I. Green, "The Uptake of Cadmium and Zinc by the Birdcherry Oat Aphid *Rhopalosiphum Padi (Homoptera:Aphididae)* Feeding on Wheat Grown on Sewage Sludge Amended Agricultural Soil," *Environmental Pollution* 96, no. 1 (1997):111–114.

8. Karola Sakekel, "Egg Substitutes Range in Quality," *San Francisco Chronicle,* 10 February 1993, p. 8.

Answers to Selected Exercises

Chapter 1

1.1 a. the student   **b.** the exam   **c.** the patient
d. the plant   **e.** the car

1.3 a. discrete   **b.** continuous   **c.** continuous
d. discrete

1.5 a. vehicles   **b.** type (qualitative); make (qualitative); carpool (qualitative); distance (quantitative
continuous); age (quantitative continuous)
c. multivariate

1.7 The population is the set of voter preferences for
all voters in the state. Voter preferences may
change over time.

1.9 a. score on the reading test; quantitative
b. the student   **c.** the set of scores for all deaf
students who hypothetically might take the test

1.11 a. a pair of jeans   **b.** the state in which the
jeans are produced; qualitative   **e.** 8/25
f. California   **g.** The three states produce
roughly the same numbers of jeans.

1.13 a. no, add a category called "Other"

1.15 a. no   **b.** not quite   **c.** the bar chart

1.17 answers will vary

1.19 a. eight to ten class intervals
c. 43/50   **d.** 33/50   **e.** yes

1.21 b. .30   **c.** .70   **d.** .30
e. relatively symmetric; no

1.25 b. centered at 75; two peaks (bimodal)
c. Scores are divided into two groups
according to student abilities.

1.27 a. pie chart, bar chart

1.29 b. the Pareto chart   **c.** number of
passengers served, size of the airline

1.31 a. skewed right; several outliers

1.33 b.
```
Stem-and-leaf of Ages   N = 37
   Leaf Unit = 1.0
      2      4  69
      3      5  3
      7      5  6678
     13      6  003344
     (6)     6  567778
     18      7  0111234
     11      7  7889
      7      8  013
      4      8  58
      2      9  00
```
relatively symmetric   **c.** Kennedy,
Garfield, and Lincoln were assassinated.

1.35 b. 0.1

1.37 a. number of hazardous waste sites (discrete)   **b.** skewed right   **c.** size of
the state; amount of industrial activity

1.39 a. skewed **b.** symmetric **c.** symmetric
d. symmetric **e.** skewed **f.** skewed

1.41 a. continuous **b.** continuous **c.** discrete
d. discrete **e.** discrete

1.43

7	8 9
8	0 1 7
9	0 1 2 4 4 5 6 6 6 8 8
10	1 7 9
11	2

1.45 c. skewed right

1.49 a. no **b.** roughly mound-shaped

1.51 a. skewed right **c.** yes; large states

1.53 a. Popular vote is skewed right; Percent vote is relatively symmetric. **b.** yes **c.** Once the size of the state is removed, each state will be measured on an equal basis.

1.55 d. horizontal axis is not a true time line

1.57 a. yes

1.59 b. bimodal distribution, outliers; different kiln sites **c.** yes

1.63 a. Stem-and-Leaf Display: Percent

```
Stem-and-leaf of Percent    N = 51
Leaf Unit = 1.0
   1    0  7
   2    0  8
   3    1  0
   4    1  3
   6    1  45
  12    1  666777
  20    1  88888999
 (11)   2  00000001111
  20    2  22222333
  12    2  44444555
   4    2  677
   1    2  9
```

b. skewed left **c.** two unusually low—Arkansas and Georgia

1.65 use a pie chart or a bar chart

1.67 a. approximately mound-shaped **b.** bar centered at 100.8 **c.** slightly above the center

1.69 a. somewhat mound-shaped **b.** .2

1.73 a–b. skewed right **c.** Yahoo!, Microsoft, Time Warner

Chapter 2

2.1 b. $\bar{x} = 2$; $m = 1$; mode $= 1$ **c.** skewed

2.3 a. 5.8 **b.** 5.5 **c.** 5 and 6

2.5 a. slightly skewed right **c.** $\bar{x} = 1.08$; $m = 1$; mode $= 1$

2.7 2.5 is an average number calculated (or estimated) for all families in a particular category.

2.9 The median, because the distribution is highly skewed to the right.

2.11 a. $\bar{x} = 2.688$; $m = 2$; mode $= 1$ **b.** skewed right **c.** yes

2.13 a. 2.4 **b.** 2.8 **c.** 1.673

2.15 a. 3 **b.** 2.125 **c.** $s^2 = 1.2679$; $s = 1.126$

2.17 a. 1.11 **b.** $s^2 = .19007$, $s = .436$ **c.** $R \approx 2.5s$

2.19 a. $s \approx 1.67$ **b.** $s = 1.75$ **c.** no **d.** yes **e.** no

2.21 a. approximately .68 **b.** approximately .95 **c.** approximately .815 **d.** approximately .16

2.23 a. $s \approx .20$ **b.** $\bar{x} = .76$; $s = .165$

2.25 a. approximately .68 **b.** approximately .95 **c.** approximately .003

2.27 a. ≈ 4.5 **b.** ≈ 2.25 **c.** $\bar{x} = 4.586$; $s = 2.892$

2.29 a. skewed right **b.** 0 to 104 days

2.31 b. $\bar{x} = 7.729$ **c.** $s = 1.985$

k	$\bar{x} \pm ks$	Actual	Tchebysheff	Empirical Rule
1	(5.744, 9.714)	.71	At least 0	Approx. .68
2	(3.759, 11.699)	.96	At least 3/4	Approx. .95
3	(1.774, 13.684)	1.00	At least 8/9	Approx. .997

2.33 a. 42 **b.** $s \approx 10.5$ **c.** $s = 13.10$ **d.** 1.00; 1.00; yes

2.35 a. $s \approx .444$ **b.** $s = .436$

2.37 a–b. $\bar{x} = 1.4$; $s^2 = 1.4$

2.39 a. $\bar{x} = 2.04$; $s = 2.806$

b–c.

k	$\bar{x} \pm ks$	Actual	Tchebysheff	Empirical Rule
1	(−.766, 4.846)	.84	At least 0	Approx. .68
2	(−3.572, 7.652)	.92	At least 3/4	Approx. .95
3	(−6.378, 10.458)	1.00	At least 8/9	Approx. .997

2.41

Sorted Data Set	Position of Q_1	Adjacent Values	Q_1	Position of Q_3	Adjacent Values	Q_3
1, 1.5, 2, 2, 2.2	1.5	1 and 1.5	1.25	4.5	2 and 2.2	2.1
0, 1.7, 1.8, 3.1, 3.2, 7, 8, 8.8, 8.9, 9, 10	3	none	1.8	9	none	8.9
.23, .30, .35, .41, .56, .58, .76, .80	2.25	.30 and .35	.3125	6.75	.58 and .76	.7150

2.43 min $= 0$, $Q_1 = 6$, $m = 10$, $Q_3 = 14$, max $= 19$; IQR $= 8$

2.45 lower and upper fences: -2.25 and 15.25; $x = 22$ is an outlier

2.47 a. min $= 1.70$, $Q_1 = 130.5$, $m = 246.5$, $Q_3 = 317.5$, max $= 485$ **b.** lower and upper fences: -150 and 598 **c–d.** No, but there are 4 extremely small observations, not identified by the box plot as outliers.

2.49 a.
```
Variable Minimum    Q1 Median    Q3 Maximum
Favre     10.00 14.75  21.50 23.00   26.00
McNabb     9.00 15.00  18.00 21.00   27.00
```
b. *Favre:* lower and upper fences: 2.375 and 35.375 *McNabb:* lower and upper fences: 6 and 30; no outliers; *Favre:* skewed left; *McNabb:* relatively symmetric

2.51 a. skewed right **b.** $\bar{x} = 38.4$; $m = 24.4$; mean $>$ median implies skewed right **c.** lower and upper fences: -26.6 and 94.2; 3 outliers

2.53 Female temperatures have a higher center (median) and are more variable; three outliers in the female group.

2.55 a. *Generic:* $m = 26$, $Q_1 = 25$, $Q_3 = 27.25$, IQR $= 2.25$; *Sunmaid:* $m = 26$, $Q_1 = 24$, $Q_3 = 28$, IQR $= 4$ **b.** *Generic:* lower and upper fences: 21.625 and 30.625; *Sunmaid:* lower and upper fences: 18 and 34 **c.** yes **d.** The average size is nearly the same; individual raisin sizes are more variable for Sunmaid raisins.

2.57 a. R $= 32.1$ **b.** $s \approx 8.025$ **c.** $s = 7.671$

2.59 $m \doteq 6.35$, $Q_1 = 2.325$, $Q_3 = 12.825$; lower and upper fences: -13.425 and 28.575; one outlier ($x = 32.3$).

2.61 a, b.

k	$\bar{x} \pm ks$	Tchebysheff	Empirical Rule
1	(.16, .18)	At least 0	Approx. .68
2	(.15, .19)	At least 3/4	Approx. .95
3	(.14, .20)	At least 8/9	Approx. .997

c. No, distribution of $n = 4$ measurements cannot be mound-shaped.

2.63 68%; 95%

2.65 a. 27; 20.2; 6.8 **b.** slightly skewed left **c.** 23.96; 1.641 **d.** largest $x = 27$, z-score $= 1.85$; smallest $x = 20.2$, z-score $= -2.29$; no **e.** 24.3 **f.** 22.95 and 24.85

2.67 a. $s \approx 7.75$ **b.** $\bar{x} = 59.2$; $s = 10.369$ **c.** $m = 60$, $Q_1 = 51.25$, $Q_3 = 69.75$; lower and upper fences: 23.5 and 97.5; no outliers.

2.69 $\sigma \approx 100$

2.71 a. 16% **b.** 81.5%

2.73 a. .9735 **b.** .16

2.75 a. .025 **b.** .84

2.77 a. At least 3/4 have between 145 and 205 teachers. **b.** .16

2.81 a. 8.36 **b.** 4 **c.** skewed right **d.** lower and upper fences: -24.375 and 42.625; no; yes

2.83 b. yes **c.** more than 2 or 3 standard deviations from the mean

2.85 a. 2.5, 3.75, 4.2, 4.75, 5.7 **b.** lower and upper fences: 2.25 and 6.25 **c.** no **d.** mound-shaped; yes

2.87 b. the sample mean gets smaller **d.** $5 \leq m \leq 10$

2.89 c. the standard deviation when dividing by $n - 1$ is closer to σ

2.91 b–c. skewed left with one outlier to the right of the other observations ($x = 520$)

Chapter 3

3.3 a. comparative pie charts; side-by-side or stacked bar charts **c.** Proportions spent in all four categories are substantially different for men and women.

3.5 a. *Population:* responses to free time question for all parents and children in the United States. *Sample:* responses for the 398 people in the survey. **b.** bivariate data, measuring relationship (qualitative) and response (qualitative)

c. the number of people who fall into that relationship-opinion category   **e.** stacked or side-by-side bar charts

3.9

x	y	xy	Calculate:	Covariance
1	6	6	$n = 3$	
3	2	6	$s_x = 1$	$s_{xy} = \dfrac{\Sigma xy - \dfrac{(\Sigma x)(\Sigma y)}{n}}{n-1} = -2$
2	4	8	$s_y = 2$	Correlation Coefficient
$\Sigma x = 6$	$\Sigma y = 12$	$\Sigma xy = 20$		$r = \dfrac{s_{xy}}{s_x s_y} = -1$

3.11 **b.** As x increases, y increases   **c.** .903
  d. $y = 3.58 + .815x$; yes

3.13 **b.** As x increases, y decreases   **c.** $-.987$

3.15 **a.** $y = 6.11 + 23.83x$ **c.** $149.06; no

3.17 **b.** slight positive trend   **c.** $r = .760$

3.19 **a.** price = dependent variable; size = independent variable   **b.** possibly, but not strong

3.21 **b.** The professor's productivity appears to increase, with less time required to write later books; no.

3.23 **a.** age of the soldier (quantitative), status of soldier—enlisted or officer (qualitative), branch of service (qualitative)   **b.** the population of responses for all soldiers in the U.S. Army and the U.S. Marine Corps; population at a fixed moment in time   **c.** side-by-side bar charts; stacked bar charts

3.27 **a.** .9108   **b.** x = first weekend gross; y = total gross   **c.** $y = 10.06 + 2.987x$
  d. 99.67 million dollars

3.29 **a.** no   **b.** $r = -.032$; yes   **c.** Large cluster in lower left corner shows no apparent relationship; seven to ten states form a cluster with a negative linear trend   **d.** local environmental regulations; population per square mile; geographic region

3.31 **a.** aluminum oxide (quantitative), site (qualitative)   **b.** higher levels of aluminum oxide at Ashley Rails and Island Thorns

3.33 **a.** Year (quantitative), number of home networks (quantitative), type of network (qualitative)   **c.** Wireless networking will increase and wired networks will decrease.

3.35 **a.** strong positive linear relationship
  b. .946   **c.** $b \approx 1$
  d. $y = 12.221 + .815x$

3.37 **b.** strong positive linear relationship

3.39 **b.** strong positive linear relationship
  c. yes; Southwest Airlines   **d.** better

3.43 **a.** .8544

3.45 **a.** 0.5   **b.** increase   **c.** 2.0; the y-intercept   **d.** 3.25; 4

Chapter 4

4.1 **a.** ,$\{1, 2, 3, 4, 5, 6\}$ **c.** 1/6
  e. $P(A) = 1/6$; $P(B) = 1/2$; $P(C) = 2/3$; $P(D) = 1/6$; $P(E) = 1/2$; $P(F) = 0$

4.3 $P(E_1) = .45$; $P(E_2) = .15$; $P(E_i) = .05$ for $i = 3, 4, \ldots, 10$

4.5 **a.** $\{NDQ, NDH, NQH, DQH\}$ **b.** 3/4
  c. 3/4

4.9 **a.** .58   **b.** .14   **c.** .46

4.11 **a.** randomly selecting three people and recording their gender   **b.** $\{FFF, FMM, MFM, MMF, MFF, FMF, FFM, MMM\}$ **c.** 1/8
  d. 3/8   **e.** 1/8

4.13 **a.** rank A, B, C
  b. $\{ABC, ACB, BAC, BCA, CAB, CBA\}$
  d. 1/3, 1/3

4.15 **a.** .467   **b.** .513   **c.** .533

4.17 80

4.19 **a.** 60   **b.** 3,628,800   **c.** 720   **d.** 20

4.21 6720

4.23 216

4.25 120

4.27 720

4.29 **a.** 140,608   **b.** 132,600   **c.** .00037
  d. .943

4.31 **a.** 2,598,960   **b.** 4   **c.** .000001539

4.33 $5.720645 \times (10^{12})$

4.35 **a.** 42   **b.** 1/42   **c.** 2/7

4.37 1/56

4.39 $\dfrac{4!(3!)^4}{12!}$

4.41

$P(A)$	$P(B)$	Conditions for Events A and B	$P(A \cap B)$	$P(A \cup B)$	$P(A\|B)$
.3	.4	Mutually exclusive	0	.7	0
.3	.4	Independent	.12	.58	.3
.1	.5	Independent	.05	.55	.1
.2	.5	Mutually exclusive	0	.7	0

4.43 **a.** 3/5   **b.** 4/5

4.45 **a.** 1   **b.** 1/5   **c.** 1/5

4.47 a. 1 **b.** 1 **c.** 1/3 **d.** 0 **e.** 1/3
f. 0 **g.** 0 **h.** 1 **i.** 5/6

4.49 a. .08 **b.** .52

4.51 a. .3 **b.** no **c.** yes

4.53 a. no, since $P(A \cap B) \neq 0$ **b.** no, since $P(A) \neq P(A|B)$

4.55 a. .14 **b.** .56 **c.** .30

4.59 a. $P(A) = .9918; P(B) = .0082$
b. $P(A) = .9836; P(B) = .0164$

4.61 .05

4.63 a. .99 **b.** .01

4.65 a. 154/256 **b.** 155/256 **c.** 88/256
d. 88/154 **e.** 44/67 **f.** 23/35
g. 12/101 **h.** 189/256

4.67 a. .7225 **b.** .50 **c.** .005625

4.69 a. .23 **b.** .6087; .3913

4.71 .38

4.73 .012

4.75 a. .6585 **b.** .3415 **c.** left

4.77 .3130

4.79 a. $P(D) = .10; P(D^C) = .90; P(N \mid D^C) =$ .94; $P(N \mid D) = .20$ **b.** .023 **c.** .023
d. .056 **e.** .20 **f.** false negative

4.81 a. continuous **b.** continuous
c. discrete **d.** discrete **e.** continuous

4.83 a. .2 **c.** $\mu = 1.9; \sigma^2 = 1.29; \sigma = 1.136$
d. .3 **e.** .9

4.85 1.5

4.87 a. {S, FS, FFS, FFFS}
b. $p(1) = p(2) = p(3) = p(4) = 1/4$

4.89 a. $p(0) = 3/10; p(1) = 6/10; p(2) = 1/10$

4.91 a. .1; .09; .081 **b.** $p(x) = (.9)^{x-1}(.1)$

4.93 a. 4.0656 **b.** 4.125 **c.** 3.3186

4.95 $1500

4.97 a. $p(x) = (.81)^{x-1}(.19)$ **b.** nonresponse; truthfulness on a sensitive subject

4.99 $2050

4.101 .0713

4.103 $P(A) = 1/2; P(B) = 2/3; P(A \cap B) = 1/3;$
$P(A \cup B) = 5/6; P(C) = 1/6; P(A \cap C) = 0;$
$P(A \cup C) = 2/3$

4.105 2/7

4.107 $p(0) = .0256; p(1) = .1536; p(2) = .3456;$
$p(3) = .3456; p(4) = .1296; .4752$

4.109 a. .4565 **b.** .2530 **c.** .3889

4.111 3/10; 6/10

4.113 a. .73 **b.** .27

4.115 .999999

4.117 8

4.119 a. .3582 **b.** .4883 **c.** .4467

4.121 a. 1/8 **b.** 1/64 **c.** Not necessarily; they could have studied together, and so on.

4.123 a. 5/6 **b.** 25/36 **c.** 11/36

4.125 a. .8 **b.** .64 **c.** .36

4.127 .0256; .1296

4.129 .2; .1

4.131 a. .5182 **b.** .1136 **c.** .7091
d. .3906

4.133 a. .0625 **b.** .25

4.135 a.

x	0	1	2
p(x)	6/15	8/15	1/15

b. $\mu = 2/3; \sigma^2 = 16/45$

4.137 a. $p(2) = p(12) = 1/36, p(3) = p(11) = 2/36, p(4) = p(10) = 3/36, p(5) = p(9) = 4/36, p(6) = p(8) = 5/36, p(7) = 6/36$

4.139 a. $p(0) = .5, p(1) = .5$

Chapter 5

5.1

k	0	1	2	3	4	5	6	7	8
$P(x \leq k)$	.000	.001	.011	.058	.194	.448	.745	.942	1.000

The Problem	List the Values of x	Write the Probability	Rewrite the Probability	Find the Probability
Three or less	0, 1, 2, 3	$P(x \leq 3)$	not needed	.058
Three or more	3, 4, 5, 6, 7, 8	$P(x \geq 3)$	$1 - P(x \leq 2)$	.989
More than three	4, 5, 6, 7, 8	$P(x > 3)$	$1 - P(x \leq 3)$	.942
Fewer than three	0, 1, 2	$P(x < 3)$	$P(x \leq 2)$	.011
Between 3 and 5 (inclusive)	3, 4, 5	$P(3 \leq x \leq 5)$	$P(x \leq 5) - P(x \leq 2)$	.437
Exactly three	3	$P(x = 3)$	$P(x \leq 3) - P(x \leq 2)$	.047

5.3 not binomial; dependent trials; p varies from trial to trial

5.5 a. .2965 **b.** .8145 **c.** .1172
d. .3670

5.7 a. .097 **b.** .329 **c.** .671 **d.** 2.1
e. 1.212

5.9 $p(0) = .000; p(1) = .002; p(2) = .015;$
$p(3) = .082; p(4) = .246; p(5) = .393;$
$p(6) = .262$

5.11 a. .251 **b.** .618 **c.** .367 **d.** .633
e. 4 **f.** 1.549

5.13 a. .901 **b.** .015 **c.** .002 **d.** .998

5.15 a. .748　　**b.** .610　　**c.** .367　　**d.** .966
e. .656

5.17 a. 1; .99　　**b.** 90; 3　　**c.** 30; 4.58
d. 70; 4.58　　**e.** 50; 5

5.19 a. .9568　　**b.** .957　　**c.** .9569
d. $\mu = 2$; $\sigma = 1.342$　　**e.** .7455; .9569;
.9977　　**f.** yes; yes

5.21 No; the variable is not the number of successes
in n trials. Instead, the number of trials n is
variable.

5.23 a. 1.000　　**b.** .997　　**c.** .086

5.25 a. .098　　**b.** .991　　**c.** .098　　**d.** .138
e. .430　　**f.** .902

5.27 a. .0081　　**b.** .4116　　**c.** .2401

5.29 a. $\mu = 10$　　**b.** 4 to 16　　**c.** If this
unlikely value were actually observed, it might
be possible that the trials (fields) are not
independent.

5.31 a. .0001406　　**b.** .0001　　**c.** .9999

5.33 a. .107　　**b.** .762

5.35

Probability	Formula	Calculated Value
$P(x = 0)$	$\dfrac{\mu^k e^{-\mu}}{k!} = \dfrac{2.5^0 e^{-2.5}}{0!}$	.082085
$P(x = 1)$	$\dfrac{\mu^k e^{-\mu}}{k!} = \dfrac{2.5^1 e^{-2.5}}{1!}$	.205212
$P(x = 2)$	$\dfrac{\mu^k e^{-\mu}}{k!} = \dfrac{2.5^2 e^{-2.5}}{2!}$	.256516
$P(2$ or fewer successes)	$P(x = 0) + P(x = 1)$	.543813

5.37

k	0	1	2	3	4	5	6	7	8	9	10
$P(x \le k)$	.055	.199	.423	.647	.815	.916	.966	.988	.996	.999	1.000

The Problem	List the Values of x	Write the Probability	Rewrite the Probability	Find the Probability
Three or less	0, 1, 2, 3	$P(x \le 3)$	not needed	.647
Three or more	3, 4, 5, . . .	$P(x \ge 3)$	$1 - P(x \le 2)$	.577
More than three	4, 5, 6, . . .	$P(x > 3)$	$1 - P(x \le 3)$	.353
Fewer than three	0, 1, 2	$P(x < 3)$	$P(x \le 2)$	.423
Between 3 and 5 (inclusive)	3, 4, 5	$P(3 \le x \le 5)$	$P(x \le 5) - P(x \le 2)$	.493
Exactly three	3	$P(x = 3)$	$P(x \le 3) - P(x \le 2)$	.224

5.39 a. .135335　　**b.** .27067　　**c.** .593994
d. .036089

5.41 a. .677　　**b.** .6767　　**c.** yes

5.43 a. .0067　　**b.** .1755　　**c.** .560

5.45 a. .271　　**b.** .594　　**c.** .406

5.47 $P(x > 5) = .017$; unlikely.

5.49 a. .6　　**b.** .5143　　**c.** .0714

5.51 a. $p(0) = .36$; $p(1) = .48$; $p(2) = .15$;
$p(3) = .01$　　**c.** $\mu = .8$, $\sigma^2 = .50286$
d. .99; .99; yes

5.53 $p(0) = .2$; $p(1) = .6$; $p(2) = .2$

5.55 a. hypergeometric　　**b.** .1786　　**c.** .01786
d. .2857

5.61 a. $p(0) = .729$; $p(1) = .243$; $p(2) = .027$;
$p(3) = .001$　　**c.** .3; .520　　**d.** .729; .972

5.63 a. .234　　**b.** .136　　**c.** Claim is not
unlikely.

5.65 a. .228　　**b.** no indication that people are
more likely to choose middle numbers

5.67 d. Either the sample is not random or the
60% figure is too high.

5.69 a. 20　　**b.** 4　　**c.** .006　　**d.** Psychiatrist
is incorrect.

5.71 a. $\mu = 50$; $\sigma = 6.124$
b. The value $x = 35$ lies 2.45 standard
deviations below the mean. It is somewhat
unlikely that the 25% figure is representative
of this campus.

5.73 a. .5　　**b.** $\mu = 12.5$; $\sigma = 2.5$　　**c.** There is a
preference for the second design.

5.75 a. yes; $n = 10$; $p = .25$　　**b.** .2440
c. .0000296　　**d.** Yes; genetic model is not
behaving as expected.

5.77 a. yes　　**b.** $1/8192 = .00012$

5.79 a. hypergeometric, or approximately binomial
b. Poisson　　**c.** approximately .85; .72; .61

5.81 a. .015625　　**b.** .421875　　**c.** .25

5.83 a. $p = 1/3$　　**b.** .3292　　**c.** .8683

5.85 a. 14　　**b.** 2.049　　**c.** no; $x = 10$ is only
1.95 standard deviations below the mean

5.87 a. .135335　　**b.** .676676

5.89 .655

5.91 a. .794　　**b.** .056　　**c.** -0.82 to 3.82 or
0 to 3

5.93 a. 36　　**b.** 4.8
c. Yes, since $x = 49$ lies 2.71 standard
deviations above the mean

5.95 a. .00006　　**b.** .042　　**c.** .0207
d. .5948　　**e.** 1

5.99 a. .0176　　**b.** .9648　　**c.** .9648

Chapter 6

6.1

The Interval	Write the Probability	Rewrite the Probability (if needed)	Find the Probability
Less than -2	$P(z < -2)$	not needed	.0228
Greater than 1.16	$P(z > 1.16)$	$1 - P(z \leq 1.16)$	.1230
Greater than 1.645	$P(z > 1.645)$	$1 - P(z \leq 1.645)$	.0500
Between -2.33 and 2.33	$P(-2.33 < z < 2.33)$	$P(z \leq 2.33) - P(z \leq -2.33)$	.9802
Between 1.24 and 2.58	$P(1.24 < z < 2.58)$	$P(z \leq 2.58) - P(z \leq 1.24)$	.1026
Less than or equal to 1.88	$P(z \leq 1.88)$	not needed	.9699

6.3 a. .9452   **b.** .9664   **c.** .8159
d. 1.0000

6.5 a. .6753   **b.** .2401   **c.** .2694
d. .0901   **e.** ≈ 0

6.7 a. 1.96   **b.** 1.44

6.9 a. 1.65   **b.** -1.645

6.11 a. 1.28   **b.** 1.645   **c.** 2.05   **d.** 2.33

6.13 a. .1596   **b.** .1151   **c.** .1359

6.15 58.3

6.17 $\mu = 8$; $\sigma = 2$

6.19 a. .1949   **b.** .4870   **c.** no   **d.** yes;
$y = 17$ lies 4.19 standard deviations above the mean

6.21 a. .4586   **b.** .0526   **c.** .0170

6.23 .1562; .0012

6.25 a. .0475   **b.** .00226   **c.** 29.12 to 40.88
d. 38.84

6.27 a. .0062   **b.** .1056

6.29 .0475

6.31 63,550

6.33 a. .3085   **b.** .2417   **c.** .0045

6.35 a. 15; 10   **b.** yes   **c.** 15; 2.449
d. 10, 11, . . . , 15   **e.** 10; 9.5
f. $z = -2.25$ **g.** -2.25; .0122; .9878

6.37 a. yes   **b.** $\mu = 7.5$; $\sigma = 2.291$ **c.** .6156
d. .618

6.39 a. .2676   **b.** .3520   **c.** .3208
d. .9162

6.41 a. .178   **b.** .392

6.43 a. .245   **b.** .2483

6.45 a. .0869   **b.** .3264   **c.** .0244

6.47 .9441

6.49 a. .3594   **b.** They do not consider height when casting their ballot.

6.51 a. .0055   **b.** .8324   **c.** .8289   **d.** yes; *Pepsi's* market share is higher than claimed.

6.53 a. 31   **b.** 3.432   **c.** no; $x = 25$ is only 1.75 standard deviations below the mean

6.55 a. .3227   **b.** .1586

6.57 $z_0 = 0$

6.59 $z_0 = .67$; the 25th and 75th percentiles

6.61 no

6.63 5.065 months

6.65 .0401

6.67 85.36 minutes

6.69 no; $x = 184$ lies only 1.26 standard deviations below the mean.

6.71 7.301 ounces

6.73 a. 141   **b.** .0401

6.75 .9474

6.77 .3557

6.79 a $Q_1 = 269.96$; $Q_3 = 286.04$ **b.** Yes; $x = 180$ lies 8.17 standard deviations below the mean

6.81 a. ≈ 0 **b.** .6026   **c.** Sample is not random; results will be biased.

6.83 a. $\pm.52$ standard deviations
b. ± 1.28 standard deviations

6.85 a. .3085   **b.** 99.92 degrees

6.87 a. .9544   **b.** .0561

6.89 a. $z_0 = -1.96$ **b.** $z_0 = .36$

6.91 a. .9651   **b.** .1056   **c.** .0062

6.93 a. .0177   **b.** .0178

6.95 a. 1.273   **b.** .1016

6.97 .1244 (exact probability $= .1236$)

Chapter 7

7.1 1/500

7.11 a. convenience sample   **b.** Yes, but only if his patients behave like a random sample from the general population.

7.13 a. first question   **b.** Percent favoring the program decreased, perhaps due to the "spending billions of dollars" wording in the question.

7.15 normal; 53; 3

7.17 normal; 100; 3.16

7.19 a. $\mu = 10$; $\sigma/\sqrt{n} = 1.5$
b. $\mu = 5$; $\sigma/\sqrt{n} = .2$
c. $\mu = 120$; $\sigma/\sqrt{n} = .3536$

7.21 c. .5468

7.23 Increasing the sample size decreases the standard error.

7.25 b. a large number of replications

7.29 a. .3758 **b.** no

7.31 a. 1890; 69.282 **b.** .0559

7.33 a. ≈ 0 **b.** Yes; the value $\bar{x} = 98.25$ is almost 5 standard deviations below the assumed mean, $\mu = 98.6$.

7.35 normal; .7; .0648

7.37 a. $p = .3$; $SE = .0458$ **b.** $p = .1$; $SE = .015$ **c.** $p = .6$; $SE = .0310$

7.39 b. .9198

7.41 a. .0099 **b.** .03 **c.** .0458 **d.** .05 **e.** .0458 **f.** .03 **g.** .0099

7.43 a. yes; $p = .46$; $SE = .0498$ **b.** .2119 **c.** .9513 **d.** The value is unusual because $\hat{p} = .30$ lies 3.21 standard deviations below the mean $p = .46$.

7.45 a. approximately normal with mean .13 and standard deviation .0453 **b.** .9382 **c.** ≈ 0 **d.** .04 to .22

7.47 a. approximately normal with mean .75 and standard deviation .0306 **b.** .0516 **c.** .69 to .81

7.49 a. LCL = 150.13; UCL = 161.67

7.51 a. LCL = 0; UCL = .090

7.53 a. LCL = 8598.7; UCL = 12,905.3

7.55 LCL = .078; UCL = .316

7.57 LCL = .0155; UCL = .0357

7.59 mean too large at hours 2, 3, and 4

7.63 a. ≈ 12.5 **b.** .9986 **c.** They are probably correct.

7.65 c. no

7.71 a. cluster sample **b.** 1-in-10 systematic sample **c.** stratified sample **d.** 1-in-10 systematic sample **e.** simple random sample

7.73 a. 131.2; 3.677 **b.** yes **c.** .1515

7.75 a. LCL = 0; UCL = .0848 **b.** $\hat{p} > .0848$

7.77 yes

7.81 a. approximately normal with mean 288 and standard deviation .9798 **b.** .0207 **c.** .0071

7.83 UCL = .2273; LCL = $-.0273$

7.85 a. 3.5; 1.208

7.87 a. 3.5; .854

7.89 a. .4938 **b.** .0062 **c.** .0000

Chapter 8

8.3 a. .160 **b.** .339 **c.** .438

8.5 a. .554 **b.** .175 **c.** .055

8.7 a. .179 **b.** .098 **c.** .049 **d.** .031

8.9 a. .0588 **b.** .0898 **c.** .098 **d.** .0898 **e.** .0588

8.11 $\hat{p} = .728$; margin of error (MOE) = .029

8.13 $\bar{x} = 39.8$; MOE = 4.768

8.15 $\bar{x} = 7.2\%$; MOE = .776

8.17 a. $\hat{p} = .78$; MOE = .026 **b.** no; $p = .5$

8.19 a. no **b.** nothing; no

8.21 Point estimate is $\bar{x} = 19.3$ with margin of error = 1.86.

8.23 a. (.797, .883) **b.** (21.469, 22.331) **c.** Intervals constructed in this way enclose the true value of μ 90% of the time in repeated sampling.

8.25 (.846, .908)

8.27 a. 3.92 **b.** 2.772 **c.** 1.96

8.29 a. 3.29 **b.** 5.16 **c.** The width increases.

8.31 (3.496, 3.904); random sample

8.33 a. (.932, 1.088) **c.** no; $\mu = 1$ is a possible value for the population mean

8.35 a. (.106, .166) **b.** Increase the sample size and/or decrease the confidence level.

8.37 a. $98.085 < \mu < 98.415$ **b.** no; perhaps the value 98.6 is not the true average body temperature for healthy people.

8.39 a. (4.61, 5.99) **b.** yes

8.41 (15.463, 36.937)

8.43 a. (17.676, 19.324) **b.** (15.710, 17.290) **c.** (.858, 3.142) **d.** yes

8.45 a. $\bar{x}_1 - \bar{x}_2 = 2206$; MOE = 902.08 **b.** yes

8.47 a. (3.333, 16.667) **b.** $(-22.040, -7.960)$ **c.** no **d.** yes; yes

8.49 a. $(-.528, -.032)$; yes, since $\mu_1 - \mu_2 = 0$ is not in the interval.

8.51 a. $(-.203, -.117)$ **b.** random and independent samples from binomial distributions

8.53 a. $(-.221, .149)$ **b.** no

8.55 a. $(-.118, -.002)$ **b.** yes, since $p_1 - p_2 = 0$ is not in the interval.

8.57 a. $(.095, .445)$ **b.** yes

8.59 $(.061, .259)$

8.61 a. $(-.082, .022)$ **b.** no, since $p_1 - p_2 = 0$ is in the interval.

8.63 $1.96\sqrt{\dfrac{pq}{n}}$; $1.96\sqrt{\dfrac{.5(.5)}{n}}$; 385;

quantitative; one; $1.96\dfrac{10}{\sqrt{n}}$; 97

8.65 a. $\mu < 76.63$ **b.** $\mu < 1.89$

8.67 $\mu_1 - \mu_2 < 4$

8.69 505

8.71 $n_1 = n_2 = 1086$

8.73 9604

8.75 $n_1 = n_2 = 360$

8.77 97

8.79 $n_1 = n_2 = 136$

8.81 $n_1 = n_2 = 98$

8.83 a. $\bar{x} = 29.1$; MOE $= .9555$
b. $(28.298, 29.902)$ **c.** $\mu > 28.48$
d. 234

8.85 $n_1 = n_2 = 224$

8.87 1083

8.89 $n_1 = n_2 = 925$

8.91 a. $\hat{p}_W = .5$; $\hat{p}_M = .75$ **b.** $-.313 < p_W - p_M < -.187$ **c.** There is a difference in the two proportions.

8.93 $(8.087, 11.313)$

8.95 97

8.97 $(33.41, 34.59)$

8.99 a. MOE $= .029$ **b.** 6147

8.101 a. $(.499, .561)$ **b.** $(.110, .230)$
c. $(.320, .520)$

8.103 at least 1825

8.105 .3874; .651

8.107 a. $(2.837, 3.087)$ **b.** 276

8.109 $(638.946, 665.054)$; no

8.111 $(2.694, 2.716)$

8.113 $(.161, .239)$

8.115 at least 97

8.117 b. widths are the same

8.119 a. 9.702   **b.** $(746.298, 765.702)$ **c.** yes

8.121 b. The standard error and the width of the interval decrease.

9.1

Test Statistic	Significance Level	One or Two-Tailed Test?	Critical Value	Rejection Region	Conclusion
$z = 0.88$	$\alpha = .05$	Two-tailed	1.96	$\lvert z\rvert > 1.96$	Do not reject H_0
$z = -2.67$	$\alpha = .05$	One-tailed (lower)	1.645	$z < -1.645$	Reject H_0
$z = 5.05$	$\alpha = .01$	Two-tailed	2.58	$\lvert z\rvert > 2.58$	Reject H_0
$z = -1.22$	$\alpha = .01$	One-tailed (lower)	2.33	$z < -2.33$	Do not reject H_0

9.3 a. $z > 2.33$ **b.** $\lvert z\rvert > 1.96$
c. $z < -2.33$ **d.** $\lvert z\rvert > 2.58$

9.5 a. Do not reject H_0; results are not statistically significant.   **b.** Reject H_0; results are highly significant.   **c.** Reject H_0; results are statistically significant.

9.7 a. .0207   **b.** Reject H_0; results are statistically significant.   **c.** yes

9.9 p-value $= .0644$; do not reject H_0; results are not statistically significant.

9.11 a. H_0: $\mu = 1$; H_a: $\mu \neq 1$
b. p-value $= .7414$; do not reject H_0
c. There is no evidence to indicate that the average weight is different from 1 pound.

9.13 a. H_0: $\mu = 80$ **b.** H_a: $\mu \neq 80$
c. $z = -3.75$; reject H_0

9.15 a. $z = -2.31$; p-value $= .0104$; reject H_0 at the 5% level of significance   **b.** Average score improvement is less than 110 at the 5% level of significance.   **c.** There is no evidence to indicate that the average score improvement is less than 110 at the 1% level of significance.

9.17 yes; $z = 10.94$

9.19 no; $z = -1.334$ with p-value $= .0918$; do not reject H_0

9.21 a. H_0: $\mu_1 - \mu_2 = 0$; H_a: $\mu_1 - \mu_2 > 0$; one-tailed   **b.** $z = 2.074$; reject H_0

9.23 a. $z = -2.26$; p-value $= .0238$; reject H_0
b. $(-3.55, -.25)$ **c.** no

9.25 a. H_0: $\mu_1 - \mu_2 = 0$; H_a: $\mu_1 - \mu_2 \neq 0$
b. yes; $z = 8.77$ **c.** p-value ≈ 0

9.27 a. yes; $z = -3.18$; p-value $= .0014$
b. $(-3.01, -.71)$; yes

9.29 a. $z = -2.22$ with p-value $= .0264$
b. significant at the 5% but not the 1% level.

9.31 H_0: $p = .4$; H_a: $p \neq .4$ **b.** p-value $= .093$;
not statistically significant **c.** no

9.33 a. H_0: $p = .6$; H_a: $p \neq .6$ and H_0: $p = .5$;
H_a: $p < .5$ **b.** no; $z = -.58$
c. no; $z = .94$

9.35 a. H_0: $p = 2/3$ **b.** H_a: $p > 2/3$
c. yes; $z = 4.6$ **d.** p-value $< .0002$

9.37 no; $z = -.90$

9.39 no; $z = -1.06$

9.41 no; $z = -.71$

9.43 a. H_0: $p_1 - p_2 = 0$; H_a: $p_1 - p_2 < 0$
b. one-tailed
c. do not reject H_0; $z = -.84$

9.45 a. yes; $z = -2.40$ **b.** $(-.43, -.05)$

9.47 Do not reject H_0; $z = -.39$; there is insufficient evidence to indicate a difference in the two population proportions.

9.49 a. yes; $z = 2.67$ **b.** $p_1 - p_2 > .001$; the risk is at least 1/1000 higher when taking *Prempro*.

9.51 Reject H_0; $z = 3.14$ with p-value $= .0008$; researcher's conclusions are confirmed.

9.55 The power increases.

9.57 a. p-value $< .0002$ **b.** Reject H_0; $z = 4.47$

9.59 a. H_0: $\mu = 7.5$; H_a: $\mu < 7.5$
b. one-tailed **d.** $z = -5.477$; reject H_0

9.61 a. H_0: $\mu_1 - \mu_2 = 0$; H_a: $\mu_1 - \mu_2 \neq 0$
b. two-tailed **c.** no; $z = -.954$

9.63 no; do not reject H_0; $z = 1.684$

9.65 a. p-value $= .1038$ **b.** do not reject H_0

9.67 yes; $z = 4$; reject H_0

9.69 yes; $z = 4.00$

9.71 a. yes; $z = 4.33$ **b.** $(7.12, 18.88)$

9.73 a. yes; $z = -3.33$ **b.** no; $z = 1$ **c.** no

9.75 no; $z = 2.19$

9.77 no; $z = -.56$

9.79 yes; $z = 3.32$

9.81 a. $(1447.49, 4880.51)$ **b.** between 1500 and 5000 more meters per week; they only have one stroke to practice.

9.83 a. H_0: $\mu = 94$; H_a: $\mu \neq 94$
b. $z = -1.331$ **c.** $.1832$
d. do not reject H_0 **e.** no

9.85 a. $.7422$ **b.** $.9783$ **c.** power increases

Chapter 10

10.1 a. 2.015 **b.** 2.306 **c.** 1.330
d. 1.96

10.3 a. $.02 < p$-value $< .05$
b. p-value $< .005$ **c.** p-value $> .20$
d. p-value $< .005$

10.5 a. $\bar{x} = 7.05$; $s = .4994$ **b.** (7.496)
c. Reject H_0; $t = -2.849$ **d.** Yes.

10.7 no; $t = -1.195$

10.9 a. yes; $t = -3.044$ **b.** 98.316

10.11 $(3.652, 3.912)$

10.13 a. Reject H_0; $t = -4.31$ **b.** $(23.23, 29.97)$ **c.** The pretreatment mean looks smaller than the other two means.

10.17 $(233.98, 259.94)$

10.19 a. 3.775 **b.** 21.2258

10.21 a. H_0: $\mu_1 - \mu_2 = 0$; H_a: $\mu_1 - \mu_2 \neq 0$
b. $|t| > 2.771$ **c.** $t = 2.795$
d. p-value $< .01$ **e.** Reject H_0

10.23 a. Yes; larger s^2/smaller $s^2 = 1.36$
b. $t = .06$ with p-value $= .95$ **c.** 19.1844
d. Do not reject H_0 **e.** $(-5.223, 5.503)$;
yes

10.25 a. No; $t = -1.16$ **b.** p-value $= .260$
c. yes; larger s^2/smaller $s^2 = 2.88$

10.27 a. No; larger s^2/smaller $s^2 = 16.22$
b. Yes; $t = -2.412$; $.02 < p$-value $< .05$

10.29 a. yes **b.** no; larger s^2/smaller $s^2 = 3.72$
c. Do not reject H_0; $t = .10$ with
p-value $> .20$

10.31 Do not reject H_0; $t = 0.24$ with p-value $> .10$

10.33 a. Do not reject H_0; $t = 1.13$
b. $(-1.487, 5.153)$, using $df = 29$; yes

10.35 a. Reject H_0; $t = 2.372$ with $.02 < p$-value $< .05$ **b.** $(.014, .586)$ **c.** 62 pairs

10.37 a. Do not reject H_0; $t = 1.177$
b. p-value $> .20$ **c.** $(-.082, .202)$
d. random sample from normal distribution

10.39 a. Do not reject H_0; $t = 1.984$; $(-7.28, 170.94)$ **b.** Reject H_0; $t = 2.307$; $(6.867, 208.433)$ **c.** Reject H_0; $t = 4.38$ **d.** $(-1.6, 8.2)$; yes

10.41 b. Yes; $t = 9.150$ with p-value $< .01$ **c.** $(80.472, 133.328)$

10.43 a. yes; $t = -4.326$; reject H_0 **b.** $(-2.594, -.566)$ **c.** at least 65 pairs

10.45 a. yes; $t = 2.82$; reject H_0 **b.** 1.4875 **d.** yes

10.47 Do not reject H_0; $t = 1.03$ with p-value $> .10$; do not proceed with the installation.

10.49 $(.190, .685)$

10.51 reject H_0; $\chi^2 = 22.449$

10.53 a. no; $t = -.232$ **b.** yes; $\chi^2 = 20.18$

10.55 a. no **b.** yes; $z = 3.262$

10.57 no; $\chi^2 = 29.433$

10.59 $(.667, 4.896)$

10.61 $F = 1.059$ with p-value $> .20$; do not reject H_0: $\sigma_1^2 = \sigma_2^2$

10.63 $(1.544, 4.003)$

10.65 Rest: $F = 1.03$ with p-value $> .20$; 80% maximal O_2: $F = 2.01$ with p-value $> .20$; maximal O_2: $F = 14.29$ with p-value $< .01$; use the unpooled t test for maximal O_2.

10.71 $(9.860, 12.740)$

10.73 yes, $t = 5.985$; reject H_0; $(28.375, 33.625)$

10.75 yes, $F = 3.268$

10.77 72

10.79 $(22.578, 26.796)$

10.81 at least 136 **b.** lower the confidence level; redesign the experiment as a paired difference test.

10.83 a. yes **b.** $F = 19.516$; there is a difference in the population variances.

10.85 a. random independent samples from normal distributions with equal variances; no **b.** yes; $t = 3.237$ with p-value $< .01$ **c.** yes; $t = 60.36$ with p-value $< .01$

10.87 no; $t = 2.2$ with p-value $> .10$

10.89 no; $t = 1.586$ with p-value $> .10$

10.91 no, $t = -1.712$

10.93 Unpaired: $(-1.69, .19)$; paired: $(-1.49, -.01)$; paired interval is slightly narrower

10.95 a. no, $t = 2.571$ **b.** $(.000, .020)$

10.97 a. two-tailed; H_a: $\sigma_1^2 \neq \sigma_2^2$ **b.** lower-tailed; H_a: $\sigma_1^2 < \sigma_2^2$ **c.** upper-tailed; H_a: $\sigma_1^2 > \sigma_2^2$

10.99 a. no, $\chi^2 = 7.008$ **b.** $(.185, 2.465)$

10.101 reject H_0; $t = 2.425$; drug increases average reaction time

10.103 yes, $t = -2.945$

10.105 no

10.107 no, $t = 1.86$ with p-value $= .112$

10.109 Use pooled t test; $t = -1.82$ with p-value $> .10$; results are nonsignificant.

10.111 a. $(5.814, 7.886)$ **b.** random sample; sampled population is normal.

10.113 Reject H_0; $t = 4.57$; yes

10.115 a. Reject H_0; $t = 4.38$; yes **b.** p-value $< .01$; yes

10.117 a. $t > 1.8$ **b.** $|t| > 2.37$ **c.** $t < -2.6$

10.121 Do not reject H_0; $t = -1.438$ with p-value $= .1782$

10.123 yes; $t = -3.33$ with p-value $= .0030$

Chapter 11

11.1

Source	df
Treatments	5
Error	54
Total	59

11.3 a. $(2.731, 3.409)$ **b.** $(.07, 1.03)$

11.5 a.

Source	df	SS	MS	F
Treatments	3	339.8	113.267	16.98
Error	20	133.4	6.67	
Total	23			

b. $df_1 = 3$ and $df_2 = 20$ **c.** $F > 3.10$ **d.** yes, $F = 16.98$ **e.** p-value $< .005$; yes

11.7 a. CM $= 103.142857$; Total SS $= 26.8571$ **b.** SST $= 14.5071$; MST $= 7.2536$ **c.** SSE $= 12.3500$; MSE $= 1.1227$ **d.**

```
Analysis of Variance
   Source DF      SS    MS    F      P
   Trts    2   14.51  7.25  6.46  0.014
   Error  11   12.35  1.12
   Total  13   26.86
```

f. $F = 6.46$; reject H_0 with $.01 < p$-value $< .025$

11.9 a. $(1.95, 3.65)$ **b.** $(.27, 2.83)$

11.11 a. (67.86, 84.14)   **b.** (55.82, 76.84)
c. (−3.629, 22.963)   **d.** no, they are not independent

11.13 a. Each observation is the mean length of ten leaves.   **b.** yes, $F = 57.38$ with p-value $= .000$ **c.** reject H_0; $t = 12.09$
d. (1.810, 2.924)

11.15

Analysis of Variance for Percent

Source	DF	SS	MS	F	P
Method	2	0.0000041	0.0000021	16.38	0.000
Error	12	0.0000015	0.0000001		
Total	14	0.0000056			

11.17 a. Completely randomized design

b.

Source	DF	SS	MS	F	P
State	3	3272.2	1090.7	26.44	0.000
Error	16	660.0	41.3		
Total	19	3932.2			

c. $F = 26.44$; reject H_0 with p-value $= .000$.

11.19 sample means must be independent; equal sample sizes

11.21 a. $1.878s$ **b.** $2.1567s$

11.23 $\bar{x}_1$ $\bar{x}_2$ $\bar{x}_3$ $\bar{x}_4$

11.25 a. no; $F = .60$ with p-value $= .562$
b. no differences

11.27 a. Yes; $F = 8.55$, p-value $= .005$
b. (−157.41, −28.59)   **c.** $\bar{x}_3$ $\bar{x}_1$ $\bar{x}_2$

11.29

Source	df	SS	MS	F
Treatments	2	11.4	5.70	4.01
Blocks	5	17.1	3.42	2.41
Error	10	14.2	1.42	
Total	17	42.7		

11.31 (−3.833, −.767)

11.33 a. yes; $F = 19.19$ **b.** yes; $F = 135.75$
c. $\bar{x}_1$ $\bar{x}_3$ $\bar{x}_4$ $\bar{x}_2$ **d.** (−5.332, −2.668)
e. yes

11.35 a. 7   **b.** 7   **c.** 5   **e.** yes; $F = 9.68$
f. yes; $F = 8.59$

11.37

Two-way ANOVA: y versus Blocks, Chemicals

Analysis of Variance for y

Source	DF	SS	MS	F	P
Blocks	2	7.1717	3.5858	40.21	0.000
Chemical	3	5.2000	1.7333	19.44	0.002
Error	6	0.5350	0.0892		
Total	11	12.9067			

11.39 a. yes; $F = 10.06$ **b.** yes; $F = 10.88$
c. $\omega = 4.35$ **d.** (1.12, 5.88)

11.41

Two-way ANOVA: Cost versus Estimator, Job

Analysis of Variance for Cost

Source	DF	SS	MS	F	P
Estimator	2	10.862	5.431	7.20	0.025
Job	3	37.607	12.536	16.61	0.003
Error	6	4.528	0.755		
Total	11	52.997			

11.43 a. Blocks are items; treatments are stores.
b. Yes. $F = 11.87$; p-value $= .000$
c. Yes. $F = 33.24$; p-value $= .000$.

11.45 a. 20   **b.** 60   **c.**

Source	df
A	3
B	4
AB	12
Error	40
Total	59

11.47 (−1.11, 5.11)

11.49 a. strong interaction present
b. $F = 37.85$ with p-value $= .000$; yes
d. no

11.51 b. yes   **c.** Since the interaction is significant, attention should be focused on means for the individual factor-level combinations.   **d.** Training: $.05 < p$-value $< .10$; ability: p-value $< .005$; interaction: $.01 < p$-value $< .025$

11.53 a. 2×4 factorial; students; gender at two levels, schools at four levels   **c.** no; $F = 1.19$ **e.** main effect for schools is significant; $F = 27.75$; Tukey's $\omega = 82.63$

11.55 a.

Source	F	SS	MS	F	P
Training	1	4489.00	4489.00	117.49	0.000
Situation	1	132.25	132.25	3.46	0.087
Interaction	1	56.25	56.25	1.47	0.248
Error	12	458.50	38.21		
Total	15	5136.00			

b. No. $F = 1.47$; p-value $= .248$.   **c.** No. $F = 3.46$; p-value $= .087$.   **d.** Yes. $F = 117.49$; p-value $= .000$.

11.57 significant differences between treatments A and C, B and C, C and E and D and E

11.59 a. significant difference in treatment means; $F = 27.78$ **b.** Tukey's $\omega = .190$
c. yes; $F = 6.59$

11.61

One-way ANOVA: Sales versus Program

Analysis of Variance for Sales

Source	DF	SS	MS	F	P
Program	3	1385.8	461.9	9.84	0.000
Error	23	1079.4	46.9		
Total	26	2465.2			

11.63 a. no; $F = 1.40$ **b.** p-value $> .10$
c. yes; $F = 6.51$ **d.** yes; $F = 7.37$

11.65 a. 2×3 factorial experiment   **b.** no;
$F = .45$ with p-value $= .642$
d. $(-22.56, -5.84)$

11.67 a. randomized block design

b.

Two-way ANOVA: Total versus Week, Store
Analysis of Variance for Total

Source	DF	SS	MS	F	P
Week	3	571.7	190.6	8.27	0.003
Store	4	684.6	171.2	7.43	0.003
Error	12	276.4	23.0		
Total	19	1532.7			

c. yes; $F = 7.43$ **d.** $\omega = 10.81$

11.69 a. randomized block; blocks are types of
crash, treatments are cars.   **b.** yes;
$F = 5.72$ **c.** no; $F = 2.11$
d. $\omega = 1298.12$

11.71 a. Completely randomized design.
b. Yes, there is a significant difference.
$F = 126.85$, p-value $= .000$

Source	DF	SS	MS	F	P
Site	2	132.277	66.139	126.85	0.000
Error	21	10.950	0.521		
Total	23	143.227			

11.73 There is no evidence of non-normality. There
appears to be slightly larger error variation for
the smaller values compared to the larger values of y.

Chapter 12

12.1 y-intercept $= 1$, slope $= 2$

12.3 $y = 3 - x$

12.7 a. $\hat{y} = 6.00 - .557x$ **c.** 4.05
d. Analysis of Variance

Source	DF	SS	MS
Regression	1	5.4321	5.4321
Residual Error	4	0.1429	0.0357
Total	5	5.5750	

12.9 a. $S_{xx} = 21066.82$; $S_{yy} = .374798$;
$S_{xy} = 88.80003$ **b.** $\hat{y} = .0187 + .00422x$
d. .44

e. Analysis of Variance

Source	DF	SS	MS
Regression	1	0.37431	0.37431
Residual Error	7	0.00049	0.00007
Total	8	0.37480	

12.11 a. $y = $ API; $x = $ ELL   **b.** yes
c. $\hat{y} = 731.277 - 3.040x$ **d.** yes

12.13 a. yes   **b.** $\hat{y} = -11.665 + .755x$
c. $\hat{y} = 52.51$

12.15 a. strong positive linear relationship
b. approximately 1
c. $\hat{y} = 12.221 + .815x$ **d.** $\hat{y} = 62.75$

12.17 a. yes, $t = 5.20$ **b.** $F = 27.00$
c. $t_{.025} = 3.182$; $F_{.05} = 10.13$

12.19 a. yes, $F = 152.10$ with p-value $= .000$
b. $r^2 = .974$

12.21 a. $y = $ cost, $x = $ distance
b. $\hat{y} = 128.58 + .12715x$
d. $t = 6.09$; $r^2 = .699$

12.23 a. yes; $t = 3.79$ and $F = 14.37$ with
p-value $= .005$ **b.** no   **c.** $r^2 = .642$
d. MSE $= 5.025$ **e.** $(.186, .764)$

12.25 a. yes   **b.** $\hat{y} = -26.82 + 1.2617x$
c. yes; $t = 7.49$ and $F = 56.05$ with
p-value $= .000$ **d.** $(.7768, 1.7466)$

12.27 a. yes; reject H_0, $t = 7.15$
b. $(.5362, 1.0944)$ **c.** yes; the value
$\beta = 1$ is contained in the interval

12.29 plot residuals versus fit; random scatter of
points, free of patterns

12.31 no

12.33 a. slight curve   **b.** 95.9% of overall variation explained by the straight-line model
c. strong curvilinear pattern indicates relationship may be curvilinear

12.35 a. fairly strong positive relationship
b. yes   **c.** relationship is curvilinear

12.37 a. $(4.6006, 5.1708)$ **b.** $(4.2886, 5.4829)$
c. $x = 8$; extrapolation

12.39 a. slight curve   **b.** 95.7% of overall variation explained by the straight-line model
c. pattern indicates relationship may be curvilinear

12.41 a. $(2.01, 3.74)$ **b.** $(-.77, 3.02)$
c. $\bar{x} = 0$

12.43 a. $(207.744, 321.902)$
b. $(134.680, 394.966)$ **c.** no

12.47 a. positive   **b.** $r = .9487$; $r^2 = .9000$

12.49 b. $r = .982$ **c.** 96.5%

12.51 a. positive   **b.** $r = .760$; yes, $t = 2.615$

12.53 yes; $t = 3.158$ with p-value $< .01$

12.55 a. correlation analysis   **b.** $r = .981$

12.57 a. no   **b.** $r = -.018$; no, $t = -.044$

12.59 a. yes   **b.** $\hat{y} = 80.85 + 270.82x$
c. yes; $t = 3.96$ with p-value $= .003$
d. $(112.1, 157.9)$

12.61 a. $r = .980$ **b.** $r^2 = .961$
c. $\hat{y} = 21.9 + 15.0x$ **d.** variance is not constant for all x

12.63

Regression Analysis: API versus ELL
The regression equation is
API = 731 - 3.04 ELL

Predictor	Coef	StDev	T	P
Constant	731.28	22.81	32.06	0.000
ELL	-3.0399	0.7551	-4.03	0.007

S = 33.72   R-Sq = 73.0%   R-Sq(adj) = 68.5%

Analysis of Variance

Source	DF	SS	MS	F	P
Regression	1	18422	18422	16.21	0.007
Residual Error	6	6820	1137		
Total	7	25242			

12.65 a. no; $t = 2.066$ with p-value $> .05$
b. $r^2 = .299$
12.67 No; variance is not constant for all x.
12.69 a. weak negative relationship
b. $r^2 = .044$; 4.4%
c. $\hat{y} = 3091.4 - 97.7x$; no, $t = -.91$
d. no
12.71 at the extremes of the experimental region
12.73 a. $\hat{y} = 20.47 - .758x$
b.

Source	DF	SS	MS	F	P
Regression	1	287.28	287.28	493.40	0.000
Residual Error	8	4.66	0.58		
Total	9	291.94			

c. Reject H_0, $t = -22.21$
d. $(-.86, -.66)$ **e.** $(9.296, 10.420)$
f. $r^2 = .984$

12.77 a. y-intercept $= 3$; slope $= -0.5$

Chapter 13

13.1 b. parallel lines
13.3 a. yes, $F = 57.44$ with p-value $< .005$
b. $R^2 = .94$
13.5 a. quadratic   **b.** $R^2 = .815$; relatively good fit   **c.** yes, $F = 37.37$ with p-value $= .000$
13.7 a. $b_0 = 10.5638$ **b.** yes, $t = 15.20$ with p-value $= .000$
13.9 b. $t = -8.11$ with p-value $= .000$; reject H_0: $\beta_2 = 0$ in favor of H_a: $\beta_2 < 0$
13.11 a. $R^2 = .9955$ **b.** $R^2(\text{adj}) = 99.3\%$
c. The quadratic model fits slightly better.
13.13 a. Use variables x_2, x_3, and x_4 **b.** no
13.15 a. $\hat{y} = -8.177 + 292x_1 + 4.434x_2$
b. Reject H_0, $F = 16.28$ with p-value $= .002$.

The model contributes significant information for the prediction of y.   **c.** Yes, $t = 5.54$ with p-value $= .001$ **d.** $R^2 = .823$; 82.3%
13.17 a. quantitative   **b.** quantitative
c. qualitative; $x_1 = 1$ if plant B, 0 otherwise; $x_2 = 1$ if plant C, 0 otherwise   **d.** quantitative   **e.** qualitative; $x_1 = 1$ if day shift, 0 if night shift
13.19 a. x_2 **b.** $\hat{y} = 12.6 + 3.9x_2^2$ or $\hat{y} = 13.14 - 1.2x_2 + 3.9x_2^2$
13.21 a. $y = \beta_0 + \beta_1 x_1 + \beta_2 x_2 + \beta_3 x_1 x_2 + \epsilon$ with $x_2 = 1$ if cucumber, 0 if cotton   **c.** No, the test for interaction yields $t = .63$ with p-value $= .533$.   **d.** yes
13.23 $y = \beta_0 + \beta_1 x_1 + \beta_2 x_1^2 + \beta_3 x_2 + \beta_4 x_1 x_2 + \beta_5 x_1^2 x_2 + \varepsilon$
13.25 a. $\hat{y} = 8.585 + 3.8208x - 0.21663x^2$
b. $R^2 = .944$ **c.** yes; $F = 33.44$
d. yes; $t = -4.93$ with p-value $= .008$
e. no
13.27 b. $\hat{y} = 4.10 + 1.04x_1 + 3.53x_2 + 4.76x_3 - 0.43x_1 x_2 - 0.08x_1 x_3$ **c.** yes; $t = -2.61$ with p-value $= .028$ **d.** no; $F = 3.86$; consider eliminating the interaction terms.
13.29 a. $y = \beta_0 + \beta_1 x_1 + \beta_2 x_2 + \beta_3 x_1^2 + \beta_4 x_1 x_2 + \beta_5 x_1^2 x_2 + \epsilon$
b. $F = 25.85$; $R^2 = .768$
c. $\hat{y} = 4.51 + 6.394x_1 + .1318x_1^2$
d. $\hat{y} = -46.34 + 23.458x_1 - .3707x_1^2$
e. no; $t = .78$ with p-value $= .439$
13.31 a. curvilinear relationship
b. $y = \beta_0 + \beta_1 x + \beta_2 x^2 + \epsilon$
c. $\hat{y} = 4114749 - 4113.4x + 1.02804x^2$
d. yes; $F = 542.11$ with p-value $= .000$
e. $R^2 = .997$; very good fit
13.35 a. 99.9%
b. yes; $F = 1676.61$ with p-value $= .000$
c. yes; $t = -2.65$ with p-value $= .045$
d. yes; $t = 15.14$ with p-value $= .000$
e. Linear: $R^2(\text{adj}) = 91.9\%$, Quadratic: $R^2(\text{adj}) = 99.8\%$, quadratic term is significant.   **f.** Quadratic term is missing.

Chapter 14

14.3 a. $X^2 > 12.59$ **b.** $X^2 > 21.666$
c. $X^2 > 29.8194$ **d.** $X^2 > 5.99$
14.5 a. H_0: $p_1 = p_2 = p_3 = p_4 = p_5 = 1/5$
b. 4   **c.** 9.4877   **d.** $X^2 = 8.00$
e. Do not reject H_0.

14.7 yes, $X^2 = 24.48$; drivers tend to prefer the inside lanes.

14.9 no, $X^2 = 3.63$

14.11 no, $X^2 = 13.58$

14.13 Yes; do not reject H_0, $X^2 = 1.247$

14.15 Yes; reject H_0, $X^2 = 28.386$

14.17 8

14.19 reject H_0, $X^2 = 18.352$ with p-value $= .000$

14.21 **a.** yes; $X^2 = 7.267$
b. $.025 < p$-value $< .05$

14.23 **a.** yes; reject H_0; $X^2 = 20.937$
b. no; $X^2 = 1.255$

14.25 **a.** No; do not reject H_0, $X^2 = 6.447$
b. p-value $> .10$; yes

14.27 **a.** $X^2 = 10.597$ **b.** $X^2 > 13.2767$
c. Do not reject H_0.
d. $.025 < p$-value $< .05$

14.29 yes; $X^2 = 24.31$

14.31 **a.** Each income category represents a multinomial population in which we measure education levels **b.** yes; $X^2 = 19.172$

14.33 Yes, reject H_0, $X^2 = 36.499$

14.35 no, $X^2 = 4.4$ with p-value $> .10$

14.37 no, $X^2 = 1.89$ with p-value $> .10$

14.39 **a.** Yes, reject H_0, $X^2 = 15.752$ **b.** Those who attend church more often are more likely to vote for Bush.

14.43 No; do not reject H_0, $X^2 = 1.311$

14.45 **a.** reject H_0; $X^2 = 18.527$ **b.** reject H_0;
$z = 4.304$; yes

14.49 yes, $X^2 = 7.488$ with $.005 < p$-value $< .01$

14.51 yes, $X^2 = 6.190$ with $.025 < p$-value $< .05$;
$(.347, .483)$

14.53 **a.** no; $X^2 = 3.259$ with p-value $= .196$
b. no; $X^2 = 1.054$ with p-value $= .788$
c. yes

14.55 no, $X^2 = 3.953$ with p-value $= .139$

14.57 **a.** do not reject H_0; $X^2 = 3.660$ with
p-value $= .454$; yes

14.59 **a.** yes; $X^2 = 11.690$ with p-value $= .003$
b. The susceptibility to a cold seems to decrease as the number of relationships increases.

14.61 Yes, reject H_0, $X^2 = 16.535$

14.63 **a.** 27.69 **b.** 5.99

14.65 consumers have a preference; $X^2 = 18.5$ with
p-value $= .0001$

14.69 no; $X_2 = 2.87$ with p-value $= .2378$

Chapter 15

15.1 **a.** T_1^* **b.** $T \leq 31$ **c.** $T \leq 27$

15.3 **a.** H_0: population distributions are identical;
H_a: population 1 shifted to the left of population 2. **b.** $T_1 = 16$; $T_1^* = 39$
c. $T \leq 19$ **d.** yes; reject H_0

15.5 do not reject H_0; $z = -1.59$

15.7 do not reject H_0; $T = 102$

15.9 yes; reject H_0; $T = 45$

15.11 yes; reject H_0; $T = 44$

15.13 **b.** $\alpha = .002, .007, .022, .054, .115$

15.15 One-tailed: **n = 10:** $\alpha = .001, .011, .055$;
n = 15: $\alpha = .004, .018, .059$; **n = 20:**
$\alpha = .001, .006, .021, .058, .132$; two-tailed:
n = 10: $\alpha = .002, .022, .110$; **n = 15:**
$\alpha = .008, .036, .118$; **n = 20:** $\alpha = .002, .012,$
$.042, .116$

15.17 **a.** H_0: $p = \frac{1}{2}$; H_a: $p \neq \frac{1}{2}$; rejection region:
$\{0, 1, 7, 8\}$; $x = 6$; do not reject H_0 at
$\alpha = .07$; p-value $= .290$

15.19 $z = 3.15$; reject H_0

15.21 **b.** $T = \min\{T^+, T^-\}$ **c.** $T \leq 137$
d. Do not reject H_0.

15.23 Do not reject H_0; $z = -.34$.

15.25 **a.** Reject H_0; $T = 1.5$ **b.** Results do not agree.

15.27 **a.** no; $T = 6.5$

15.29 **a.** Do not reject H_0; $x = 8$ **b.** Do not reject H_0; $T = 14.5$.

15.31 **a.** paired difference test, sign test, Wilcoxon signed-rank test **b.** reject H_0 with both tests; $x = 0$ and $T = 0$

15.33 yes, $H = 13.90$

15.35 **a.** no; $H = 2.63$ **b.** p-value $> .10$
c. p-value $> .10$

15.37 no; $H = 2.54$ with p-value $> .10$

15.39 **a.** Reject H_0; $F_r = 21.19$.
b. p-value $< .005$ **d.** $F = 75.43$
e. p-value $< .005$ **f.** Results are identical.

15.41 **a.** Do not reject H_0; $F_r = 5.81$.
b. $.05 < p$-value $< .10$

15.43 a. $r_s \geq .425$　　**b.** $r_s \geq .601$

15.45 a. $|r_s| \geq .400$　　**b.** $|r_s| \geq .526$

15.47 a. $-.593$　　**b.** yes

15.49 a. $r_s = .811$　　**b.** yes

15.51 yes

15.53 yes, $r_s = .9118$

15.55 a. Do not reject H_0; $x = 2$　　**b.** Do not reject H_0; $t = -1.646$

15.57 a. Do not reject H_0; $x = 7$　　**b.** Do not reject H_0; $x = 7$

15.59 Do not reject H_0 with the Wilcoxon rank sum test ($T = 77$) or the paired difference test ($t = .30$).

15.61 Do not reject H_0 using the sign test ($x = 2$); no

15.63 yes; $r_s = -.845$

15.65 reject H_0; $T = 14$

15.67 a. Reject H_0; $F_r = 20.13$　　**b.** The results are the same.

15.69 a. Reject H_0; $H = 9.08$
b. $.025 < p$-value $< .05$　　**c.** The results are the same.

15.71 a. no　　**b.** significant differences among the responses to the three rates of application; $F_r = 10.33$ with p-value $= .006$.

15.73 $T = 19$. $T_{.05} = 21$ ($T_{.01} = 18$.) Reject H_0.

15.75 $z = 1.285 < z_{.05} = 1.645$. Lighting not effective.

15.77 $H = 7.43$ $df = 3$ p-value $= 0.059$. No significant difference.

15.79 a. $r_s = .738$.　　**b.** p-value $= .025 < .05$. Yes, positive correlation.

Index

Credits

Introduction. © **Royalty-Free/CORBIS**

Chapter 1. © **Picture Arts/Royalty-Free/CORBIS**

Chapter 2. © **Photodisc/Royalty-Free/Getty Images**

Chapter 3. © **Photodisc/Royalty-Free/Getty Images**

Chapter 4. © **Getty Images**

Chapter 5. © **Photodisc/Royalty-Free/Getty Images**

Chapter 6. © **AFP/Getty Images**

Chapter 7. © **PictureNet/Royalty-Free/CORBIS**

Chapter 8. © **AP/Wide World Photos**

Chapter 9. © **Getty Images**

Chapter 10. © **CORBIS/Sygma**

Chapter 11. © **James Leynse/CORBIS**

Chapter 12. © **Getty Images**

Chapter 13. © **Will & Deni McIntyre/CORBIS**

Chapter 14. © **Dave Bartuff/CORBIS**

Chapter 15. © **Royalty-Free/Getty Images**